UNDERGROUND OPERATORS CONFERENCE 2023

Premier underground mining conference

27–29 MARCH 2023
BRISBANE, AUSTRALIA

The Australasian Institute of Mining and Metallurgy
Publication Series No 01/2023

AusIMM

Published by:
The Australasian Institute of Mining and Metallurgy
Ground Floor, 204 Lygon Street, Carlton Victoria 3053, Australia

ISBN 978-1-922395-21-4

ORGANISING COMMITTEE

Alex Campbell
MAusIMM(CP)
Conference Co-Chair

Anne-Marie Ebbels
MAusIMM(CP)
Conference Co-Chair

Chris Carr
FAusIMM(CP)

Katrina Crook
MAusIMM(CP)

Alastair Grubb
MAusIMM(CP)

Peter Hills
FAusIMM(CP)

Erin Langworthy

Joe Luxford
FAusIMM(CP)

Adrian Pratt
FAusIMM(CP)

Iain Ross
MAusIMM

Farell Simanjuntak

John Stanton
MAusIMM

AUSIMM

Julie Allen
Head of Events

Fiona Geoghegan
Manager, Events

Samara Brown
Conference Program Manager

REVIEWERS

We would like to thank the following people for their contribution towards enhancing the quality of the papers included in this volume:

Mark Adams

Alex Campbell

Geoff Capes

Chris Carr

Katrina Crook

Geoff Dunstan

Chris Elliott

David Finn

Andrew Glastonbury

Ed Gleeson

Donald Grant

Alastair Grubb

Rodney Hocking

Evan Jones

Marie Jones

Joe Luxford

Anthony Malavisi

Maddie Merrett

Adrian Penney

John Player

Adrian Pratt

Iain Ross

Clint Scott

Delia Sidea

Uday Singh

John Stanton

Iain Thin

John Tucker

Claudia Vejrazka

FOREWORD

On behalf of the Organising Committee, we are delighted to welcome you to the Underground Operators Conference 2023.

The conference brings together more than 1000 mining professionals and leading experts to connect, learn, share and collaborate on issues facing underground operations and to find new solutions.

The Organising Committee has put together an impressive range of speakers and topics, with technical sessions, panel discussion and keynote speakers spread over three days, showcasing leading innovations and research. The conference provides delegates numerous opportunities to share their expertise, meet new contacts and rejuvenate past connections.

Finally, we would like to thank the Organising Committee, the AusIMM events team, authors, paper reviewers, attending delegates and all our sponsors and exhibitors including:

- Major Conference Sponsor: BHP
- Sustainability Partner: Caterpillar
- Blasting Technology Partner: Dyno Nobel
- Collaboration Partner; Coffee Cart and Breaks Sponsor: Epiroc
- Innovation Partner; Name Badge and Lanyard Sponsor: Orica
- Life Cycle Partner and Networking Hour Sponsor: Sandvik
- Destination Partner: Tourism and Events Queensland
- Gold Sponsors: Barminco, Mining Plus and Polymathian
- Silver Sponsors: DSI Underground and Master Builders Solutions
- Conference Dinner Sponsor: Byrnecut
- Conference Mobile App Sponsor: Jennmar
- Notepad and Pen Sponsor: MineGeoTech
- Delegate Bags Sponsor: Macmahon

We welcome you to Underground Operators Conference 2023 and trust that you will find it an enjoyable and rewarding event.

Yours faithfully,

Alex Campbell MAusIMM(CP) & Anne-Marie Ebbels MAusIMM(CP)
Underground Operators Conference Organising Committee Co-Chairs

SPONSORS

Major Conference Sponsor

BHP

Sustainability Partner

CATERPILLAR®

Blasting Technology Partner

DYNO
Dyno Nobel

Collaboration Partner; Coffee Cart and Breaks Sponsor

Epiroc

Innovation Partner; Name Badge and Lanyard Sponsor

ORICA

Life Cycle Partner and Networking Hour Sponsor

SANDVIK

Destination Partner

TOURISM & EVENTS Queensland

Gold Sponsors

Barminco

MP
MINING PLUS

Polymathian
Industrial Mathematics Software

Silver Sponsors

DSI
UNDERGROUND
A SANDVIK COMPANY

MASTER®
BUILDERS
SOLUTIONS

Conference Dinner Sponsor

BYRNECUT

Conference Mobile App Sponsor

JENNMAR

Notepad and Pen Sponsor

MINEGEOTECH

Delegate Bags Sponsor

Ω
MACMAHON

CONTENTS

Keynote

Case studies and operating practice

Collaboration between suppliers, operations and alternative industries

Feasibility studies and mine design

Health and safety; and navigating social licence

Technology and innovation

Keynote

The successful deployment of a fatal risk system, designed for the front-line

C J Pitzer[1] and J Nel[2]

1. CEO, Safemap International Inc, Vancouver, BC, V6C 3P6. Email: corrie.pitzer@safemap.com
2. Operations Manager, Debmarine Namibia, Windhoek. Email: jan.nel@debeersgroup.com

ABSTRACT

The prevention of fatal risks is a key objective for the global resources industry, especially considering poor performance with regard to fatality rate reduction and the continued occurrence of mine disasters globally. However, the industry has a credible history, over several decades, of the development and deployment management systems to control potential Serious Injury, Fatal and Catastrophic (P-SIFC) risks.

While these systems have had a significant impact on the thinking and approaches to the prevention of fatal risks, fatal accident rates have not shown the desired reduction. Recent fatal and serious injury events in Australia have led to the establishment of several enquiries to further identify improvements that could be considered. Globally, fatal accident rates remain stubbornly high or plateaued in many jurisdictions, and several catastrophic events have occurred, such as tailing dams disasters in Brazil and South Africa, and the recent coalmine disaster in Turkey.

Apart from the apparent lack of progress, it is evident that the deployment of critical risk systems has introduced more bureaucracy, slower response systems and additional burdens on already overloaded middle and supervisory levels of management. Organisations are rarely able to effectively tap into the risk source of risk knowledge of the frontline worker and supervisor, for various reasons discussed in this paper.

To establish a valid and valuable channel of critical risk information flow from the frontline employee, Debmarine Namibia instigated the development and deployment of a system called SafeSENTRY, where the focus is on the 'dynamic identification of dynamic risks'.

This paper outlines the system, the principles of the design, the range of risk metrics that can be derived from the observations of the frontline operators, and demonstrates how a range of posited shortcomings in critical risk management is addressed how features of the High Reliability Organization (HRO) have been developed.

INTRODUCTION

Safety performance in the global mining industry continues to be under scrutiny, especially considering the highly visible and severe mine disasters of the past decade. In 2019, 287 mining deaths were reported, with most of those deaths from one event, the Brumadinho Dam failure, where 270 people died, including members of the community living downstream. This disaster followed another disaster in Brazil, at the Mariana Dam where 19 people died. More recently, a gas explosion at a Turkish mine killed 41 miners underground. Mine safety performance as measured by fatality rates is less publicised and is difficult to validly compare with other types of industries, although it is generally accepted that mining occupations are at the higher end of exposure to serious injury and fatal risk.

Obtaining all mining industry data in all countries is an impossible task, with countries like China and Russia not readily sharing information in public forums. For jurisdictions where such information is more readily available, such as in the USA, Canada, Australia and South Africa, the trends are mixed in some cases, and concerning in others. A large group of mining companies is represented in the data of the International Council for Mining and Metallurgy (ICMM, 2022), shown in Figure 1, and trends globally in fatality rates seem to show a significant reduction of about 50 per cent, since 2012, when they started to publish data, but the rate of improvement seems to be slowing in the latter years since 2019.

Fatality Rates Global - ICMM

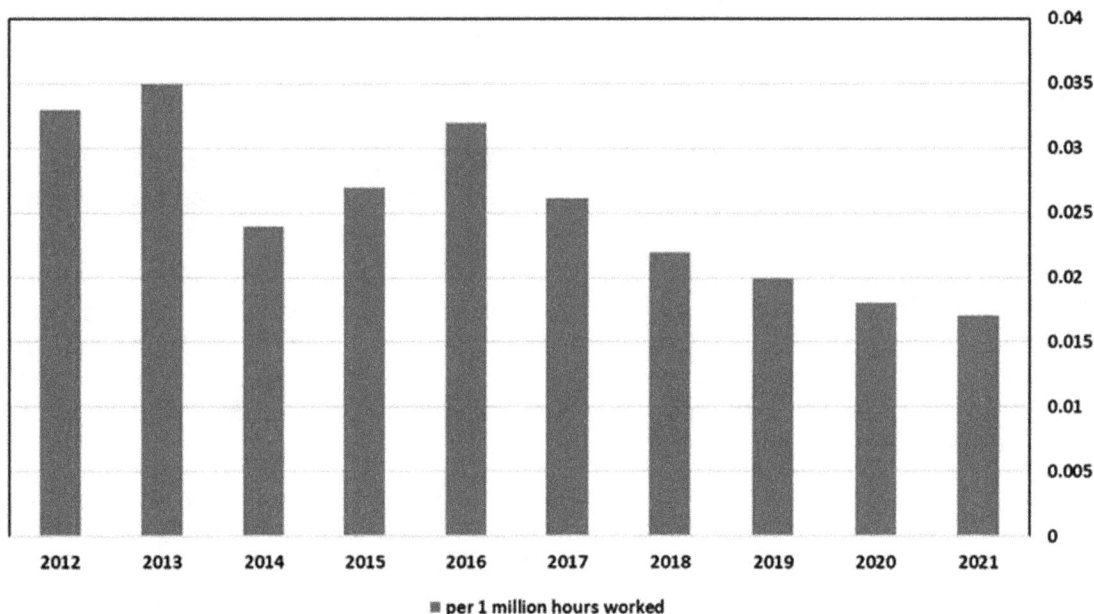

FIG 1 – Global Mining Fatality Rates 2012–2021 from ICMM members (the high death toll in the Brumadinho tailings dam disaster has been excluded from the 2019 rate calculation). Source: ICMM (2022).

However, in some jurisdictions, such as the USA mining industry, very little improvement is evident over two decades, as reported on the Mine Health and Safety Administration website (2021), represented in Figure 2.

Fatalities USA (MHSA)

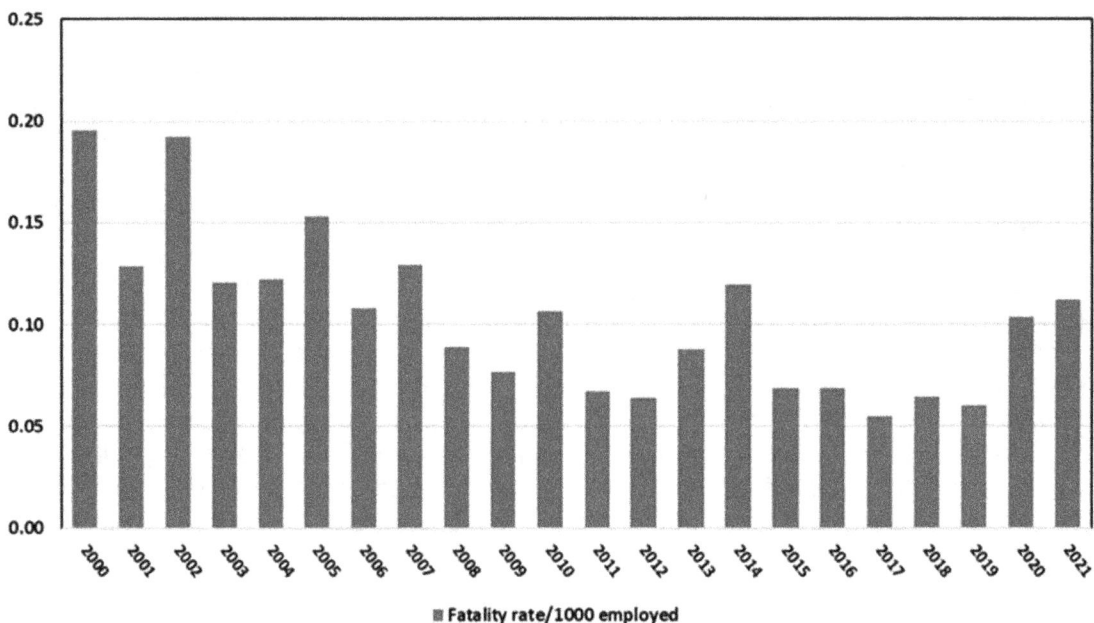

FIG 2 – Fatalities in the USA between 2000–2021.

It is clear from the table above that despite the gains made between 2000 and 2008 in the annualised fatality rates in the USA, the rate has plateaued since then, and even shows significant increases over the past five years (especially in 2020 and 2021).

Similar trends are evident in Australia, where the mining fatality rate, while significantly lower than Construction, Electricity / Gas / Water / Waste, Transport / Postal / Warehousing and especially

Agriculture Forestry / Fishing, remains stubbornly plateaued at over two deaths per 100 000 workers. Shown in Figure 3 from the web database of Safework Australia.

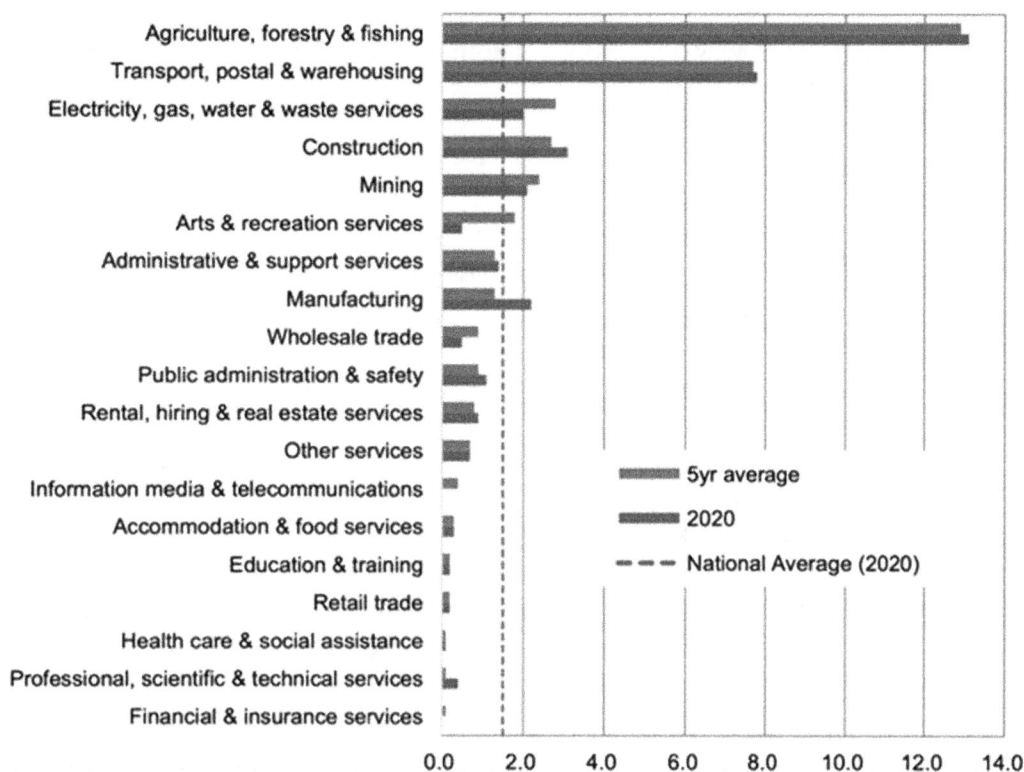

FIG 3 – Australian workplace fatality rates 2020 (Safework Australia, 2022).

A review of mining fatalities in Queensland (Brady, 2019), for the period of 1999–2022 shows no significant decrease in actual fatalities, with these remaining at an average of approximately two fatal events per annum, as shown in Figure 4.

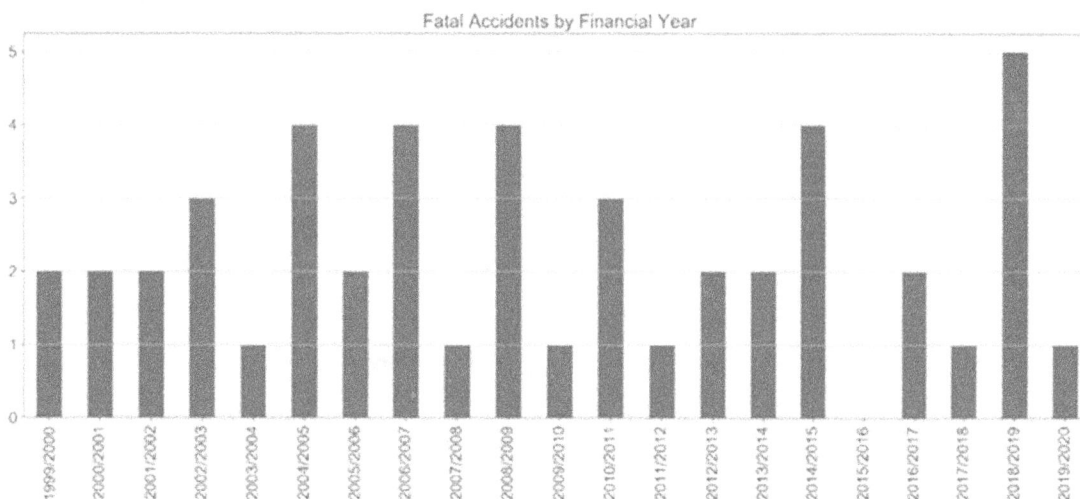

FIG 4 – Queensland fatalities 1999–2022 (Brady, 2019, p 18).

From the above information, it is suggested that claimed gains in safety performance, as measured with injury rates, are not supported by trends in serious injury and fatal rates – and that the mining industry is in need of a 'step change' in approach and prevention methodology.

THE MANAGEMENT OF SIFC (SERIOUS INJURY, FATAL, CATASTROPHIC) RISK

The management of fatal risks through formal risk management systems has been in place for several years, even decades now, with the original formalised deployment in the nuclear industry in the USA, through the work of the Institute of Nuclear Power Operations and the US Department of Energy.

In the mining industry, Australia has very much led the way since the early 1990s, resulting in the publication initially of 'Fatal Risk Protocols', which have gradually developed into more comprehensive systems. In 2015, the ICMM developed and published a manual on deploying critical control management systems, which is now widely followed in mining, and other industries, globally.

While these systems and protocols are generally well-received and effective, there is still a need for a greater degree of sophistication, or precision, in the application and impact, given the continuing disturbing trends in fatal and catastrophic events previously mentioned.

Catastrophic events, outside of the mining industry, have especially led to much being published about the emergence of the 'complex' organisation, and the apparent inability to reach the ultimate levels of controls. This had been foreseen already in the 1980s, with the publication in 1984 of a seminal work by Charles Perrow: 'Normal Accidents: Living in high-risk technology'. He illustrates how the growth of *complexity* of interactions and *coupling* as two axes of a matrix, creates ever-higher risks and greater difficulties in preventing catastrophic events, as shown in Figure 5.

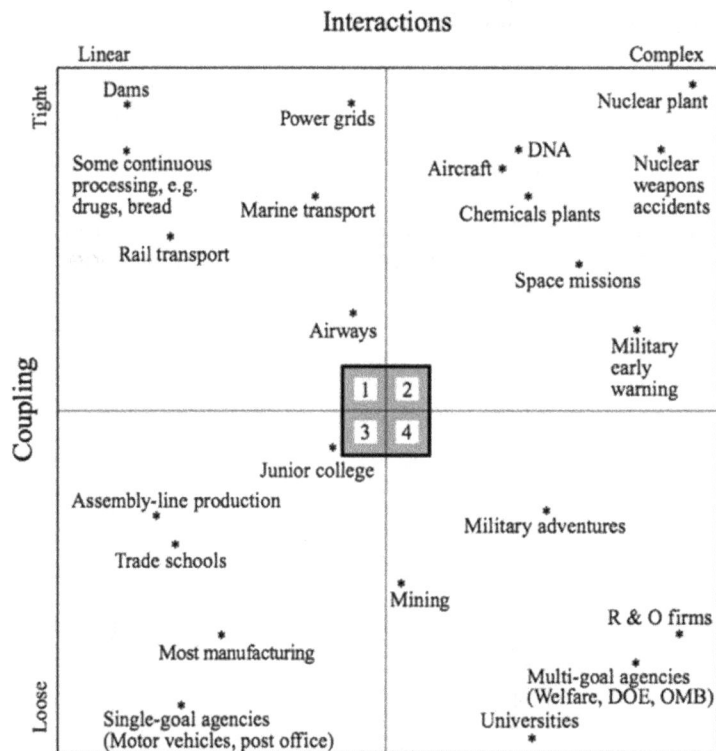

FIG 5 – Perrow's matrix (Perrow, 1984, p 96).

While these two axes are typically 'structural' in nature, there is also a sociological dimension to the management of risks, or more precisely, in the breakdown of control of risks.

It is beyond the scope of this paper to delve deeply into this dimension, other than to provide context for the broader organisational culture issues that often underpin the management of risk. The work of several academics is relevant here, but most importantly is the phenomenon of 'risk homeostasis', developed by Prof Gerard Wilde (1994), and later redeveloped and labelled by Prof John Adams (2001) as 'risk compensation'.

Essentially, the basic tenet is that humans (and organisations) have a 'target level of risk' that they are willing to tolerate, in order to gain the benefits they want to derive from that activity.

If people perceive that the level of risk is less than their target, they may modify their behaviour to increase their exposure to risk. Conversely, if they perceive a higher than acceptable risk, they will compensate by exercising greater caution. This balancing act occurs often and by everyone, eg driving a car on a clear and sunny day will be done at higher speeds than on a rainy day with slippery roads. The broader and more sinister implications of this notion is that safety interventions designed to make workplaces safer, will conversely attract higher levels of risk-taking behaviours, and offset any gains that were intended.

An example of this behavioural shift was noted by the author during the implementation of a Proximity Alert System (PAS) in Brisbane, Australia, in 2003. The system was developed for a road repair crew, responsible for re-surfacing a road, by spreading out bitumen with 'brooms', in front of a multi-roller. A previous fatal accident involving the roller hitting one of the workers prompted the request to develop the system. It was noted that during the development and deployment of the system, the workers were gradually reducing the distance between themselves and the approaching roller, until it reached a point where they were simply waiting for the alarm (at about 2–3 metres) to sound before they moved forward. Before, with the PAS system in place, they maintained a distance of 6–7 metres, and were constantly watching the roller. The behavioural shift illustrated the process of risk compensation, as the workers perceived higher levels of protection. Needless to say, the implementation of this system was not recommended.

At an organisational level, the same perceptual distortions are noticed, as organisations improve their safety performance, to 'near zero' levels, and believe that the safety defences in place are effective. This creates a range of 'delusions' (false beliefs) which were explored by Dekker and Pitzer (2016).

In this paper, they explore the impact on an organisation's exposure to catastrophic events because of applying simplistic, linear models and safety techniques to complex processes. Of consequence is the emergence of a culture of cult-like compliance, and the application of inappropriate techniques (such as safety audits, safety cases, risk matrices etc) that create a misconception that risk is 'under control, when nothing goes wrong'.

These 'near zero' organisations uphold several other false beliefs: human error as a cause of accidents, safety performance as objectively measurable, and that accidents and incidents are indicators of flaws and failures of the production process which can be systematically eliminated. The net effect is that risk control is often superficial, reactive and tenuous, until the next serious, fatal or catastrophic event shatters the delusions.

SHORTCOMINGS IN CURRENT RISK MANAGEMENT APPROACHES

In view of the apparent poor performance on improving fatality rates, and continued occurrence of catastrophic events, it is contended that the current risk management models fall short in six key areas, and each are stated below, as a framework to evaluate the safeSENTRY against, see the *Discussion* section.

Risk is seen as 'static', yet is inherently 'dynamic'

Over time, the safety profession developed and deployed a range of 'risk assessment tools' to guide the effective control of risk, with the 'hierarchy of control' and the 4 × 4 risk matrix the most common. These tools have good practical application and value for the practitioner in the field, but have also contributed to misconceptions about the 'nature' of risk. The risk matrix, especially, created the impression that risk can be readily quantified as the product of likelihood and severity, and neatly placed on a chart, often with three colours of red, yellow and green, to emulate a simple traffic-light system. However, the very essence of probability, the random variability in likelihood, exposure and consequences, would strongly suggest that risk is a highly dynamic and emerging characteristic of circumstances, and cannot, therefore, be quantified in categorical local terms.

Risk 'control' is over-emphasised, at the expense of risk 'discovery'

The typical risk management system, as described by the well-known ICMM model (2015) that is widely applied in the resources industry, is highly biased towards controls, and much less effort is given to the identification of risks. By its own admission, the list of Major Unwanted Events (MUE's) is accepted as a given and most mining risks can be described by these known critical risks. While this is essentially true, the unintended consequence of this is that other potential unwanted events are not actively *searched* for in day-to-day routines and are assumed to be 'caught' by many layers and levels of critical controls. As a further consequence, many risks outside of the risk register (MUE's) and outside the control net have a higher potential of occurring, with more severe consequences, as there are fewer control systems in place for these risks.

Control hierarchies create complex defences-in-depth and 'false confidence'

The logic of layered defences seems incontrovertible, to create the maximum level of protection for the operations and personnel. This principle was famously illustrated by the Swiss Cheese model of Prof James Reason and became one of the most widely used graphics in safety management. However, these defences are designed based on the known causation of known local events and cannot and do not represent the full gambit of possible failures, while creating an illusion of control. The mere fact that serious events are extremely rare, suggests that the amount of information about such events will be extremely limited, and very little will be readily known and available about causation, and possible failure modes in the system. Additionally, the more layers there are, the more confident the operators will be that the system is fail-safe. Natural or normal levels of attention and awareness may well be lessened – what Vaughan (1996) called 'redundancy-induced overconfidence'.

Learning from incidents creates 'risk myopia'

Risk myopia is not a common term in safety or risk management but is offered here to describe a phenomenon that the author has widely observed in industry. It can be described as a natural bias or heuristic. The availability heuristic refers to the tendency to assess the probability of an event based on the ease with which instances of that event come to mind. Thus, there is a tendency to overestimate the likelihood of events with greater 'availability' or recency in (corporate) memory. The focus, and effort required, to study and learn from local or recent events will most likely create a bias to focus on similar events as potential risks.

Additionally, as is often the case in a large corporation or a closely linked industry such as mining, there is often an enthusiasm to learn from own and/or other companies' events, resulting in the dissemination of accident reports or 'briefs'. However, the consequence of this may be that weight is given to certain (dramatic) events that have no or little realistic relevance for a particular site or group, but create an unnecessary or wasted effort to identify similar risks related to those incidents – in essence, risk myopia.

Management processes are 'disconnected'

The concept of 'connectedness' is not commonly used in safety management, or other circles, as a measure of performance. It is hypothesised here as an alternative measurement of organisational effectiveness and as a 'measure' of safety.

A classic example of this could be the well-known principle of 'Just In Time', as originally practised in Japanese car manufacturing, particularly Toyota, (see Ashburn, 1977). A just-in-time (JIT) inventory system is a management strategy that has a company receive goods as close as possible to when they are actually needed, avoiding the need to hold large inventories. If the same principle of connectedness is applied in the field of risk management, it can be argued that it is not the actual and separate parts of the system that define the capacity to maintain a safe operation (or with risk under control) but rather the quality, speed and clarity of the connectedness between the parts. As an example, it is not the quality or quantity with which hazard reports are being completed, but the speed with which the hazards so identified are addressed – and the speed of response is only optimal if the two parts of the systems are tightly connected. As such, safety is defined by the author as the degree of *connectedness* of a variety of key risks and organisational processes. The following elements of connectedness in a risk management system should be considered:

- *Feedback loops* are lines of information exchanged between parties who need these, in order to ensure the flow of operations. A feedback loop on risks associated with certain high-risk tasks, needs to provide the required information (eg procedures) to the person who will execute the job, and feedback from that person back to the supervisor.

- *Hierarchical alignment* refers to the extent to which various job levels act in unison, towards a common goal and can be measured through perception analysis.

- *Integration of various* functional activities into the operational processes is a key measure of the degree to which risk identification and control become an upstream focus, or, adversely, is a downstream reactive focus.

- *Forward focus* is associated with integration in some respects, with the major balance of the risk management process towards the dynamic identification of dynamic risk, or towards the control of risks that are, retrospectively, linked with incidents. An example is whether the organisation is reactively responding to near misses, or proactively responding to potential risks.

- *Dynamic verification* of critical controls is a pivotal component of effective risk management, which seldom happens in practice (inspection-driven oversight by control owners or assigned leaders/personnel is not a dynamic process and will miss failures that occur in the 'real world').

The management of critical risk lacks 'reliability'

Most organisations attempt to control their critical risks through a focused system, that is often given a label to create attention and buy-in, such as Cardinal Rules, Golden Rules and more recently Life Saving Rules. These programs have varying and limited success, as it is often associated with harsh responses from management, such as Zero Tolerance of non-compliance, and as such, effectively shutting the life blood of the control system: transparency about potential hazards and problems, and creates risk secrecy. A significant and positive evolution in the field of critical risk management was the High Reliability Organization (HRO), which started with the work of researchers at the University of California around 1995, and has since become very popular in many organisations and industries, essentially describing a type of organisation that 'should have lots of accidents, but don't.' The characteristics of the High Reliability Organization (Weick and Sutcliffe, 2001, p 10) are that:

- They anticipate mistakes and failures, *so an open climate for reporting errors* is important.

- They favour interaction between employees with different points of view that describe the observed events as accurately as possible.

- They are *sensitive to operational processes*, and learn from near misses.

- The *pursuit of resilience*, by being ready to respond to risks.

- The *deference to expertise*, whereby problems are identified and solved by those with the knowledge and skills to do it best.

The focus will be to assess whether the SafeSENTRY system counteracts these gaps and weaknesses.

DEPLOYMENT OF A FATAL RISK MANAGEMENT SYSTEM

Given the outline of multiple impediments, potential failures and limitations of typical risk management systems described above, the challenge to design and deploy a system to address all these issues was conceptualised at Debmarine Namibia, a submarine resource recovery organisation in Namibia.

Background

Debmarine Namibia is the country's leading marine recovery company and is a recognised world leader in marine diamond exploration and recovery technology. The company is a joint venture marine diamond prospecting and recovery company, owned in equal shares by the government of the Republic of Namibia and the De Beers Group, itself part of the Anglo American group of companies. Debmarine Namibia is a wholly owned subsidiary of Namdeb Holdings (Pty Ltd).

Recovery operations take place off the south-west coast of Namibia in the Atlantic 1 license area of Namdeb Holdings. Debmarine Namibia's head office is in Windhoek, the capital of Namibia. Logistics support services are based in Oranjemund, which serves as the transit base for employees between the recovery vessel fleet and head office.

Debmarine Namibia owns, operates and maintains six production vessels, namely: mv Debmar Atlantic, mv Debmar Pacific, mv Grand Banks, mv Gariep, mv Mafuta and mv Benguela Gem. The seventh member of the fleet, mv SS Nujoma, is a diamond exploration and sampling vessel.

No human hands touch the diamonds during the entire production process at sea. The company operates on a 24/7, 365 basis and the crew onboard work in rotating teams, 28 days on board followed by 28 days shore leave. Crew members are transported to and from Oranjemund by fixed wing aircraft, and flown to and from the vessel by helicopter.

Method and design

Prior to the deployment of the SafeSENTRY system, Anglo American deployed a global program called Elimination of Fatalities, which consisted of a core team of safety and risk specialists, joined by internal staff of Anglo American in a team of approximately 10–15 personnel. Teams visited each of the 42 mine sites to conduct a week-long fatal risk review, and give feedback to the local management teams on the findings. Debmarine Namibia participated in the activity and expressed the need for an in-house system that could achieve the same focus on fatal risk exposures. The risk management system that was developed at Debmarine Namibia is called the SafeSENTRY system, to align with typical terminologies that would be custom in a marine environment. In this case, traditionally, ships would have a 'lookout' to warn ship commanders of potential dangers, such as debris, other ships, icebergs, navigation marks or land.

The SafeSENTRY system is a rotational role that is shared in a natural work team each day and supported by the SafeSENTRY application, installed on a commercial tablet device. Prior to deployment, a series of training programs are conducted, to train 'MasterSENTRIES', who are responsible for the training of all other employees on risk observation and recording to the SafeSENTRY application. Training sessions for supervisors and leadership levels on the managing of the overall risk system are also provided. At the beginning of 2020, the system started off as a paper-based system but soon ran into significant challenges with capturing and data analysis. An app was then developed to improve the overall effectiveness of the system.

The observation process (guided by the application) requires the allocated SafeSENTRY to visit other workplaces, for about 45–60 minutes at a time, and to engage with other available team members for a risk discussion. They use a 'risk lens' tool, embedded in the application, to identify risks that may be overlooked, underestimated, shortcut taken, work pressured, inherent or tolerated. This is known as the '6Y' lens tool. Each observed risk is then analysed, using an embedded graphic RiskCalculator to assess the risk level of each observation, as determined by the potential impact, exposure and likelihood, in that order. This results in a rating on a scale ranging from minor to major threat. The application is user-friendly with most of the sections of the application completed by taking photographs of situations, recording voice commentary, or simple typing in text boxes.

Mostly, the observation would be of a high-risk task – working at heights, lifting, hot work, confined spaces etc, (but is not limited to these) – and will require the SafeSENTRY to assess and verify simple critical controls associated with that activity, rating them as Pass or Fail. The SafeSENTRIES also observes 'other' activities and can add them as critical risks, as well as adding critical controls that may be needed or not available. There is also an opportunity for the SafeSENTRY to record any 'good practices' observed, as well as to record what actions he/she took to reduce risk or rectify a failing. The app then requires the assessment of four statements on safety culture, as part of a continuous pulse perception survey.

Before the SafeSENTRY signs off for the day/shift (typically after completing two to three observations), they complete an assessment of the leadership of that location, by answering whether they noticed any of six (pre-defined) leadership practices by any manager or supervisor.

The final step is for the SafeSENTRY to do a handover on the SafeSENTRY application to the supervisor of the area where the observation took place, and after review, the supervisor signs off by assessing the quality of the observations and records any comments and follow-up actions, if required.

The process described above generates a variety of data sets as recorded on the application. The SafeSENTRY observation process described above, operates with and without wi-fi connection depending on availability and uploads its data whenever connection to the internet is established.

RESULTS AND ANALYSIS

The key data sets that are collected as part of the SafeSENTRY process include the following.

Participation rates

The participation of employees in the SafeSENTRY process is an obvious key performance indicator (KPI) for the entire system of engagement of employees, and a target of 36 sessions per week in the entire fleet has been set, implying that on each vessel, at least one SafeSENTRY session is completed each day, six days per week.

The participation rates are reported in Figure 6 (since inception in August 2021 to October 2022, at the time of the writing of the paper).

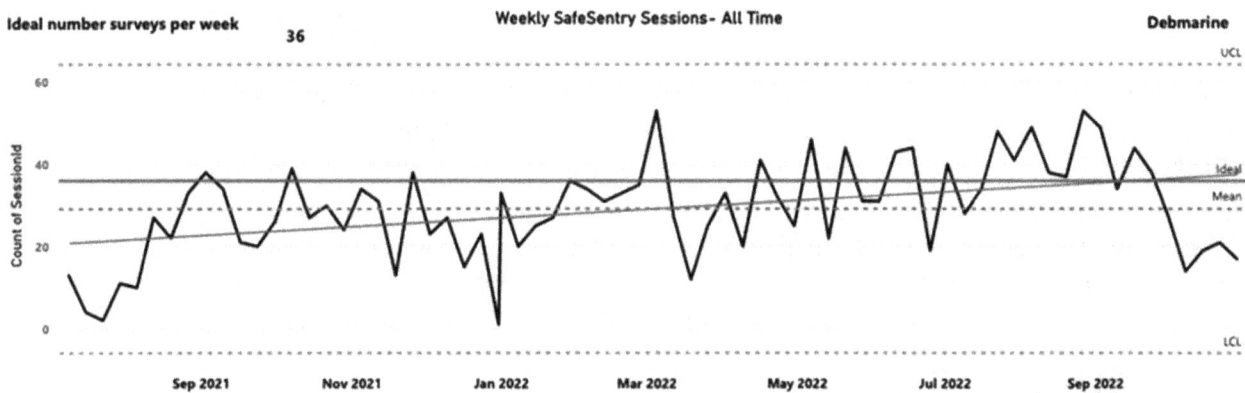

FIG 6 – SafeSENTRY participation rates per week.

It is evident from the data above that the organisation saw a gradual increase in participation rates since inception, and started to consistently hit the target of 36 sessions per week around May of 2022, and has been mostly above the target for the three months of August, September and October, although participation rates dropped slightly below target in the month of October 2022.

From a participation rate perspective, the system appears to be functioning effectively.

Systemic/cognitive ratio

The ratio of systems versus cognitive is a function of the 6Y lens tool that is mentioned before and is linked to the classifications that the SafeSENTRY observers used when they first observed the critical risks. The lens has six dimensions, with three of those typically systemic and the other three typically cognitive.

It was argued that risks classified as 'Tolerated', 'Pressured' (by workloads) or Inherent (in the operational process of design) are typically 'Systemic', that the 'origin' of the risks is outside the immediate focus of the frontline employees.

If the risks were classified as 'Overlooked', 'Misjudged' or a 'Shortcut' (Rewarded), one can argue that the risks resulted from the immediate actions and behaviours of the frontline employees, and are therefore 'Cognitive/Behavioural'.

It was an assessment from early risk classifications in the Anglo American Elimination of Fatalities process (Anglo American, 2019), and during the design and piloting period of the SafeSENTRY system, that the classifications of 'Systemic' typically occur in organisations that focused more on downstream control of risks. Conversely, 'Cognitive/Behavioural' classifications occur more in organisations where the focus is on upstream elimination or substitution of risks. If viewed in the terminology of the Hierarchy of Control, it is the higher order versus lower order of controls. From this assessment a 'ceiling' (Threshold) was set, that ideally, the Systemic observations should be less than 40 per cent, and the Cognitive/Behavioural should make up more than 60 per cent of classifications. In essence, this will imply that the organisation is less prone to generating risks upstream.

The control charts on this parameter are presented in Figure 7.

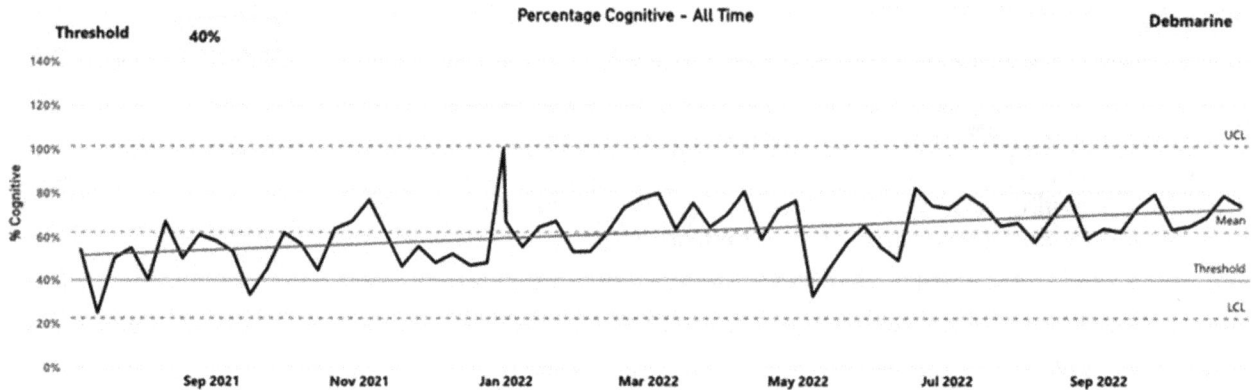

FIG 7 – Control chart of Systemic/Cognitive ratio.

It is evident from the data above that the organisation has gradually improved the ratio since inception in August 2021, and is consistently above the 40 per cent threshold on 'Systemic' classifications.

From the perspective of this parameter, the quality of management's response to failures appears to be improving, by increasingly focusing on higher-order controls of risks. Many examples have been recorded of the effort by management to improve designs, eliminate risks or reduce the likelihood of events (as opposed to mitigating impact after the incident).

Risk level rating

Each risk observed by the SafeSENTRY is given a risk level rating, by using a Critical RiskCalculator, as shown in Figure 8, with scales for the dimensions of Impact, Exposure and Likelihood. Note that the Impact scale starts at Lost Time Injury as the lowest impact rating, to force a focus on serious injury and higher risks, and that the Likelihood scale uses behavioural anchors that can be readily applied by the frontline operator, as against the usual (non-sensical) definitions of rarely, frequently, often etc).

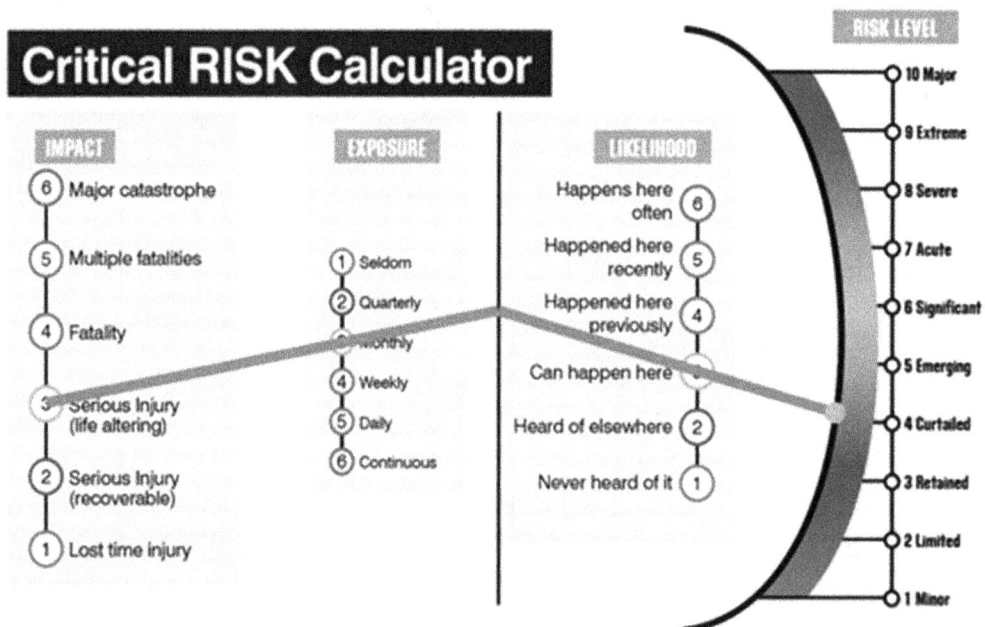

FIG 8 – Graphic of the RiskCalculator.

This parameter is reviewed on two different dimensions:

The **first dimension** is to assess the 'normal distribution' of the risk levels, as presented in Figure 9.

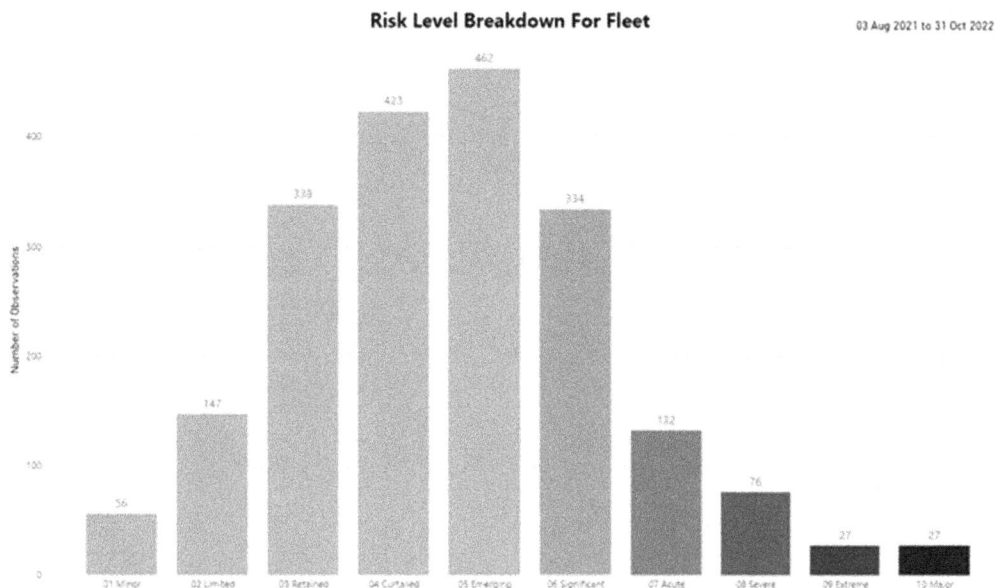

FIG 9 – Distribution of risk level ratings.

The chart in Figure 9 shows a skew to the left, towards the lower end of the risk levels. This indicates that the risks are generally seen as of lower likelihood and exposure levels Ideally, the focus should be more on the higher-risk classifications, but it will be shown further in this paper that the focus on fatal and higher-impact risks is adequate.

The **second dimension** is to assess the risk levels, on a control chart, with a 'ceiling' set at six, or at 'significant level' as presented in Figure 10.

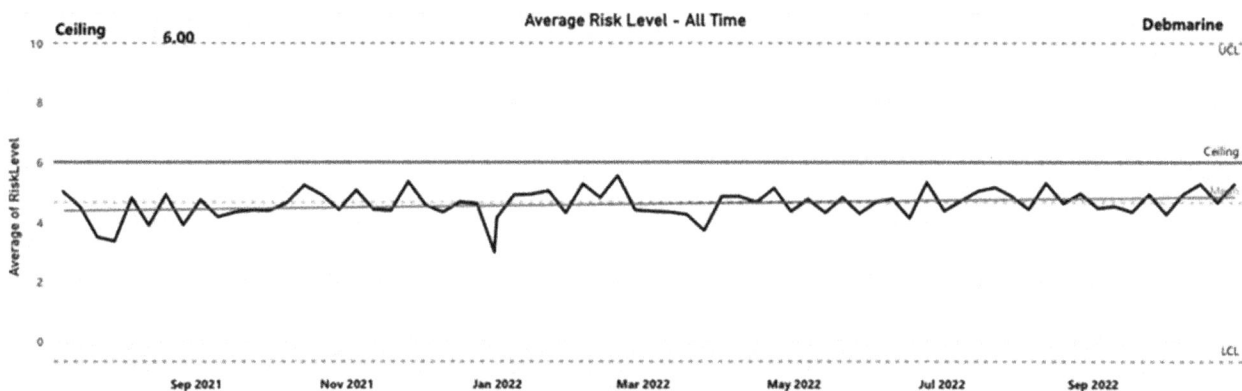

FIG 10 – Control chart of Risk Level ratings.

It is evident from the data above that the organisation has continued to retain the risk levels at below the 'significant level' albeit with a (very) gradual increase in ratings apparent in the overall period.

Focus on impact

The breakdown of the Impact scale, shown in Figure 11, on the Risk Calculator (as assessed by the observers) shows the proportions of each, with the biggest proportion being that of Serious Life Altering Injuries, at 37 per cent of all observations.

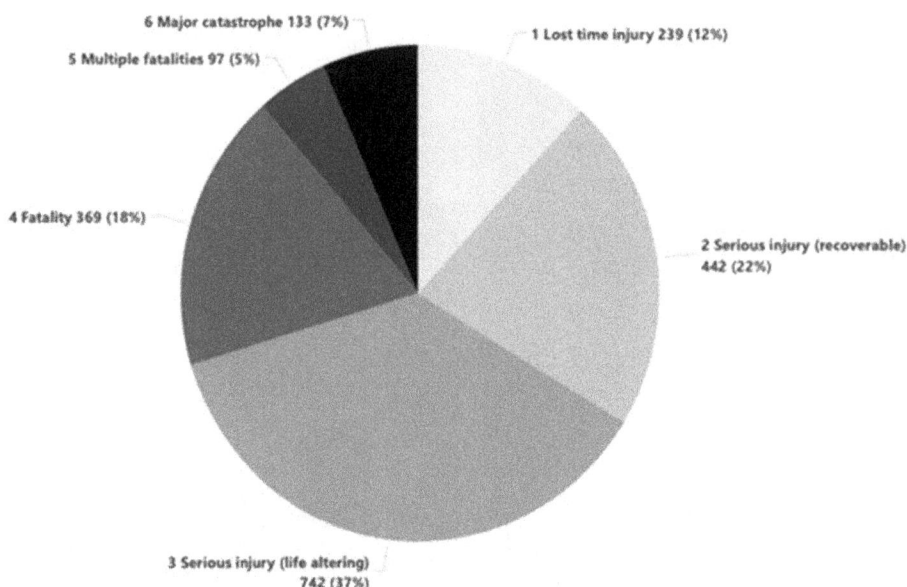

FIG 11 – Pie chart of Impact breakdown.

Of more value is the overall proportion that high Impact events (fatal, multiple fatal and catastrophic events) have, which in the overall period August to October 2022 amounts to 30 per cent of all observations, with fatal risks at 18 per cent.

The data presented in Figure 12 shows the overall trend of the proportion that High Impact (fatal/multiple fatal/catastrophic) risks constitute of overall observations, since inception.

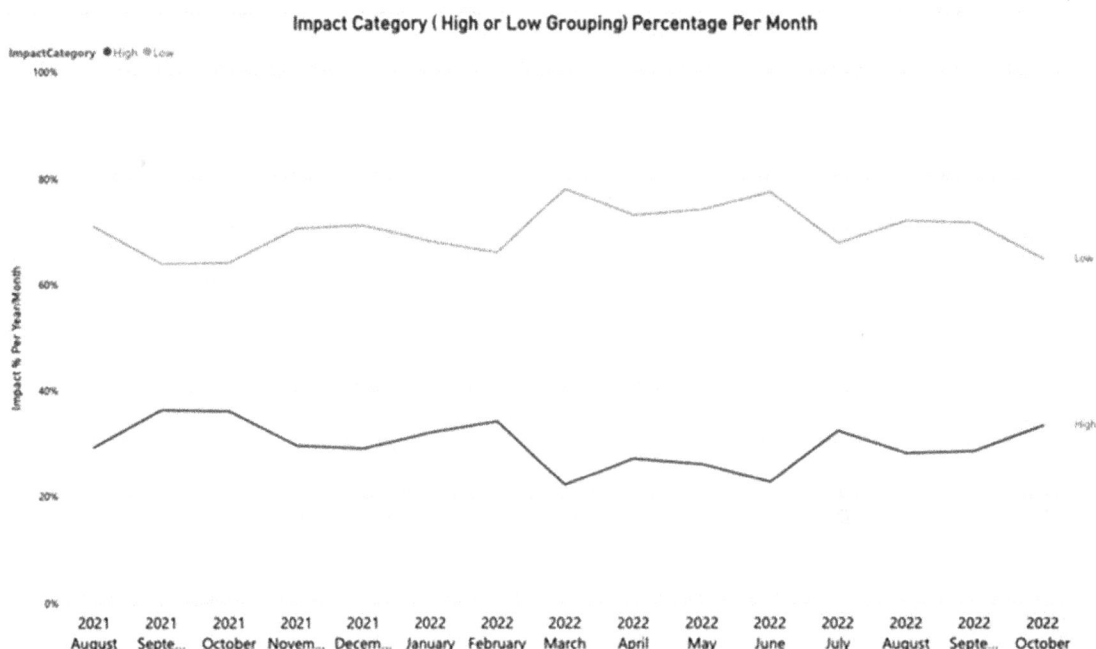

FIG 12 – Line graphs of High Impact risk classifications.

The line graph of the High Impact proportion is of particular focus, indicating that High Impact risks remained on average about 30 per cent of all risks identified, which, arbitrarily, is a targeted level of all observations. The fundamental focus of the system is on high impact risks (the Impact rating scale used in the risk calculator starts at Lost Time Injury) and that lower injury risks and other non-compliance issues are to be collected under the existing systems of HSE observations.

Ratio of Priority Unwanted Events (PUE's)

The focus of the SafeSENTRIES is naturally directed towards the various high-risk activities that are performed in the vessels, of which there is daily a high variety being performed.

The application is pre-populated with the selected PUE's of Debmarine Namibia, and provision is made for observations outside of those risks to be also identified and recorded.

Figure 13 shows the proportion of PUE's observed by the frontline employees, of which 18 per cent (354) of the PUE's observed was of Lifting, 11 per cent (217) of Working at Heights, and 7 per cent (147) of Equipment safeguarding. The highest proportion however was labelled as 'Other', with 45 per cent (887) of the observations.

Critical Risk (PUE) Observations   03 Aug 2021 to 31 Oct 2022

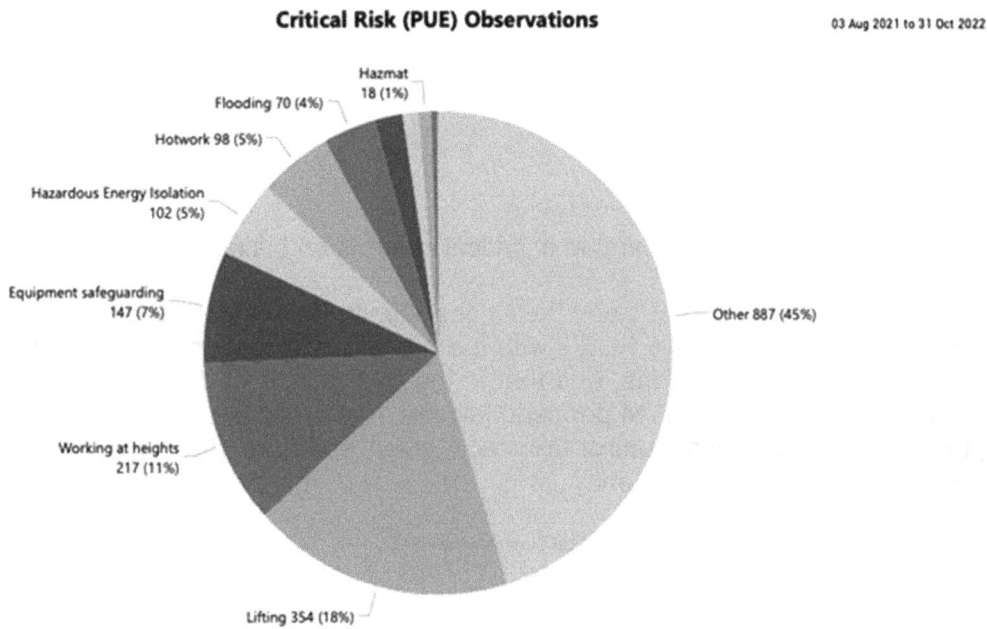

FIG 13 – Proportion of PUE's observed.

Critical control failures

The verification of the performance of critical controls is a very important aspect of the SafeSENTRY system, because the SafeSENTRY arrives unannounced at a location where a high-risk activity is being performed, and the testing of the controls happens dynamically as part of the overall observation process. The application is pre-populated with the various PUE's (WAH, Hot work etc) and also with the associated critical controls for each of those. However, the controls are limited to a maximum of four ('Super') critical controls, which are within the direct reach or capability of a frontline employee to actively verify, at that location or activity. The controls are also selected based on their active prevention of an incident. For example, they would not verify if a person is qualified to operate a crane by checking the relevant certificates (which would anyway not be available) but whether the lifting zone is barricaded/access controlled. The SafeSENTRIES would review the controls during the observation by selecting 'Pass' or 'Fail' as seen in Figure 14.

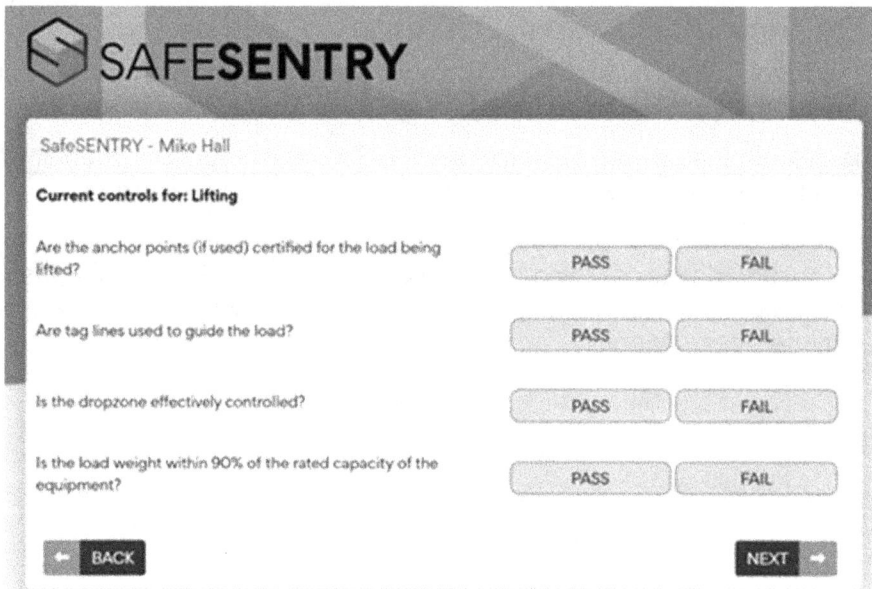

FIG 14 – Example of critical controls on Lifting.

In each review period report, the PUE's and the Critical Control failures are shown graphically as in Figure 15a, which shows that the two PUE's with the highest observations, Lifting and Working at Heights have similar failure proportions, and that in the total period of deployment, 4083 controls were tested/verified, and with 700 (17.14 per cent) failures. A more detailed analysis of each of those PUE's and their associated critical control failures is given in Figure 15b. This provides valuable information on follow-up and improvement.

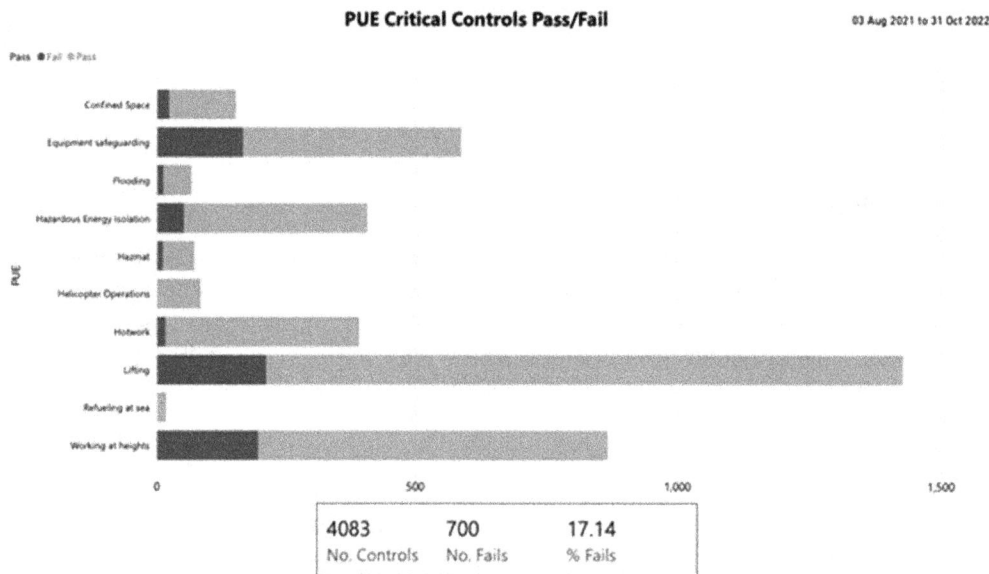

4083	700	17.14
No. Controls	No. Fails	% Fails

FIG 15a – PUE observations and failures.

Critical Controls Pass/Fail   03 Aug 2021 to 31 Oct 2022

CriticalRisk	Fail	Pass	CriticalRisk	Fail	Pass
Confined Space	15.79%	84.21%	**Hotwork**	4.34%	95.66%
Are all people involved with the entry listed on the permit?	18.42%	81.58%	Are fuels protected from ignition?	5.10%	94.90%
Is access to the confined space controlled?	18.42%	81.58%	Are ignition sources from the hotwork contained to the protected area?	7.14%	92.86%
Is there a specific rescue plan and equipment?	13.18%	86.84%	Is there a firewatch, who knows how long to stay after the hot work is completed?	2.04%	97.96%
Is there a spotter/lookout/guard/to check the area?	13.16%	86.84%	Is there an approved permit for the task?	3.06%	96.94%
Equipment safeguarding	28.40%	71.60%	**Lifting**	14.99%	85.01%
Are all machine guards are in place?	42.86%	57.14%	Are tag lines used to guide the load?	15.82%	84.18%
Emergency stops are within easy access of the related machine?	11.56%	88.44%	Are the anchor points (if used) certified for the load being lifted?	15.36%	84.64%
Equipment is locked, tagged and tried before use?	16.33%	83.67%	Barriers and rails	85.71%	14.29%
There are no gaps in the guards?	42.86%	57.14%	Is the dropzone effectively controlled?	54.16%	45.84%
Flooding	19.70%	80.30%	Is the load weight within 90% of the rated capacity of the equipment?	9.86%	90.14%
Classified doors are closed and locked.	19.70%	80.30%	Proper PPE	54.55%	45.45%
Hazardous Energy Isolation	12.75%	87.25%	Proper Procedures	45.45%	54.55%
Does the operator or mechanic have control of their personal key?	13.73%	86.27%	Warning Signs	55.56%	44.44%
Has the "Try" test been applied to ensure there is no energy?	9.80%	90.20%	**Refueling at sea**		100.00%
Is the equipment locked with a signed and dated tagged?	14.71%	85.29%	Fuel rate and tank level monitored?		100.00%
Is there an approved Permit To Work (PTW) for the task?	12.75%	87.25%	Is a radio link established?		100.00%
Hazmat	16.67%	83.33%	**Working at heights**	22.93%	77.07%
Are hazardous materials stored correctly in an appropriate cabinet or designated area?	27.78%	72.22%	Is the worker using an approved anchor point?	41.47%	58.53%
Are ventilation fans working?	5.56%	94.44%	Is the worker using lanyards to prevent falling objects?	19.35%	80.65%
Do all containers have labels identifying their contents?	11.11%	88.89%	Is the worker wearing an approved and inspected harness?	17.51%	82.49%
Is access controlled to hazardous material?	22.22%	77.78%	Is there an approved permit to work for the task?	13.36%	86.64%
Helicopter Operations	1.19%	98.81%			
Are landing conditions confirmed as green (Check with CNO or bridge).		100.00%			
Ensure no loose items/articles/debris on deck.	4.76%	95.24%			
Is an HAD at the CO2 lance and dressed in complete fire fighting suit?		100.00%			
The foam system is lined up and ready for use and the valve is open, then closed after		100.00%			

FIG 15b – Critical controls pass/fail.

An (arbitrary) maximum fail rate of 25 per cent had been set during the initial design and piloting of the system, which has proved to provide a level that has practical value. The data presented in Figure 16 shows the average fail rate for the various periods of review.

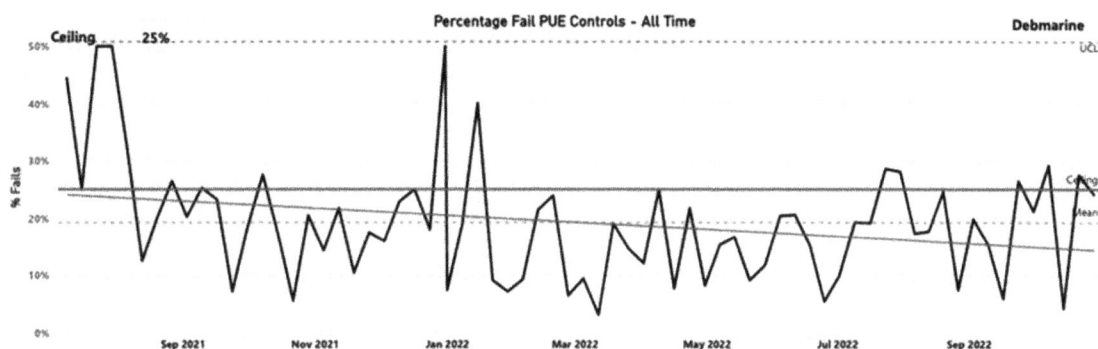

FIG 16 – Control charts of critical control failures.

The data suggests that the organisation has been able to improve (decrease) the critical control failure rate over time, with a sharp increase above the 25 per cent level in the last month of the review period – which prompted further investigation and follow-up by the management of the appropriate vessel where it occurred.

High impact control failure rates

A more in-depth analysis is made of the failure of critical controls as they are associated with high impact (fatal/multiple/catastrophic) and low impact (injury-related) risks. This comparison is made based on the understanding that failures of high impact risks' critical controls are obviously less desirable and therefore require more detailed and quicker responses. The chart in Figure 17 shows the proportion of these two types of control failures in each month.

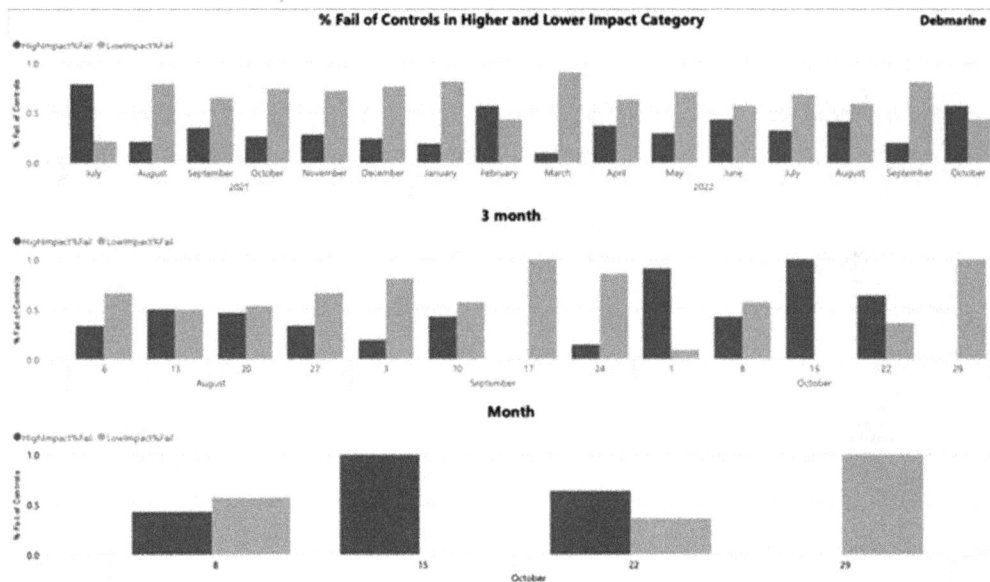

FIG 17 – Percentage of critical control failures of high versus low impact risks.

The above data is represented as a ratio for each period, as follows in Figure 18.

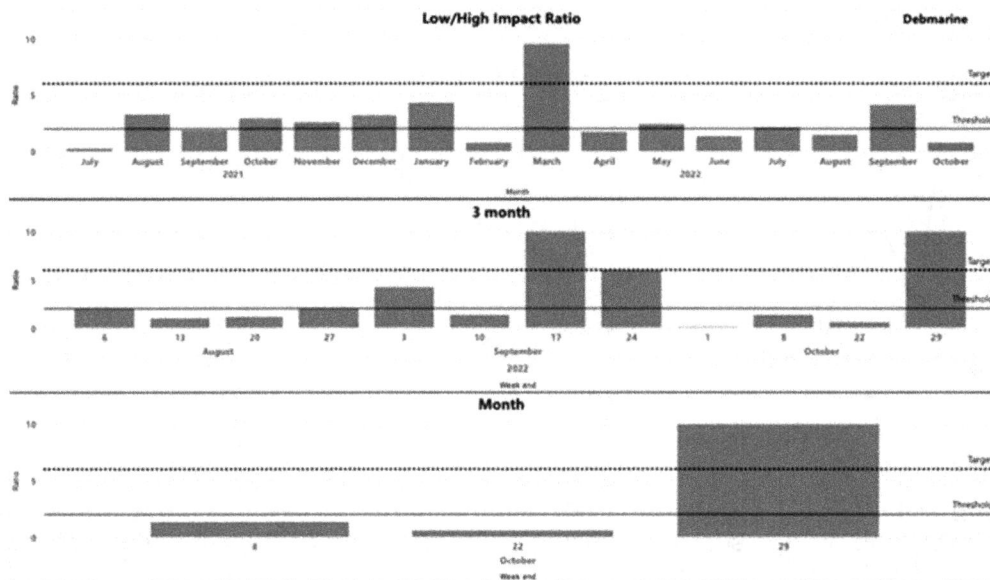

FIG 18 – Ratios of high versus low impact control failures.

The chart in Figure 18 depicts the ratios in each month, displayed against an ideal (target) of six, where the proportion of low impact control failures is significantly higher than the proportion of high impact failures, which would suggest that the risk of a fatal event is lower. When the ratio approaches or goes below the threshold level (two) the risk of a fatal event occurring is *increasing*, and will prompt a quick response to arrest the trend, as it happened in the first three weeks of the month of October.

Safety culture factors

The SafeSENTRY observations include a pulse safety culture perception survey, conducted by the SafeSENTRY when they engage with fellow employees. It consists of seeking the agreement/ disagreement on just four statements (which refresh daily to display four different statements), as shown in Figure 19. The overall measurement of perceptions is based on eight factors, with each one associated with a positive and a negative statement, in total 16 statements. In this way, the eight factors are surveyed on a rolling basis every eight days.

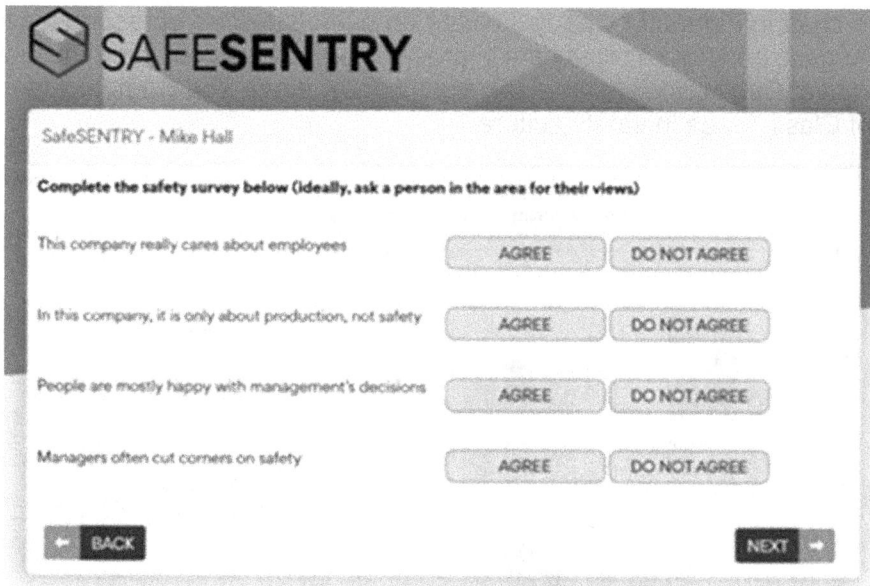

FIG 19 – Example of pulse survey statements.

The responses are shown as the Net Positive Response (NPR), calculated by subtracting the negative response levels from the positive response levels, and displaying this against the Top Company response levels (as derived from the Safemap Safety Culture database of over 1000 companies and 1.5 million employees, globally), shown in Figure 20.

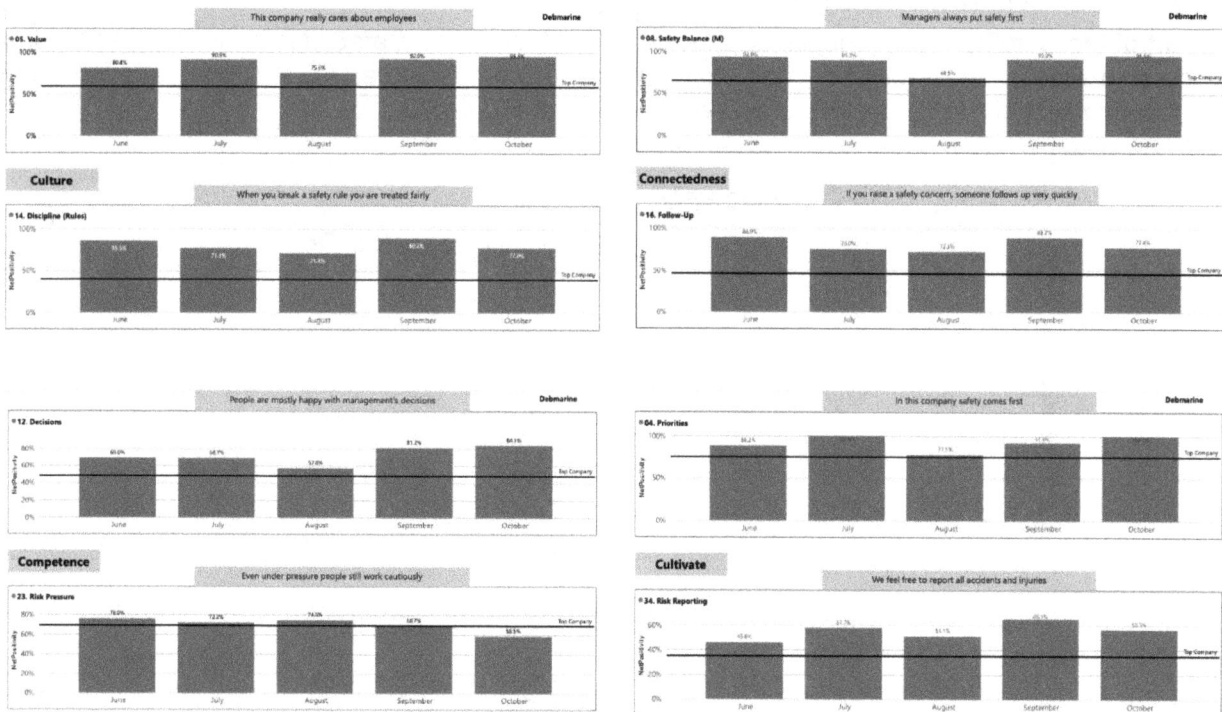

FIG 20 – NPR response levels on pulse culture surveys per indicator.

The charts above show the response levels on four indicators of the De Beers strategies for achieving their vision of Pioneering Brilliant Safety: Culture, Connectedness, Competence and Cultivate, over a five-month period. One factor, Risk Pressure, shows a declining rate towards 'I know people have to break safety rules to get jobs done', although still overall at Top Company response levels, while there is a significantly improving perception on Risk Reporting of 'We feel free to report all accidents and injuries'.

The data on the various factors is aggregated for each month, as below, and reviewed by the management teams on a continuous basis.

The trends in the chart in Figure 21 clearly indicate an overall decline of responses in the period of mid-July to mid-August, and then a significant improvement in the period after that. However, overall, the response levels at Debmarine Namibia are significantly above Top Company levels, suggesting it to be in 'Best of Class' range in safety culture.

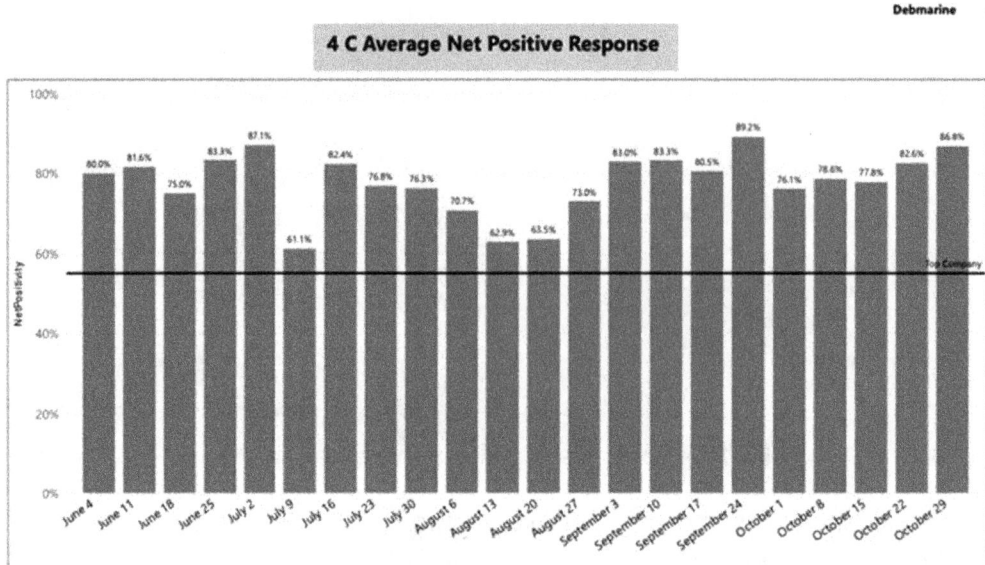

FIG 21 – NPR response levels on pulse culture surveys, all indicators combined.

Leadership practices ratings

As part of the SafeSENTRY process, at the end of each shift, the SafeSENTRY is required to do a 'leadership assessment', based on their observations, by completing an Agree/Do Not Agree review of observing six defined leadership practices, as listed in Figure 22.

FIG 22 – Leadership statements included in the SafeSENTRY assessment.

The aggregated responses are presented on control charts as below in Figure 23, with an 'Ideal' set at 90 per cent.

FIG 23 – Leadership statements as done by SafeSENTRIES.

The results show a gradual improvement in the ratings given to leadership by the SafeSENTRIES. In the initial period of the deployment, between August 2021 and March 2022, Positive leadership responses were consistently below the target level, but have been consistently above target performance since then to October 2022. This is strong evidence that the system has positively impacted the behaviours of the leaders in Debmarine Namibia and can also be seen in the results of the pulse culture surveys, discussed above.

DISCUSSION

Six key issues (gaps and limitations) of current risk management approaches were posited in the section *Shortcomings in Current Risk Management Approaches*. The purpose of this discussion is to illustrate how the SafeSENTRY system has succeeded (or not) in addressing these:

- Risk is seen as 'static', yet is inherently 'dynamic'.
- Risk 'control' is over-emphasized, at the expense of risk 'discovery'.
- Control hierarchies create complex defences-in-depth and 'false confidence'.
- Learning from incidents creates 'risk myopia'.
- Management processes are 'disconnected'.
- The management of critical risk lacks 'reliability'.

Risk is seen as 'static', yet is inherently 'dynamic'

What is strikingly evident from the observations of the SafeSENTRIES is that, while the total tally of observations has now exceeded 2000 over a period of 11 months, to date very few 'repeat' observation has been noticed, and even in these few cases (three or four), the same hazard was observed, but different risks associated with them. While the categorisation into PUE's has been available throughout, the observations within each is not repeated and illustrates both the complexity and dynamic nature of the risks on the vessels.

Additionally, the fact that critical controls are randomly checked and verified, strongly contributes to the understanding that the safety and production processes are non-static and to the understanding (and effort) of the generative nature of safety as an 'emergent property' (Leveson, 2017, p 101) that it is to be constantly 'created and recreated' and is a function of the interaction of many components.

It can be concluded that the SafeSENTRY system is effective in capturing the dynamic range of risks, as reflective of the constant shifts in location, likelihood, exposures and probability of consequences/impact.

Risk 'control' is over-emphasized, at the expense of risk 'discovery'

The SafeSENTRY redresses the disbalance of the risk management system, towards risk identification. While there is a significant focus on the effectiveness of the critical controls of each observed risk, the SafeSENTRIES. would spend most of the time in the field on searching for and identifying risks. By estimation, about 50 per cent of the engagement between employee in any session is on discovering risks, and about 40 per cent on the analysis, classification and controls verification, with 10 per cent of effort (time) on the pulse culture survey, leadership assessment and good practices commentary.

Each week, the senior executives of the enterprise meet to discuss the weekly report and best observations during the that period, and also review topics to improve the correct balances in responding to and managing the observations. Also on a weekly basis, each vessel's management team uses the observations of the week to review with crews, as a continuous training and improvement process, and also to conduct further reviews of findings to improve safety management systems.

On a monthly basis, the management teams of each vessel and corporate managers meet to review the range of analyses (such as provided in this paper) in order to take appropriate actions, close out of risks, and to identify opportunities to improve the overall risk management system. The SafeSENTRY system is a dynamic, random risk discovery process, much in line with the very characteristics of risk. It essentially 'combats randomness with randomness'.

Various business units participate in these reviews to ensure wide sharing of information, increasing risk transparency in the company and that the basic principles of managing the system are maintained.

Control hierarchies create complex defences-in-depth and 'false confidence'

The value of control systems, curtailing both the likelihood and impact of risks (so both preventative and mitigative) is unquestionable. The deployment of the hierarchy of control in fatal risk management in the mining industry is wide and effective, but has also contributed to the complexity, with many examples of controls interacting and actually initiating failures. While the *intended* consequence is the increase of redundancy, there is an additional, *unintended*, consequence of induced over-confidence, or risk homeostasis.

The SafeSENTRIES are continuously, and again randomly, challenging the controls' performance, in real time, in-field and 'live', which is then also testing the defences-in-depth. However, with the selection of the critical controls to populate the application, care was taken to only select the most critical of the critical controls (referred to as 'super' controls). As such the complexity (too many layers of controls) is significantly reduced, at least as its prioritised at the frontline.

The converse of risk homeostasis is chronic unease (Fruhen and Flinn, 2015) and it's quite evident from the experiences of the management team, that the process is creating a range of alerts that flows through on a daily, weekly and monthly basis. Chronic unease is firstly induced at the operational level, as a consequence of the risk discussions between SafeSENTRIES and fellow employees. It is estimated that for the identification of over 2000 critical risks, there would have been 4000 engagements (if the SafeSENTRY and the fellow employee are considered two engagements for each observation). These would clearly be risk discussions that would otherwise not have taken place.

Secondly, the handover of the daily observations to the direct supervisors at the end of each shift is another engagement and another risk discussion. Provision is made for supervisors to do a review of the observations, to add comments and to 'sign off' on the observations. The Sentries are also required to include comments on specific actions they have taken to reduce risks observed, where it was possible for them to act.

The middle and senior management of the organisation have access to the database on a daily basis, and can actively review information and take actions where necessary. The same database is also accessed by the executive/corporate management, and by the Vessel and Executive Management, on a regular basis. The data is available in numerous formats and empowers management to self-discover new patterns and trends in the data.

It is clear that the organisation is making progress towards what has been described as 'mindfulness', '*the capability to see the significant meaning of weak signals, and to give strong response to weak signals...which holds the key to manage the unexpected*' (Weick and Sutcliffe, 2001, p 3).

The analysis of the failure rates of fatal/high impact risks versus injury/low impact risks (see Figure 18) also proved to be of important value, as it created a sharp focus on 'risks that can kill', and the extent to which the controls for these risks are intact – and is a typical example of 'weak signals, to which the management give strong responses'.

Learning from incidents creates 'risk myopia'

The database contains a rich source of risks that are continually identified and evaluated, and a practice has evolved on the vessels to use the weekly reports, and selected observations, as risk skills training. Because of the nature of the vessel rotations (28 days at sea), the crews are available for extended training once per week as part of routine crew meetings.

The observations on the application are done with photographs, and supplemented with voice recordings, which can readily be used, directly from the database, during crew meetings to be included as training content and for practical analysis and illustrations. This has proved to be a valuable process, and also to counteract the problem of risk myopia, which results from learning only from past accidents. As previously mentioned, accidents are rare events, so the training content is very limited, and if expanded to include accidents from other vessels or other companies, the training content is not very relevant.

The amount of information (SafeSENTRY observations) that can be used for training is vastly higher, very relevant to their immediate circumstances, and of practical value. The training content is literally generated by the frontline employees themselves and in reviewing the voice recordings and photos, it has become very obvious that the observations and discussions as recorded are often of extremely high quality and valuable.

It can also be noted from the diagram in Figure 13 that nearly the majority of observations are made outside the range of defined PUE's, with 45 per cent of risk identified to date, labelled as 'Other' risks (not the defined PUE's). This indicates that the system is delivering observations outside of the obvious PUE focus, and counteracting the effect of myopia.

Management processes are 'disconnected'

The 'disconnection' of management process is addressed in the SafeSENTRY system in a variety of ways:

Feedback loops, where relevant and often urgent information about fatal and other serious risks, are strongly re-enforced with the SafeSENTRY system, between frontline workers themselves, with the regular engagements, and between frontline workers and supervisors, through the formalised handover at the end of shifts. There is also a 'high speed' connection between the daily observations of the SafeSENTRIES and the various levels of management, often directly to the top of the organisation. Numerous examples exist of observations that were identified and resolved with rapid responses.

Hierarchical alignment is achieved through the participation of frontline employees in the critical risk management system, and through the constant 'pulse' culture surveys that allows management to respond early on trends and deteriorations of perceptions. Additionally, the leadership assessment by the SafeSENTRIES allows for a continuous review of the impact that leaders make, especially as a result of their interaction with frontline employees. The improvement in the impact of leadership is reflected in the increasingly more positive leadership ratings shown in Figure 23.

The SafeSENTRY also engenders a greater degree of decentralisation and provides the frontline with more ownership and authority about their work environment.

Integration of risk management into the operational processes is achieved by constantly searching for upstream solutions to risk reduction, and numerous examples exist of improvements, modifications and additional controls that followed the reporting of identified deficiencies by the SafeSENTRIES. In many cases, the SafeSENTRIES would recommend improvements, which proved to be very practical, innovative and valuable.

A forward focus on risk has certainly replaced a reactive focus on accidents. As mentioned above, a practice evolved whereby crews are trained using the observations as content, instead of using accidents. One of the Vessel Captains, Capt Harold Haoseb (2021), expressed it as follows in a video-recorded interview:

> In the past we have made the mistake to fix problems without really understanding the problem and with this (SafeSENTRY system) we are not waiting for something to go wrong and then do an incident investigation. We are proactively looking for ways to improve and to eliminate the risks.

The *dynamic verification* of controls has proved to be an effective continuous improvement system, especially because the performance of the controls are reviewed and tested in 'real-time', and by frontline employees who have a greater knowledge of the validity of these controls. As reflected in Figure 15b, the failure rates of each PUE and each of the 'super controls' are continuously available for follow-up and improvement by the line managers, and the failure rate of critical controls has consistently been improved upon, as shown in the decreasing failure rate in Figure 16.

This process is able to counteract the spawning and growth of what Reason (1997, p 10) termed 'pathogens' (undiscovered errors and latent conditions) in the organisation, to act on these and to create a deep-seated readiness to respond to risk.

The management of critical risk lacks 'reliability'

The creation of 'chronic unease' or mindfulness in an organisation has long been stated as an ideal condition, or state of mind to be in. The HRO, as stated above, *anticipate mistakes and failures, so an open climate for reporting errors is important*. However, such a condition could lead to a state of constant fear, which is impossible to sustain in any normal work environment. Many other attempts in companies to create a reporting culture, by encouraging the reporting of near misses, has failed because of the 'unnatural' flow of information directly from the frontline to a database managed by the safety or risk department. This often creates dissent amongst the direct supervisors and middle managers, who are at the 'receiving end' of corrections to be made.

In the SafeSENTRY system, the natural hierarchies and authority of the supervisor and middle management are respected and reinforced, and it seems that the SafeSENTRY system is able to strike a balance, by creating a 'random routine' of risk reviewing and giving feedback to the front-line supervisor in the first instance. The increased participation rates of the employees in the system are also supporting the fact that risk transparency *(and open climate for reporting errors)* is developing and supported by the high levels (above Top Company responses) of employees indicating that they 'feel free to report all accidents and injuries', as shown in Figure 20.

There is clearly a significant focus on 'the anticipation failure' and the potential for it. Each of the over 2000 observations is a potential serious Injury, Fatal or Catastrophic event, and it has been demonstrated how observations have clearly prevented imminent failures, by being identified and rectified in time.

The *aversion to over-simplifying* can be seen in the encouragement of all employees to freely report and rectify issues and concerns, allowing different perspectives on risks. Feedback from supervisors and operational managers indicates that employees have shown an increasing willingness to share and give transparent feedback, and that risk discussions occur more frequently. The range of personal interactions, estimated to be over 4000 additional ones since the inception of the system, spans the full scope of SafeSENTRY-Employee, SafeSENTRY-Supervisor, Supervisors actively engaged in the process, and all management levels involved in responses and follow-up.

The *sensitivity to operational processes* has been demonstrated by the fact that the organisation has gone beyond 'learning from accidents' and beyond 'learning from near misses', to learning from dormant and latent risks ('far misses'), as illustrated by the recorded interview with Capt Harold Hoaseb (2021) 'We are proactively looking for ways to improve and to eliminate the risks', as quoted above.

The *commitment to resilience* is demonstrated on a daily and continuous basis by the various levels of management, constantly reviewing the data as it flows into the database, but an important by-product is the evolution in frontline leadership. The system seems to add the important 'characteristics' of intrinsic motivation as defined by Deci (1971) and Pink (1995) of *mastery, authority, purpose (and belonging)* to the roles of frontline operators, as reflected in comments from operators and supervisors participating in the system. This is also suggested by the high and sustained response levels (currently at 95.2 per cent NPR, which is significantly above Top Company benchmark) on the pulse survey on 'Value' ('This company really cares about employees').

Probably the most outstanding feature of the SafeSENTRY system is how it *defers to the expertise* of the front-line operators and the supervisors – also captured by a Debmarine Namibia axiom of: 'Humans are the strongest link in the safety chain'. There is also strong evidence that the SafeSENTRIES often take immediate actions to rectify issues and deficiencies that they have observed, which fosters a responsiveness to problems, rather than a reporting of problems. As such, it contributes to an organisation that can respond quickly and flexibly to deviations.

CONCLUSION

It is concluded that the SafeSENTRY has proved to be a valuable process of employee engagement, the proactive identification and management of risks and an embedded process of (self) challenging the integrity the risk control systems. It was reported from members of the management team that, as a result of the participation of employees in the SafeSENTRY observations, crew engagement and critical control verifications, they became more risk competent – and the reasons for working safely became more personal and understood. It creates a culture of psychological safety, and risk transparency between the various hierarchical levels of the organisation, that has led to, in the case of Debmarine Namibia, the discovery and containment of several 'multiple fatal' and 'catastrophic' exposures.

The SafeSENTRY system has progressed through a 12-month period of design, deployment and 'proof of concept' was successfully demonstrated in the Debmarine Namibia company. Indications are that the system is highly effective in generating valuable data on Serious Injury, Fatal and Catastrophic risks, that drives increased levels of risk control, and at the same time, increased levels of 'chronic unease'.

A key question remains… whether the system can have application beyond the company and context where it was designed. While there currently is wider applicability, having been deployed in the associated Namdeb land operations, and commended in the open cut and underground diamond operations of the Venetia Mine in South Africa, there still needs to be deployment in say, non-mining environments to answer that question definitively. The SafeSENTRY system is currently in design and being piloted in the utility industry of USA, and a future review of this will be made.

REFERENCES

Adams, J, 2001. *Risk* (Routledge: Abington).

Anglo American, 2019. Elimination of Fatalities Report, Anglo American South Africa, unpublished.

Ashburn, A, 1977. Toyota's 'famous Ohno system', *American Machinist*, July, pp 120–123.

Brady, S, 2019. Review of all fatal accidents in Queensland mines and quarries from 2000 to 2019, prepared for Department of Natural Resources, Mines and Energy, Queensland, Australia.

Deci, E L, 1971. Effects of externally mediated rewards on intrinsic motivation, *Journal of Personality and Social Psychology*, 18:105–115.

Dekker, S and Pitzer, C, 2016. Examining the asymptote in safety progress: a literature review, *International Journal of Occupational Safety and Ergonomics*, 22(1):57–65.

Fruhen, L and Flin, R, 2015. Chronic Unease: A State of Mind for Managing Safety, Centre for Safety, University of Western Australia, Perth.

Haoseb, H, 2021. Recorded interview on board the Debmarine Atlantic vessel.

ICMM, 2015. Critical Control Management, Implementation Guide (International Council for Mining and Metallurgy: London).

ICMM, 2022. Safety Performance: Benchmarking Progress of ICMM Company Members in 2021, London, UK.

Leveson, N G, 2017. Engineering a Safer World: Systems Thinking Applied to safety, (MIT Press: Cambridge).

Mine Safety and Health Administration, USA, 2021. Available from: <https://arlweb.msha.gov/stats/centurystats/mnm stats.asp>

Perrow, C, 1984. *Normal Accidents: Living with High-Risk Technologies* (1st ed.) (Princeton University Press: Princeton).

Pink, D, 1995. *Drive* (Penguin Books: New York).

Reason, J, 1997. *Managing the Risk of Organizational Accidents* (Ashgate: Hampshire).

Safework Australia, 2022. Report on Work-related Traumatic Injury Fatalities, Australia. Available from: <https://www.safeworkaustralia.gov.au>

Vaughan, D, 1996. *The Challenger Launch Decision* (University of Chicago Press: Illinois).

Weick, K E and Sutcliffe, K M, 2001. *Managing the Unexpected: assuring high performance in an age of complexity* (Jossey-Bass: San Francisco).

Wilde, G J S, 1994. *Target risk: Dealing with the danger of death, disease and damage in everyday decisions* (PDE Publications: Castor & Columba).

Case studies and operating practice

DeGrussa Copper Mine – mine closure experience

M Bellamy[1] and H Shedden[2]

1. Mine Engineering Superintendent, Sandfire. Email: mattyb_302@hotmail.com
2. Senior Mining Engineer, AMC Consultants. Email: hshedden@amcconsultants.com

ABSTRACT

The DeGrussa copper deposit was discovered by Sandfire Resources in April 2009. Construction of the 1.6 Mt/a concentrator, open pit and underground mining operations commenced in March 2011 with first concentrate production achieved in October 2012. Over the next ten years, DeGrussa grew to include both the DeGrussa and Monty underground mines, establishing itself as one of the world's premier copper producers with total production of 663 000 t of contained copper by the time the mine closed.

Final ore was trucked from Degrussa on 5 October 2022. This paper examines the mine closure experience from both a planning and operational perspective.

The paper provides an overview of the site's mine planning process including the factors involved in determining the end of the mine life and how this was tracked and communicated in the lead up to the end of mining operations. It will also look at some of the specific operational aspects involved in maintaining steady state production all the way through to the last day while facing increasingly challenging conditions.

These factors, along with the external factors faced by most of the industry such as labour shortages and COVID related disruptions, contributed to making it an interesting journey to the end of operations at DeGrussa.

INTRODUCTION

Sandfire Resources made the initial discovery of the DeGrussa deposit in April 2009. Following the delivery of the maiden resource estimate in February 2010 and pre-feasibility in March 2011, construction of the 1.6 Mt/a concentrator, open pit and underground mining operations commenced. First concentrate was produced in October 2012.

By April 2013 mining of the open pit was completed with the operation transitioning to a wholly underground operation. In 2015 the Monty deposit was discovered 10 km east of the DeGrussa operation. Development of the Monty underground commenced in October 2017 with the first ore produced in March 2019. Monty ore was trucked to the DeGrussa concentrator for processing via a dedicated haul road.

Prior to closure it was one of the world's premier copper producers, with total production of 663 000 t of contained copper as at the end of Q1 FY23. During its operation DeGrussa became highly regarded for its high-grade and high-quality concentrate. It also became known for its ability to consistently deliver in line with budget production and financial return: during its operation Sandfire never missed a production guidance target.

Underground mining at DeGrussa was primarily conducted using longhole open stoping methods with paste fill. Monty utilised Modified Avoca and Avoca methods with a combination of Cemented Aggregate Fill (CAF) and rock fill. Mining activities were conducted by Byrnecut. The overall layout of DeGrussa and Monty is shown in the long sections in Figures 1 and 2.

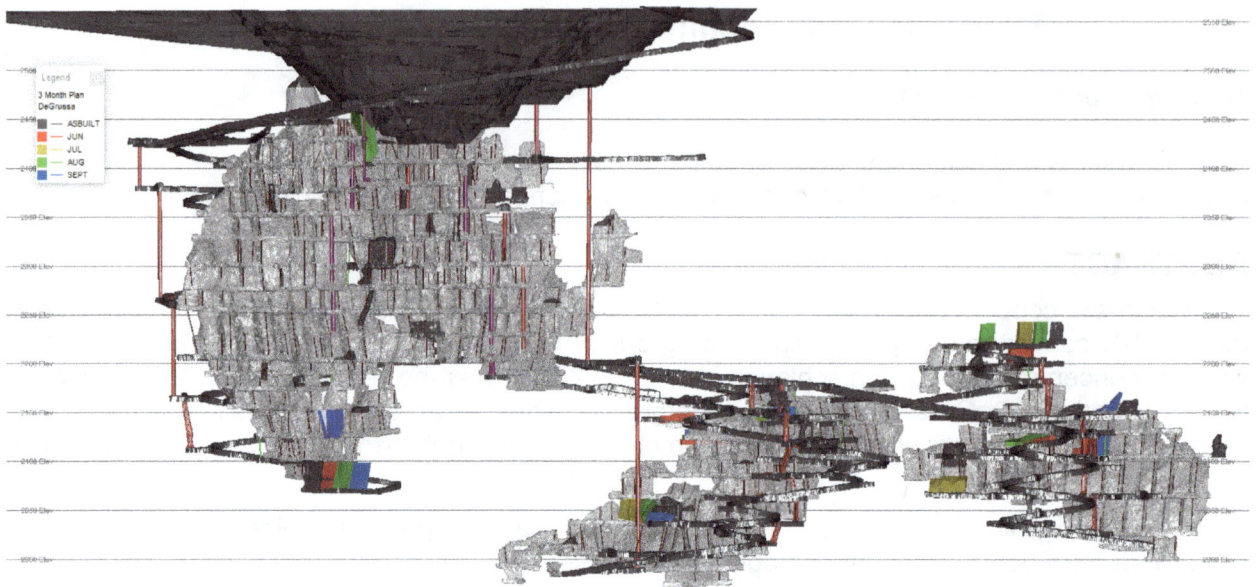

FIG 1 – DeGrussa FY23 budget long section.

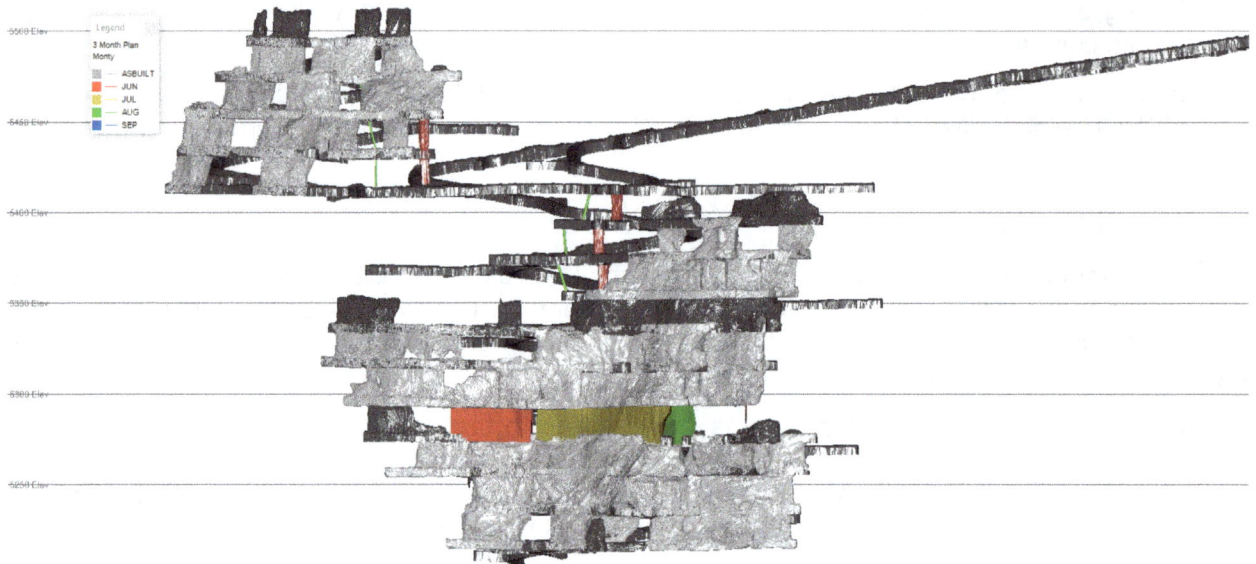

FIG 2 – Monty FY23 budget long section.

DEGRUSSA PLANNING SYSTEM

All aspects of the mine planning for DeGrussa were handled on-site; this includes all planning and scheduling from the life-of-mine (LOM) down to weekly scheduling. This helped ensure alignment across all levels of planning from the LOM to the weekly and shift level plans.

The LOM was reviewed and updated annually following the release of the updated geology resource model in December. This then flowed into the annual Financial Year (FY) budget and down to the rolling three-month plan and 7/21 weekly plan.

As part of the process, a single live EPS schedule was used for all levels of scheduling, ensuring that all schedules were in line with the LOM plan. It also allowed critical path activities to be monitored daily where necessary, highlighting issues early and allowing resources to be re-allocated where required.

Once the updated geology resource model was handed over all stope and development designs were reviewed and updated. This process also considered any variations to the geotechnical structural model and incorporated any prior learnings from stoping completed over the previous 12 months. Due to the level of detail required these design updates were completed manually.

The LOM update also included a review of all the modifying factors and scheduling assumptions using historical stoping data. For detailed stope designs stope dilution was determined individually based on the Equivalent Linear Overbreak/Sloughing (ELOS) method where specific overbreak values were nominated for each face of the stope by the geotechnical team. For the LOM, dilution was estimated based on a regression formula that captured historical stoping dilution plotted against stope size. Each lode had its own regression. Recovery was also estimated using a regression formula. An example regression is shown in Figure 3.

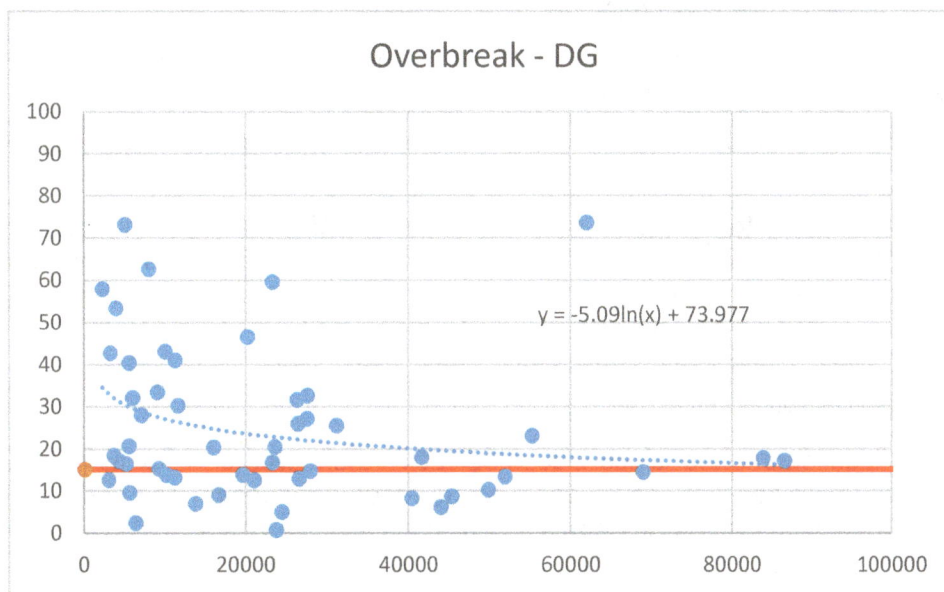

FIG 3 – Example regression curve used to estimate mining dilution.

Similarly, all mining assumptions including production rates and drilling factors were reviewed and updated against the previous 12 months' actuals. Following this process aided in ensuring that the Budget that derived from the LOM was both realistic and achievable. The authors have experienced operations where the long-term plan is managed off-site. At times this leads to a disconnect between schedules and expectations – situations which can be avoided with a site-based approach.

This approach along with close collaboration with all departments on-site resulted in DeGrussa and Sandfire earning a reputation for delivering and often exceeding market guidance.

When would the mine end?

A review of the last five LOM plans shows there was little movement in the planned end date. Based on the December 2017 geology resource model the end date was expected to be the end of Q3 FY22. Table 1 summarises the changes since then.

TABLE 1

Last five LOM plan forecast end dates.

LOM plan	Forecast end date
December 2017	Q3 FY22
December 2018	End FY22
December 2019	Q1 FY23
December 2020	30 September 2022
December 2021 / Budget FY22	30 September 2022

During 2018 there was a significant amount of engineering work completed on the extraction of the crown pillar stopes to the DeGrussa open pit. This review highlighted some additional geotechnical issues concerning the pit stability. As part of this, the planned large double lift stopes were split into smaller single lift stopes so that the open pit exposure and stability could be better managed. This contributed to the end date being revised to the end of FY22 as published in the December 2018 LOM update.

Underground diamond drilling performed during 2019 allowed a resource upgrade and detailed design on an upper extension to the C5 orebody. This work together with an increase to the reserve following some localised strike extensions added a further three months production. By the end of December 2019, the end date was projected to be the end of Q1 FY23.

Since the release of the December 2019 LOM report the completion date remained stable with mining progressing largely in line with each years LOM plan.

At the release of the December 2020 LOM the end date for mining was stated as 30 September 2022.

In March 2021 the geology team undertook a further review of the resource to identify any further remnant areas that could be included into the LOM. Engineering analysis of these remnant areas resulted in a small increase in the mining inventory however all identified targets could still be extracted within the existing LOM time frame.

Based on this the December 2021 LOM and FY22 Budget both confirmed that all mining would be complete by 30 September 2022.

What this review highlights is that since December 2017 there was only six months added to the planned end date. This extension can be attributed to reserve extension and updates to the crown pillar extraction. There was minimal 'schedule creep' in the last five years. This is a testament to how a robust planning system together with consistent execution can deliver reliable results.

EMPLOYEE RETENTION AND ENGAGEMENT

As the mine plan developed and the end date became firmer, it was recognised that employee retention would be critical to ensure uninterrupted production could be maintained till the last day.

To address this, Sandfire implemented a retention package that was rolled out to all employees in July 2020. Staff turnover following the rollout of the package was reduced significantly. From a team of 11 mining engineers there were only two who chose to leave prior to the completion date; anecdotally a turnover rate far lower than most sites over the same time frame. The retention package was also offered to new starters as they signed on, aiding the attraction of high-quality replacements without having to rely heavily on short-term contractors.

This trend was also seen across the whole technical team; there were no contractors at all working within the team for the final three months of production.

The table below shows the impact the retention package had on turnover of site-based Sandfire personnel: dropping from over 21 per cent to below 9 per cent (Hana, October 2022, personal communication).

TABLE 2

Sandfire site based personnel turnover.

Year	Number of resignations	Total headcount	% Turnover
2019	41	180	22.78
2020	38	175	21.71
2021	15	171	8.77
2022	4	165	2.42

Several 'Life After DeGrussa' workshops were also held commencing in November 2021. Topics covered included Finance, Mental Health – Adapting to a Change, and Career Transition.

Discussion within the underground technical team showed there was a high level of confidence in finding work once mining was complete given the employment conditions in the mining.

Once the redundancy process commenced and personnel were notified of their end date, there were several other initiatives implemented including assistance with interview skills and CV writing through an external provider. The Employee Assistance Program (EAP) service was also extended to all personnel for one month after their last day.

There was also continued communication through a series of 'Celebrating the success of DeGrussa' newsletters and YouTube videos as production came to an end.

CONTRACTOR MANAGEMENT

As the end of operation approached, the mining services contract with Byrnecut was varied to move away from a schedule of rates towards more of a fixed and variable component. This provided both parties with more certainty given the possibility of some fluctuations in production rates as the number of available mining areas decreased. For Sandfire, this reduced the risk of personnel being allocated to other sites in a 'skills shortage' environment. For Byrnecut, it reduced the risk of not recovering costs as available mining fronts and mining physicals reduced.

In April 2022 Sandfire and Byrnecut agreed to a Sandfire sponsored retention package for all Byrnecut personnel on-site to further safeguard the final months of production. Retention of key operators was also assisted by Byrnecut being proactive in ensuring that everyone who stayed would have work within the group once the mine finished. By the start of August 2022, of a workforce of approximately 120 the vast majority knew what site they would be transferring to. This reduced the risk of distraction emanating from uncertainties.

COUNTDOWN TO CLOSURE

The following section will provide a month by month run down of the events leading up to the final truckload of ore and preparing the mine for closure. From a mine planning and operational perspective, the lead up to June was focused on delivering the FY22 production targets and ensuring that the final FY23 Budget provided a realistic end date for production.

This work enabled discussions to commence early with employees and contract partners so that all parties could begin to plan what the final months would look like. The overriding concept was aimed at maximising production from both mines right up to the last truckload of ore. This would be followed by a planned retreat from both mines with the aim of full closure of the underground workings as opposed to a care and maintenance approach.

June

The primary focus for June was ensuring that the operation achieved the FY22 production targets and corporate guidance, which it did.

As part of the FY23 Budget process, work commenced on a more detailed closure plan for DeGrussa. This started by including all major plant items that were planned to be recovered from underground into the mine schedule and linking them to production activities. This enabled a staged approach to the retreat, with infrastructure removal occurring simultaneously with production. Table 3 shows the items added to the schedule to assist with closure planning.

TABLE 3

Major plant items to recover.

Item	Number
Secondary Fan	28
Primary Fan	4
Sub Station	10
Refuge Chamber	8
Distribution Board	31
Pump Station	3
Mono Pump	3

Some items, such as ladderways and vent doors, were omitted as the cost removing them was unlikely to be recovered. It was also agreed that services poly pipe, paste reticulation pipe and electric cable would not be recovered. This was based on both the low salvage value and the high potential for manual handling injuries undertaking this removal on such a large scale.

This enabled more detailed planning to commence and gave a prediction for how long it would take to completely strip out all the underground infrastructure. This work showed that DeGrussa would be stripped of all salvageable infrastructure by 24 October.

July

In early July the FY23 Budget was finalised and approved. This further confirmed that all production activities would be complete by 30 September. Full retreat from the underground workings and closure works would be complete by the end of October.

During the month it was also identified that there was an opportunity to utilise Byrnecut to complete surface earthworks required for mine closure: capping Potentially Acid Forming (PAF) dumps, levelling waste dumps, drill site rehabilitation and minor tailings dam wall earthworks. Using the existing on-site capability fitted in with the mining schedule and eliminated the requirement to mobilise additional resources to site.

Activities underground continued in line with the budget. However, the team continued to look for opportunities to recover more of the resource and improve the efficiency of the operation. One of these opportunities was to review an area of the mine where longhole stoping could be used instead of cut-and-fill development. Analysis completed in 2015 determined a block of ore close to a major fault zone could not be extracted through stoping and recommended a series of cut-and-fill development lifts be used. Following completion of a number of these lifts the ground conditions were re-assessed and it was agreed that a series of longhole stopes could be safely extracted against the fault zone. This approach was adopted successfully in the last two cut-and-fill development levels, eliminating the need for 100 m of jumbo development.

Redundancies also commenced this month within the Sandfire technical team.

August

Mining activities in August progressed well and were in line with the budget. All indications were that production was on track for completion. Planning work continued on how to best undertake the retreat of the underground infrastructure, with more detailed planning of the pump station removal undertaken.

During the month there were a couple of 'Eat Street' celebration nights held on-site. These events involved Perth staff travelling to site to celebrate end of operations.

September

September production started strong with all indications that production would be complete by the end of the month. On 14 September the last truckload of ore was mined at Monty and removal of all underground and surface infrastructure was completed shortly after on 22 September, with the portal gates shut and the boxcut backfilling commenced.

At DeGrussa the progressive shutdown of the underground pumping infrastructure commenced early in the month as mining was completed from the lower levels. These activities all proceeded largely in-line with the EPS closure schedule and in most cases the removal of infrastructure was able to be completed faster than originally estimated.

Production activities were impacted late in the month by equipment downtime which resulted in mining of the last two stopes pushing into the first week of October. Despite this slippage work to remove underground infrastructure continued with levels progressively closed throughout the month.

This process was assisted by having a closure checklist completed for all levels, with the final step being a barricade at the level access. Through this process we were able to ensure that nothing was left behind and all potential hazards such as open holes around vent rises and escapeways were managed and/or demarcated. Despite knowing we would be fully sealing the mine; the overriding principal was to ensure everything was left with the possibility that the mine may be re-entered at a later date.

By the start of September all technical staff had been advised of their redundancy and departure date. The technical team including Engineering, Geology and Survey had dropped from 33 at the start of August to 12 at the end of September.

October

As noted above it was not until October 5th that the last truckload of ore was hauled to the surface, marking the end of production from DeGrussa.

FIG 4 – Last load of ore from DeGrussa.

With the last truckload of ore came the shift in focus to removing all remaining infrastructure ready to completely close the underground. The removal of the underground infrastructure continued to be undertaken by Byrnecut with several operators remaining to assist. The primary ventilation fans for

DeGrussa were originally installed by Mine and Tunnel Ventilation (MTV). As part of the closure, MTV were engaged to remove the fans and prepare them for sale and transport off-site.

In consultation with the site environmental team, works were also undertaken to remove any surface infrastructure that would no longer be required. This included the removal of redundant dewatering infrastructure, surface compressors and water tanks. There was also a focus on ensuring all laydown yards and workshop areas were left empty and all hydrocarbon contaminated areas were contained and remediated.

By October 19th all works underground were completed and the portal entrance was blocked off with waste.

FIG 5 – DeGrussa portal sealed.

On October 20th MTV completed the removal of all primary fans, with steel plates bolted over all fresh and return air rises.

Paste filling of the last remaining crown pillar stope was also completed on October 20th. Although this paste filling was not required for mining purposes, it was agreed that it would be best for the long-term stability of the open pit and pit ramps while also ensuring that no large water flow path remained between the pit and the underground.

WHAT NEXT FOR DEGRUSSA?

It was announced in September that the Sandfire board had approved the 'DeGrussa Processing Expansion Project'. This would see processing continue till January 2023 utilising heavily transitional and mineralised waste stockpiles. There was also test work carried out in September on the viability of processing oxide stockpiles; however, at the time of writing no decision had been made on this project.

By the time the author had finished on-site on October 25th all Byrnecut mining personnel and equipment had been demobilised from site. A few workshop personnel remained as the last stores were loaded. All entrances to the DeGrussa underground were sealed and a significant amount of rehabilitation and clean-up of the boxcut and open pit had been completed. The remaining rehabilitation works have been handed over to the environmental team on-site to coordinate.

All infrastructure had been placed in laydown yards ready for auction, with a professional auctioneer engaged to manage the disposal of the assets.

FIG 6 – Laydown yard ready for auction.

CONCLUSION

The closure of the DeGrussa and Monty underground mines marks the end of mining at one of the best copper projects in recent memory. Both mines have now been sealed with the remaining rehabilitation works handed over to the on-site environmental teams.

Key learnings from this process were:

- The effectiveness of incentivising the workforce – early to stay to the end of mine life. Operations are at significant risk of losing personnel for more secure employment approaching a closure. The ability to recruit new labour with a short mine life remaining is also reduced. Considering the reduction in turnover rates that were observed following the announcement of the retention package, the cost of the package was likely well below the combination of the cost that would have been incurred recruiting staff and the opportunity cost that turnover represents.

- Early engagement of stakeholders (staff, contractors, auctioneers etc) is key.

- Integrating closure tasks into the mine schedule simplified organising the removal of mine infrastructure.

From the authors' viewpoint the end of production, retreat of infrastructure and final sealing of the underground workings exceeded expectations, with all activities completed ahead of schedule and without incident.

ACKNOWLEDGEMENTS

The authors would like to thank the following for their assistance. Sandfire; Greg Peden (General Manager Operations), for approving the submission of this paper; the Sandfire technical team for all the hard work ensuring the closure was completed in a safe and efficient manner; the contracting partners, including Byrnecut and MTV, for their support right through to the end of operations.

A global review of productivity and operational performance in sublevel caving mines

A D Campbell[1]

1. Principal Engineer, Beck Engineering, Brisbane Qld 4000.
   Email: acampbell@beckengineering.com.au

ABSTRACT

The sublevel caving (SLC) mining method is becoming more widely adopted as an underground mining method as mines become deeper and target lower grade mineralisation. SLC has been successfully adopted in increasingly complex and wide-ranging conditions in the past 10–20 years, demonstrating the flexibility of the method. Various challenges encountered at SLC mines around the world have led to significant differences in operating practices, mine performance and mine productivity. Unfortunately, aspects related to productivity, operational performance and best practice are not widely published and remain with individual companies. Limited information is publicly available to compare operational performance of SLC mines in the various conditions where the mining method is now being applied. To overcome this shortcoming, a global review of SLC operations was undertaken over a three-year period as part of the Cave Mining 2040 research consortium.

The purpose of the review was to document technical and operational practices, hazard management strategies and key benchmarking metrics at SLC mines around the world. A total of 21 mines spread over four continents and 17 different mining companies were included in the project. The main outcome of the project was a manual of operational and technical practices, benchmarking of key metrics and documentation of the current state-of-the-art for each aspect of SLC mining. This paper describes the main findings from the benchmark in terms of mine productivity, including mine development, production and mining costs. The implications of mine scale, depth and ground conditions on operational performance at each of the benchmarked mines is described and charts are provided comparing key productivity metrics.

INTRODUCTION

Sublevel caving (SLC) is a top-down mining method that relies on gravity flow of ore that is fragmented by blasting (Kvapil, 1965). Blasted ore is extracted in accordance with a prescribed tonnage or grade shutoff. As the ore is extracted, the overlying waste material caves naturally as mining progresses.

SLC is a low capital and early production mining alternative to block caving, with greater geometric and sequence flexibility (Bull and Page, 2000). It is a potential mining method when longhole open stoping is not financially viable and when block caving is not possible due to orebody geometry, rock mass conditions or capital investment constraints. The global benchmark found that the SLC mining method is being increasingly adopted in a wide range of orebody geometries and mining conditions that are far more novel than text-book adaptations of the method. The flexibility of the SLC mining method is an advantage of the method. However, there is limited information, benchmarking and operating guidelines available in the literature for the more recent and novel adaptations of the SLC mining method. This benchmark includes a wide variety of SLC operations, including various commodities, mining scale and production rates, blind caves and those with open pits above, longitudinal and transverse SLCs as well as modified SLC methods, such as sublevel shrinkage (SLS) and sublevel retreat (SLR).

The benchmark identified a wide range of operational performance and mine productivity due to factors such as mine scale, operating depth, rock mass conditions, haulage and materials handling systems and mine layout. The range of some productivity measures vary from mine to mine by up to a factor of ten for the benchmarked mines. Mine development rates were also found to vary significantly from site to site, with most variability caused by the rock mass conditions as well as differences in rates achieved in developed versus developing countries.

This paper also discusses the uptake of tele-remote and electric load-haul-dumps (LHDs), as well as the production ramp up rates achieved at recently constructed SLC mines. Mining costs for a number of SLC operations are also compared.

GLOBAL BENCHMARK OF SLC MINES

The global benchmark of SLC mines and operational review was carried out from 2019 to 2022 as part of the Cave Mining 2040 research consortium. The objective of the project was to:

- Document the current state of SLC mines around the world including technical and operational practices, hazard management, and current challenges at each mine.

- Benchmark and compare mines in a wide range of operating conditions, countries, commodities, mining depths and adaptations of the SLC mining method.

- Document advances in technology and operating practices over the past 10–20 years.

- Benchmark mines with unique conditions and challenges, for the benefit of future mines.

- Develop an extensive document to serve as a manual of operational practices around the world and identify the current state-of-the-art for each aspect of SLC mining.

The project was conceptualised following from the widespread use of the benchmark of five SLC operations as part of the Mass Mining Technology group (Power and Just, 2008). A total of 21 mines spread over four continents participated in this project and an additional three mines which have ceased operations were benchmarked from the literature. The majority of SLC mines are located in Australia and South Africa as shown in Figure 1. Four (known) large scale SLC mines were not included in the benchmark due to access limitations and other constraints. The author acknowledges that there are other SLC mines around the world that were not included in the project. Details of the benchmarked mines is summarised in Table 1.

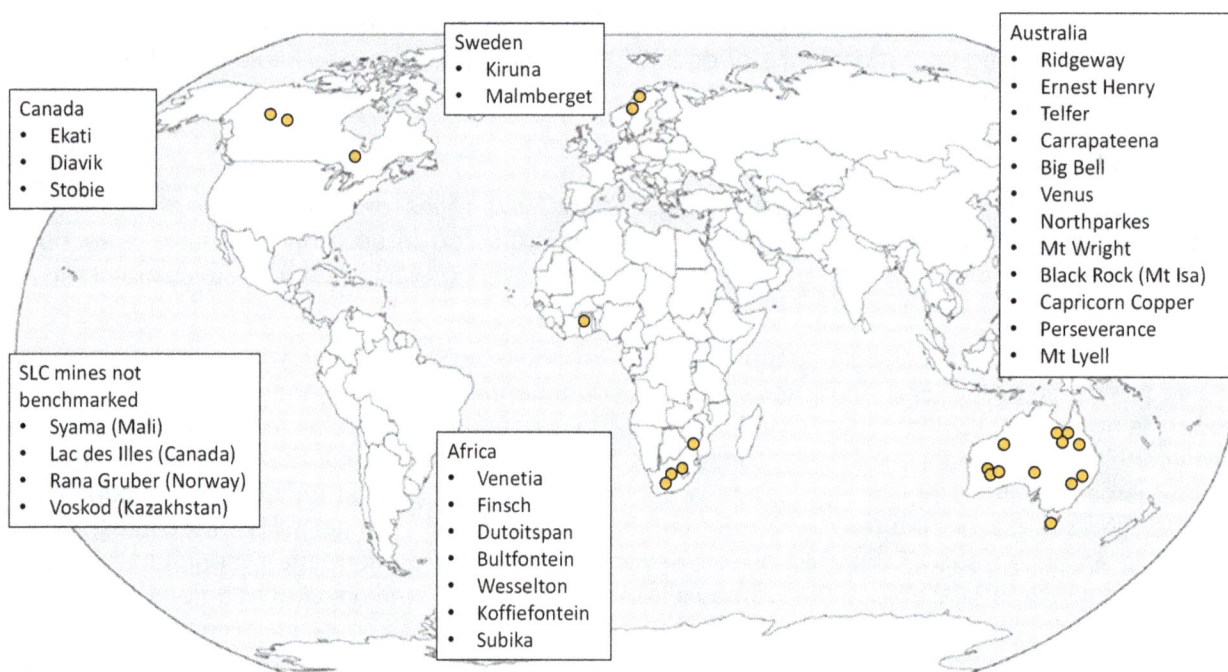

FIG 1 – The location of sublevel caving mines included in the global review.

TABLE 1

Mining parameters for each of the benchmarked mines.

Mine	Peak production rate (Mtpa)	Commodity	Level layout	Sublevel spacing (m)	Cross-cut spacing (m)	Orebody rock mass conditions
Ernest Henry	6.825	Copper Gold	Transverse	25	15	Good
Telfer	5.6	Gold	Transverse	25	14	Good
Northparkes	1	Copper Gold	Transverse	25	15	Good
Mt Wright	1.4	Gold	Both	25	12	Poor to fair
Koffiefontein	1.1	Diamond	Transverse	23	19	Poor to fair
Bultfontein SLC	0.5	Diamond	Transverse	20	19	Poor to fair
Dutoitspan NWC SLC	0.2	Diamond	Longitudinal	15	18	Poor to fair
Finsch	3.2	Diamond	Both	25	21	Fair
Carrapateena	4.25	Copper Gold	Transverse	25	15	Good
Ridgeway	6	Gold	Transverse	25 and 30	14	Good
Perseverance	1.5	Nickel	Transverse	25	14.5	Poor to fair
Stobie	1.825	Nickel	Transverse	21.5 and 30.5	12.2	Fair
Kiruna	27	Iron Ore	Transverse	28.5	24.75	Very good
Malmberget	17	Iron Ore	Both	25	22.5	Good to very good
Mt Lyell	2.5	Copper Gold	Both	20 and 25	15	Fair
Big Bell	1.55	Gold	Longitudinal	25	15	Good
Ekati (Koala SLC)	1	Diamond	Transverse	20	14.5	Poor to fair
Ekati (Panda SLR)	1.2	Diamond	Transverse	20	14.5	Poor to fair
Diavik (A154S Pipe)	0.5	Diamond	Transverse	25	15	Poor to fair
Diavik (A418 Pipe)	1.1	Diamond	Transverse	20 and 25	15	Poor to fair
Black Rock (Mt Isa)	0.4	Copper	Longitudinal	20 and 25	15	Poor
Capricorn Copper	1	Copper	Both	25	15	Fair
Venetia K01	4.5	Diamond	Transverse	25	17.5	Good
Venetia K02	1.5	Diamond	Transverse	25	17.5	Fair to good
Venus	0.7	Nickel	Transverse	25/30/35	20	Poor to fair
Subika	2.5	Gold	Transverse	25	15	Very good

The project included a site visit and underground tour for most of the benchmarked mines. Each mine provided access to technical and operational information, mine plans, procedures, and management plans to enable a review of current practices. A total of 120 parameters were recorded for each mine ranging from production, development, equipment fleet, infrastructure, drill and blast and recovery. The information provided was used to write a comprehensive report for each mine, including details of geology, mine design and planning, mine operations, drill and blast, geotechnical, infrastructure and material handling, and hazard management practice. These reports provide context for the comparisons made in the benchmarking metrics and charts, as each of the mines has similarities and differences to the other mines. The reports also outline practices at each mine and act as reference sources and operating guidelines.

The range of operating depths for the benchmarked mines is illustrated in Figure 2. Some SLC mines operate close to surface, particularly sublevel retreat (SLR) mines, which start production directly below a previously mined open pit. SLR mines include Black Rock, Ekati, Diavik, Venetia K01 and Koffiefontein. The deepest SLC mines are Perseverance, which had a final mining depth of 1100 m, and the Mt Lyell and Telfer mines with final mining depths of 1085 m below surface. Kiruna and

Malmberget in Sweden currently have mining depths of 1000 m and 900 m, respectively, but have plans for future mining depths of up to 1380 m below surface and studies for additional depth extensions.

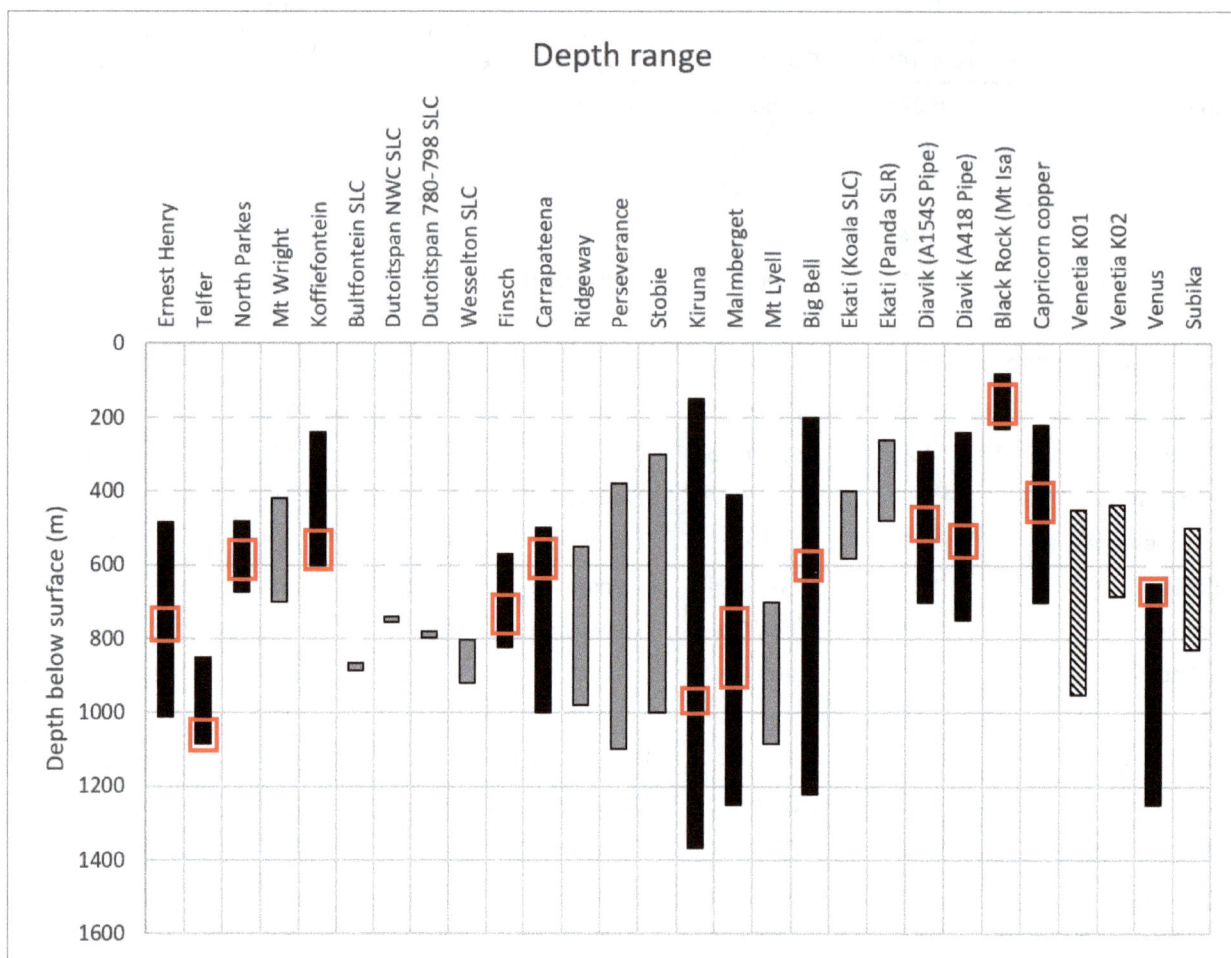

FIG 2 – Depth range for each of the benchmarked mines. Dark bars indicate mines currently in operation, light bars indicate mines that have ceased SLC mining and the hatched bars are mines that are yet to start production. The red box indicates the depth of operating production levels at the time of the benchmark.

Development

Development metrics for each of the benchmarked mines are summarised in Table 2 and Figure 3. Mines with favourable and more competent ground conditions (ie Carrapateena, Ernest Henry, Capricorn Copper, Ridgeway) typically achieve development rates in the order of 200 to 300 m/month per jumbo with an advance 4 m to 4.3 m per development round. Mines in less favourable conditions, particularly the kimberlite mines, typically achieve slower development rates in the order of 50 m to 150 m/month per jumbo. The factors that most impact development rates are the ground conditions, which dictates the ground support and rehabilitation requirements as well as the drive size and cut length. Transverse SLC's typically have higher development rates compared to longitudinal SLC mines as there are more available headings on each production level. Mines that were benchmarked where development had been completed prior to the site visit are noted accordingly in Table 2.

Most of the benchmarked mines use split feed jumbos and complete face drilling and ground support with the same machines to reduce tramming and set-up time. Carrapateena uses a mixed fleet of fixed long rail jumbos for face boring and split feed jumbos for ground support installation. Kiruna and Malmberget use jumbos for face boring only and specialised equipment for mechanical scaling as well as bolting and meshing. The use of secondary development and ground support equipment enables the two Swedish mines to achieve development rates that are slightly higher than the Australian mines on the basis of metres per month per face drill.

Many of the benchmarked mines also use shotcrete or fibrecrete as part of the ground support regime. The use of shotcrete at each mine varies for multiple reasons and this is not listed in the benchmark or the charts due to limitations in being able to explain the nuances of each site in a paper format.

TABLE 2

Development parameters for each of the benchmarked mines.

Mine	Development per month (m)	Number of Jumbos (face drills)	Number of support rigs (bolters and scalers)	Development rate per jumbo per month	Development cut length
Ernest Henry	630	3	0	210	4.3
Telfer	Dev completed	NA	0	NA	NA
Northparkes	Dev completed	NA	0	NA	NA
Mt Wright	Dev completed	NA	0	NA	NA
Koffiefontein	120	3	0	40	3
Bultfontein SLC	30	1	0	30	2
Dutoitspan SLC	25	1	0	25	2
Finsch	750	6	0	125	3
Carrapateena	1350	6	0	225	4.5
Ridgeway	675	3	0	225	4.3
Perseverance	550	5	0	110	3.5
Stobie	Dev completed	NA	NA	NA	NA
Kiruna	2000	7	7	285*	4.5
Malmberget	1330	4	8	333*	4.5
Mt Lyell	Dev completed	3	0	NA	NA
Big Bell	620	4	0	160	4
Ekati (Koala SLC)	Dev completed	NA	NA	NA	NA
Ekati (Panda SLR)	Dev completed	NA	NA	NA	NA
Diavik (A154S Pipe)	10	1	0	10	3
Diavik (A418 Pipe)	Dev completed	NA	NA	NA	NA
Black Rock (Mt Isa)	280	1	0	280	4
Capricorn Copper	330	1	0	330	4
Venetia K01	1400	7	0	200	4.2
Venetia K02	800	4	0	200	4.2
Venus	250	4**	0	120	4
Subika	1000	5	0	200	4

*Development metres calculated per face drill only, excluding supporting bolting and scaling rigs.

**Not all jumbos used for development.

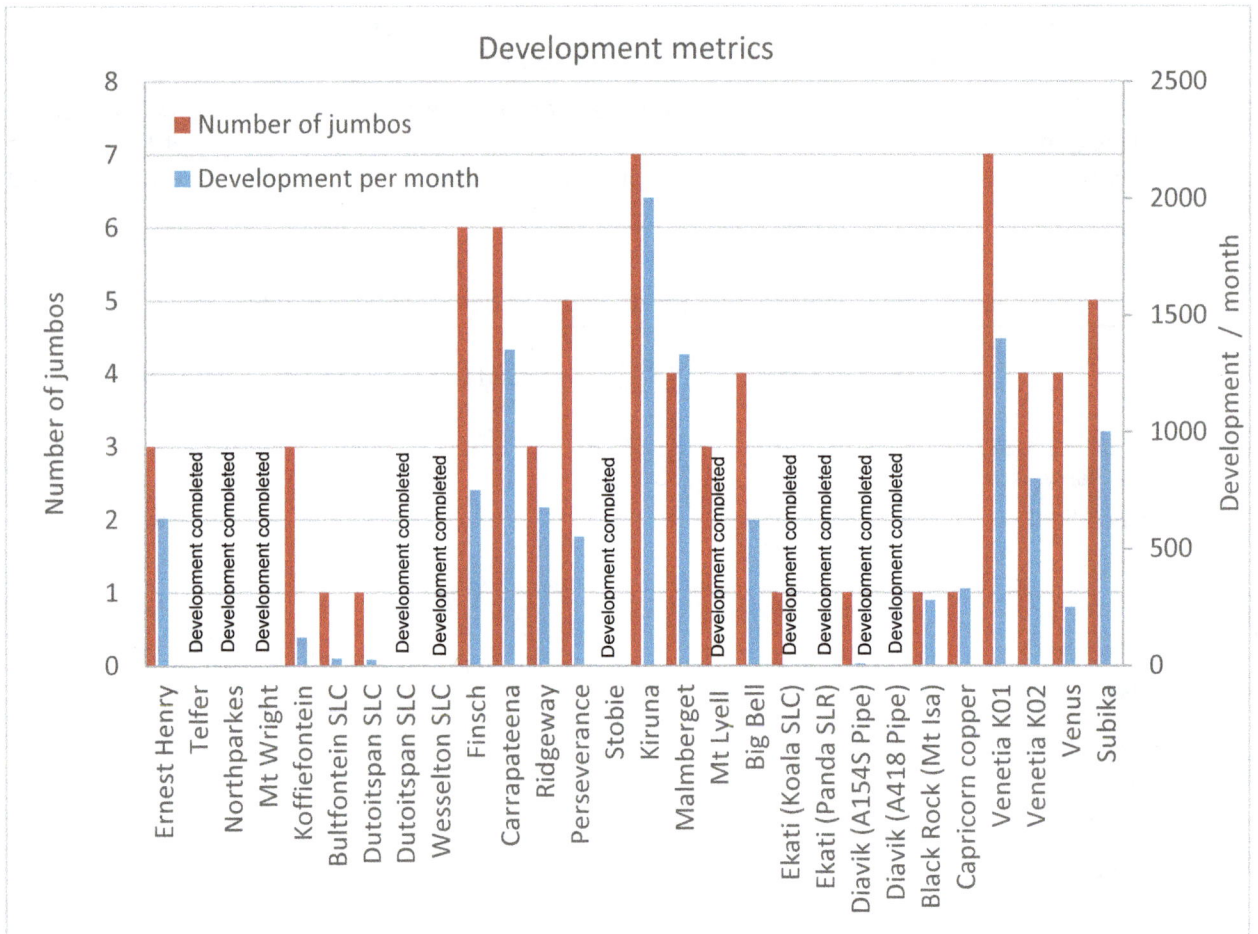

FIG 3 – Development metrics for number of jumbos and total development metres mined per month.

Production

Benchmarked parameters for production drilling and LHD's are provided in Table 3. Typical production drilling rates achieved in mines with strong competent rock are in the order of 8000 to 10 000 m/month per drill rig. Many of the kimberlite (diamond) mines achieve slower drilling rates, which is due to a combination of dry drilling requirements and higher frequency of redrilling. Big Bell and Capricorn Copper also fall into this classification because of issues with hole stability and redrilling requirements. Many of the mines in developing countries also achieve lower productivity rates, which is common for many of the production metrics benchmarked. This is partly due to many of the diamond mines in South Africa operating three 8-hour shifts per day and the mines not operating on weekends. A clear outlier in the chart shown in Figure 4 is the production drilling rate achieved at Mt Wright. The comparably high drilling rate at Mt Wright is mostly attributed to the relatively soft rock mass, which enabled high drilling rates to be achieved.

TABLE 3

Production parameters for each mine.

Mine	Number production drills	Total production drilling per month	Number LHDs in production levels	Number LHDs in rehandle/crusher level	Average t/day /drawpoint	LHD type	LHD bucket size (tonnes)
Ernest Henry	3	34 500	6	4	420	Diesel	21
Telfer	2	11 000	7	1	350	Diesel	21
Northparkes	1	10 000	2	0	250	Diesel	17.8
Mt Wright	1	18 000	3	0	250	Diesel	17.8
Koffiefontein	3	8000	5	2	300	Diesel	8.1
Bultfontein SLC	1	750	3	0	200	Diesel	7
Dutoitspan NWC SLC	1	750	3	0	200	Diesel	7
Finsch	6	25 000	9	4	320	Diesel	10
Carrapateena	3	27 000	5	0	400	Diesel	21
Ridgeway	2	20 000	6	2	390	Diesel	17.2
Perseverance	4	-	5	2	420	Diesel	17.5
Stobie	-	-		-	190	Diesel	-
Kiruna	10	80 000	29	0	700	Diesel and Electric	25
Malmberget	6	50 000	14	0	665	Diesel and Electric	25
Mt Lyell	2	-	4	2	280	Diesel	21
Big Bell	2	10 000	4	0	270	Diesel	17
Ekati (Koala SLC)	2	-	3	1	180	Diesel	10.5
Ekati (Panda SLR)	1	-	2	0	260	Diesel	10.5
Diavik (A154S Pipe)	0.5*	1900	1	0	135	Diesel	13.4
Diavik (A418 Pipe)	1	3500	1	0	250	Diesel	13.4
Black Rock (Mt Isa)	1	6500	2	0	250	Diesel	12.5
Capricorn Copper	2	11 635	2	0	275	Diesel	21
Venetia K01	5	26 580	7	4 (shared by K01 and K02)	200	Diesel	21
Venetia K02	2	8620	2		200	Diesel	21
Venus	2	13 200	2	0	190	Diesel	21
Subika	3	11 000	2	0	200	Diesel	15

*Production drill shared between underground operations.

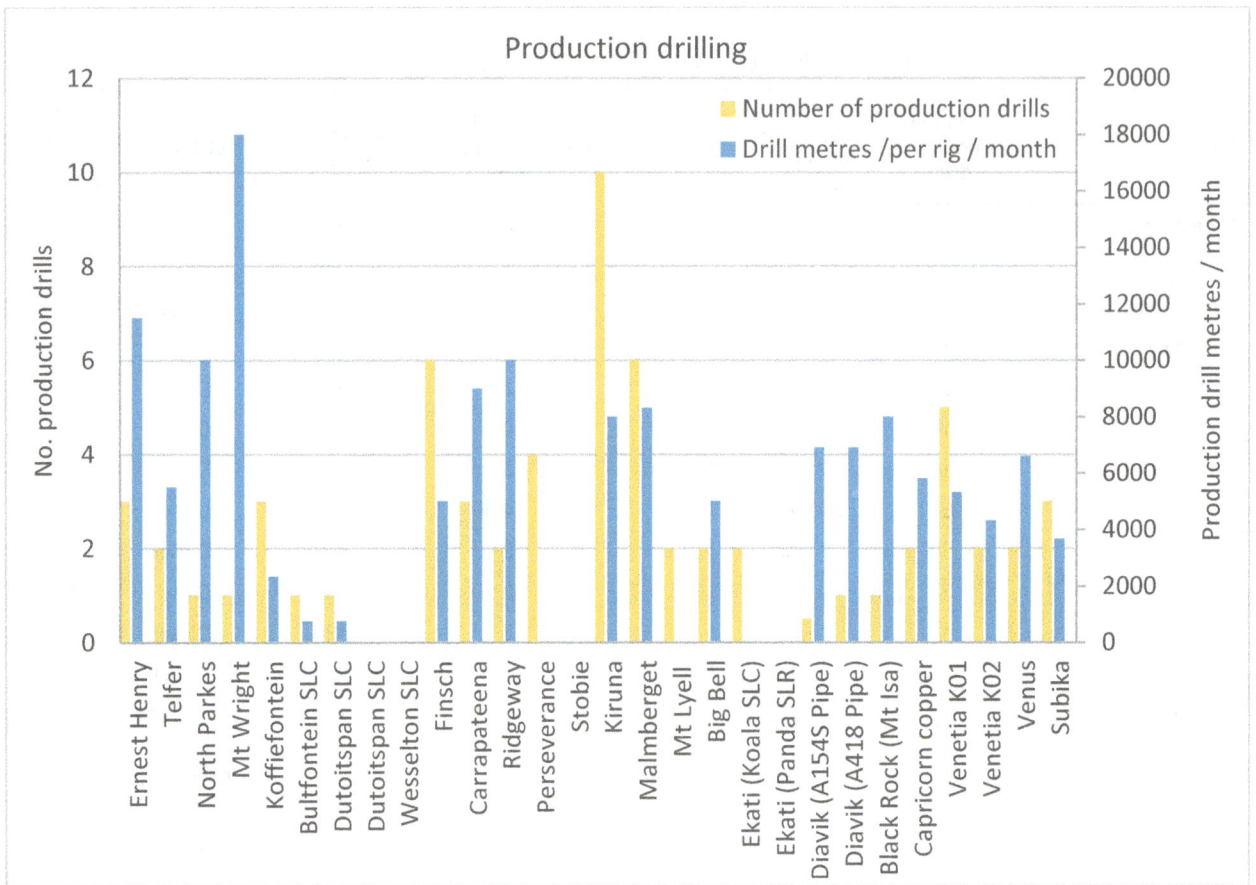

FIG 4 – Number of production drill and monthly production drill metres.

Productivity for SLC mines is typically measured as a function of production tonnes and the number of available drawpoints for production (Power, 2021). This metric, measured in tonnes extracted per day divided by the number of available drawpoints is also used in mining studies to estimate mine production for a SLC operation given the size of the level footprint and number of available production cross-cuts, as well as a means to back calculate the number of production cross-cuts and active production levels that are required to achieve a specific production rate. Transverse SLCs generally have a lower productivity rate for each active drawpoint as the production intensity and utilisation is lower compared to a longitudinal SLC. Transverse SLCs are generally larger operations and have higher production rates due to the number of available drawpoints for production. SLC mines typically mine one ring in each cross-cut every 7 to 10 days, on average. Mines with lower production rates and more difficult conditions have a slower ring turn-over rate compared to those with more favourable conditions. Production ring turnover rate is also impacted by the draw strategy and requirements for pre-charging and maximum explosive sleep time.

LHD productivity rates from the benchmarked mines are provided in Figure 5 and in Table 2. SLC mines in developed countries with favourable ground conditions typically achieve productivity rates of 350–420 t/day/drawpoint. These mines include Carrapateena, Ridgeway, Telfer and Ernest Henry. The highest productivity rates are achieved by Kiruna and Malmberget. These two mines achieve high production rates due to the favourable ground conditions and production cross-cut widths of up to 7 m, which in turn allows the largest LHDs of any SLC mine to be used. These loaders have bucket capacities of over 25 t. The iron ore being mined also has a density of 4.2 t/m^3 on average, which means less volume of material needs to be moved (per tonne) compared to the other benchmarked mines with typical ore densities of 2.6 to 3.0 t/m^3. The mines with more difficult ground conditions that are subject to more frequent delays due to drill and blast problems, rehabilitation and redrilling typically achieve lower productivity rates in the order of 200 to 250 t/day/drawpoint.

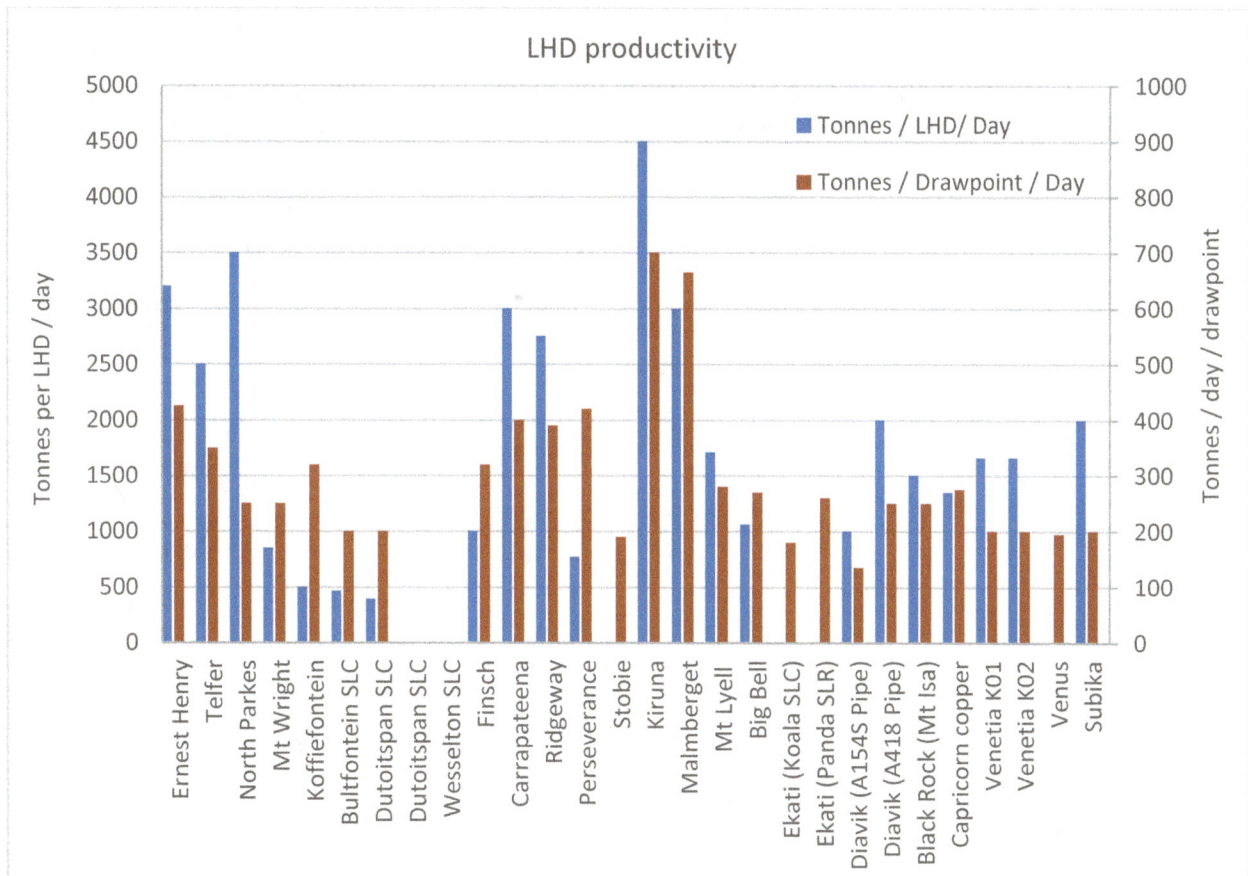

FIG 5 – Production loader productivity and drawpoint productivity.

LHD productivity in terms of tonnes per day per loader was found to vary significantly from mine to mine. The rates varied between 400 t/LHD/day to 4500 t/LHD/day. Multiple factors were found to influence this parameter. Most important was the LHD size, bucket capacity and the loading methodology (into trucks or into orepasses). As expected, mines with larger LHDs tipping to multiple orepasses significantly outperform trucking operations with smaller LHDs. It was unexpected that the impact to productivity was up to 10 times higher in the most extreme comparison, and 2–3 times higher for mines operating ~20 t LHDs tipping to orepasses compared to mines with 10–14 t LHDs tipping to trucks. The reasons for the differences are somewhat obvious, though the differences in productivity identified in the benchmark was more profound than expected. Other factors, such as mine layout, tramming distances and number/length of shifts (two shifts versus three shifts per day) were also found to impact LHD productivity. The mining depth was not found to impact productivity metrics in the benchmarked data, and this is most likely due to some of the more productive mines being also some of the deepest SLCs in the study.

The benchmark found a weak correlation between orebody rock strength and rock mass rating, where mines with higher quality rock mass generally were also the mines with the higher productivity rates in terms of tonnes per day per drawpoint. However, there is considerable scatter and variability in the data indicating that this is a generalisation only, but not the only factor that strongly influences productivity.

The footprint area of each production level is a significant factor influencing the annual production rate of the benchmarked mines, as shown in Figure 6. This is expected, given that most SLCs operate with three levels in production and the footprint area of a level dictates the number of available production cross-cuts per level. Mines with orepasses and high-capacity materials handling systems such as Ridgeway, Ernest Henry, Telfer, Malmberget and Kiruna have the highest production rates. Telfer, Malmberget and Kiruna are all outliers in terms of footprint area. Both Telfer and Kiruna have large orebodies (in horizontal area) with a strike length of around 800 m and 4 km respectively. Each level has many production cross-cuts. Kiruna is required to break each level up into separate operating blocks due to the scale of the mining footprint and strike length of the orebody. The Malmberget mine operates in 14 separate orebodies of varying depth, size and geometry. Although the total operating footprint is very large, there is a loss in productivity compared to the nearby Kiruna mine due to the additional complexity operating and hauling from multiple orebodies concurrently. It should also be noted that the Kiruna and Malmberget iron ore mines have high density ore, which also contributes to the high annual tonnage produced.

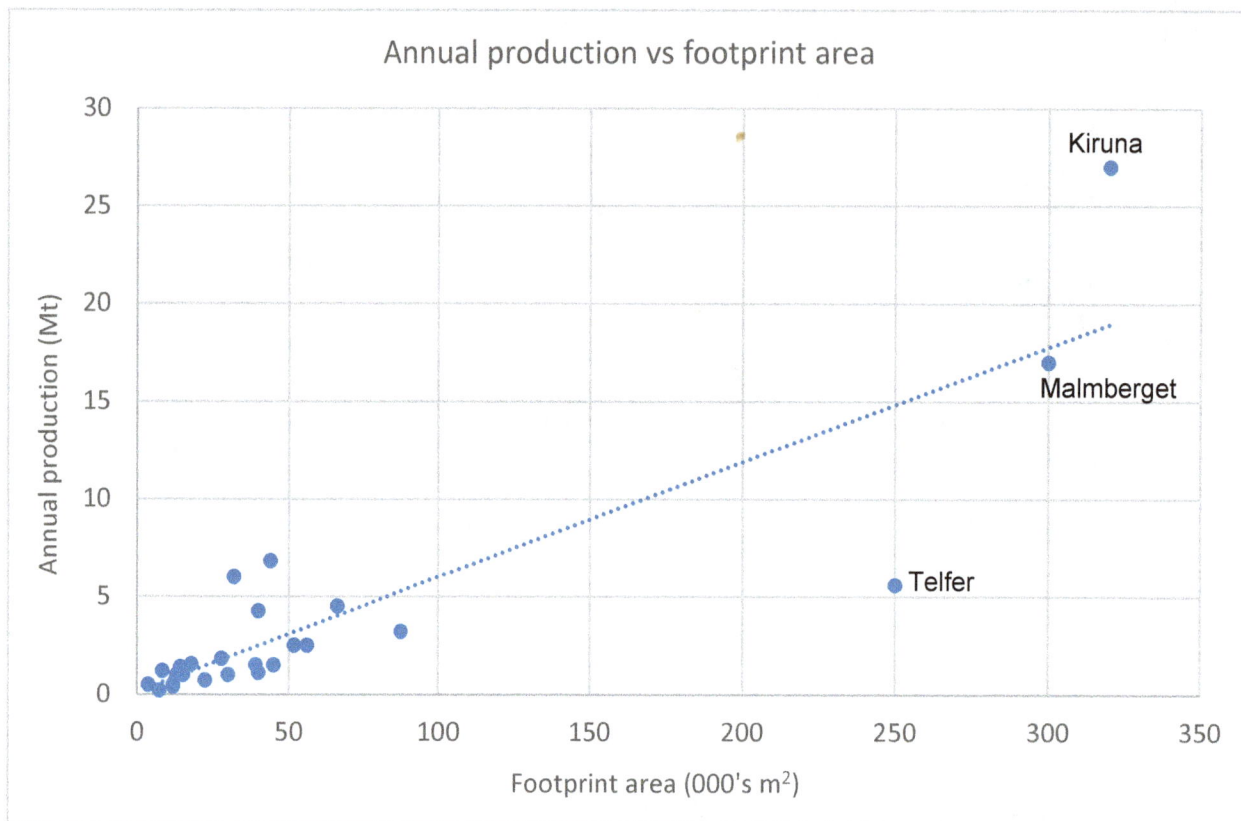

FIG 6 – Annual mine production tonnes as a function of production level footprint area.

Only the Kiruna and Malmberget mines use tele-remote loaders for SLC production. These machines are operated remotely from an underground mining control centre. LHD tramming is automated and the dig and tip cycles controlled by an operator in the control centre. Despite tele-remote capability, up to 80 per cent of mine production is completed using manual operations (ie people in machines) at the two mines. Despite widespread uptake of tele-remote loading in the global mining industry, it was somewhat surprising to find that only a small amount of global mine production from SLC mines was conducted via remote and automated loading operations. This is also particularly interesting given the regular layout of an SLC mines is highly suited to automated loading operations.

The Kiruna and Malmberget mines were also the only two mines to utilise tethered electric loaders. The use of tethered LHDs in an SLC is generally not practical due to the layout of production levels, with the production cross-cuts and orepasses (or truck tipping location) on the other side of the main footwall access drive. Access requirements for other equipment on the level generally means that the trailing cable would be hit or would need to be constantly disconnected to allow access. The 'crows foot' layout (Mooney *et al*, 2021) developed for the Carrapateena mine negates this problem and enables tethered LHDs to operate unrestricted. There were no mines in the benchmark that use battery powered LHDs.

PRODUCTION RAMP UP

Information for the production ramp up at Carrapateena, Ernest Henry and Ridgway was provided as part of the study (see Figure 7). These three contemporary SLC mines in Australia are of comparable mining depths and level geometry. The three mines reached a production rate close to nameplate after approximately four years from first production. The ramp up rate was an increase in annual production of 1 Mt to 1.5 Mt per annum. This time does not include the initial construction and development period for each mine, which varied significantly due to the length of decline development required. The notable step-change in the production rate at Ernest Henry was due to the change over from trucking operations to shaft hoisting. This ramp-up information is provided for the purpose of benchmarking for future mining assessments and studies. Data from more mines was not available due to mines not keeping such records and because the production ramp up had occurred over many years and even multiple decades at some mines.

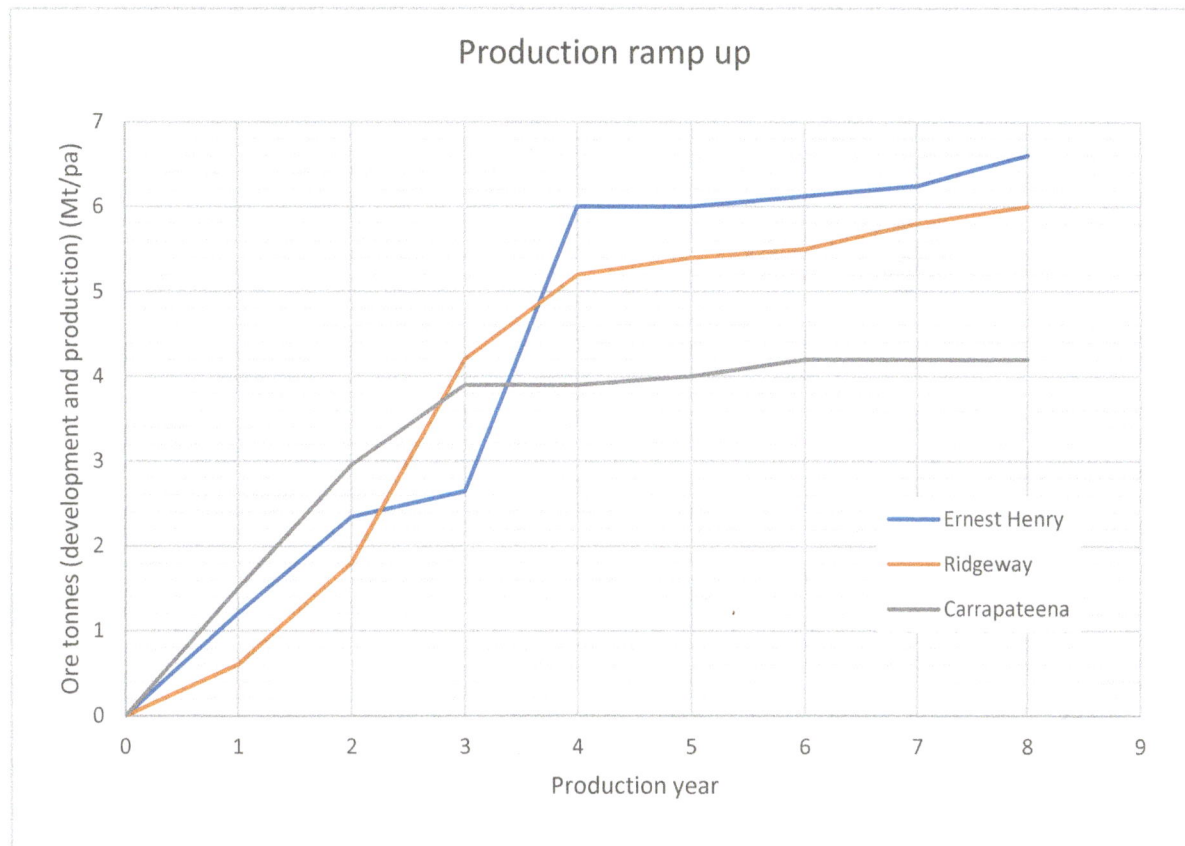

FIG 7 – Production ramp up for Ernest Henry, Ridgeway and Carrapateena.

MINING COSTS

Mining costs as a function of the annual production rate are illustrated in Figure 8. Mining cost information was not provided by some mines and the names of individual mines with information shown in Figure 8 are not provided for confidentiality reasons. The line of best fit for the mining costs provided rather than individual data points to ensure individual mines remain unidentifiable. The range provided is for ±20 per cent. The mining costs have been split into two groups: mines in developed countries and those in developing counties. The mining costs have been converted to USD and the mining costs includes underground operational costs only, including development, production, drill and blast, haulage to surface and mine services. Costs associated with ore processing and treatment, transport, smelting are not included. CAPEX is also not included.

Mining costs in developed counties were found to vary considerably more as a function of the annual production rate compared to the mines in developing countries. Mining costs for the mines in developing counties was also considerably lower for the mines with low production rates which is most likely due to these mines operating on a low cost scavenge basis that is focused on overdrawing the sides of historic caving operations. At higher production rates, the mining costs for developing and developed countries are more comparable. This is due to the cost of labour being offset and almost inversely proportional to the productivity rates achieved per person.

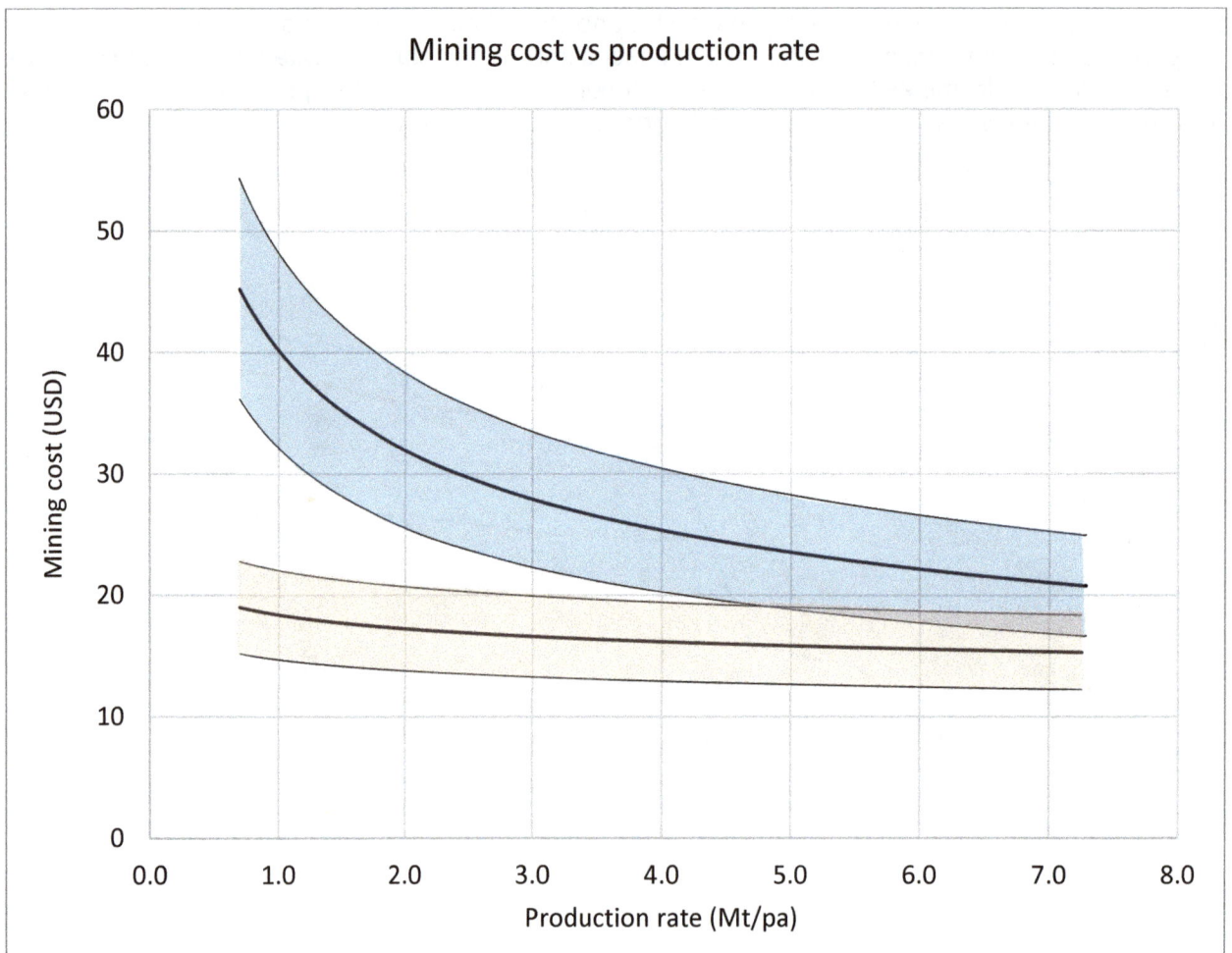

FIG 8 – Mining cost versus production rate for developed countries (blue) and developing counties (orange).

CONCLUSIONS

This paper provides the main findings from a global benchmark of SLC mines for mine productivity, including mine development, production and mining costs. A total of 21 mines spread over four continents and 17 different mining companies were included in the project. The development, production and productivity measures were found to be affected by a wide range of variables including mine scale, depth, ground conditions, mine layout, trucking versus orepasses and shift scheduling. The range of some productivity measures was found to vary by up to a factor of 10 for the benchmarked mines.

Typically, development rates in the order of 200 to 300 m/month were recorded per jumbo for large scale operations with favourable mining conditions. Smaller mines with less favourable conditions often achieve half this rate. Productivity rates of 350–420 t/day/drawpoint were achieved by large scale SLCs in favourable ground conditions. The smaller operations with less favourable ground conditions were found to achieve considerably lower productivity rates in the order of 200 to 250 t/day/drawpoint.

Despite widespread uptake of tele-remote loading in the global mining industry, it was found that only a small amount of global production from SLC mines was conducted via remote and automated loading operations. This is despite the regular layout of many SLC mines, which is highly suited to automated loading operations. All the SLCs benchmarked continue to rely on diesel LHDs, and no operations were using battery powered LHDs at the time of the site reviews.

ACKNOWLEDGEMENTS

The author would like to acknowledge the Cave Mining 2040 research program sponsor companies (AngloGold Ashanti, BHP, OZ Minerals, Newcrest and Vale) for their continued support of cave mining research. The global SLC benchmark and best practices review project would not have been possible without the in-kind support, site visits and access information by the various mines who participated in the project. The author must personally thank the following people for their assistance during the project:

- Big Bell (Westgold) – Jogi Samosir, Wei Duan, Tim Green

- Black Rock (Glencore) – Andrew Shiels, John McConnell, Alastair Grubb, Albert Young

- Carrapateena (Oz Minerals) – Glen Balog, Daniel Hronsky, Matthew Fargher, Daniel Bruce, Chad Parken, David Cox, Mollie Poulter, Tessa Ormerod, Claire Chauvier, Ming Xia, Peter Burns

- Capricorn Copper (29 Metals) – Albert Sta Ana, Alonso Gonzales, Josh Moran

- Diavik (Rio Tinto) – Paul Duplancic, Steffan Herselman, Rob Atkins

- Ernest Henry (Glencore/Evolution) – Michael Corbett, Aaron Harrison

- Finsch and Koffiefontein (PetraDiamonds) – Refilwe Mafiri, Johan Langenhoven, Douglas Baxter, David Bailey, Mongezi Magwaza

- Kimberley Mines (Bultfontein, Dutoitspan, Wesselton) (KEM JV) – Derick Du Plessis, Brent Atling

- Kiruna and Malmberget mines (LKAB) – Gurmeet Shekhar, Yunus Hazar, Matthias Wimmer, Steve Belohlawek

- Mt Wright (Resolute) – Stuart Long

- Northparkes (CMOC) – Ellie Hawkins, Sarah Webster, Zecheng Li

- Subika (Newmont) – Frans Basson

- Telfer and Ridgeway (Newcrest) – Paul Kline, Otto Richter, Lino Manca, Robbie Lowther

- Venetia (DeBeers and Anglo American) – Freddie Breed, Denver D'Angelo, Tshepo Mokele

- Venus (BHP) – Michael Hopkins, Rigo Rimmelin

REFERENCES

Bull, G and Page, C H, 2000. Sublevel caving – today's dependable low-cost 'ore factory', *MassMin 2000* (ed: G Chitombo) pp 537–56.

Kvapil, R, 1965. Gravity flow of granular materials in hoppers and bins-Part 2 Coarse material, *International Journal of Rock Mechanics and Mining Sciences*, 2(3):277–292.

Mooney, A, Grosser, H, Marsden, J and Dunstan, G, 2021. Carrapateena Sub Level Cave Crows Foot Design – Safety in design to support efficiency and production automation, *Underground Operators Conference 2021* (The Australasian Institute of Mining and Metallurgy: Melbourne).

Power, G R, 2021. An analysis of productivity in sublevel cave mines, *Underground Operators Conference 2021* (The Australasian Institute of Mining and Metallurgy: Melbourne).

Power, G R and Just, G, 2008. Mass Mining by Sublevel Caving, Internal report, Mass Mining Technology Research Group, University of Queensland.

Challenges and solutions for deformation management of Cannington's hoisting shaft

A Clarkson[1], C Hall[2] and M Sandy[3]

1. Senior Project Geotechnical Engineer, South32, Cannington Mine, Mckinlay Qld 4823.
   Email: andrew.clarkson@south32.net
2. Geotechnical Superintendent, South32, Cannington Mine, Mckinlay Qld 4823.
   Email: christopher.hall@south32.net
3. Principal Geotechnical Engineer, AMC Consultants, Perth WA 6005.
   Email: msandy@amcconsultants.com

ABSTRACT

Cannington Mine is an underground silver lead zinc operation located in north-west Queensland which has been in production since 1997. Ore is primarily moved through an underground material handling system comprised of a crusher, conveyor and hoisting shaft. The Fowler Shaft was constructed in 1997 using a strip and line method with the intent for it to provide a means to hoist ore from the loading station 600 m below surface. The shaft also serves as the route of key services including high voltage power, mine dewatering and fresh air.

Despite the adherence to a mining exclusion zone, large-scale change in geotechnical conditions around the shaft were first observed in 2015. Ongoing ground movement around the shaft lead to cracking and delamination of the concrete liner.

Shaft management required the on-site team to measure deformation rates and magnitudes, forecast change, manage rockfall risk and deforming pipework, and ultimately, hoisting skip clearance.

Monitoring techniques to observe and measure deformation were required to evolve with the magnitude of movement. Some techniques leveraged off modern technology while some of the most useful techniques were based on logic and first principles. All shaft deformation management strategies relied on the co-operation and alignment between Cannington's Infrastructure Maintenance, Survey and Geotechnical Departments.

INTRODUCTION

The Fowler Shaft was constructed in 1997 using a methodology which Benecke, Hancock and Weber (2016) summarised as:

- Pilot hole drilled from surface to 645 m.

- Raise bore shaft to a diameter of 1.8 m.

- Developed vertically, using bench drill and blast methods to strip out to full diameter, with the fired material mucked into the 1.8 m raise.

- Line the shaft by using internal steel formwork and concrete to establish the nominal 5.6 m internal diameter. The thickness of the concrete liner is assumed to be 300 mm and thicker in zones where shaft wall overbreak occurred.

The Fowler Shaft uses rope-guided skips with a payload of 15 t each to provide a means to hoist ore from the loading station 600 m below surface. The shaft serves as the route of key services including high voltage power, mine dewatering and fresh air. Geotechnical conditions have deteriorated in response to mining activities and progressive stoping as the mine has matured to its current remnant state. Progressive extraction of the orebody and stoping in proximity to the shaft exclusion zone has progressively resulted in deformation since 2015.

GEOLOGICAL SUMMARY

The location of the shaft was selected to minimise potential ore sterilisation and reduce the likelihood of intersecting adverse ground conditions. Two exploratory boreholes were drilled from surface to collect geology and structural information. Geological interpretation has been studied and refined over the subsequent 25 years, based on underground mapping and diamond drill information. Figure 1 shows a summary of the current geological interpretation, relative to the shaft infrastructure.

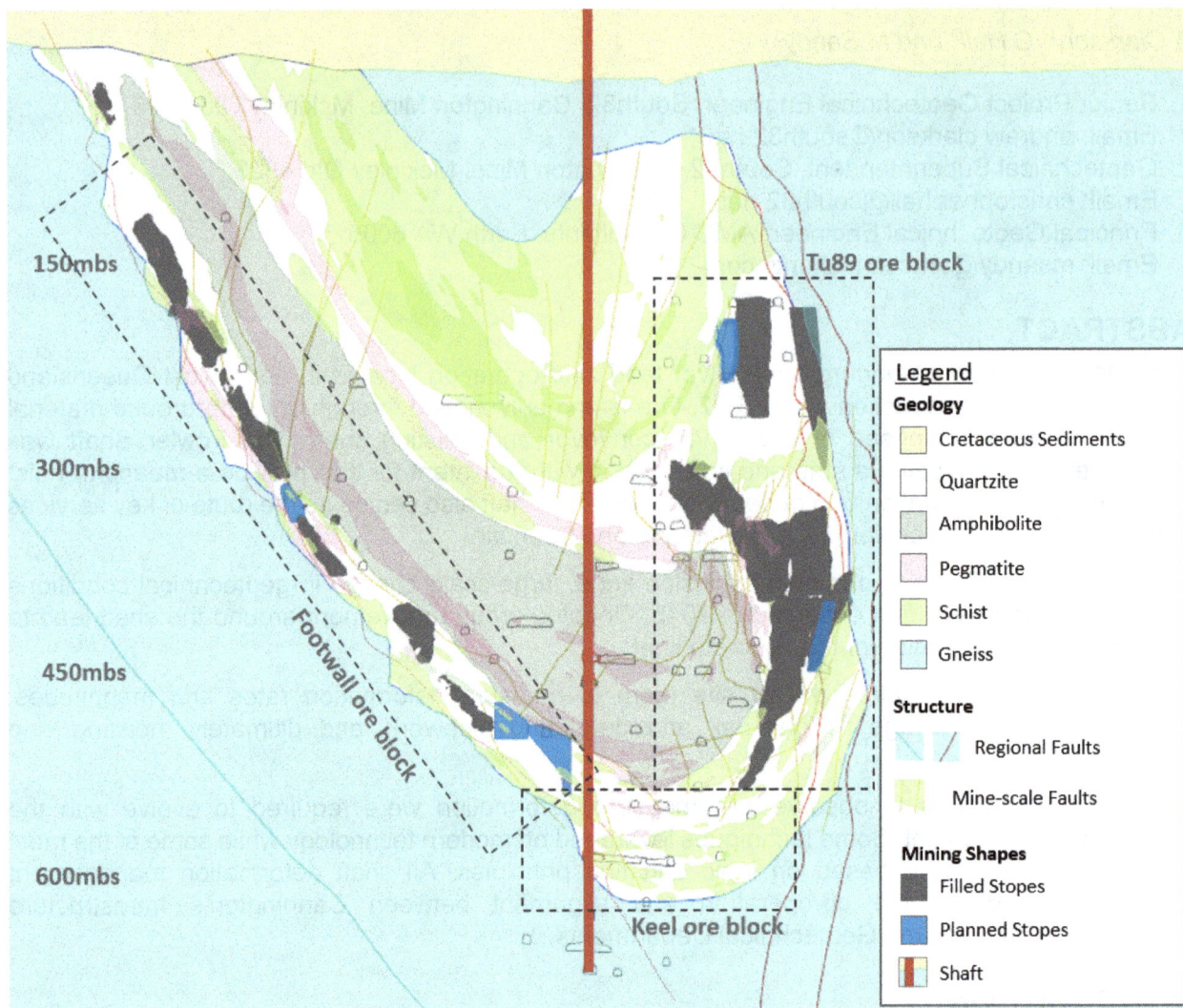

FIG 1 – Cross-section of shaft location, relative to geological interpretation, looking north.

Although Cannington has complex geological conditions with significant variability between lithologies, the host rock surrounding the shaft can be broadly summarised by the following rock types: cretaceous sediments which includes weathered mudstones and clays, gneiss which is hard and foliated, quartzites which are competent and strong, schist which is soft and jointed and amphibolite which is blocky and jointed.

Water inflow to the shaft is limited, largely because of mine dewatering which has dropped the pre-mining water table above the mining footprint from 50 mbs to 200–300 mbs (Hall, 2015).

MECHANISM OF SHAFT DEFORMATION

Complex mining induced stresses interact with the different lithological units that host the Fowler shaft. As Cannington has matured, various mining sequences have been extracted which have led to strain on the Fowler shaft. As the surrounding conditions change, variable deformation characteristics are apparent.

Mine scale subsidence occurs as a result Cannington's mining activities. The ore is found in mineralised quartzites which are typically very competent with strengths generally around 200 MPa. This ore is extracted by a longhole open stoping method and backfilled with paste or unconsolidated rock fill materials. Cannington's paste fill typically has a UCS of around 1 MPa and a modulus of 0.35 GPa which are about two orders of magnitude less than the ore. Extraction of ore results in stress redistribution and subsidence. The backfill has little effect on the stress redistribution. It should nonetheless limit but not totally prevent subsidence and deformation.

The footwall and keel mining sequences progressively triggered subsidence within the shaft and surrounding rock mass. While this undercut mining was occurring, the Tu89 stoping block sequence to the east was progressing, causing a stress shadow effect, resulting in reduced confinement of the rock mass surrounding the shaft.

Different scale deformations have caused different issues. There are large sections of host rock surrounding the shaft mobilising as intact units. This is evident in areas of the liner that are mostly undamaged above and below horizontal shear planes and damage to service pipelines 'bowing' due to shaft large scale vertical deformation. Mine wide deformations cause damage to mechanical items and services, movement on hoisting infrastructure and affect skip clearances. This mine wide deformation has ultimately led to skip clearances falling below acceptable limits and the requirement to plan for shaft decommissioning.

A large portion of the shaft has experienced localised deformation. This is where mine wide stress changes have led to changing stress on structure, and formed abutment stress concentrations in some locations and relaxation in others. The liner material is ~300 mm thick concrete reinforced with 20 mm rebar. The concrete in the shaft barrel is brittle and deformation manifests from cracking to spalling and eventually total failure of the liner. This damaged liner is a rockfall hazard with maintenance personnel and infrastructure in the line of fire of potential liner failure.

Numerical modelling

The mining induced stress impact on the shaft was first investigated by Beck Engineering in 2017. Plastic strain modelling results were looked at to explain the mechanisms of shaft damage and the forecast magnitudes of this damage.

The strain mechanisms of angular distortion, shaft barrel extension and horizontal strain were combined to assess current and future total strain. In-field observations and survey data were used to calibrate the inputs for a revised model in late 2019. An example of the 2019 model results with a focus on the Fowler Shaft is shown in Figure 2.

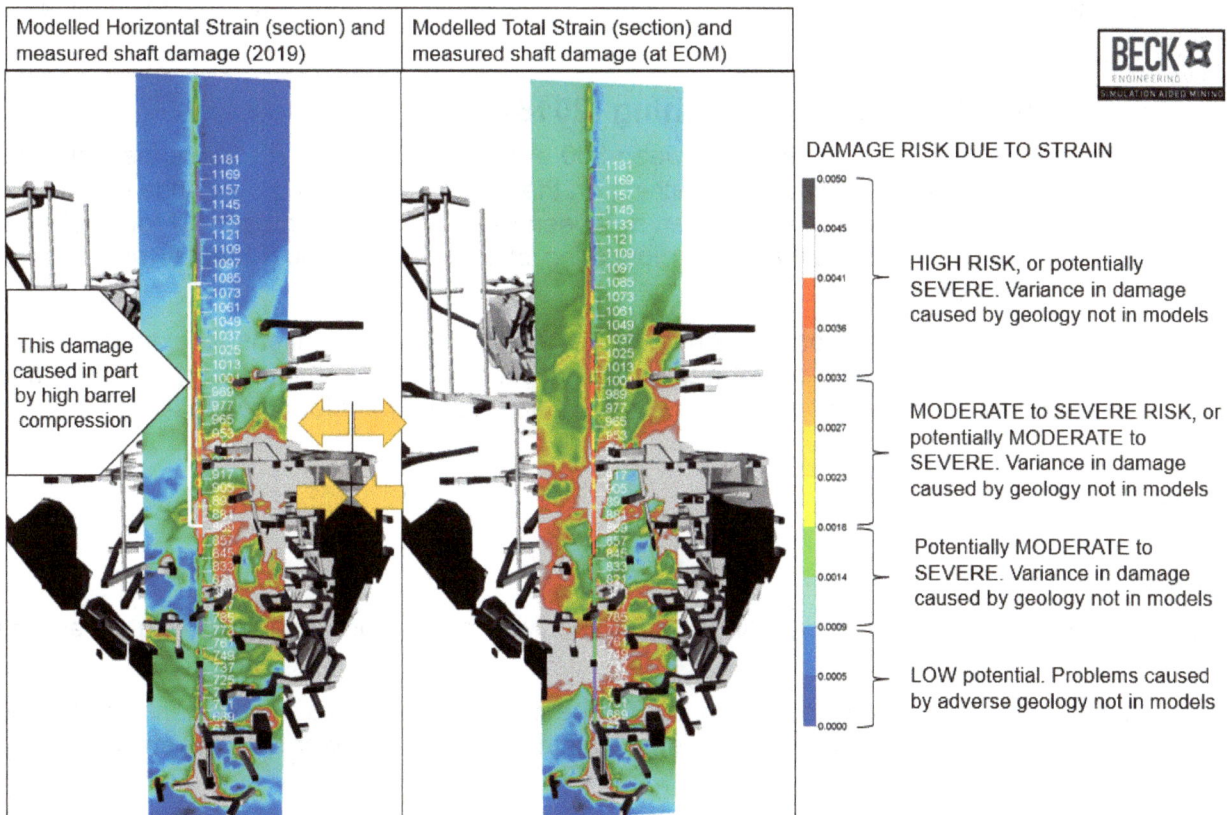

FIG 2 – Example of plastic modelling results (Beck, 2020).

TIMELINE OF SHAFT DEFORMATION

Deformation of the rock mass at specific locations of the shaft and the subsequent damage to the concrete liner progressed over time, which introduced two key risks:

1. Localised rockfall-type hazard, where the concrete liner crushed or delaminated from the rock mass and falls from height onto people working on shaft infrastructure below.

2. Damage to skip infrastructure, where the clearance between the hoisting skip and the shaft wall reduced to the point that the skip could catch on the wall and cause an interruption to safe production.

Throughout the life of the shaft, physical inspections have been conducted by Fixed-plant Maintainers and Geotechnical Engineers to ensure that any signs of change are identified and escalated. While these inspections were a constant feature in the shaft's management, a raft of alternative controls and monitoring techniques were trialled and implemented throughout its life.

Initial indication of shaft concrete liner movement (2006–2014)

The first signs of shaft liner deterioration were reported in 2006 when the area of concern was focused on the interface between the weathered surface sediments and the top of fresh rock. In response an external study was commissioned to assess the condition of the concrete liner and advise on the necessary steps to maintain reliable hoisting for the remaining mine life, which at that time was a further seven years, until 2013.

Love (2006) identified that concrete liner cracking coincided with concrete pour points around the perimeter of the shaft, which would prove to be a weak point in the concrete liner design in the years to come. Recommended passive controls which included visual inspections and crack monitoring were implemented until the step-change in damage eventuated.

Acceleration of movement (2014–2018)

A step-change in deformation rate and magnitude throughout the lower South Zone of the mine (below the 280 mLv) was identified in 2014/2015. This change triggered rehabilitation campaigns in squeezing underground development drives and impacted the conditions in the shaft. Cracks in the concrete liner opened to the point that detached slabs could be manually, hand-scaled as a mitigating control.

Final years and shaft decommissioning (2018–2022)

Deformation at a consistent rate change between 350 and 400 metres below surface inevitably led to the west quadrant of the shaft liner encroaching on the minimum operating clearance of the west skip. By this time, the critical areas of the shaft had been rehabilitated with mesh and rock bolts, so the focus was to establish a short-interval monitoring strategy which could identify any sudden changes to west skip clearance.

During this time a project was commissioned to identify and assess solutions to extend hoisting operations. The outcome was the decision to transition Cannington to an all-trucking operation and re-route mine services away from the shaft to eliminate any requirement for personnel access in the future.

MONITORING

In-field observations

The Fowler Shaft contains counterbalanced east and west skips, each equipped with an access platform above the skip.

Geotechnical and fixed plant inspections occur frequently from both skips. Fixed plant inspections occur weekly and are carried out prior to weekly maintenance on the lower load out stations at the bottom of the Fowler Shaft.

Geotechnical inspection frequency has been variable based on the observed conditions in the shaft. Prior to 2015, these geotechnical inspections occurred on an annual basis. From 2019 due to the high rates of deformation in the shaft, these inspections were carried out on a weekly basis.

The observed damage included: torsional displacement, evident in spiral cracking, shear displacement, evident in shear movement along the liner's cold joints and stress break out, evident by axial 'pinching' deformation, generally along the south-east and north-west parts of the shaft barrel as shown in Figure 3.

FIG 3 – Radial compression damage on south-east quadrant of shaft perpendicular to induced primary stresses.

Extensometers

During the November 2017 – February 2018 shaft liner rehabilitation campaign, 8 × 8 m extensometers and 8 × 6 m SMART cables were installed to monitor liner convergence. Following this rehabilitation work and upon recommencing hoisting activities, these instruments were heavily damaged due to ore scats spilling from skip unloading movement at the surface and overhanging ground support elements carried by the skips, also tearing mesh liner. These instruments were repaired and fitted with Newtrax Bluetooth modules, either stored in the instrument hole collar or protected by covering with bolted conveyor belt sheet.

After a lot of difficulty maintaining these instruments the collected data showed minimal convergence, supporting the mechanism of the shaft host rock largely subsiding with the shaft as solid units. These instruments were decommissioned, and the Bluetooth units retrieved for use with other instruments located throughout Cannington's underground tunnel accesses.

Quantify scaled material

In 2019, regional fault mobilisation and subsequent horizontal strain where structures intersect the shaft led to increased liner damage. The quantities of scaled material during shaft inspections were recorded. These quantities were estimated from material brought up with personnel on the skip and that which was scaled and dropped. The rate of scaling was significant during summer 2019/2020 which led to bringing forward additional rehabilitation out of schedule in January 2020. These quantities are presented in Figure 4 which shows mass of scaled concrete plotted against time. The dates and tonnages of stope firings throughout the mine are also plotted.

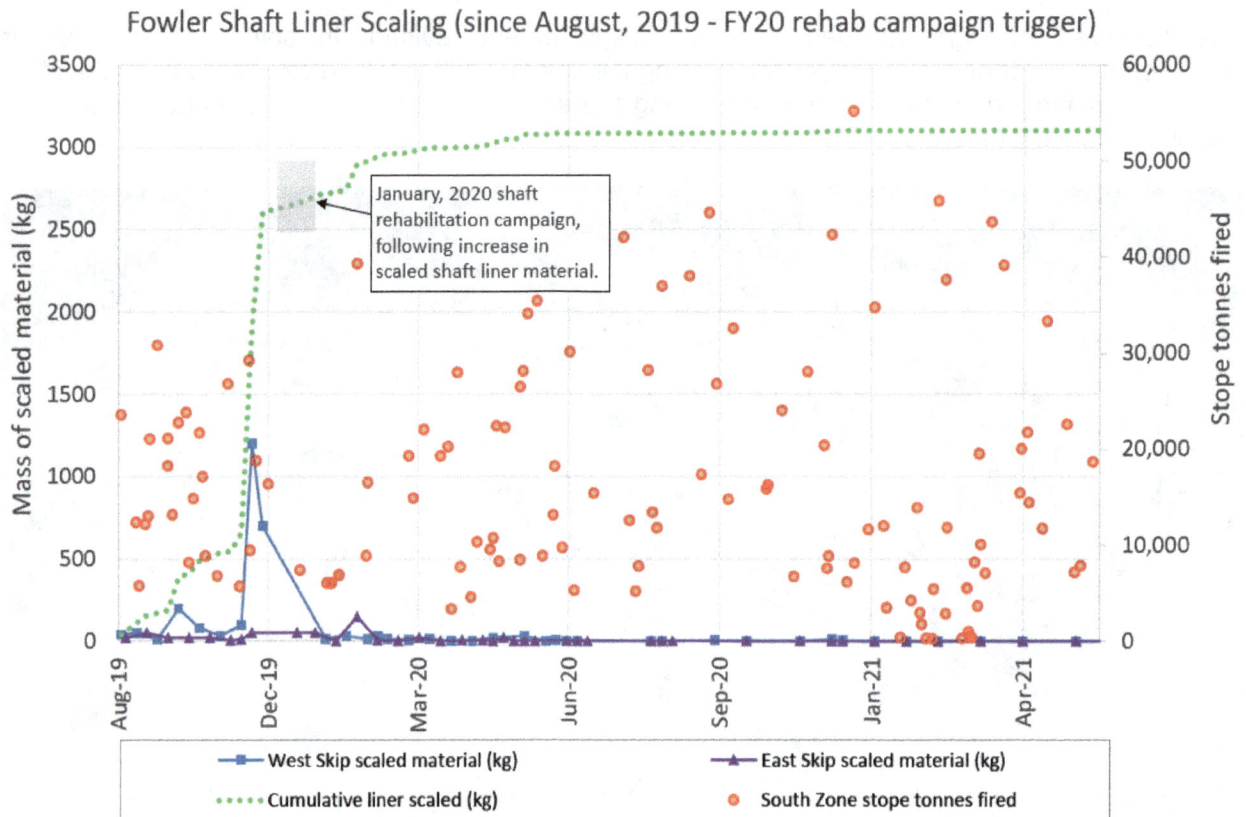

FIG 4 – Mass of concrete liner scaled over time versus South Zone stope firings.

Specialised survey results

In 2016, shaft survey specialists DMT were approached to perform a kinematic laser survey of the shaft between the surface and load out station approximately 645 m below surface. This survey methodology proved successful in providing an accurate shaft inclination plot (see Figure 5). DMT performed further scans in 2017, 2019 and 2021 whilst the shaft was still operational. These scans proved the most useful tool in providing an accurate and complete profile of the shaft deformation relative to the surface hoisting frame.

These shaft profile scans highlighted directional deformation trends that would continue until the shaft was decommissioned for hoisting ore.

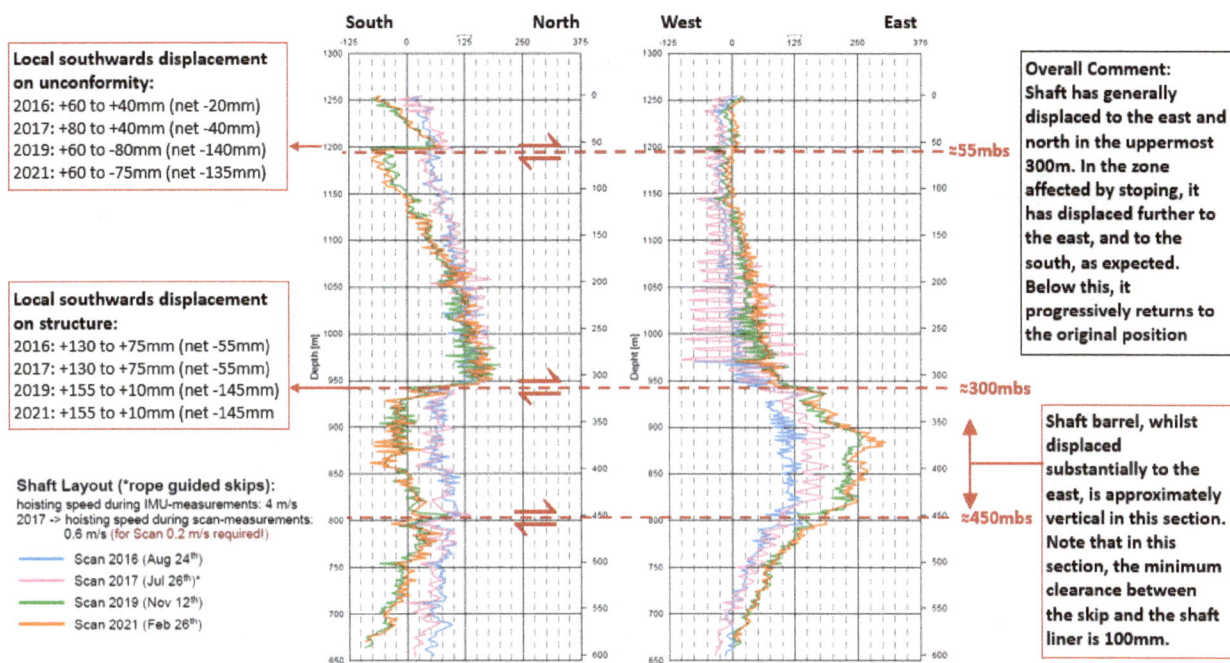

FIG 5 – Progression of mine-scale shaft deformation over time (adapted from Sandy, 2020).

LiDAR scanning and point cloud comparison

In early 2021, the collection of surveyed LiDAR scans for the purpose of ground movement was introduced to Cannington. Initially LiDAR techniques were focused on scanning horizontal development drives and proved successful for conducting point-cloud comparisons to determine rates and magnitudes of deformation.

Handheld LiDAR scanning was trialled in the shaft with the intent to replicate the style of monitoring used in horizontal development. Whilst LiDAR technology provided an additional layer of movement verification, it ultimately proved too limited for permanent skip clearance monitoring application and was discontinued after a six month trial.

Physical measurements of skip clearance

Some of the earliest measurements taken in the shaft were basic tape-measure or distometer (hand-held laser) measurements. Measurement points were marked-up on the shaft liner and a safe work instruction was developed to ensure a repeatable process to measure skip clearance was followed. Despite the error introduced by the sway of the skip, the practice of manually measuring the clearance between the west skip and the west wall was sustained throughout the shaft's hoisting life. Figure 6 demonstrates how manual measurements were plotted against survey data to verify survey accuracy and provide short-interval (weekly) monitoring.

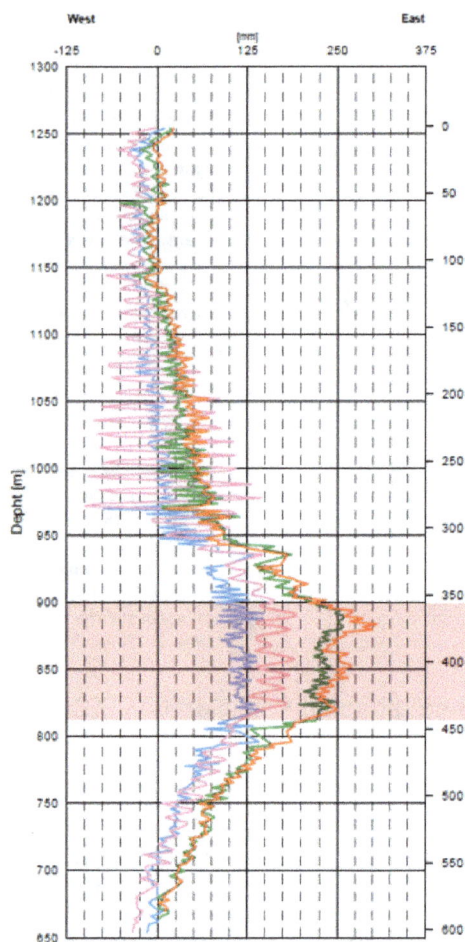

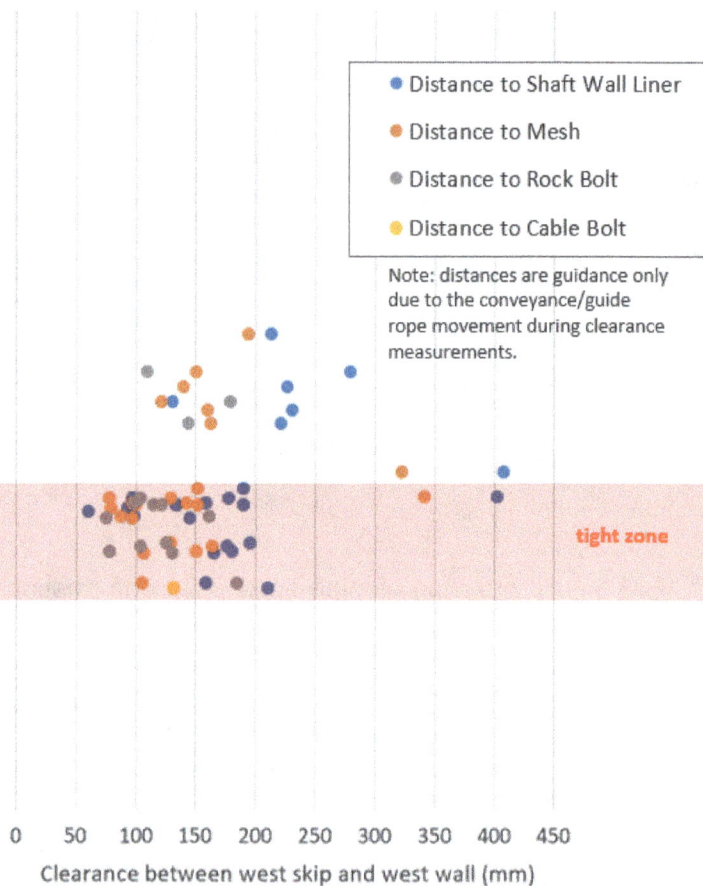

FIG 6 – Example of correlation between manual distometer measurements and MDT shaft survey.

Video recordings and review of live hoisting runs

The risk of 'skip interaction with shaft wall' continued to be called-out and escalated throughout the final 12 months of hoisting. Trouble-shooting workshops led to the realisation that the skip may behave differently at hoisting speed (14 m/s) than during inspections (≤2 m/s) In response, several fittings were welded to the standard skip platform:

- A single fitting on the canopy was designed to hold a small camera with a wide-angle lens.

- Several clamp-type fittings were welded to the handrail of the inspection platform (Figure 7).

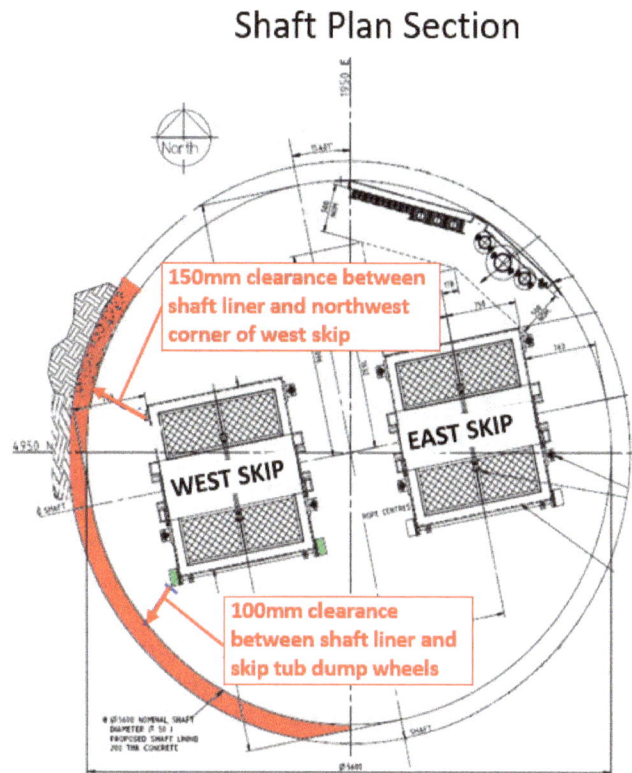

FIG 7 – Configuration of plastic 'strikers' used to monitor clearance between skip and concrete liner.

Flexible plastic 'strikers' were attached to the welded clamps and set to pre-determined lengths. The flexibility of the strikers meant that when a striker contacted the wall, it could be easily identified when reviewing the video footage. This monitoring design allowed for recordings of the west skip's behaviour and clearance to the west wall under full operating conditions (both loaded and unloaded) to be assessed.

The wide-lens camera was often deployed prior to a physical inspection with the intent to gain an understanding and isolate personnel from any step change in shaft damage that may have occurred.

MITIGATING CONTROLS

Scaling

Scaling loose liner material prior to weekly maintenance work at the bottom of the shaft was a simple and effective control in ensuring safe access. Scaling equipment is readily available at Cannington and the hoist downtime required for rehabilitation is not excessive. Throughout the period of accelerated movement and into the shaft's final operational years, many areas of damaged and loose liner material were visually identified and safely removed by scaling. Figure 8 provides some examples of scaled shaft liner material.

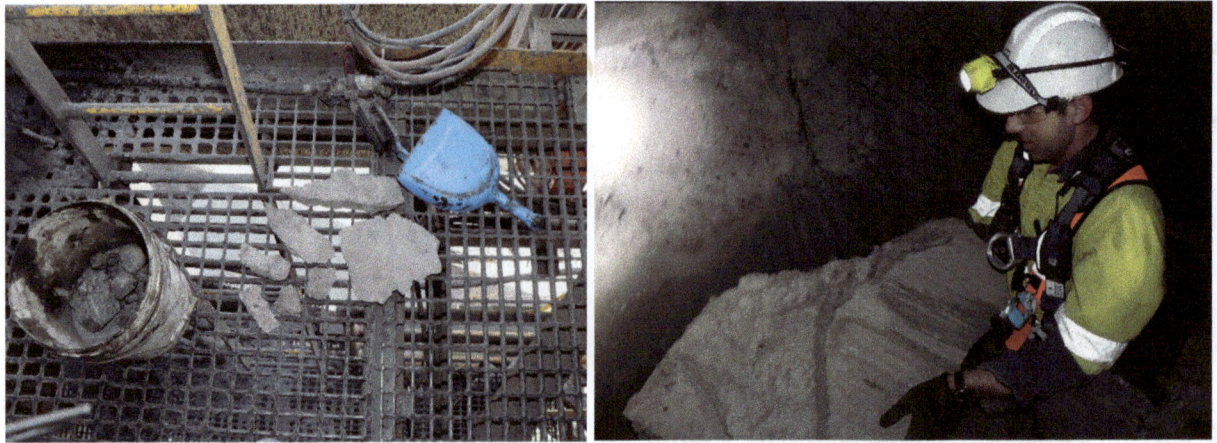

FIG 8 – Examples of scaled shaft liner material.

During most of the shaft's operational life, deformation and liner damage were non-existent and there was initial resistance from maintenance crews to perform scaling as any falling material either liner or ore scats during hoisting can potentially cause to damage to infrastructure at the bottom of the shaft. As awareness of the geotechnical issues was shared amongst the maintenance teams and likewise awareness of the personnel exposure to maintain the shaft infrastructure to the geotechnical department. This method became accepted and required prior to any access below the hoisting section of the shaft.

As more of the damaged shaft liner became contained with additional ground support, less shaft barrel liner was required to be scaled. In the later stages of the shaft's operational period, scaling actives included bleeding and repairing bagged ground support.

Rehabilitation

Platform design/construction

Prior to major rehabilitation being conducted a purpose-built platform was designed, constructed, and intermittently installed on a Cannington Shaft Skip in order to facilitate ground support activities in the shaft (Hendry and Moshos, 2016). This skip platform allowed sufficient space for rehabilitation to be conducted. This platform included storage areas for rehabilitation equipment such as airleg drills and a grout bowl as well as consumables storage which included rock bolts and chain link mesh rolls.

The skip top platform included an overhead wing roof for operator protection from scat fall off from above and foldable north and south wings to allow 180° of shaft barrel access from either the east or west skip. Installation of the skip top platform took approximately three shifts to both install and uninstall.

Ground support design

The shaft underwent various rehabilitation periods between 2016 and 2021. The ground support methodology evolved with site experience and purpose. The overall intention of the installed ground support was to line the brittle shaft barrel with ductile steel mesh to contain the failing liner material. The methodologies at each rehabilitation period are outlined below.

During all rehabilitation and other activities in the shaft including inspections, a requirement was made to check scale any loose material above the work areas from both skips prior to rehabilitation work being undertaken.

October 2016

This rehabilitation campaign included the underground tunnel breakthroughs (shaft plats) at 325 mLv, 440 mLv and 450 mLv. These areas were originally supported with 50 mm shotcrete and split sets. As this was intended to restrain potentially large wedges, high capacity underground tunnel support was used.

This rehabilitation included steel weldmesh (2.4 m × 3.0 m × 5.4 mm R10 mesh), grouted solid rock bolts (2.4 m × 17 mm GP cement grouted) and cable bolts (twin strand 6 m × 15.2 mm bulbed strand).

November 2017 – February 2018

The intention of this rehabilitation campaign was to contain damaged shaft barrel liner material throughout the shaft. The damaged areas were prioritised, and work was undertaken from areas of highest priority.

This rehabilitation included galvanised chain-link mesh rolls (2.1 m × 10 m × 3.15 mm wire with 50 mm aperture), bolted with 2.4 m × 20 mm GP grouted solid bolts (Keenan, 2017).

During this rehabilitation period, ~200 m vertically of shaft liner was contained in steel mesh (Marschke, 2018).

July 2019

This rehabilitation period was a continuation of the liner coverage methodology used previously and largely was infilling areas between 325 mLv and 450 mLv that were not completed previously.

It was found that the GP cement grouting was taking a lot of time to complete in the support cycle compared with pinning the mesh rolls and drilling bolt holes. This was heavily time consuming due to the size of the grout bowl on the tight work platform and need to continually retrieve cement bags from the surface at a slow skip movement rate. A decision was made to utilise resin capsules as used widely in Cannington's underground workings. Each solid bolt was anchored with 2 × 1.2 m fast setting resin capsules.

December 2019

This was unscheduled rehabilitation. Due to a significant increase in damage noted during inspections and an upward trend in material scaled (as shown in Figure 4), further liner meshing was brought forward.

This rehabilitation predominantly included mesh liner coverage above the 325 mLv, extending the portion of mesh liner as the damage progressed higher. The same methodology of chain-link mesh rolls and resin encapsulated solid bolts was continued. This rehabilitation covered ~80 m vertically of shaft barrel over three weeks.

2020 – June 2022

Throughout the Fowler Shaft's final hoisting operational period, various minor rehabilitation activities were conducted, these were minor works, typically including 'spot bolting' areas of convergence and mechanical damage.

Localised convergence occurring throughout the previously rehabilitated areas was typically identified with failed bolt plates and bagged mesh; where the mesh was holding material to the point that it was encroaching on the skip clearances and required contained material to be removed.

Mechanical damage on the installed mesh was caused by hoisted ground support elements overhanging the skips and spilled ore. This torn mesh was often stitched together with d-shackles and spot bolted where required.

When the shaft was decommissioned for hoisting activities in June 2022 approximately 55 per cent of the shaft liner had been supported with mesh and bolts.

Winder operations

Adjustment of winder hoisting speeds was trialled in the final 12 months of hoisting. The adjustment was based on the hypothesis was that the skips swing less when running at slower speeds. As a result, winder operations: modified maximum hoisting speed from 14 m/s to 12 m/s to reduce pendulum effect.

This reduced speed also marginally reduced the damage that could have occurred by contact between the hoisting skips and shaft barrel. Whilst the risk reduction may have been marginal, in the final months of hoisting when skip clearance was critical, all controls were under consideration.

Regaining clearance

One of the main oversights of the rehabilitation program was the failure to recognise that rock bolts and cablebolts had potential to protrude from the concrete liner and impede on skip clearance if they are installed in unfavourable locations. Several rock bolts and cablebolt were installed in the vicinity of the north-west and south-west corners of the west skip. This issue was remedied by cutting the tails of the rock bolts, where possible, and the cablebolts, but taking care that the cablebolt strands had enough tail to ensure that they would not pull through the plating system.

Similarly, sagging mesh often required local re-pinning tight against the wall using a dynabolt and complementary plate system as shown in Figure 9, to reduce the likelihood of a skip catching the mesh and tearing it away.

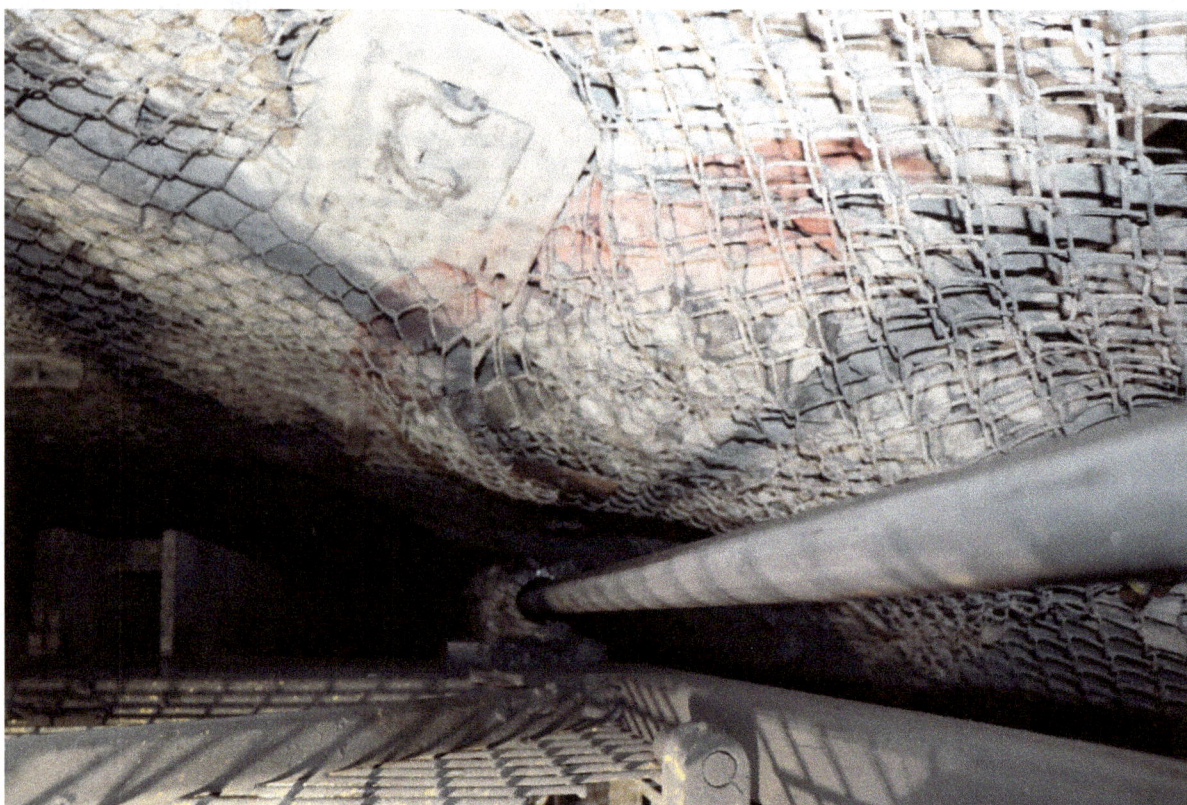

FIG 9 – Limited clearance between west skip and west wall in early 2022.

CONCLUSIONS

Risk management of a deforming shaft is complex. The complexity is exacerbated when regular personnel access inside the shaft is required to maintain the hoisting infrastructure. Successful strategy to control associated hazards in these environments is similar to many complex work areas that require monitoring over the life of the infrastructure:

- Maintain an on-site presence and invite fresh eyes to inspect.

- Be willing to adjust if historical controls fail to be effective.

- Identification of subtle warning signs and escalating to levels of management so strategic decisions can be made in a timely manner.

- Conduct tasks in-house whenever possible but arrange for expert support for specialised mining tasks.

The last skip of ore was hoisted from Cannington's Fowler Shaft in June 2022. The Transition-to-Trucking Project has been in progress since February 2021 with the intent to convey Cannington into its next phase of life as an all-truck operation.

ACKNOWLEDGEMENTS

The authors would like to thank South32 for allowing this paper to be shared with the mining community. They would also like to thank the past and present members of Cannington's Geotechnical, Survey and Fixed Plant Maintenance Departments, in particular Lance Gatty who has fulfilled the role of Specialist Winding Operations at Cannington for 13 years.

REFERENCES

Beck, D, 2020. Cannington Global Stability, Beck Engineering report P2019097

Benecke, N, Hancock P and Weber, M, 2016. Latest developments in the practice of shaft inspection, in *Proceedings of the Proceedings of the XVI Congress of the International Society of Mine Surveying* (Australian Institute of Mine Surveyors: Singleton).

Hall, J, 2015. Hydrogeological Review of Cannington Open Pit Project, RPS report 1757B-010a.

Hendry, S and Moshos, V, 2016. MCA Engineering Document No: 008021-Pre-0001 Installation and Safe Use of Skip Top Platform.

Keenan, S, 2017. GE-314 – Shaft Rehabilitation Ground Support Change, Internal South32 Cannington Report.

Love, A, 2006. Fowler hoisting shaft assessment of cracking concrete lining, Coffey report HZ00048/01-AA.

Marschke, V, 2018. Shaft Rehabilitation Close Out Report, Internal South32 Cannington Report.

Sandy, M, 2020. Shaft Decommissioning Geotechnical Considerations, AMC Consultants Project 220057.

Impact testing of raise bore protection mesh

M Doran[1], X Hatch[2], T Mahoney[3] and G Mungur[4]

1. Geotechnical Engineer, BHP Olympic Dam, Roxby Downs SA 5725.
   Email: meggie.doran@bhp.com
2. Senior Mine Production Engineer, BHP Olympic Dam, Roxby Downs SA 5725.
   Email: xavier.hatch@bhp.com
3. Lead Engineer Geotechnical, BHP Olympic Dam, Roxby Downs SA 5725.
   Email: ted.mahoney@bhp.com
4. Superintendent Geotechnical Operations, BHP Olympic Dam, Roxby Downs SA 5725.
   Email: glenton.mungur2@bhp.com

ABSTRACT

The Olympic Dam mine utilises 1.1 m to 1.4 m diameter raise bores for generating the initial slot void for most production stopes. Slot raise bores generally span two development levels ranging from 30 m to 50 m in height. Raise bores are reamed upward and typically leave a cap of rock equivalent to two times the raise bore diameter on the drill level (minimum 3 m). This provides safe access over the cap for subsequent production activities. It is therefore critical to ensure robust falling object protection for personnel accessing underneath, and in the vicinity of the raise bore void on the lower level.

Recently, a number of mesh failures have occurred at Olympic Dam as a result of falling rock from the underside of the cap and raise walls. In addition, as Olympic Dam transitions to deeper and more challenging ground conditions, rockfall-off onto the raise bore protection mesh is anticipated to increase in frequency.

This paper discusses the evolution of raise bore protection mesh used at Olympic Dam and presents results from recent site-based drop testing. This impact testing was carried out using a built for purpose object to replicate point loading and shear failure of mesh wires. Testing was also carried out on an improved mesh design which utilises two layers of Geobrugg high-tensile MINAX mesh. Results from the drop testing suggest an improvement in performance of the raise bore protection mesh is realised with the updated design.

INTRODUCTION

Olympic Dam mine is Australia's largest underground mine, located approximately 570 km NNW of Adelaide, South Australia (Figure 1). The mine hosts one of the world's most significant deposits of copper, gold, silver and uranium. The mine was first operated by Western Mining Corporation (WMC) and was later purchased by BHP Billiton in 2005.

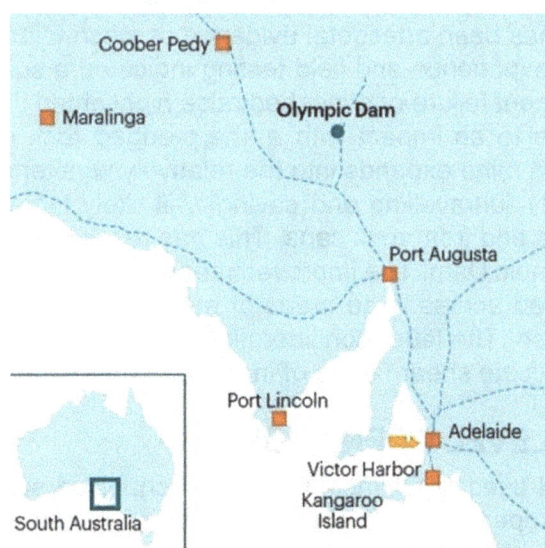

FIG 1 – Olympic Dam location in South Australia.

Since 2005, Olympic Dam's Northern Mining Area (NMA) has mined at least 738 stopes, equivalent to approximately 6.5 Mt of stope ore per annum. Mine planning for the next 50 years will include extensive development into the mine's Southern Mining Area (SMA) which is a relatively new mining front for Olympic Dam (Figure 2). The SMA has greater structural and geological complexity which have implications on the design of stopes and expected stability of the ground. Challenging mining conditions have already been encountered in the SMA where one of the area's first mined stopes effectively doubled in size with more than 55 per cent stope overbreak (weak rock mass conditions relating to Mashers Fault, Geotech Department – (Olympic Dam Block Note Addendum Tangerines, 2021)).

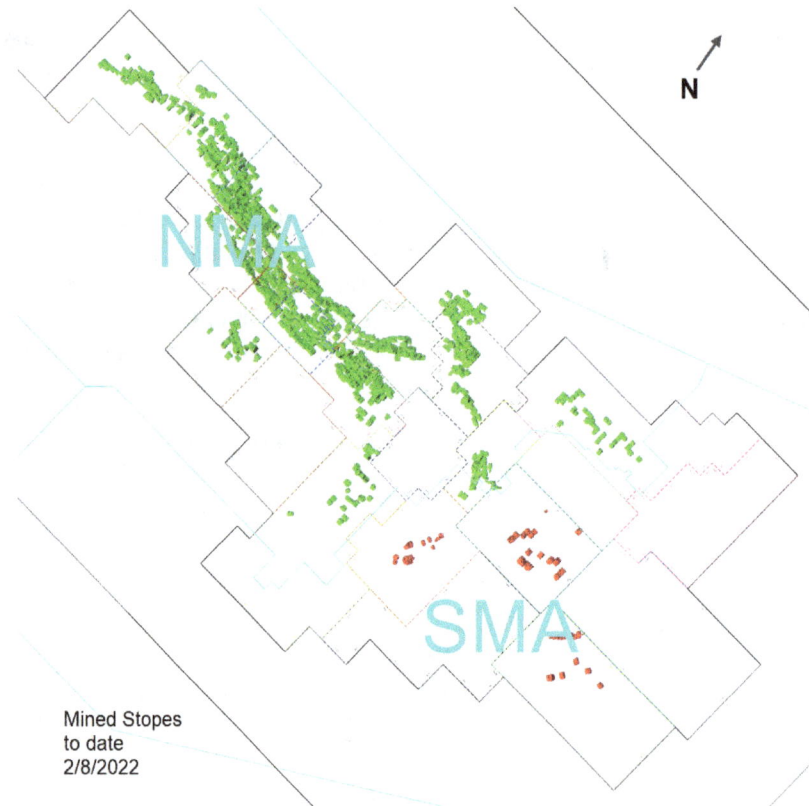

Mined Stopes
to date
2/8/2022

FIG 2 – Mined Stopes in the NMA (738 stopes) and SMA (44 stopes) on record.

As is common in the mining industry, raise bores are routinely used at Olympic Dam for a variety of purposes including ventilation, backfilling and production blasting (Yao, Sampson-Forsythe and Punkkinen, 2014). In order to protect personnel from falling rocks whilst working directly under and in proximity to raise bore holes, Olympic Dam utilises high-tensile steel wire mesh at the bottom of raise bores. Although there has been anecdotal evidence of mesh failures due to falling rocks from within raise bores, previous experience and field testing indicated a suitable standard of protection was provided. However, a recent failure of a meshed raise highlighted that shear failure of individual wires from point loading due to an impact with a sharp-edged rock is a failure mechanism that requires consideration. As the mine expands into the relatively weaker and altered rock units of the SMA, gravity-driven instability (unravelling and caving) will likely increase the frequency of rocks falling from within raise bores and from rock caps. This has resulted in a redesign to the raise bore mesh protection used at Olympic Dam. The improved mesh protection not only considers impacts in which the energy is distributed across large areas of mesh but also considers point loading on a relatively smaller area of mesh. The latter consideration was tested on-site with a built for purpose impactor in an attempt to replicate shear failure of individual wires.

RAISE BORES IN SUBLEVEL OPEN STOPING

The standard mining method used at Olympic Dam is mechanised sublevel open stoping (SLOS) illustrated in Figure 3. The stopes are filled shortly after being mined with cemented aggregate fill (CAF) in order to maintain stable stress fields in the mine and limit rock mass deterioration around large open voids. This method is preferred for extracting high-grade ore with low dilution and it is also effective for minimising the risk of subsidence. Mining levels are spaced approximately 30 m to 60 m vertically to suit orebody geometries, geotechnical conditions, and longhole drilling capabilities.

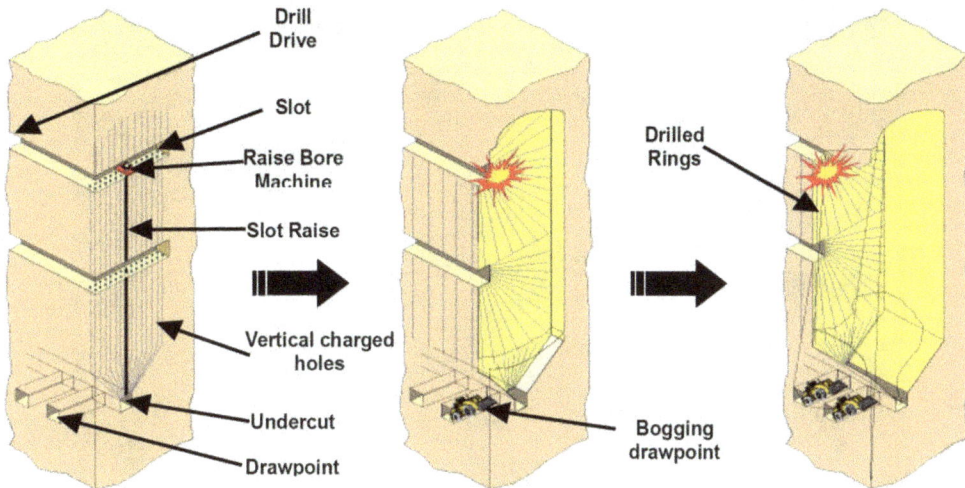

FIG 3 – Sublevel open stoping method used at Olympic Dam (Ehrig, McPhie and Kamenetsky, 2012).

Within the footprint of each stope a slot raise is drilled to create the initial void required for the production rings to effectively fragment ore. At Olympic Dam, production raise bores typically have a diameter of either 1.1 m or 1.4 m and span between two development levels. The raise walls are unsupported during the excavation process and there are limited ways of rectifying any significant instability that may develop during the reaming process. Depending on the raise bore dimensions, there are several options available to protect personnel working in the vicinity of open raise holes including high-tensile steel mesh, grouted pentice plugs, separation controls, or a combination of the above. For personnel working in the vicinity of a 1.1 m or 1.4 m production raises, Olympic Dam typically installs high-tensile steel mesh to minimise the risk of rockfall related injuries or damage to equipment which is common in the mining industry (Garant, Roberge and White, 2009). The investment in a raise can be substantial and the consequences of a major failure can be highly disruptive to operations (Penny, Stephenson and Pascoe, 2018).

PREVIOUS SINGLE LAYER MESH STANDARD

Historically, raise bore protection consisted of a single panel of Geobrugg® TECCO® G65/4 mesh installed over the raise on the lower level as shown in Figure 4. The panel consists of the G65/4 mesh which is manufactured with 4 mm diameter high-tensile steel wires arranged in a rhomboidal pattern with a 65 mm diameter minimum aperture. For added capacity a 14 mm diameter boundary rope and with brake rings intersecting over the middle of the raise are woven through the G65/4 mesh. The panel is attached to the back with split set bolts.

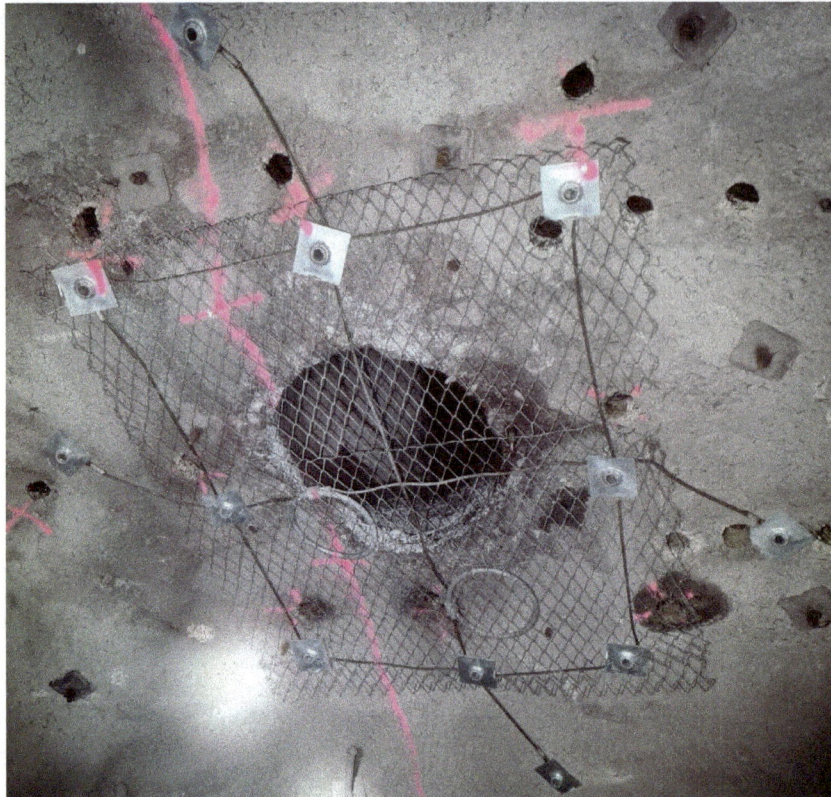

FIG 4 – High-tensile G65/4 steel mesh installed over the breakthrough of a raise bore.

In 2016 an impact test was carried out at Olympic Dam to verify the suitability of the G65/4 mesh panel. The test consisted of dropping three angular rocks weighing approximately 115 kg each approximately 50 m from the collar of a raise onto the G65/4 mesh. The calculated kinetic energy of a 115 kg rock falling 50 m is 55 kJ at impact. The rocks were dropped consecutively without being removed from the raise bore prior to the next rock being dropped. Observations from this test showed that the mesh was able to withstand three repeated impacts without compromising the ability of the mesh to retain the rocks. Minor damage was noted after the third impact however the mesh was still able to retain the cumulative mass of all three rocks (Figure 5).

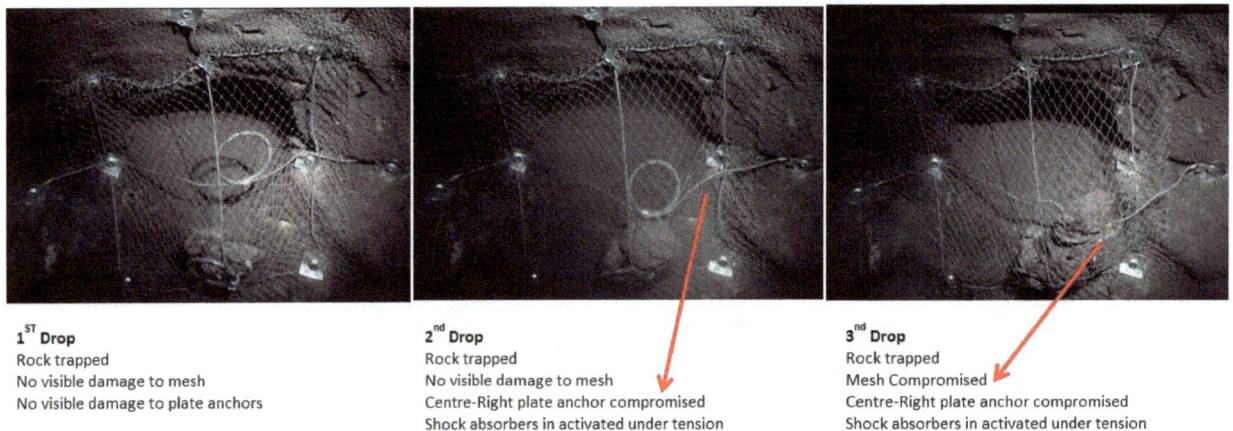

1ST **Drop**
Rock trapped
No visible damage to mesh
No visible damage to plate anchors

2nd **Drop**
Rock trapped
No visible damage to mesh
Centre-Right plate anchor compromised
Shock absorbers in activated under tension

3nd **Drop**
Rock trapped
Mesh Compromised
Centre-Right plate anchor compromised
Shock absorbers in activated under tension

FIG 5 – Images from internal report showing the condition of the G65/4 mesh after each rock impact from 2016 site-based test.

SHEAR FAILURE OF INDIVIDUAL STRANDS IN HIGH TENSILE STEEL MESH

A recent incident involving the failure of the G65/4 mesh has required a re-assessment of the design for raise bore protection. In January 2021, upon re-entry to a production site, a wedge-shaped rock weighing approximately 30 kg was observed on the floor after dislodging from the raise and penetrating the mesh. Given the distinct etchings on the wedge-shaped rock, is believed to have fell from the cap of the raise bore near the 52 L and penetrated through the protection mesh installed on the 57 L, a height of approximately 40 m. This equates to 13 kJ of kinetic energy at impact, significantly less than both the 55 kJ of capacity demonstrated in the 2016 site-based drop tests and test results for G65/4 mesh provided by Geobrugg®. Upon inspection of the mesh, it was observed that individual wires were sheared as a result of the impact. Discussion with the mesh manufacturer indicated the impact was high velocity, and 100 per cent of the impact energy was concentrated on a few individual wires creating enough force to shear through (Bucher, January 2021, personal communication).

As a result of this failure a new mesh panel was designed in consultation with Geobrugg® and the mine production workforce (Figure 6). In order to account for the point loading and shear failure mechanism a mesh with higher tensile capacity wires was incorporated into the panel. Geobrugg® MINAX® 80/4.6 high-tensile mesh was selected for further testing which provides a thicker diameter wire – 4.6 mm and higher tensile wire resistance relative to the G65/4 mesh. The MINAX® 80/4.6 mesh has an 80 mm aperture which was considered more favourable for drilling production holes through the mesh without damaging wires. This is historically a problem with the smaller aperture of the G65/4 mesh. However, the larger aperture of the MINAX® 80/4.6 mesh would also permit small scats to pass through the mesh. As such, the redesigned panel incorporates two layers MINAX® 80/4.6 mesh with an offset in the overlap to reduce the effective aperture size. The double layer was restricted to the area under the raise and did not extend to the extents of the full panel. This allowed the production team to take advantage of drilling holes through the single layer of the larger aperture MINAX® 80/4.6 mesh.

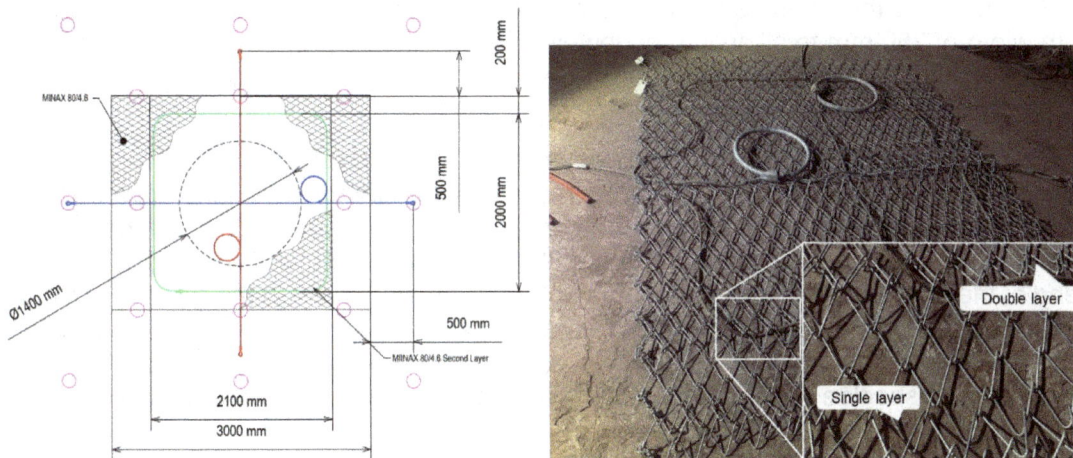

FIG 6 – MINAX® 80/4.6 double mesh panel design with production holes indicated in pink (left) and photograph of MINAX® 80/4.6 double mesh panel used in testing.

SITE-BASED DROP TESTING ON MINAX® 80/4.6 DOUBLE MESH PANEL

In order to satisfy management of change requirements and demonstrate the effectiveness of the re-designed panel with 80/4.6 mesh a series of tests were devised. The objective of the testing was to replicate shear failure of the G65/4 panel under similar conditions to that observed in the January 2021 incident. Secondly, a successful test will also demonstrate that the double layer of 80/4.6 mesh withstands the same impact which caused shear failure in the G65/4 mesh.

Impactor drop test device

In order to replicate the point loading mechanism required for shear failure observed in the failure, a built-for-purpose impactor was engineered and manufactured for use in the drop tests. An impactor was designed based on the shape and dimensions of the rock which penetrated the G65/4 mesh in the January 2021 incident (Figure 7). The impactor was manufactured with the ability to change the mass by adding or removing weight plates. This accounts for the range of possible rock masses which may fall within a raise. The drop tests were also constrained to production raises in the current mine schedule at the time which resulted in variability in drop heights across the battery of tests and variability from the failure height of 40 m in the original incident.

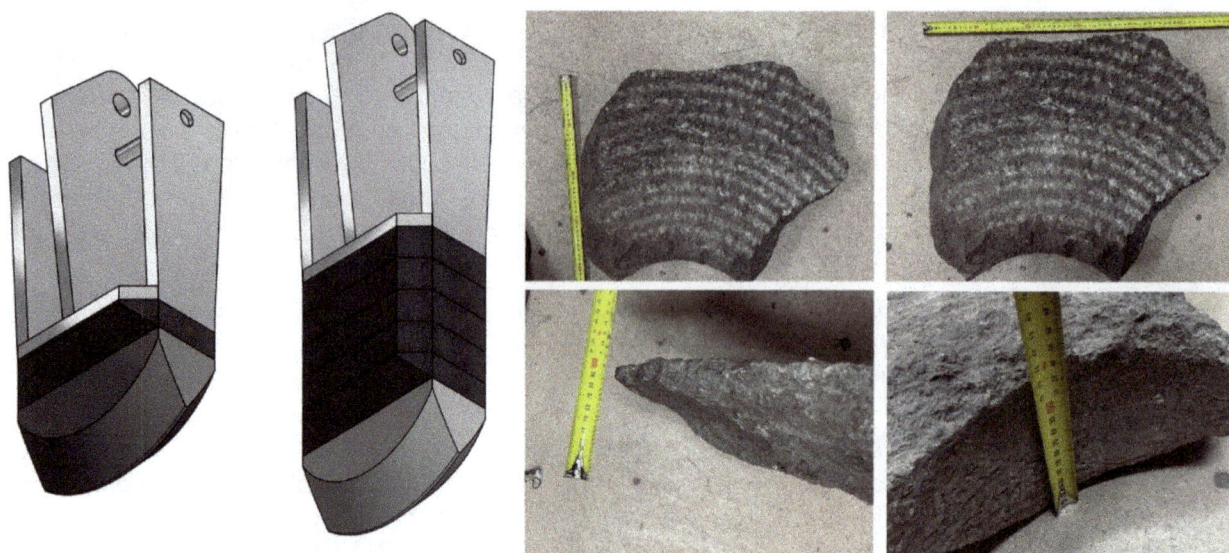

FIG 7 – Impactor with 40 kg and 100 kg weight configurations (left) to simulate 30 kg rock which penetrated G65/4 mesh in January 2021 (right). Approximate dimensions of the 30 kg rock are 440 mm × 600 mm × 130 mm × 10 mm.

A degree of conservatism is built into the impactor design as the radius of curvature of the sharp edge is 3 mm, whilst the sharp edge of the rock is approximately 10 mm. The impactors had a maximum width of 250 mm to allow it to be lowered though the 300 mm hole in the raise cap on the upper level. A safety rated eyelet was installed on top of the impactor in order to lower and suspend the impactor prior to being dropped.

A number of safety considerations were implemented in the rig used to manoeuvre and release the impactor down the raise. Firstly, to manage the mass of the impactor an engine hoist was positioned over the raise cap. Using the articulation of the engine hoist it was possible to lift the impactor from the tray of a light vehicle and position it over the 300 mm raise cap hole. Secondly, a winch installed on the end of the engine hoist allowed the impactor to be lowered through the raise cap to the desired height in a controlled manner. Both of these features enabled the elimination of a significant amount of manual handling of the impactor. In order to release the impactor once positioned in the raise a quick release snap shackle attached between the winch line and impactor was activated by pulling on a lanyard.

Observation areas with portable lighting systems for video recording the impactor striking the mesh were set-up outside of exclusion zones on the breakthrough level.

Test results

A total of six drop tests were completed using the designed impactors – three tests using the old Geobrugg meshing standard G65/4, and two using the new meshing standard G80/4.6 (Figure 8). An additional test (Test 3) was completed without mesh to assess the trajectory of the impactor. This test confirmed the impactor was upright as it fell through the raise. Tests 1–4 were conducted in two different locations. Test 1 (40 kg from a 29 m height) and test 2 (70 kg from 29 m height) did not result in mesh failure. However, when the height of fall and mass was increased to 42 m and 55 kg, with an impact energy of 22.6 kJ, the old meshing standard proved ineffective at stopping the projectile. This mesh failure was at an impact energy 59 per cent less than previously successful distributed load testing of 55 kJ.

Tests 5 and 6 trialled the new G80/4.6 meshing standard to support the impact of a projectile weight of 100 kg falling from 30 m. The new standard of mesh did not fail. The mesh was re-installed at the same location so test 6 was not affected by the previous drop test 5 impact. The point of impacts for each of these tests were adjacent, but not in contact, to the bisecting break rings. The mesh stretched but did not tear (Figure 9).

Protect raisebore as per existing mesh standard with G65/4 single layer mesh				
No.	Height (m)	Weight (kg)	Calculated Energy (kJ)	Mesh failure?
Location 1 - remesh after each test				
1	29	40	11.4	no
2	29	70	19.9	no
Location 2 - remesh after each test				
3	42	55	n/a	n/a - no mesh
4	42	55	22.6	yes
Protect raisebore as per new mesh standard with G80/4.6 double layer mesh				
Location 3 - remesh after each test				
5	30	100	29.4	no
6	30	100	29.4	no

FIG 8 – Drop-test results.

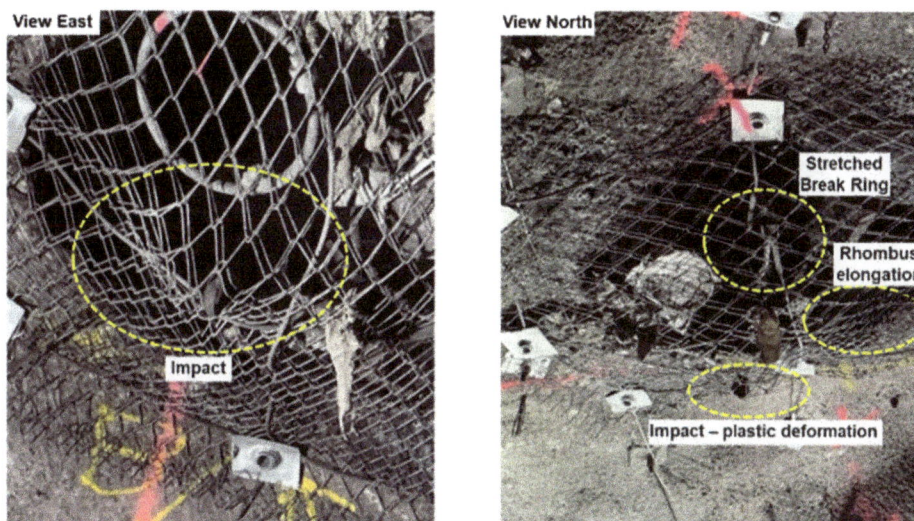

FIG 9 – Impact location of test 6 – mesh stretched but not torn (break ring stretched)

CONCLUSIONS

The stability of stope raise bores at Olympic Dam are influenced by a range of factors related to stress, structure, and rock mass conditions. It is therefore paramount a suitable level of protection is installed beneath a raise bore which will also account for shear failure in localised areas of the mesh from sharp angled rocks. Drop testing suggests meshing raises with two sheets of mesh with thicker diameter wires provides more robust support. Drop testing using distributed load testing does not accurately consider point load failure. Point loading results in failure at lower impact energy than mesh which has been tested using a distributed load scenario.

The double layered, thicker mesh has been tested to be able to support a 100 kg projectile falling from a 30 m height whereby the previous single layer meshing technique was unable to sustain 55 kg from 42 m, a 23 per cent reduction in impact energy.

ACKNOWLEDGEMENTS

The authors wish to thank BHP for permission to publish this paper. The authors would also like to thank all those who have contributed and been involved in the drop testing program conducted at Olympic Dam.

REFERENCES

Ehrig, K J, McPhie, J and Kamenetsky, V, 2012. Geology and mineralogical zonation of the Olympic Dam iron oxide Cu-U-Au-Ag deposit, South Australia, *Geology and Genesis of Major Copper Deposits and Districts of the World: A Tribute to Richard H Sillitoe* (eds: M Harris, F Camus and J W Hedenquist), Special Publication 16: 237–268 (Society of Economic Geologists, Littleton: CO).

Garant, J, Roberge, S and White, R, 2009. Working Safely Under Inverse Raise Boreholes at Brunswick Mine – The story of The 'Trampoline', in *Proceedings the 43rd US Rock Mechanics Symposium and 4th U.S.-Canada Rock Mechanics Symposium* (American Rock Mechanics Association: Asheville).

Olympic Dam Block Note Addendum Tangerines, 2021.

Penny, A, Stephenson, R and Pascoe, M, 2018. Raise bore stability and risk assessment empirical database update, in *AusRock 2018: The Fourth Australasian Ground Control in Mining Conference* (The Australasian Institute of Mining and Metallurgy: Melbourne).

Yao, M, Sampson-Forsythe, A and Punkkinen, A R, 2014. Examples of ground support practice in challenging ground conditions at Vale's deep operations in Sudbury, in Deep Mining 2014: Proceedings of the Seventh International Conference on Deep and High Stress Mining (eds: M Hudyma and Y Potvin), pp 291–304 (Australian Centre for Geomechanics: Perth), https://doi.org/10.36487/ACG_rep/1410_19_Yao

Managing high volume water inflows of 300 L/s in the planning and development of a large scale underground Gold Mine – the Didipio story

J P Evert[1] and S Pearce[2]

1. Dewatering Consultant, Mineright, Brisbane Qld 4006. Email: minerightau@gmail.com
2. Project Director, OceanaGold, Kershaw S Carolina 29067, USA. Email: sean.pearce@oceanagold.com

ABSTRACT

Didipio gold-copper mine operation is situated in the island of Luzon, approximately 270 km north of Manila in the Philippines, in an area of high seasonal rainfall. The hydrogeological water flow model implied that there would be high connectivity between regional structures and underground operations. This model predicted that underground mine groundwater inflows were expected to quickly rise as the decline progressed to a peak of 510 L/s.

Designing a capital pump station to deal with 510 L/s has its own challenges, however developing the decline down to the required level, always dealing with 300 L/s of water and constructing the pump station is extremely challenging and required a robust water management plan and mine design. When dealing with expected high volumes of water, underground mine design and scheduling plays a key role in protecting electrical infrastructure, equipment, secondary pumps, and the decline from flooding. A good drainage system and controlled dewatering plan is essential for safety and progress.

This paper details the dewatering techniques and mine designs required to accomplish decline development with high water inflows as used in the Didipio project. It will describe the learnings and outcomes that are achievable with a flexible design, as new structures are mapped, and water intersections are increased. It also highlights the potential for adaptability in mine planning to provide evolving improvements or to evade problems that are expected from the constant reconciliation of data recorded during development.

INTRODUCTION

Didipio gold-copper mine operation is situated in the island of Luzon, approximately 270 km north of Manila in the Philippines, in an area of high seasonal rainfall.

OceanaGold acquired the high-grade underground gold and copper mine in 2006 through a merger with Climax Mining Ltd and commenced commercial production as an open pit operation in 2013. In 2015, the mine transitioned to underground operation, with production from the underground commencing in early 2017.

FIG 1 – Location map: Didipio Mine.

The hydrogeological groundwater flow model implied that there would be high connectivity between regional structures and underground operations.

Evidence from mining the open cut, rainfall events and monitoring bores, allowed for a comprehensive groundwater model to be developed. This model was then used as a template for mine design.

The principal question asked when you are to decline into an area of ground when there is an uncertainty of volume is 'what is the best way to prepare for it?'

The key is to have an overriding underground water management plan. This plan will dictate the ethos of how to control and manage the water inflows into the mine.

Didipio was required to pump 450–500 L/s constantly from underground for life-of-mine and to allow access to the orebody. Designing a capital pump station to cater for this volume was not without its own challenges. However, declining 2800 m down to the location to construct the capital pump station with the expectation of large volumes of water was much more complicated. If inflows exceeded the pumping capacity and reached electrical infrastructure all will be lost, so the development of a robust water management plan was required.

From the start of Portal #1 at 2665RL there was always an uncertainty on how much water and when it would be intersected. The open pit was still in operation with an expected bottom of 2460RL so assumptions were made that water intersection volumes would increase when mining got below the RL of the pit. This assumption was reasonably accurate, total water inflows were around 50 L/s in this top section of decline.

The Didipio mining method was large stope and paste fill so there was an expectation that stopes had to be reasonably dry prior to paste filling to satisfy inrush risks. Therefore, dewatering of the orebody was imperative for the safe management and extraction of the resource. It also implied that a top-down method of mining was the likely scenario required as dewatering was likely to dry out the top section of the resource first.

WATER MANAGEMENT KEY PRINCIPLES

In the case of Didipio, the delicate balance between dewatering and production was acknowledged from the onset and integral to the development of their key principles for water management.

At Didipio as large volumes of groundwater ingress was added to the mining cycle it was then prudent to hold off development in areas of known water flow until interim or capital infrastructure was commissioned. Scheduling programs used in modern day mining cannot schedule to the unknown so it was likely that schedules would 'slip' due to water inrush and recovery pumping. It was important that this distinction was recognised from the beginning. Getting to the capital pumps was the highest priority for the Didipio project and to get there required a solid plan. This plan needed to be outlined from the start of the operation to all stakeholders so that focus was created and maintained on water management outcomes.

Figure 2 shows the cycle of water management which is used to describe the steps involved with achieving a successful water management outcome at Didipio.

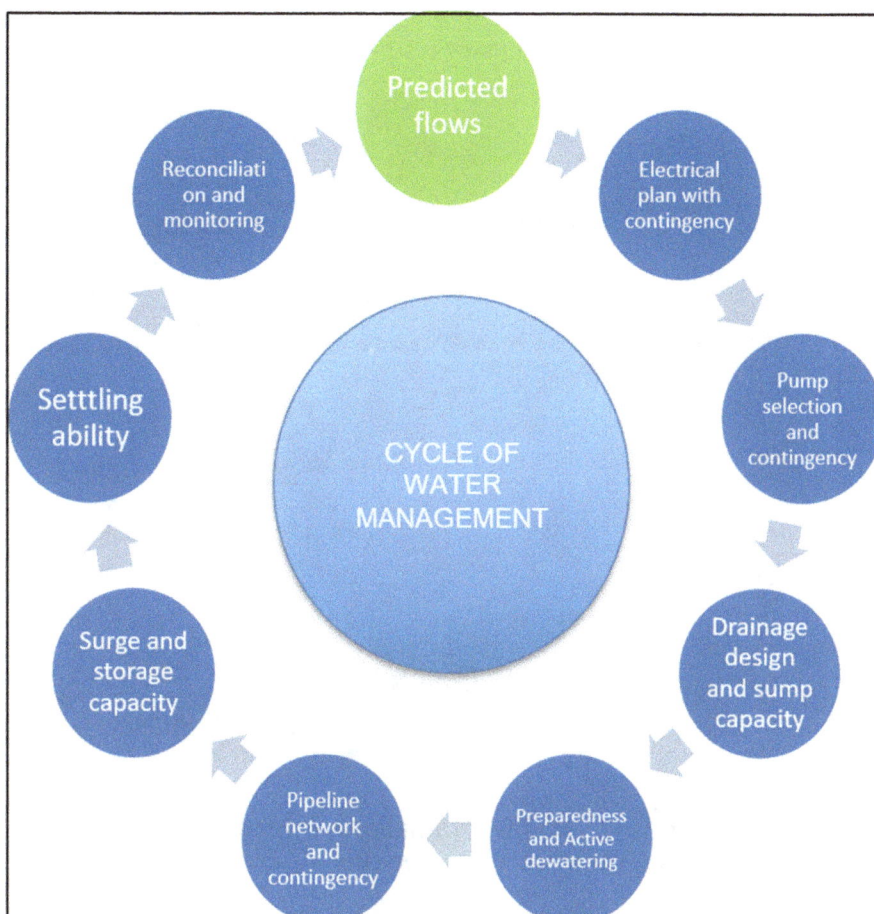

FIG 2 – Cycle of water management.

Didipio water model predictions

The initial predicted groundwater flow model at Didipio was expected to be 400 L/s at the full depth of the decline decreasing after three years to a significantly lower figure of 150 L/s. As the decline was mined and the model was recalibrated the peak predictions increased to between 440–510 L/s with a 400–470 L/s flow rate being sustained over life-of-mine.

Hydrogeologists produced a groundwater model which was a guide to how much water would be expected over the life-of-mine. This model was recalibrated several times as the underground mine progressed in the first two years. Underground inflows were monitored and recorded for comparison against the model and an ongoing review and calibration of the model was undertaken to refine inflow predictions as the mine developed.

Having a predictive tool of this nature allowed for recalibration of pumping capacities that were required. The Hydrogeological department at Didipio collected data and recalibrated the model to give operations up to date information. Figure 3 shows four iterations of the recalibrated groundwater inflow model. The dramatic difference between 2017 model and 2018 model implied a life-of-mine reliance on Capital pumps and a sustained higher power cost.

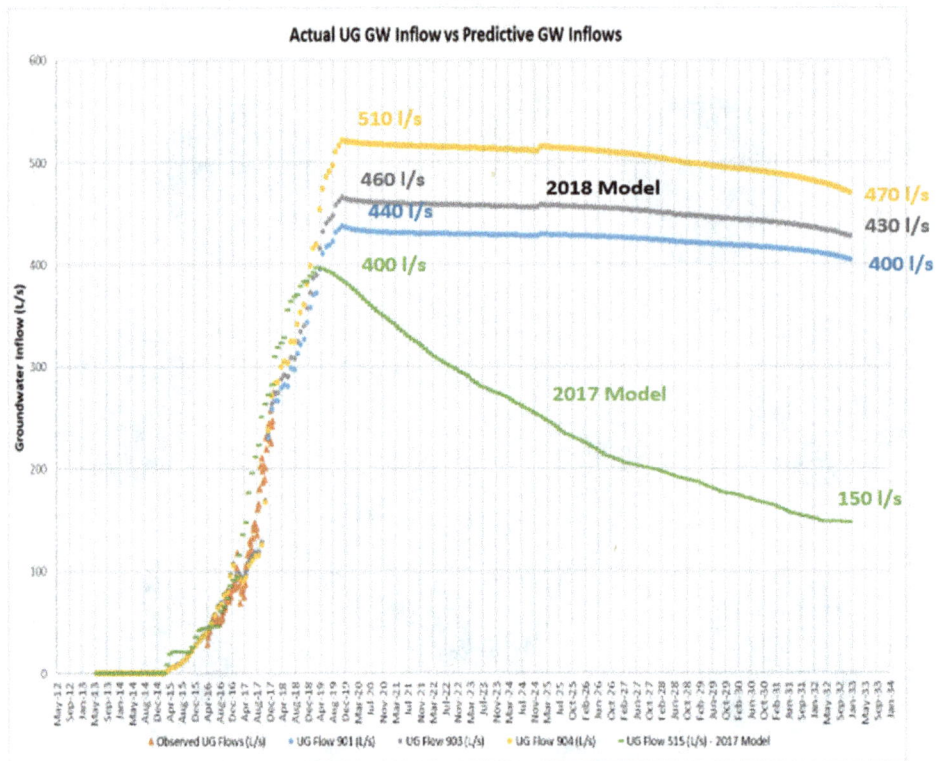

FIG 3 – Didipio groundwater model, over time, dramatically changed implying a reliance on capital pumps does not abate over the life-of-mine.

The model was a guide for initial pump and pipeline selection for the interim stage of the development as well as the baseline for all underground capital pump station discharge requirements. With this predictive model capital pumps were selected for total ex-mine water flow of 650 L/s.

Interim pumping infrastructure was designed with significant contingency and in a manner that was modular to cope with greater or significantly lower pumping volumes than predicted by the groundwater model.

It is important to recognise from the start that the peak inflows were likely to be significantly decreased if the mining schedule included active dewatering and increased flow rates early in the mine life.

Electrical plan

The Didipio power supply consisted of a mains overhead line supplied from a domestic power station as well as an on-site bank of diesel generators as backup. The underground was designed as a ring main unit so that there was always a way to get power reinstated to the underground pumps.

The site ethos was that if power was lost from the domestic supply the gensets were started and the underground was first to be reinstated. Depending on conditions, first the pumping system then the ventilation was energised. The mill and all surface infrastructure were then brought online if power supply permitted.

In Didipio mine with its high groundwater inflows, electricity to power pumps was the most important requirement to sustain production. Having a solid electrical backbone with built in contingency was critical. This was particularly significant during the development phase of the mine before the Capital pump station was constructed. If power was lost, then all groundwater would invariably overflow from the catch sumps and report to the decline face within a specific time period dictated by the level sump volumes. If power wasn't restored within a specific time period the Jumbo and auxiliary face pump would need to be pulled back from the face until power was restored. Recovery of the decline could be onerous at times depending on the cycle and length of time of the power outage and schedules would be pushed out.

Even though all efforts were made to prioritise reinstatement of lost power, there were several power outages during the project that caused the schedule to slip and extended the construction and commissioning of the main Capital pump station underground at 2270 RL. It is important to note that power outages were recognised as a significant risk to the operation. The design of electrical infrastructure placement underground was successfully chosen to ensure flooding damage to essential electrical components did not occur nor compromise the operation.

Pump selection and contingency

From the Portal (2665RL) to the first production level was 1650 m of development at 1 in 7 down. As the decline progressed the volume of water required to pump increased requiring decline catch sumps so that water could be intercepted prior to running to the active face.

Once the inflow into the decline face approached 15 L/s a catch sump was cut on the side of the decline at an appropriate place and a submersible pump was installed to handle the water. Therefore, 15 L/s was chosen because water flows above this would require a pump cycle during the bogging cycle and extend the cycle time dramatically. The submersible was either a 26 kw or a 42 kw depending on the flow required. The 42 kw pumps were initially daisy chained up the decline to dewater the decline.

It is important to note that every sump had two pumps hanging in the sump plumbed up with 100 mm quick release flexible hose. One pump was operational, and the other was on standby.

These sumps were positioned approximately 50 m vertically apart or 350 m along the decline.

- J 405 42 kw submersible Sulzer pump: estimated flow rate of 50 L/s with 50 m of head pumping through 160 mm ID poly pipe.

- J 205 26 kw submersible Sulzer pump: estimated flow rate of 30 L/s with 50 m of head pumping through 160 mm ID poly pipe.

Once the decline had reached approximately 2565RL (100 m vertically below the portal) an Interim pump station was installed on the decline to cater for the water. This pump station, shown in Figure 4, pumped out through the portal through the two × 160 mm ID poly lines. The Interim pump station was designed as a modular system that could be relocated or added too as required. Electrically, the starter equipment for the pumps were set-up with a Variable Frequency Drive, (VFD), allowing for the pumps to be run at an optimal speed and to also minimise high start-up currents. Importantly, the hopper size could be minimised to reduce the mining footprint. The VFD's allowed the pumps to maintain the water level in the hopper. This was to avoid running the pumps at full speed emptying the hopper very quickly during low volume inflows and thus causing many start/stops as well as potential overflows. Soft starters will only allow as few as eight starts per hour. Minimising the start/stops per hour is critical to maintain pump rates. The interim pumps also had vacuum canisters so they could pull directly from a sump or decline as built-in contingency, but this was never utilised other than set-up in the decline cuddy as shown in Figure 4.

At Didipio significant groundwater inflows into the decline appeared to be structurally orientated. As each structure was intersected more water was added to the system. This allowed for detailed mapping to occur on the structures and predictive measures were able to be employed such as planning of catch sumps on the downside of the expected structure to maximise the sumps effectiveness. It is also important to note that probe drilling was always conducted during this stage of decline development either by a diamond drill rig for long straights or with production and development Jumbos for short sections.

FIG 4 – IPS Pump Station located on the decline at 2564 m RL.

Drainage design and sump capacity

As the decline progressed to the crown pillar top mining level (2430RL) all sumps were then designed to be off the decline and on the level access. The level access was specifically designed to cater for the sump and appropriate drain holes to the levels below and an IPS was located on the footwall drive of this level. This then allowed for the management of all pumping systems at Didipio to be separate from the decline and as such overflows or pump failure would not have a direct impact on the decline development.

To control the water expected from the orebody all catch sumps at Didipio were designed on the level access vertically below each other so the all the water was centralised and drainage between levels was easily achievable. Access from the decline to the level sump was limited to a maximum fall of 900 mm. From the sump the level access inclined to intersect the footwall drive slightly higher than the decline access. This ensured that if the level sump pumps failed or power loss occurred, the level access would flood until it overflowed into the decline and entrapment of miners on the level was never possible. This shallow depth on all accesses allowed for Integrated tool carriers with man baskets to access the level or sump to change out pumps without any safety or mechanical concerns. It also allowed for the maximum amount of time to reset power or change out a pump.

All sumps between levels were interconnected with large drain holes and initially two DN250 rising main holes were designed between sumps so that as the mine was developed the water was able to be pumped through the shortest path possible. However, drill rig availability and access to sumps was difficult early in the operation so separate rising main holes were drilled to bypass three levels to interconnect the Interim pump stations as the mine progressed.

As each level was cut dual submersible pumps were installed in the sumps to pump water 30 m vertically to the level above via the twin 160 mm poly decline drain lines. At the 2340 Level another IPS Pump Station was installed and two 90 m rising mains were drilled and installed to bypass three levels. Predrilled drain holes were then opened between levels, and all water from all levels above reported to the 2340 Level sump. All auxiliary pumps in the 2430, 2400 and 2370 sumps then became redundant. Decommissioning sumps that required submersible pumps allowed resources and costs to be reallocated.

Submersible pump failure was a common issue at Didipio. The use of Fibrecrete as a ground support created an enormous number of issues for the longevity of the pumps. Fibres float to the sumps and mat around the suction of the pump and dramatically decrease the discharge output till eventually the pump prematurely fails by cooking or running dry. It was estimated that approximately 250 kg of fibres were floating to the sumps each month. Figure 5 displays a photograph of fibres collected from all operating sumps in one shift.

FIG 5 – When dealing with high volume water, rebound of shotcrete exacerbates the amount of fibre floating to sumps and destroying pumps. Collecting fibre from sumps and taking them out of the mine was an important maintenance program adopted at Didipio.

It is important to note that centralising the sumps in the x-cuts, off the decline, allowed for all decline development to only deal with the water encountered over the distance of 210 m between levels. This allowed development of the decline to remain a priority and limited, as much as possible, the maximum cycle time between rounds. Some major water bearing faults could not be avoided over this distance and catch sumps were required to be mined on the decline as a necessity however as soon as the level sump was commissioned then the decline catch sump was allowed to drain to the new sump.

As the mine progressed the decline design was slightly changed to avoid crossing major known faults. Figure 6 shows the drainage sequence as the mine is developed with each level. Scheduling service holes, rising main and drain holes in a timely manner is integral to maintaining water management capacity.

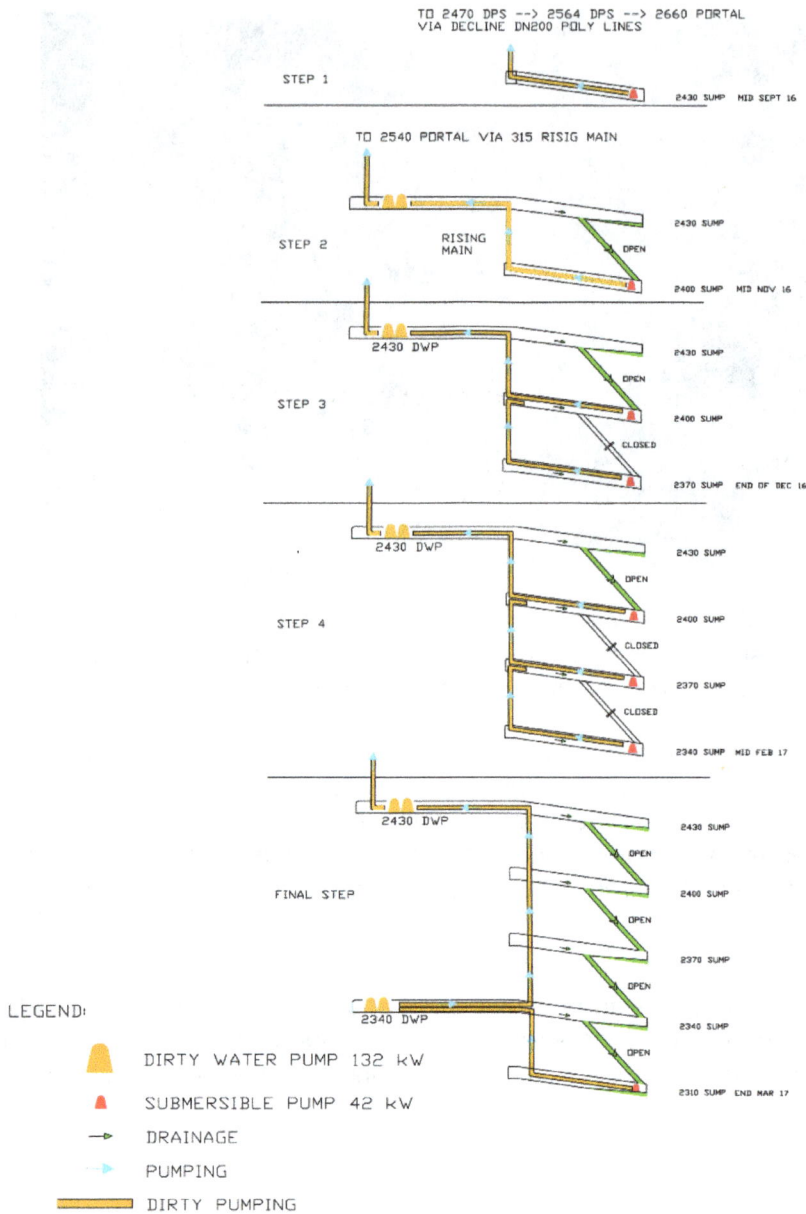

FIG 6 – Sump drainage sequence between level showing the transition between submersible sump pumps and the IPS centrifugal pumps.

Pipeline network and contingency

The 2430 RL was the top of the production levels and the access to the crown pillar. Production levels below this were 30 m apart extending to 2010RL. In total there were to be 15 levels in the mine. 2430 RL level sump was then the collection sump for the top of the orebody and the decline above this level. Two interim pump stations were placed on this level with a capacity of 300 L/s. Two × 349 mm rising main holes were drilled 110 m from the 2540 RL Portal #2 bench to the 2430 access x-cut. Two × DN250 steel pipes were then grouted into these holes. Exploration diamond drill holes that were drilled on the 2430RL were piped into the IPS to increase dewatering rates. These holes were all drilled trough valved standpipes and could be physically turned off if required during power outages.

All water collected at the 2430RL was then pumped to the 2540 Portal and discharged out of the pit via a diesel pit pump system (mid-September 2016 to mid-January 2017) then by using the 2540 Portal #2 Capital Pump station (commissioned mid-January 2017).

As mining progressed down the decline twin 160 mm ID poly lines were extended with the service in 100 m lengths flange bolted together. Once the fourth level was reached and larger rising mains were installed between these levels these twin 160 mm ID poly stayed active for contingency until the system was tested and then they were decommissioned and used further down the mine.

As each level was entered, flow rates increased so a second IPS system was installed on the 2340RL to cater for the extra inflows. Have two rising mains allowed this system to be commissioned easily. Table 1 outlines the flow rates encountered on each level prior to the commissioning of the 2270 Capital pump station. An average of 50 L/s was produced from each level. A HOLD was placed on a level once it reached 50 L/s so that flow rates could be managed out of the mine.

TABLE 1

Flow rates from each level in Feb 2018 prior to commissioning of the capital pump station.

Panel 1 Interim inflows per level Feb 2018			
Level	Location	Inflows (L/s) Feb 2018	Remarks
L2430	Level Sump	30	90% from orebody related development
L2400	Level Sump	40	95% from orebody related development
L2370	Level Sump	50	Predominantly from orebody related development
L2340	Level Sump ·	45	Predominantly from orebody related development
L2310	Level Sump	60	75% from orebody related development
L2280	Level Sump	50	95% from footwall and orebody related development
L2250	Level Sump	20	Access to orebody on hold. Predominantly from Alan Fault
Decline	Decline face	5	
Total		**300**	

Preparedness for major water intersections

As with all underground mines, exploration diamond drill holes are generally ungrouted. Decline and level design ultimately intersect these diamond drill holes. With the added risk of an underground aquifer that produces high flow rates the potential of intersecting ungrouted diamond drill holes are high and can be catastrophic if not controlled successfully, especially in the decline.

Diamond drill hole intersections with extreme water inflows of 40 to 50 L/s at Didipio were not uncommon. Increased awareness was highlighted in the decline as intersections would be known to overwhelm pumping capacities at the face. Training of jumbo operators was essential so that panic was avoided when high volume intersections occurred. All historic diamond drill holes were noted on the mining plan as potential risks and highlighted on the jumbo survey memo so the proper cautions could be made. The use of a tapered plug (Figure 7) that was permanently kept on the jumbo, allowed jumbo operators to stop the water flow coming from a diamond drill hole by simply putting it on the end of the drilling steel and pushing it in the hole to stem the flow (finger in the dyke principal). A supervisor would then be notified, and an appropriate jumbo installed groutable packer was brought to the face to be installed to stop the flow (Figure 8).

Managing the intersection of structures and diamond drill holes, with pressures exceeding 20 Bar, with this method allowed the control of the high inflow water possible.

The procurement of a high-pressure grout pump and appropriate training was also required so that intersected diamond drill holes could be grouted appropriately to allow mining to continue.

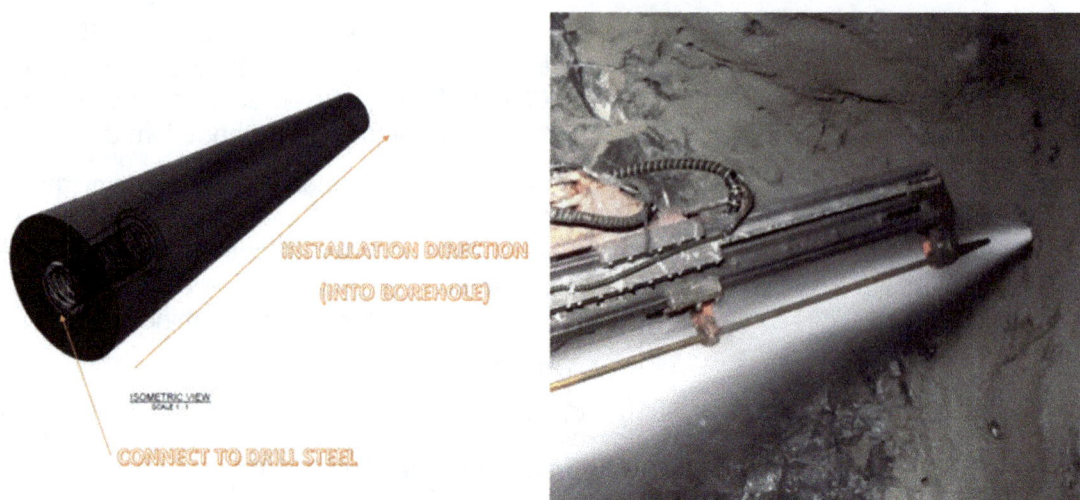

FIG 7 – Tapered plug used to stem water inflow from diamond drill hole intersection in development faces. The tapered plug allows time for appropriate equipment to be brought to face to stop and grout the water flow.

FIG 8 – After a tapered plug has stemmed the flow a Jumbo packer is installed into the hole to stop water inflow from intersected structures or diamond drill holes in development faces. Once installed the unit can be pressure grouted to make safe so mining can continue.

Active dewatering

It was recognised from the start of the Didipio Project that once the 2270 Capital pump station was constructed and commissioned that it would be possible to actively dewater the orebody. At the time it was unclear at what rate the dewatering would lower the level of the water in the orebody so mining continued on each level until approximately 50 L/s of water was intersected and all contingencies in pumping were close to being exhausted. As the total ex-mine water flow got to 300 L/s, levels were put on hold or headings were stopped until the 2270 Capital pump station was commissioned.

Simultaneously however, data showed conclusively that the water level had dropped significantly while mining the first panel. Once the 2270 Capital pump station was commissioned Didipio now had the opportunity to turn on the diamond drill holes that had been drilled across the orebody at the 2280RL, 2270RL and 2250RL and feed the Capital pump station with an extra 210 L/s of clean water to take the average flow from the mine up from 300 L/s to 510 L/s. Figure 9 shows the data collated from vibrating wire piezometers indicating the lowering water level in the mine as pumping rates increased.

Once this increased flow rate was achieved it was quickly noticed that the upper levels began to 'dry' up to an inflow regarded as negligible. This data showed that the dewatering rate could be scheduled at Didipio if flow rates were sustained.

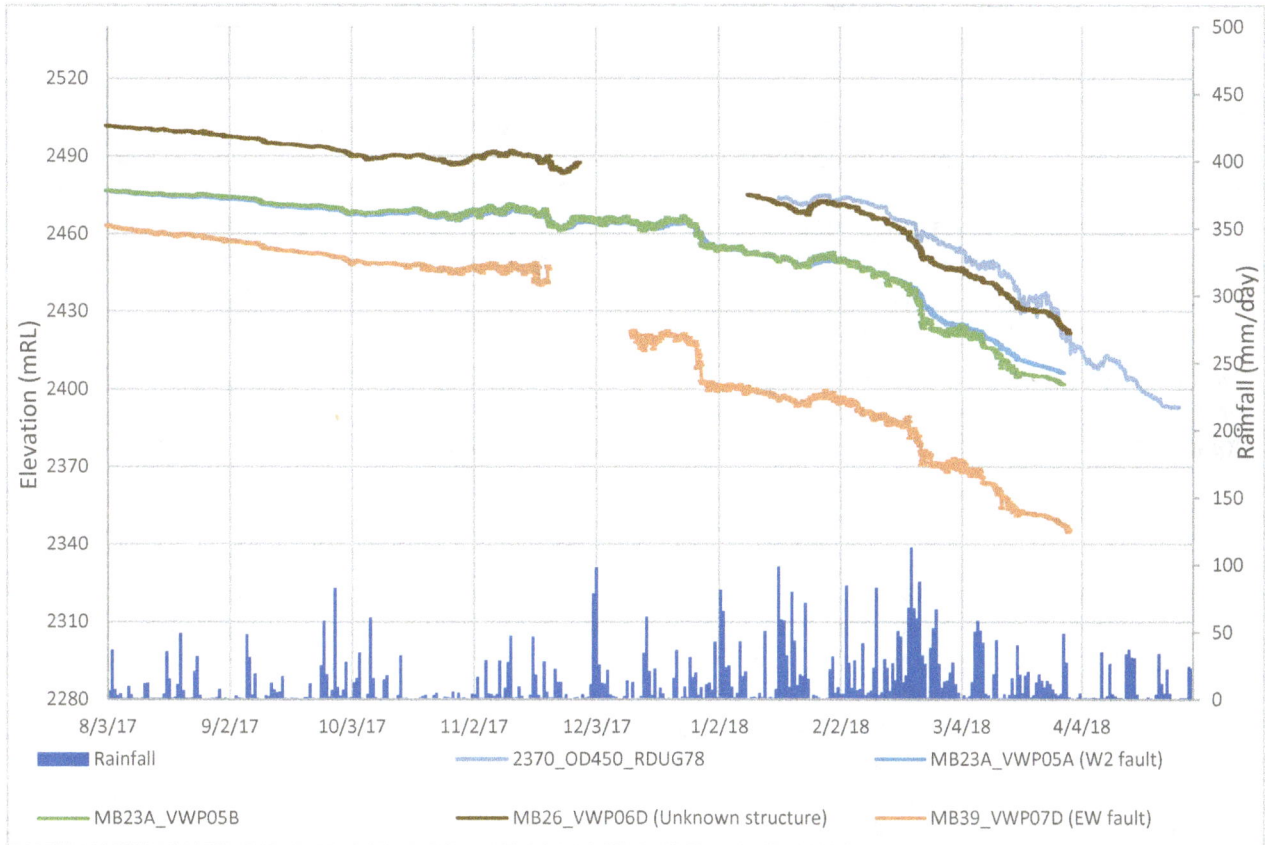

FIG 9 – Data from multiple vibrating wire piezometers showing decrease of water pressure in response to increased dewatering.

Surge and storage capacity

To deal with massive volumes of water and associated silt that comes with mining and paste filling, and the surface need for clean discharge, a purpose-built storage stope was mined to capture all water prior to entering the Capital pump station.

This water storage stope was mined between 2310 and 2250 levels in an area of marginal grade. The total volume of the void in the stope was 26.5 ML (16.5 ML for water and 10 ML for settling). The storage stopes aim was specifically to allow for surge control during power outages as well as much needed time for crucial maintenance on the 2270 Capital pumping infrastructure. Table 2 shows the maintenance time gained for different inflows. Prior to the storage stope commissioning all maintenance activities on the system invariably had a flood event associated with it. After commissioning all mine water was directed to the storage stope allowing work to be conducted on sumps, pumps and pipelines in a preventative maintenance setting. When there was scheduled or unscheduled power outages all underground water could be controlled in the storage stope thus eliminating the severity of flood events. All mine water above the 2310 level was drained into the 2310–2250 storage stope. The mine had finally been able to move from reactionary to preventative and thus development and production became a much easier prospect.

TABLE 2

Surge capacity table showing 9 hours available at 500 L/s.

Storage volume				**16.5**	**ML**		
Inflow	L/s	250	300	350	400	450	500
Surge capacity	hours	18	15	13	11	10	9

Settling ability

The bottom 10 ML of the storage stope was dedicated to *in situ* settling of fines. These fines concentrated at the bottom of the stope and were extracted at the 2250 Stope plug. This stope plug was designed to a suitable pressure to withstand the 60 m of potential head inside the stope once full. The stope plug was a single pour and had appropriate hydraulic hoses placed inside backs and wall cavities pre pour to allow post high-pressure grouting of the plug to ensure a proper seal. It was constructed with appropriate pipework and automatic valving to allow for the mud to flow out in a controlled manner to fill a specific mud truck to be taken to ore or waste stockpiles.

With an average solids content of 0.5 per cent the water entering the storage stope had a minimum silt content of 13 m^3 of solids per day. Extracting this silt prior to entering the 2270 capital pump station produced a much more reliable pumping system and cleaner discharge water to surface.

Reconciliation and monitoring

Tracking water flows with magflow flowmeters on all pumps from the beginning of the project and reconciling these flows against pressure drop data taken from strategically placed vibrating wire piezometers and Trackit loggers is an invaluable tool for high volume wet mines. It is also very important to present this data in a visual way so that all facets of operations can feel confident in progress. Didipio employed a method called seep mapping to show the difference in flows from specific areas of the mine each week. This seep mapping coupled with vibrating wire piezometer data and resource diamond drilling created a visual platform that was updated each week as the water level changed. Figure 10 shows the contours created to form a water inflow risk level map of the 2310 level. Seep data is placed on each drive to get a total from each level and this data is then reconciled against the previous week.

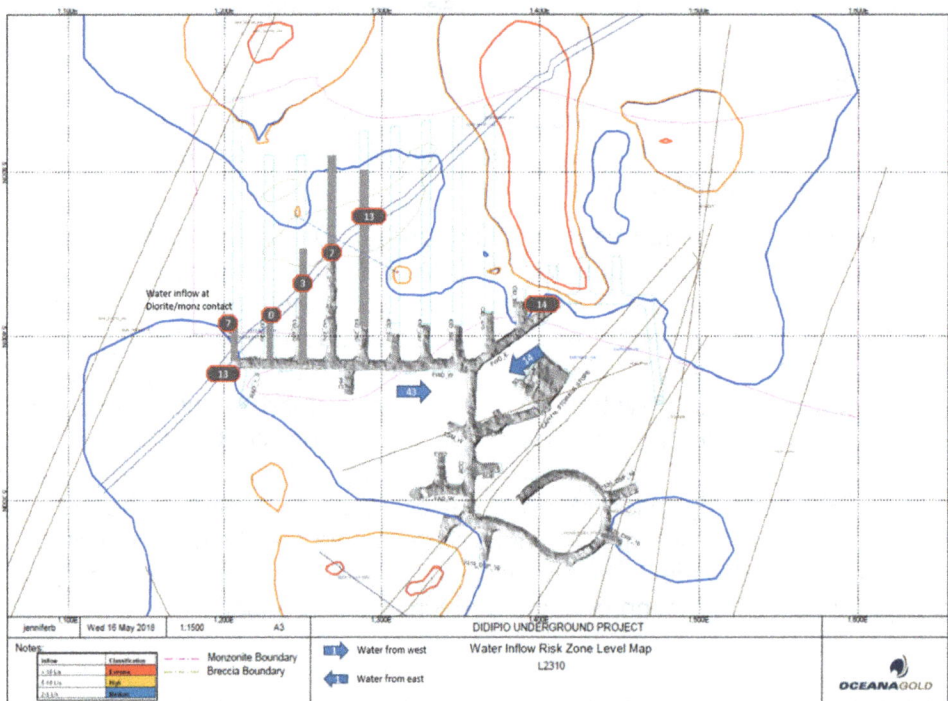

FIG 10 – Water inflow risk zone level map showing significant flow rates from all areas of the 2310 level.

LESSONS LEARNT

From historical data obtained while Mining Panel 1 decline from 2430 level to 2250 level it was evident that the Didipio construction schedule slipped dramatically due to flood events. The root cause of these flood events was fundamentally a loss of power resulting in overflow of uncontrolled water to eventually flood the decline and put capital infrastructure at risk. The lack of suitable storage for the inflowing water is the key component missing during the development stage of the decline.

Entering the orebody on six levels prior to installing the 2270 Capital pump station resulted in flow rates between 50 L/s to 120 L/s discharging from each level. This water was uncontrolled and detrimentally affected the decline development during sustained power outages. Positively however, it also allowed for the collection of a huge data set of water intersections and trends that allowed for predictive mining and design work to be carried out for each subsequent level.

This data showed that the eastern end of the monzonite and the north-west fault area on the western side of the Didipio orebody as being the greatest conduit of water. These areas were then able to be targets for active dewatering when the 2270 Capital pump station was commissioned.

As expected, once the Capital pump station was commissioned at 2270RL and an active dewatering horizon was established on the lower levels the water table was dramatically lowered in a relatively short time allowing for unencumbered mining on 2310 levels and above.

The overall system was again dramatically improved once the 2310–2250 storage stope was commissioned which allowed critical maintenance work to be done on the capital system when required by storing these large volumes of water.

The data also categorically showed that to totally dewater a production level then it was imperative to have a dewatering gallery a minimum of one level below this intended mining level.

A review was conducted after the 2270 Capital pump station was commissioned to investigate the outcomes of Panel 1 mining and what improvements could be made to increase production rates from lessons learnt for the mining of Panel 2.

From the data collected during the mining of Panel 1 it was noted that a high percentage of water inflows were associated with the orebody and major faults and it was evident from the first 2800 m of decline that the water inflows in the decline were much less, on average in the order of 20 L/s, for each level descended.

Recommendations were then made that the Panel 2 decline would have priority to be pushed down to establish the next capital pump station prior to any orebody development. Data collected indicated that inflows of up to 175 L/s would be expected. This mining could be completed much easier and quicker with less water to deal with.

A dewatering gallery would then be drilled at the bottom most level to feed the new capital pump station with maximum water that would dewater Panel 2 in a more economic and rapid fashion. With monitoring, this would then allow access to production fronts in the upper part of Panel 2 in a top-down approach.

As seen on 2340 and 2310 level a dramatic lowering of the water table was achieved by installing targeted holes into known wet regions of the orebody to dewater ahead of development. These levels saw a reduction in flows in the order of 100 L/s to negligible over a relatively short period.

From Panel 1 learnings showed that there were substantial cost/tonne increases associated with high flow rates and these costs were able to be captured during the mining of Panel 1.

These cost/tonne increases include but are not limited to:

- Stress to workforce resulting in staff turnover and lost production (Significant).
- Uncontrolled water intersections on levels of varying flow rates causing flooding issues and significant delays (Major).
- Recovery after a power outage is longer due to uncontrolled water inflows (Major).
- Increase roadwork and road re-building requirements (Significant).

- Water on electrical mining equipment both fixed and mobile resulting in premature failure and added downtime (Significant).

- Increased differential and wheel end repairs.

- High percentage of fines getting flushed to sumps and pumps shortening pump life (Major).

- Shortening of tyre life on all equipment (Major).

- Pump power consumption increase with dirty water (Significant).

- High pump wear resulting in premature failure (Major).

- Charging issues, both development and production (Significant).

- Paste filling issues both dilution and bleeding requirements (Significant).

- High degree of difficulty during remote bogging over paste resulting in sheeting requirements and extended bogging time (Significant).

- Re-bound from shotcreting exaggerated due to wash off while applying (Significant).

- Shotcreting potentially not practical resulting in Jumbo reallocated to ground support.

- Increased fibres reporting to sumps resulting in failure of both submersible and centrifugal pumps (Major).

Ultimately, with active dewatering, the same amount of water is required to be pumped but the pumping costs of having a controlled clean water pumping system is base case. Extracting water from levels below with a stable flow rate feeding into a more reliable pumping system reduces both power consumption and pump costs substantially. As the water table is reduced below levels there is also storage within the system and recharge time is extended to give extra time to get power going during times of power loss.

CONCLUSION

Didipio's Panel 1 was successfully dewatered and is in production unencumbered by water issues.

From the data and learnings and using the same water management plan from Panel 1, Didipio's Panel 2 is well placed to successfully dewater the system in a more efficient and economic manner.

ACKNOWLEDGEMENTS

The Authors would like to thank the management of OceanaGold Philippines for permission to write and publish this paper.

We would like to express our thanks and appreciation to the individuals both internally and externally who have brought their considerable expertise, patience, and unwavering commitment across several disciplines to 'stay afloat' and develop a world-class operation at Didipio.

Planning for underground mining at Savage River Iron Ore Mine

U Singh[1], P Harper[2] and N Burgio[3]

1. Underground Project Manager, Grange Resources Limited, Brisbane Qld 4000.
   Email: uday.singh@grangeresources.com.au
2. CEO and Principal Consultant, Enthalpy Pty Ltd, Melbourne Vic 3000.
   Email: pharper@enthalpy.com.au
3. Director, Stratavision, Sydney NSW 2000. Email: natburgio@outlook.com

INTRODUCTION

This paper presents the evolution of the studies and strategic plans to extend the life of operations at the Savage River iron ore project in north-western Tasmania. The current Life-of-Mine (LOM) plan for the Savage River operations is based on a series of open pits. Drilling has confirmed high-grade extensions of the magnetite resource at depths that exceed practical limits for open pit mining. Grange Resources Limited, the owner of the operation, is investigating underground options – not only to extend the LOM plan but also to maintain peak production at a reduced operating cost while reducing greenhouse gas emissions into the future.

What makes this project unique is that it requires introducing a productive and cost-effective mining method to exploit a challenging geotechnical and geometrical orebody in an area of high rainfall. To support the study efforts, the development of an exploration decline, underground drilling programs, and an exploratory drive excavated into the orebody for geotechnical investigation was key to providing the necessary data for a pre-feasibility study (PFS).

This paper describes the pragmatic approach and processes to developing the underground mining options study through data collection and then integrating these options into the open pit plans to optimise an overall strategic plan for the Savage River operations.

The learnings for operators from this experience are how the technical and pragmatic approaches were aligned to develop the strategic plans, and the importance of timely investment in gathering the necessary data.

PROJECT BACKGROUND

The Savage River mine and concentrator plant are in north-west Tasmania, approximately 100 km south-west by sealed road from Burnie. The mine has been in operation for over 55 years, extracting magnetite from a series of open pits. Grange Resources Limited (ASX: GRR) owns the mine and the downstream processing facilities, which include an on-site concentrator and a pelletising plant at Port Latta on the Bass Strait coast. The pelletising plant and dedicated port facilities at Port Latta are located 70 km north-west by sealed road from Burnie (Figure 1).

Magnetite concentrate slurry is pumped from the mine to the pelletising plant through an 85 km long pipeline. The pipeline is currently limited to annual throughput of 2.85 Mt of concentrate.

The local topography is rugged, with incised valleys and steep hills. The deposit is dissected by the west-flowing Savage River. Regional vegetation includes open heathland, and rainforest comprising wet eucalypt and acacia. Climate is wet temperate, with an average annual rainfall of 1950 mm and mean monthly temperatures ranging from 3°C to 19°C.

The mining operations comprise two active open pits, the North and Centre pits, which are elongated and have a north–south orientation. The mineralisation strike length is 4 km. Centre Pit is separated from North Pit by the Savage River, which runs through a pillar area left to retain the river channel. North Pit is approximately 360 metres deep. Future cutbacks and extensions could deepen it to approximately 450 metres below the natural land surface.

North Pit is the primary ore source, producing approximately 5 Mtpa. Mining of the current pit shell (Stage 6) is expected to be completed in 2022. Stage 6 is the North Pit design used for the PFS. Operation of the pit has been impacted by difficult geotechnical conditions; the west wall is hosted in variably altered rocks that are prone to slumping and movement. The east wall is in competent ground but is affected by significant geological structures and is subject to falls of ground due to the interaction of the wall with faults having an unfavourable orientation. A further cutback of the east wall to extend the open pit mine life is in progress.

FIG 1 – Location of Savage River mine.

During the PFS the exploration decline was developed from March 2019 to September 2020 and included 2.5 km of development, incorporating a geotechnical investigation drive through the orebody and 11 km of resource drilling. The decline portal is located in the south-east corner of North Pit, with the decline developed in the east wall of the pit and traversing from the southern end of the resource to the north. The decline has been designed with a minimum 60 m standoff from the LOM pit shell. The decline dimensions are 5.8 m (height) × 5.5 m (width), with a 1:7 gradient.

STUDY PURPOSE AND SCOPE – PFS

An underground concept study and a mining options study was commissioned in 2018 to evaluate the potential for underground mining to be implemented at Savage River. The study results suggested that underground mining methods – block caving (BC) or sublevel caving (SLC) – could be viable and economically attractive alternatives to increasing the depth of North Pit. Grange commissioned a PFS to investigate the viability of an underground mine (Grange Resources, 2021). The PFS commenced in 2019.

At the end of the scoping study and during the first phase of the PFS, a high-level independent peer review (IPR) was carried out to determine whether the current strategy was appropriate. This involved engaging with consultants and the study team and facilitating a Board strategy session. It allowed the development of a Study Charter which met Grange's corporate objectives and funding strategy and provided direction for the study team and boundaries to work within. The key learning in this pragmatic approach was to ensure all key stakeholders are aligned before investing time and money in developing a business case for the Board to consider.

A risk assessment, involving technical and operational experts with experience in caving and new mine development, was also conducted early in the project to identify key potential risks and the data required to adequately analyse and quantify these risks. This was critical for ensuring that the required data were collected early in the project and potential mitigation measures were carried forward into the mining study.

This PFS was treated as a standalone business case to investigate the viability of an underground mine and provide data for evaluating underground and open pit options. A subsequent enterprise optimisation study was conducted to identify the optimum transition point for transitioning to underground mining and this will be used to inform the feasibility study (FS). The integrated pit and underground plan will then become the LOM optimisation plan.

The exploration decline provided an opportunity to optimise and better target drilling of the underground resource and develop bulk sample drives to improve orebody knowledge. Underground exposures of the orebody provided information that could be calibrated with drilling data to define geological structures and geotechnical parameters necessary for mine design, ore recovery and production forecasting.

Strategic fit

The rationale for the underground project is to profitably extend the life of the operation. From a strategic perspective, the underground mining option is directly comparable to deepening of the existing North Pit. Current open pit designs and the underground options in the PFS address the extraction of magnetite ore defined as Mineral Resources in North Pit.

Key requirements were to:

- extend the mine life
- reduce operating costs
- minimise pre-production capital
- improve the internal rate of return
- transition to underground mining with no disruption to ore supply
- apply technology to reduce greenhouse gas emissions.

The mining options analysed were based on findings of the concept study, which recommended that low-cost caving methods be assessed. The PFS includes evaluation of three BC options and one SLC option, with each option using different systems for materials handling. Both options (BC and SLC) include a scavenge mine designed to recover 'early' ore left behind on the margins of North Pit.

The following PFS options were considered:

- Option 1a: Scavenge mine and BC – trucking to the Run-of-Mine (ROM) pad via twin declines
- Option 1b: Scavenge mine and BC – trucking to the vertical haulage shaft
- Option 1c: Scavenge mine and BC – trucking to the inclined conveyor
- Option 2: Scavenge mine and SLC – trucking to the ROM pad via twin declines.

The SLC option was the least favourable, due to the higher operating costs and the operational challenges of developing and sequencing multiple levels of production in poor ground. The BC options were more advantageous in terms of economics and operability but more challenging technically, with cave behaviour and ore recovery being key issues.

Option 1a – Scavenge mine and BC trucking to the ROM pad via twin declines – is most favourable economically when compared with the other materials handling options studied but is less amenable to scale-up and incurs a larger carbon emissions footprint. The BC options with trucking to the shaft (Option 1b) and conveyor (Option 1c) provide more scalability and have a more favourable carbon emissions footprint.

The underground mining layout for Option 1c and interaction with the open pit is displayed in Figure 2. The exploration decline (shown in brown) has already been constructed and provides a platform to access the scavenge mine (shown in green). The scavenge mine commences above the pit floor. It does not require extensive dewatering infrastructure for first production and provides access to ore remaining at the margins around the pit. This provides early production while the decline is progressed to the undercut and extraction levels (shown in blue).

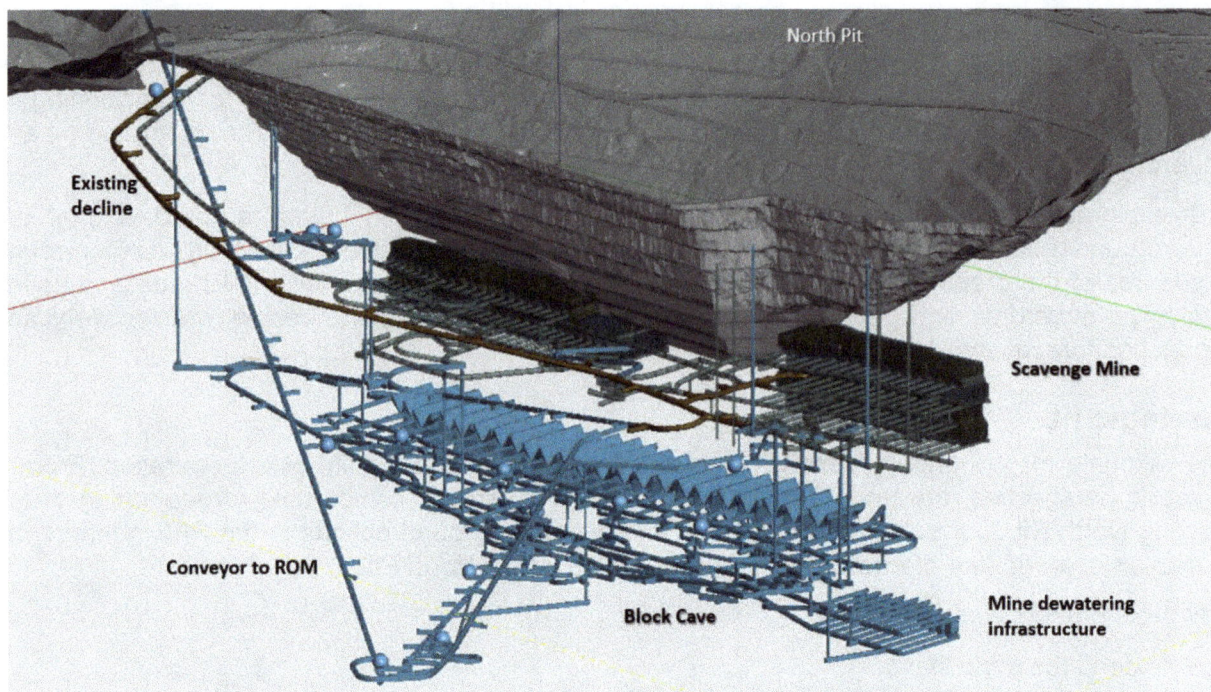

FIG 2 – Underground layout for Option 1c (looking south-west).

Study team approach

A Steering Committee consisting of senior managers and External Experts who convened monthly to review the study progress and provide advice and guidance to the study team. The External Experts team consisted of industry experts in BC and SLC mining who had operational and technical experience using these methods. The Steering Committee reviewed progress and guided the study approach, with the External Experts team providing advice and support where required.

Risk assessment workshops were conducted during Phase 1 and 2 of the PFS. These workshops assessed all key project aspects that could have an impact on the underground mine. The workshop participants included PFS study team members, operational management, industry specialists and the External Experts team.

An independent peer review (IPR) process was instigated as part of these workshops to provide an 'arm's length' review of the study and was carried out intermittently through phases of the study to final delivery to the Board. Having the IPR team involved early in the study process served to minimise re-work at the end of the study and provided opportunity to challenge the status quo and identify areas for improvement. A key advantage the independent reviewers bring is their international operational experience.

The risk assessment workshops considered all key elements of the project, including:

- geology, geotechnical, hydrogeological, infrastructure and mining risk
- OH&S risk
- project execution risk
- human resources, community and social risk
- permit and licensing risk.

Apart from the learnings and identification of study gaps, a very important aspect of these workshops was the improved alignment with the overall operational strategy for this opportunity. Robust discussions were held to challenge the inherent risks of caving this orebody – whether by BC or SLC methods. The ultimate selection of the BC method was based on this operational strategy, but it also aligned with Grange's corporate objectives (Study Charter).

The workshops also demonstrated that further work was required to establish a more-detailed Project Execution Plan, which would also provide increased confidence in the capital cost estimate.

This additional work was carried out towards the end of the PFS, when the mining option selected for advancement to FS was bedded down.

In addition to the Project Execution Plan, a Forward Work Plan was developed to ensure identified gaps and risk mitigation requirements are adequately addressed, resourced, and funded in the FS. While presenting a technically and economically viable opportunity to the Board, the study team objectives included a presentation of a well-thought-out and comprehensive budget to execute the FS.

Geology and exploration

The Savage River magnetite deposits were discovered in 1877. Systematic exploration commenced during the late 1950s with the Bureau of Mineral Resources conducting airborne and ground-based magnetic surveys to delineate the Savage River deposit and two smaller anomalies to the south (at Long Plains and Rocky River). Exploration and resource definition over recent years has involved reverse circulation and diamond drilling. The ore resource definition drilling for North Pit used in the PFS was completed in June 2020 and included the exploration decline drilling.

Over the period of the PFS (2018–2020), a total of 29 km of drill core targeted North Pit. Of this drilling, 18 km were drilled during Stage 1 and Stage 2; this included 19 drill holes drilled from the top of the east wall, and north of North Pit. Stage 3 drilling totalled 11 km from 18 drill holes collared from underground drilling platforms using decommissioned stockpile headings.

A decline was developed in the east wall of North Pit to facilitate acquisition of geological data from underground mapping, diamond drilling, structural interpretation, and a suite of geochemical analyses on pulps. Two independent Mineral Resource updates were completed – by Optiro in 2019 and Xstract in 2020.

Mineral Resource audits by both an internal Grange team and independent consultants were carried out. The audits determined the resource estimation work was consistent with acceptable industry practice.

Geotechnical data collection and analysis

The ground conditions encountered are similar to conditions expected from the drilling and geotechnical interpretations. A structural geology expert was engaged to conduct structural interpretation from the drilling, mapping and open pit data. This formed the basis for the geotechnical domain models and subsequent analysis. Laboratory tests and *in situ* measurements, including overcoring stress measurements, Schmidt hammer and Equotip rebound tests, were used to assess boundary conditions and rock mass properties.

The ground conditions at the decline are generally good and hydrogeological observations indicated that the east wall is generally tight and dry. The ground support used in the exploration decline and ancillary development (stockpiles, sumps and refuge chamber cuddies) comprised mainly fibre-reinforced shotcrete and resin bolts.

A geotechnical investigation drive (or bulk sample drive, BSD) was developed through the ore at the -60 mRL. Several breakaway headings were developed successfully from the BSD to test intersections and breakaways in the ore. The ground conditions in the ore ranged from fair to very poor. Ground stability assessments indicate a highly variable rock mass which will require specific and targeted ground support systems – particularly in weaker areas of the mining footprint. While the BSD was typically supported using a combination of mesh, fibre-reinforced shotcrete and resin bolts, steel-reinforced shotcrete arches and self-drilling anchors were also successfully trialled in areas of poorer ground conditions.

Based on the work completed, ground support recommendations have been developed for both mining options (BC and SLC). Because of poor ground conditions in ore and the predicted levels of damage, the drives will require regular rehabilitation, with the level of rehabilitation dependent on the geotechnical domain and the induced stresses. The operating cost forecasts include provision for regular rehabilitation.

A caveability assessment was conducted on the PFS mine design. The extraction level is situated at -280 mRL, has a drawpoint spacing of 34 m × 20 m, and involves two separate caving fronts that diverge from a roughly central point in the orebody footprint. For empirical caving analysis using Laubscher's methodology, the upper quartile values for the MRMR (mining rock mass rating) provided conservative results.

The minimum cave width for the North Pit underground operation ranges between 60 m and 120 m, which is narrower than most recent cave developments globally, and few benchmarking examples exist. A further complicating factor for the use of empirical minimum cave span assessment methods is the high cross-sectional variance that is interpreted to exist in the cave column. Numerical analysis studies into the caveability and flow behaviour of the rock mass were therefore important in establishing whether continuous caving could be achieved. The simulation of mining sequences also helped define an optimum mining strategy. A review of other block cave operations with similar conditions – such as New Afton, Argyle, Cullinan, Finsch and the skarns at DOZ (Deep Ore Zone, Grasberg) – was also important in helping formulate this strategy.

Hydrology and hydrogeology

Potential water inflows to the underground mining operations were investigated to develop a mine water management plan and assess the net impacts on the local hydrogeological environment. Rainfall and pit dewatering data spanning a long time frame, as well as borehole monitoring data, were used to develop hydrological and hydrogeological models. These models were verified using data from subsequent underground development.

The key water management issues are:

- significant volumes of rainfall which will then rapidly make its way down into the underground mine workings through the pit and cave zone
- groundwater inflows from the fractured basement rock aquifers which host and surround the magnetite orebody
- possible inflows from the Broderick Creek flow-through system, a high-flow channel to convey flows from upstream reaches of Broderick Creek in the north to discharge points in the south.

Simulations and trade-off studies were conducted to compare peak pumping capacities and optimum underground storage of water. The extraction level and supporting infrastructure were also designed to minimise the risk of losing or damaging critical infrastructure through flooding by carefully assessing the potential flooding scenarios and water flow paths. As a result, the mine design considers pumping capacity and incorporates surface water diversion channels and underground buffer storage voids.

Mining method and design parameters

The mining methods considered were BC and SLC, and a scavenge mine was designed to recover the ore left in the walls of the Stage 6C pit shell. The scavenge mine uses the SLC mining method and has the same layout and profiles as the SLC mine. Due to their higher mine operating costs, the PFS did not consider non-caving mining methods. This aligns with the findings from the previous mining studies.

The target annual underground mining production is rated to meet the current capacity of the pipeline that delivers concentrate slurry from the mine to the pelletising plant in Port Latta. The maximum pipeline capacity is 2.85 Mtpa of concentrate.

Detailed design parameters have been determined, including for:

- mine layouts (BC and SLC)
- access design (BC and SLC)
- production sequencing (BC and SLC)
- drawpoint design (BC only)
- undercut design and strategy (BC and SLC).

Multiple cave flow modelling runs with PGCA and FS4 software were conducted, and open pit interaction and subsidence for the SLC option were generally similar to that modelled for the BC options. Sensitivity analyses were conducted on the flow and recovery using a range of simulated draw widths and flow parameters. Simulations were also conducted on drawpoint damage, including reduced recoveries and early closure, and rehabilitation, including a second set of lintels.

The PFS considered several materials handling options with associated underground configurations and underground and surface infrastructure. For the BC (and scavenge mine) option, the following three materials handling systems were evaluated:

- trucking to the ROM pad via twin declines
- trucking to the vertical haulage shaft
- trucking to the inclined conveyor.

For the SLC mining option, only trucking to the ROM pad via twin declines (Option 2) was evaluated due to the relatively shallow depth of mining, and reduced mining inventory.

Option 1a – trucking to the ROM pad via twin declines – was considered to have significant advantages as a go forward option, including:

- best net present value (NPV) – despite having the highest operating cost
- ease of construction – it does not require specialised engineering skills or procurement of specialised equipment
- does not require a crusher – trucking to the surface removes the requirement for underground crushing, but a tramp removal system on the surface is required before the ore can be fed into the surface crusher and conveyor system (the scavenge mine also requires a surface tramp removal system)
- compatibility with the scavenge mining system and interoperability – the scavenge mine will use trucks of the same size
- potential for future upgrade to include use of an electric fleet and mine automation.

However, Option 1a had significant limitations, including:

- Trucking to surface at 6 Mtpa was considered the upper limit of productivity, representing a constraint on future expansion.
- Operating costs would be higher, which was not the preferred long-term strategy.
- The greenhouse gas emissions would be higher, which does not align with Grange's strategy for decarbonisation.

Shaft haulage and conveyor haulage offer more flexibility for increasing production and reducing diesel consumption. In addition to reducing the carbon emissions footprint, these haulage systems have lower ventilation demand. The conveyor haulage option has slightly better economics than the shaft haulage option. A detailed materials handling study will be conducted to finalise a go forward option for the FS.

Project implementation

The implementation scope of work for the FS included:

- definition of project objectives and key performance indicators (KPIs)
- project implementation planning for the FS and for project execution
- human resources, including personnel and organisational structure issues
- procurement and contracting strategy
- project scheduling for all options considered
- project control systems and IT infrastructure
- operational management.

Project objectives and KPIs

The overall objectives to be satisfied through project implementation are:

- Maintain a zero-accident rate, as measured by a zero LTI (lost time injury) rate.

- Establish a proactive safety culture and as a minimum, ensure compliance with Grange's Safety and Environmental Management System (SEMS).

- Behave in an environmentally responsible manner and ensure compliance with the Site Environmental Management Standards.

- Reduce the greenhouse gas emissions from the site in accordance with Grange's objectives.

- Establish a healthy and harmonious working environment for all personnel on-site and ensure compliance with Grange's Human Resource Management Standards.

- Ensure compliance with all internal and external stakeholder requirements, including the Savage River site, Grange Resources, local council, EPA and State regulations.

- Achieve the critical customer requirements for the project.

- Strive for best-in-class standards in project delivery.

- Recognise the role of the Operations Team as the 'customer' of the project and the need to ensure integration and commitment of the Operations Team in the transition to underground mining.

Project scheduling

The detail and execution for pre-commissioning, commissioning and handover will be covered in the FS, in conjunction from the open pit planning team, and will include:

- integration of underground production and Centre Pit ore feed

- personnel requirements, including transition and retraining of current personnel

- additional site accommodation requirements

- tender options for underground contractor and explosives supplier

- decommissioning of redundant open pit equipment and facilities

- upgrades to current site infrastructure, including workshop and power

- Environmental Impact Statement (EIS) approval status

- mobilisation requirements.

It is important to note that the open pit will be operating in conjunction with the underground operations in the preliminary stages of underground development and production, and the transition requirements will be covered during this period.

Project review, risks and opportunities

Geotechnical and hydrogeological analysis and risks have informed the mine designs for all options considered. SLC mining presents geotechnical and production risks, some of which have been mitigated in the design and others which would require ongoing operational management.

The footprint for the BC is expected to cave readily despite its narrow span but has the potential to cave preferentially along geological structures, which could affect recoveries. Secondary fragmentation is expected to be fine, which, combined with clay and water, could result in mud rushes. The assessment also indicated that the poor ground and presence of numerous faults in the ore zone would adversely affect stability on the extraction level of a block cave. This is mitigated by using wide drawbell spacings, resulting in larger pillars and limiting the life of the drawpoint. Extensive ground support and rehabilitation have also been factored, including rehabilitation of access drives and construction of new drawpoints after 250 kt of draw.

The IPR team found that the geotechnical conditions are forecast to be severe in comparison to other underground caving operations. However, various design and planning strategies have been put in place to help mitigate these impacts. Due to the ground instability predicted, the IPR team considered there remains some residual risk to the proposed mining strategy that warrants review. The purpose of the development design and draw strategy review was to address the predicted 'extreme damage to the extraction level pillar because of imposed stresses' and seek ways to reduce the damage by design and draw strategy rather than the inclusion of remediation works and discounting of the results to allow for 'poor' performance, as proposed in the PFS.

Other IPR team suggestions include:

- incorporating an assessment of transition to electrification in the FS (should trucking be the preferred materials handling option)

- assessing the risks and benefits of a faster draw strategy through fewer active/open drawpoints to minimise damage and maximise draw from active drawpoints

- reviewing or testing the sensitivity of the stability analysis for rock strength assumptions on the predicted ground conditions.

A risk assessment was also conducted. The key findings are summarised as follows:

- Several risks were classified as *Extreme*. They include:
  o mud rush for mining method options (BC or SLC)
  o asbestiform material affecting occupational health and safety
  o suboptimal value proposition in the transition from open pit to underground mining.

- A total of 16 risks were classified as *High*. They cover geotechnical, mine flooding, air blast, mining, infrastructure permitting and approvals risks.

Some of these *High* and *Extreme* risks (eg geotechnical and asbestiform minerals) are already being managed through Grange's existing safety management systems for underground development. These, and other identified risks, will need to be further evaluated and mitigated during the FS.

KEY LEARNINGS

The key learnings and actions that contributed to successful study execution include:

- Conducting workshops, including high-level risk assessments, with all key stakeholders and decision-makers before commencing the study.

- Using workshop outcomes, establishing a Study Charter that all stakeholders align with and agree to. This ensures the right project to meet corporate objectives is being designed and developed.

- Selecting an appropriate study team that applies best practice minimum standards and protocols.

- Sourcing well-qualified personnel who follow a pragmatic approach and have relevant experience to assist in the design and developing the project the right way.

- Appointing a Steering Committee consisting of external experts, experienced operators and senior management to guide and advise the study team.

- Engaging a credible IPR team early in the process to support the study team to deliver a professional study document with clear and transparent messaging (this minimises re-work).

- Articulating the business case (value proposition and risk profile) in a manner that stakeholders understand (stakeholders were engaged at key decision-making points throughout the study to ensure alignment was achieved during the process).

- Including sensitivity assessments, reliability assessments and contingency planning to manage risk and retain flexibility to accommodate changed conditions.

- Benchmarking the proposed project against similar projects to apply the most appropriate and realistic assumptions (benchmarking against projects that do not have similar geometries, as an example, can be misleading).

- Developing a comprehensive Forward Work Plan for the next phase of work (this provides an immediate gap analysis and realistic plan to advance the business case).

- Understanding the difference between the Forward Work Plan and Project Execution Plan and spending adequate time developing these plans in a comprehensive manner (this gives decision-makers a clearer appreciation of future requirements).

- Adopting a holistic approach to ensure a robust and sustainable LOM plan is achieved (the underground PFS is to be integrated in the open pit plan to become the Enterprise Optimisation LOM plan).

Other key factors contributing to the study's success were that senior management clearly articulated strategic objectives, and that the leadership fostered a culture that ensured all issues and challenges could be discussed and addressed in an open and non-threatening manner.

CONCLUSION

The Savage River mine in north-west Tasmania has been extracting magnetite from a series of open pits for over 55 years. Exploration drilling below North Pit confirmed high-grade extensions of the magnetite resource at depths that exceed practical limits for open pit mining. Grange Resources initiated a study process to evaluate options for extracting the resource at depth. This included developing an exploration decline and conducting detailed data collection and analysis.

A mining options evaluation study was conducted to identify suitable methods that would meet the technical and business requirements of the project. A high-level risk assessment was conducted early in the project to identify key risks, which then informed the data collection requirements and strategy for the project.

A Study Charter was developed which aligned the Board, senior management, and the study team in establishing a term of reference, as well as study boundaries, exclusions and hurdle rates. A Steering Committee was formed to guide, mentor, and advise the study team in meeting its objectives.

Communicating the study progress and highlighting the options, risk profile and value proposition with all stakeholders, including the IPR team throughout the process ensured there were no surprises and no requirement to undertake re-work at the end of the study.

A robust Forward Work Plan was developed for the next phase of study, including a realistic Project Execution Plan for the project delivery. The study findings were incorporated in an enterprise optimisation study, which will be used to determine the optimum point for transition to underground and the best allocation of capital for maximising value for the company. The findings of the Enterprise Optimisation LOM plan will provide the scope for the next level of evaluation and definitive feasibility study.

ACKNOWLEDGEMENTS

We wish to thank Grange Resources Limited for its support and approval to publish this paper. We also extend our appreciation to the External Experts team, IPR team and Grange senior management for their guidance and mentoring through this study process.

REFERENCES

Grange Resources, 2021. Savage River Prefeasibility Study confirms technical and financial viability of an underground operation, Savage River Operations, Tasmania. ASX Release dated 23 December 2021. Available from: <https://www.grangeresources.com.au/announcements> [Accessed: 26 November 2022].

Friction rock stabiliser QAQC and a process to investigate high pull test failure rates

R Varden[1], R Hassell[2] and J Player[3]

1. MAusIMM(CP), Senior Geotechnical Engineer, MineGeoTech, Perth WA 6000.
   Email: richardvarden@minegeotech.com.au
2. MAusIMM(CP), Manager Operations, Mining and Civil Integrity Testing, Boulder WA 6433.
   Email: rhett.hassell@minecivilit.com.au
3. MAusIMM(CP), Principal Engineer, MineGeoTech, Perth WA 6000.
   Email: johnplayer@minegeotech.com.au

INTRODUCTION

Friction rock stabilisers, commonly termed 'Split Sets', friction rock bolts, or continuous friction coupled bolts are the most widely used rock bolt in the Australian underground metalliferous industry. Historically, they were used for temporary reinforcement, mainly within ore development. However, their use as primary reinforcement has grown with the demand for higher development rates at lower costs. The introduction of galvanisation has increased their longevity and pattern bolting in conjunction with weld mesh, specifically made for use in development headings, has allowed the use of friction rock stabilisers to increase rapidly with development jumbo's undertaking the installation.

The installation of friction rock stabilisers appears to be deceptively simple, particularly when compared to resin encapsulated bolts. The friction rock stabilisers capacity is a function of the borehole parameters, rock mass quality/strength, the bolt dimensions and steel quality, and lastly and perhaps the most important the operator skill levels. However, as in industry we are still seeing common occurrences of friction rock stabilisers not reaching their design pull test capacity.

Pull testing the friction rock stabilisers is used to verify the installation of the design capacity. It involves loading the collar of the bolt with a hydraulic ram until the design load is reached or the bolt slips. Pull tests often show failure rates of 20 per cent or higher. This is despite over 40 years of use and with millions of units installed annually. Such high failure rates would suggest there remain issues with the quality of installation.

Ground support quality assurance and quality control (QAQC) programs are needed in combination with operator training to ensure that pull test failure rates are kept as low as practical. However, there are times when despite the QAQC being completed the pull test failure rates increase quite suddenly. When this occurs, it is important to conduct a thorough investigation. These investigations often show there are several factors that are not obvious to personnel installing the bolts. To complete a useful investigation, it must also be done methodically, working through the entire process that includes less obvious factors such as steel quality and the calibration of the hydraulic pull test kit.

The aim of the paper is to give mine practitioners an understanding of the complexities of friction rock stabilisers, what QAQC requirements they should be doing on-site, what can go wrong during installation and, if necessary, a process of how to conduct investigations into high friction rock stabiliser pull test failure rates.

FRICTION ROCK STABILISERS

Friction rock stabiliser (FRS) is used in this paper to refer to what the industry commonly called 'Split Sets'. This is not to be confused with other friction rock bolts such as the Swellex or Spirol Bolt.

An FRS is a continuously frictionally coupled (CFC) rock bolt (Windsor and Thompson, 1993) that consists of a 3.1–3.5 mm steel C-section made from a strip width of 107.5–120 mm. It is typically rolled into a C-section shape with a 25 mm slot along the entire length and tapers at the toe end. The outside diameter is typically 46 mm to 47 mm when being installed by a drilling jumbo. They are manufactured in a variety of lengths but typically range between 0.9 m and 3 m. The bolt is inserted into a borehole that is typically 2–4 mm smaller than bolt diameter. The C-section deforms as it is inserted into the borehole and develops a frictional force against the borehole surface.

The FRS has its origins in the USA coal mining industry. On 2 December 1975, United States Patent 3,922,867 by Scott was granted. The abstract of this invention stated:

> Friction rock stabilisers such as for example roof anchors, comprising a generally annular body which from end-to-end has a slot through its thickness and is circumferentially compressible for installation into a bore of diameter substantially smaller than the normal outer diameter of the body whereby, after such installation, the resilience of the body causes the body outer periphery to anchor by frictional engagement with the surrounding wall of the bore. Also, an anchoring method employing a stabiliser of this type.

Scott (1975) states that the modulus of steel and dimensions of the bolt are selected so that yield of the steel occurs into the plastic range.

Tomory (1997) expands on that understanding with the following statement:

> With slot closure, the Split Set bolt is deformed beyond the yield point and into the cold working portion of a stress-strain curve. Anchorage, or bond strength, is produced by the reaction of the spring-like Split Set against the walls of the drill hole. If the bolt is removed, and the steel unloaded, there will be some amount of spring-back.

On 21 November 1978 United Stated Patent 4,126,004 by Lindeboom for Ingersoll-Rand Company was granted. This patent improved the friction component and enhanced the collar section of the unit.

Since the invention by Ingersoll-Rand and the expiration of the patent there has been a number of manufactures making their own version of the FRS, with a number of adaptions to increase friction that includes closed section bolts, point anchors or grouting. These are not covered in this paper.

FRS were originally designed for short-term reinforcement in a large part due to their susceptibility to corrosion damage from their large surface area and thin steel thickness. A significant improvement has been made with galvanisation.

Galvanisation provides protection against corrosion damage and thus increases the bolt life allowing reinforcement strategies to change from temporary to permanent reinforcement. This is particularly noticeable in the Australian gold mining industry where many operations have a mine life of <5 years. Combined with a lower unit cost (compared to resin or cement grout encapsulated bolts) and speed of installation, FRS make an attractive reinforcement option.

In Australia, an estimated 15 million units are installed every year.

How friction rock stabilisers work

The FRS consists of a rolled tube (with a slot along the entire length) which is driven into a smaller diameter drilled into the rock mass. When left ungrouted it relies on friction between the tube and the rock to prevent pull out. The system does not suffer from the problems encountered with other reinforcement systems such as point anchored rock bolts where if the anchor slips or the rock around the plate fails the capacity of the bolts drop to zero and failure of the rock around an excavation can occur. This problem is less severe in the case of a fully coupled frictional bolt because even if slip does occur, or if the face plate breaks off, the remaining length of the bolt is still anchored/bonded and will continue to provide a degree of reinforcement (Villaescusa and Wright, 1997).

One critical limitation of ungrouted split set bolts is that although the bolts are simple and quick to install and stand up to blast vibrations relatively well, they have a very low initial bond strength per metre of embedment length (Villaescusa and Wright, 1997), usually in the range of 35–50 kN per metre of embedment.

Load transfer and embedment length

The load transfer concept is fundamental in understanding reinforcement behaviour. Windsor and Thompson (1993) describe load transfer in three stages:

1. Ground movement at the excavation boundary, which causes load transfer from the unstable rock, wedge, or slab to the reinforcing element.

2. Transfer of load via the reinforcing element from the unstable area to a stable interior region within the rock mass.

3. Transfer of the reinforcing element load to the rock in the stable interior zone.

For a fully coupled frictional bolt such as an ungrouted split set bolt, failure by slippage of the reinforcement element can either occur within the unstable or the stable regions in the rock mass depending on the amount of embedment length available in both regions.

The critical embedment length is the minimum length of reinforcement needed to mobilise the full bolt capacity or strength. This minimum length may need to be increased if the reinforcement element is oriented at unfavourable angles with respect to the free surface (Windsor and Thompson, 1993), see Figure 1.

Villaescusa and Wright (1997) state the strength per metre of embedment length of an ungrouted Split Set bolt is limited by the radial prestress set-up during installation. This is a function of the Split Set bolt diameter, the borehole diameter, and any geometrical irregularities occurring at the borehole wall. Slippage and failure within the unstable area due to short embedment lengths, can be minimised with the installation of a proper face plate to provide surface restraint.

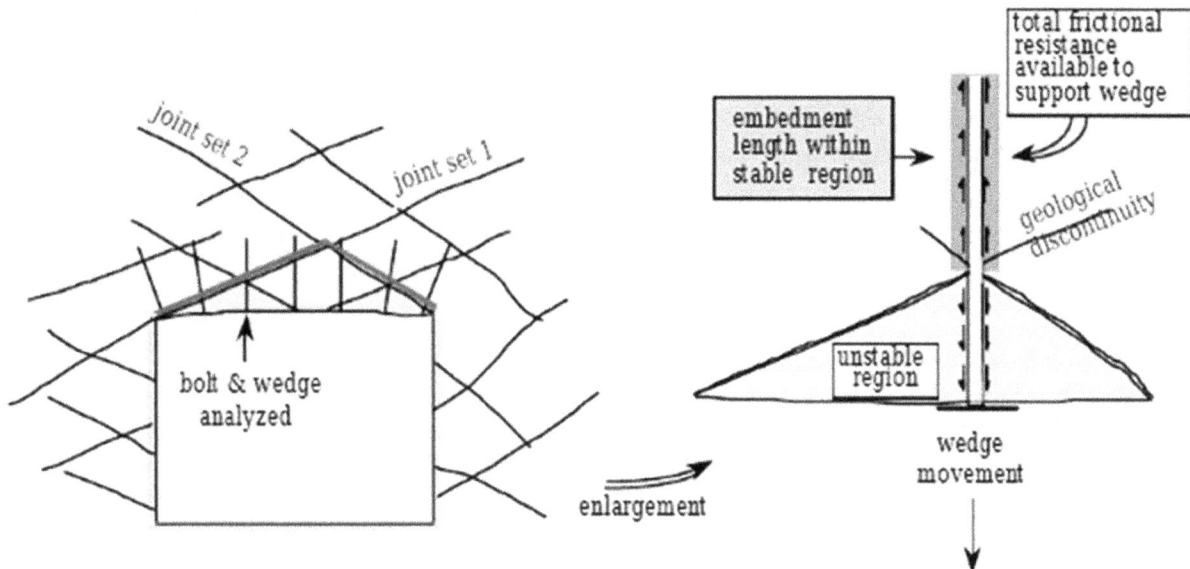

FIG 1 – Load transfer and embedment length concepts.

Embedment length and pull testing

Friction Rock Stabilisers like all reinforcement elements are pull tested from their collar. This mobilises the embedded length of frictional resistance from the FRS. This capacity is different from the design capacity of the FRS which is based on the capacity of the length of FRS that is in the stable region of the rock mass. This is usually referred to in tonnes per metre (t/m). For example, most 2.4 m length friction bolts are pull tested to a maximum load of 8–10 t. If 8 t are achieved this equates to 8 t over 2.25 m (effective embedded length) which is 3.56 t/m. If 10 t are achieved, then this increases to 4.44 t/m (25 per cent higher). Pull testing by some sites have been safely completed up to loads of 15 t.

By comparison resin grout encapsulated bolts have capacities per metre is in the order of 16–25 t/m, Villaescusa and Wright (1997).

Changes to the friction bolt design and installation practices

Since the introduction of FRS to the mining industry there has been modifications to the FRS design. There has also been changes to the equipment that is used to install the FRS. The more significant changes that the authors are aware of include:

- Change of split tube dimensions (diameter and split).

- Shape of the FRS cross-section.

- Improved ring/collar strength.

- Metallurgy (the High Strength Low Alloy Steels HSLA that are used for FRS have a yield stress range of 250–690 MPa and over 40 grades), but the original Ingersoll Rand Split Sets used two different steel grades depending on the location of manufacture either 380 or 415 MPa yield stress (Tilman, Jolly and Neumeier, 1984).

- Galvanisation.

- There have been large increases in the percussion power and drilling feed force of drifters, for example 250 Nm to 1000 Nm of torque.

- Plate design.

These changes have resulted in an increased ability to install bolts faster, which has resulted in the FRS becoming more attractive to mining operations. However, with the increase in technology it has created new challenges to the personnel installing the bolts and those designing ground support schemes and undertaking QAQC.

FRICTION BOLT QUALITY ASSURANCE AND QUALITY CONTROL

Correct installation is critical to FRS performance. There is a general understanding in the mining industry that FRS are quick and easy to install. This is correct, but also misleading as often the quality of installation is lacking. There are a considerable number of parameters that the operator needs to consider when installing bolts.

Modern underground development drilling jumbos have a considerable amount of percussion power and rotational torque, with faster drive times. There is significant pressure on operators to advance headings quickly. Contracts for development focus on metres advanced, with installation quality clauses generally being secondary if included at all.

Within this framework it is can be seen how the quality of installation may be affected. To support installation quality standards a robust QAQC program is needed. This following subsections outlines the possible QAQC options that may be used by sites. A site manager, supervisor or engineer can use the elements in Table 1 as a template for best practice FRS installation QAQC.

TABLE 1

Components of an FRS QAQC Program.

Materials	Environmental	Installation procedure	Instructions and assessment
Manufactures specifications	Evaluate rock mass conditions	Borehole length	Ground support instructions
Bit sizing	Actual borehole size	Borehole angle	Bolt and ring spacing
Certification of pull test equipment	Corrosion	Borehole straightness	Task observations
		Alignment of bolt with borehole	Pull testing
		Insertion dolly	Operator training
		Drive time and no rotation	

Materials

Confirm manufacturers specifications

The ground support manufacturer should have a quality management system that is ISO 9001 accredited. During the ground support supply contract discussions, either the mine site or the ground support supplier will specify the bolt dimensions, the required physical properties and importantly the tolerance of each.

Batch certificates are a document that attest that the provided ground support meets the product specifications. These should be provided by the supplier for every consignment and reviewed by site personnel to ensure they meet the contract requirements. Arrival and completion dates for the batch numbers should be kept, so that site personnel can figure out where different batches have been installed underground.

Site personnel can verify some specifications on-site. The key parameters to be aware of include FRS diameter, FRS thickness, FRS strip width (width of the steel when laid flat), steel tensile strength and galvanisation thickness. Steel tensile strength can be estimated in the field with a metal hardness tester and galvanisation thickness measured with a coating thickness gauge.

Drill bit types and size

There are a variety of different drill bit types commercially available. Sites use either spherical or semi-ballistic button bits of varying face geometries. For a 47 mm diameter FRS, ground support suppliers recommend a borehole size between 43 mm to 45.5 mm. Drill bit suppliers offer a 43 mm bolting bit. However, this is misleading as the tolerance is 1 mm to 1.5 mm larger meaning the actual diameter of the bit starts at 44 mm to 44.5 mm.

Re-sharp bits are previously used bits that have been re-sharpened, to increase usage. They will be a smaller diameter than the new drill bits.

To ensure the correct bit sizes are being supplied and used on the jumbo the drill bits should be confirmed with a bit gauge. These are a machined plate of steel typically with three specified sizes decided by a testing program eg 43 mm, 44 mm, 44.5 mm (Figure 2).

FIG 2 – Bit sizing template.

When measuring the bits to use in an evaluation program, bit rings are recommended (Figure 3). They are recommended over verniers because bits are often not circular, and the buttons are often proud of the bit body. Vernier measurements can be precisely wrong. Sizing the bits with closely incremented rings is more accurate. Both gauges and rings need to be of suitable thickness to minimise bits passing through by twisting.

A site may recommend that smaller diameter drill bits are used in specific weaker rock masses.

FIG 3 – Bit rings for sizing bits.

Pull test equipment

Sight certification of gauge calibration and that it is in date.

Environmental

Rock mass properties

Weaker rock masses such as those with a low UCS, significant jointing or having undergone large deformations will allow higher rates of drill penetration. This may mean the hole size is much larger than the bit. Broken rock masses may also provide less confinement to the bolt further reducing the frictional resistance.

In extremely hard, competent rock types, eg dolerites, the borehole surface may be smooth which may also reduce the amount of resistive friction generated by bolt.

Minerals such as talc may act also as a lubricant reducing friction.

All data collected during the QAQC process needs to be related back to the rock mass it was being installed in to understand if this is a factor in varying installation quality throughout the mine.

Borehole size

The size of the hole is dependent on the local geology (principally hardness and fracturing) and the bit size:

- If the hole diameter is too large, the FRS when inserted will not have sufficient compression to close the gap and produce the frictional force to resist sliding.

- If the hole diameter is too small, the FRS when inserted significantly deforms and sometimes twists which reduce the area of metal in contact with the borehole. Thus, reducing frictional resistance and significantly yielding the steel so it has less spring to push against the side of the hole.

Borehole measurements are taken with a borehole micrometre. It can produce variable results and is dependent on the operator and position of each measurement. It is recommended to take a measurement every 0.5 m along the borehole. Readings are taken 0.5 mm increments. The authors are not aware of any devices that can obtain a full 3D geometry of a rock bolt hole.

Corrosion

Corrosion of ground support systems in mines is common. FRS can be particularly susceptible to corrosion damage due their large surface area, thin walls, damage to coatings during installation and direct contact with the rock mass and water/sulfides.

In areas affected by groundwater, and in drives where the ventilation has been turned off for extended time periods steel corrosion may occur as early as six months after installation, A dry, well-ventilated drive may still show little corrosion damage after ten years.

The process to assess the severity of corrosion damage involves environmental condition mapping of the underground excavations in combination with qualitative and quantitative assessments of the condition of the reinforcement and support systems. This is discussed in more detail in Hassell, Villaescusa and Thomson (2004), Hassell (2007) and Jones, Hassell and Power (2022).

Installation procedure

Borehole length

The hole needs to be drilled at least 300 mm longer than the bolt in case surface fretting occurs before or during bolt installation, resulting in shortening of the hole. Bolts installed in holes that are too short will not have the plate sitting flush with the rock surface.

Borehole angle

Borehole angle is important to achieve the required loading of the bolt and to ensure it is installed to the maximum depth. Note a 2.4 m bolt typically only have 2.3 m of engagement with the rock allowing for the taper and surface hardware. This reduces to 2.25 m when the pull ring is installed. The effect of dump is shown in Table 2.

TABLE 2

Effect of dump on length on true height.

Bolt length	Dump	Vertical distance
2.4 m	Vertical	2.4 m
2.4 m	10 degrees dump	2.36 m
2.4 m	20 degrees dump	2.26 m
2.4 m	30 degrees dump	2.08 m
2.4 m	40 degrees dump	1.84 m

The borehole should be drilled perpendicular to the development drive surface. A 15-degree tolerance is generally acceptable.

Borehole angles can be visually assessed during inspections. If actual measurements are needed, then a long straight pole can be inserted into the inside of the FRS and the angle can be surveyed.

When bolt angle increases it becomes more likely for overdriving of the ring of the FRS through the surface fixture, see Figure 4. Increases in the relative angle of the surface fixture to the FRS will result in bending of the FRS tube near the collar of the hole.

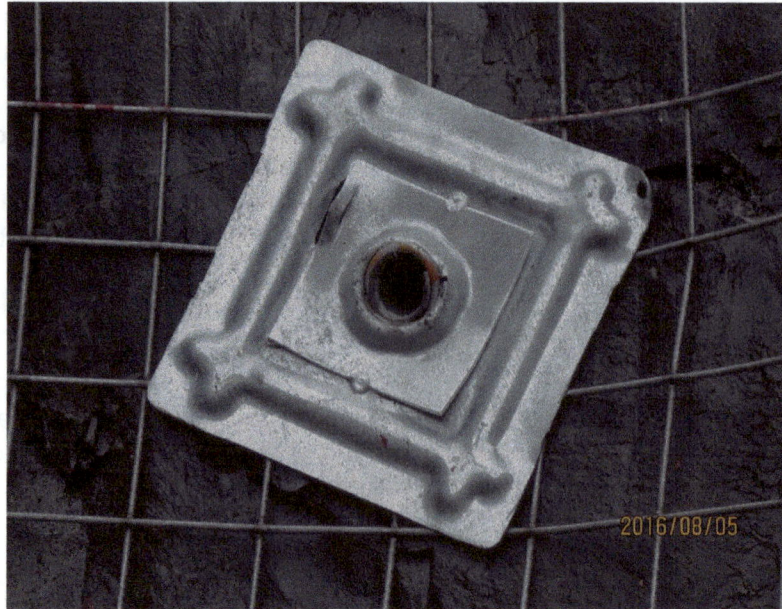

FIG 4 – Overdriven FRS through surface fixture.

Borehole straightness

In most lithological units there will be deviation of the borehole. Deviation is associated with the operator skill and the control of the feed-rate, the rock mass quality and the bit type. If the borehole is not straight, the installed FRS is not installed straight. This may prevent the bolt pulling axially and allows the FRS to be pulled at a point collapsing the FRS resulting in loss of friction. Borehole straightness can be observed by shining a light up the borehole.

Alignment of the bolt with the borehole

Once the borehole has been drilled and the drill rod is extracted the operator needs to align the FRS with borehole. This is a critical part of the installation process. If the alignment is incorrect the bolt will be forced into the borehole at an angle (Figure 5). This creates a rotational force which may result in the bolt twisting during installation and reducing the amount of metal in contact with the hole (Figure 6). Twisted or corkscrew FRS have a reduced pull-out resistance.

This practice also leads to a higher rate of FRS bending prior to full installation. These FRS must be discarded resulting in wastage.

FIG 5 – Poor alignment of the FRS with borehole.

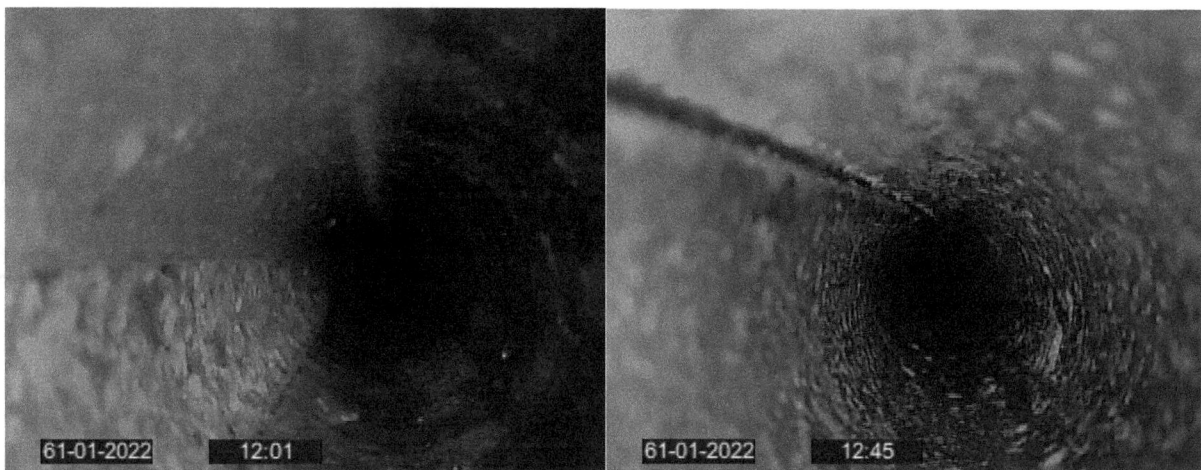

FIG 6 – Internal view of an FRS showing a well installed bolt (left) and twisted bolt (right).

Insertion dolly

There are two types of dollies in use, 39 mm dolly for installation of insert bolts and 47 mm dolly for installation of 47 mm bolts. Some operators may use the 39 mm dolly for 47 mm FRS as they find it easier to install the bolts. However, the smaller dolly has the potential to damage the collar and allow misalignment resulting in lower capacity.

Drive time and no rotation

Manufactures historically specified FRS drive times. However, these drive times may not be a good sign of installation quality due to the increased power of modern drifters. Site specific measurements should be completed. Rotation should not be used when installing FRS. Rotation increases the risk that the bolt may twist in the borehole, reducing the area of metal in contact with the borehole surface and ultimately the FRS load.

Instructions and assessment

Ground support instructions

Quality installation starts with ensuring the development drill operator has been made aware of all the installation requirements. The ground support scheme instructions are the method of transferring this information to the operator to achieve an effective ground support scheme.

Bolt and ring spacing

Ring spacing refers to the distance between each ring of bolts while bolt spacing refers to the distance between each bolt within the ring. These can be measured using laser distance measurers, counting mesh squares or by survey pick-up. Some mining software has the functionality to calculate both bolt and ring spacing from survey pick-ups or interpretation of Light Detection and Ranging (LiDAR) scans.

Task observations

Task observations or installation inspections are an especially important part of quality control. They involve watching the operator as they install the ground support scheme, comparing that against the relevant site procedure and then talking with them about any issues they may have. They can be completed by engineers or supervisors.

Items that should be included as part of the task observation include:

- Check the condition and storage of the ground support scheme consumables.

- Has mechanically scaling been completed?

- Bit size checks are performed.

- Bolts are aligned with the borehole during installation and rotation is not used.

- Installation angle and spacing of the rock bolts.

- Pull rings have been used on the minimum number of FRS.

- Any ground condition or installation quality concerns reported by the operator.

It is important to engage in a conversation with the operator during the task observations. This allows the engineer to gain feedback and a better understanding of the performance of the FRS and ground conditions and reinforces the procedure to the operator. This allows a relationship to be built up between operator and engineer. The operator gains trust in why installation quality is important, and the feedback helps both parties.

Pull testing

Pull testing refers to a process by which the FRS is loaded from the collar with a hydraulic ram connected to a calibrated gauge. Loading of the FRS continues to a pre-determined load is achieved, sometimes referred to the design load. If the FRS does not achieve the design load, either due to slippage of the FRS or damage to the retaining ring, this is referred to as a pull test failure. It is rare for displacement to also be recorded during routine pull testing.

Pull testing is needed on a routine basis to ensure that the most recently installed FRS are being pull tested. Only FRS that have had a pull ring installed during installation can be pull tested. Most sites aim for two pull rings per cut. It is important the position of the FRS with pull rings is varied between the backs, walls and shoulders so different orientations can be evaluated.

Generally, pull testing is completed monthly by the ground support supplier technicians, but some organisations complete pull testing in-house. Testing rates varying depending on the mine site but most generally aim to pull test between 1 to 2 per cent of the installed bolts. If different bolt lengths are being used ie 2.1 m and 2.4 m, it is important that the required testing percent is applied to both.

The following data should be collected for each test:

- Date and pull test operator's name.

- The location of the bolts; level, drive and location in the drive (backs, left hand shoulder, left hand wall, right hand shoulder, right hand wall).

- The FRS type and length being evaluated.

- The installation quality of the FRS and if there is evidence of twisting or poor alignment during installation, bolt twist can be estimated in 90° or 45° increments allowing plotting of pull testing data on a bolt twist curve for further analysis. If FRS collar direction is different from the rest of the FRS.

- Maximum load achieved and comments if FRS did not achieve the design load eg bolt slipped, retaining ring damaged, poor angle of installation.

- If an FRS had a pull ring but could not be tested then the reason should be given eg void behind plate, damaged bolt, angle of installation and could not attach the jack.

Best practice would be to uniquely number the test FRS, survey these FRS and review for spatial coverage across the mine.

Reviewing pull test results

Results should be reviewed once the pull testing is complete. They should be examined by calculating the number of FRS passed and failed. The total that the pass/fail mark is calculated from is the total number of FRS that were tested and must exclude the FRS that could not be tested.

It is still valuable to review the number of bolts that had pull rings installed but could not be tested as this can provide a guide to the quality of installation. If a large number of FRS are unable to tested it should be investigated as it is unlikely that required testing percentage can be achieved.

What pull load constitutes a pass or failure is site specific and is related to the sites ground support scheme design criteria. It is generally related to the 35–50 kN/m of embedment length. The following can be used a guide for some ground support designs:

- 2.1 m bolt length, 80 kN pass load.
- 2.4 m bolt length, 100 kN pass load.
- 3.0 m bolt length, 120 kN pass load.

Actions when high failures rates occur

When failures exceed a set level then actions are required by site to investigate and remediate the issue. The actions must be clearly stated in the site Ground Control Management Plan (GCMP).

It is recommended that sites prepare a Trigger Action Response Plan (TARP) included in the GCMP. A TARP defines the minimum set of actions needed in response to a deviation from normal working conditions. The actions are escalated based on the risk level. The TARP also outlines responsibility. This will vary between sites that are owner operator or contractor operator. It is good practice for the pull tester to give a verbal report to the geotechnical engineer and or the underground manager before leaving site.

Action escalation may involve the closure of a level/heading until the required action has been taken, eg reinstallation of FRS or grouting of FRS. It may also include a full investigation by the site geotechnical engineer and or external consultants.

Operator training

Observations of bolt installations show that with powerful drifters the installation times are often less than 10 seconds for 2.4 m FRS. Rotation is not recommended, but some operators use rotation as they find it easier to install the FRS. With large headings and the need to have personnel standing behind the machine jacks, operators find it hard to align to the hole when inserting the bolt leading to poorly installed bolts and damaged FRS.

Operator training to a high standard is critical to the performance of the ground support scheme. Unfortunately, there is not a standardised training program for ground support installation. The quality of training is dependent on the company, the mine site and the available experienced personnel. It is recommended that ground support installation training material and procedures be regularly reviewed by experienced jumbo operators, geotechnical engineers, and management to ensure they cover all the necessary information.

It is important that the site geotechnical engineers develop a relationship with operators. This is so they can impart their technical knowledge as well as receive the feedback from operators on installation practice and quality.

INVESTIGATING FRICTION BOLT INSTALLATION ISSUES

When there is a high failure rate that is not immediately resolved an investigation must be undertaken. This should be planned in such a manner so that the program progressively investigates what could be a complex and variable problem with no one solution or obvious reason for the failure.

Site investigations quite often involves the instruction to install more bolts with pull rings which are pull tested either by site personnel or when the bolt supplier representative arrives on-site. The trial may also include using different bit sizes or other suppliers' bolts. A problem that is often seen is little supporting information is collected such as bit size, hole size, installation practices. The investigation needs to complete a representative number of tests. Analysis of results is to be conducted statistically and to consider rock mass conditions.

A preliminary investigation is similar to the QAQC requirements:

- Review previous months pull tests. Is there a trend?
- Underground inspection of the areas of failure. Is there a geological reason behind the failures?
- Interview the operators or contract supervisors. Are they new operator? Has there been a change in supply for bits? Have they noticed a change in ground conditions? Consider if a system to track the pull tests results by individual operators can add value.

- Check bit sizes. The manufacturer may have made a change that increased bit size, or advertised size range is different to physical size, eg 43 mm bits measure as 44.5 mm to 45 mm.

- Check that the contractor has not changed supplier, or the FRS manufacturer has not changed design. This is potentially best evaluated from steel hardness and yield stress rather than metallurgical reports given the broad range of HSLA steels that can be used and that there is not a standard steel grade.

- Check if site sends bits for re-sharpening off-site and is this happening, eg bit sharpener has closed for Christmas holidays and site is using new bits only instead of re-sharps.

When the above has been completed, no clear issue has been identified and the problem still exists then bolt trials need to be conducted.

Completing friction bolts trials

The following process supplies guidance on how a trial may be set-up:

- Use a template for data to be collected. Ideally, as a minimum it should include test date, operator, drive location, rock mass conditions, bolt type and length, location installed in the drive, and the pull test load. Additional information includes bit size, drilling time, hole length, borehole diameter, bit size post drilling, drive time and operator comments.

- Consider when the bolts are to be pull tested, after installation or days or weeks later.

- Number of bolts to be trialled or pull rings to be installed (if part of development cycle).

- Are the trials going to be in situ, ie part of the development cycle or setting up a specific area with defined number of FRS. These may result in quite different outcomes and have a degree of varying control.

- Are different FRS from different suppliers being used? Understand the technical differences in each bolt, diameter, metal thickness, strip width and metal strength.

- Ensure that you can be present for the installation of the bolts and photographs are helpful to refer to.

It is preferred if testing and calculations use kN and kN/m opposed than tonnes, however this is dependent on the gauge of the pull test gear. Using kN enables a larger variation of detail to be captured. Additionally, measuring the true embedment length is necessary to calculate load per metre.

When conducting trials, it is important to make only one change to any variable at a time to fully understand the effect of that change.

In situ testing

This style of testing is attractive to management in that it causes less interruptions to development, however, is far harder to control and determine effects of individual changes. This style of testing can be useful in identifying operator error for example. Each operator can be given separate areas or tags to be attached to the pull ring plate and then results are compared by operator.

The situations where this may be beneficial are:

- To show operator variations.

- Different manufacturers bolts can be evaluated on a larger scale during the normal development cycle under general operating conditions.

- In cycle changes when varying one parameter to give a broader understanding of the problem.

- Variation in rock mass conditions and depth below surface.

The main limitation in these programs is lack of certainty in bit sizes and drive times, unless specifically recorded by a second person.

Specific testing trial area

The most effective method is to complete the trial in a specific area where variables can be controlled and measured. The recommended procedure to follow includes:

- Establishing the parameters to be evaluated.

- Understanding the material properties of the FRS to be installed.

- Selecting a trial area that has a typical rock mass for what you are trying to evaluate.

- Install enough bolts so you can get confidence from the results. A minimum of ten holes per trial variation is recommended.

- Holes should be spaced allowing pull tests to be practically conducted, 500 mm to 800 mm apart and no higher than 1.5 m off the floor. Holes to be numbered, Figure 7.

- Before drilling the hole select the bits to be used. Bits need to be measured before and after each hole completed using bit gauge with 0.25 mm graduations, Figure 3.

- Time the drilling of each hole. Holes should be angles slightly up to drain water. Measure hole length.

- If available, use a borehole camera to inspect the hole. This can be is useful to visually assess hole roughness. If borehole micrometre available measure the borehole diameter.

- Install the FRS as per site procedures. Installation time to be recorded.

- Inspection of each FRS for borehole straightness, twisting and gap closure.

- Complete the pull testing. There is value in recording deformation via the ram-extension to understand load deformation performance of the FRS.

- Review the results. Repeat trials may be required if additional variables are changed.

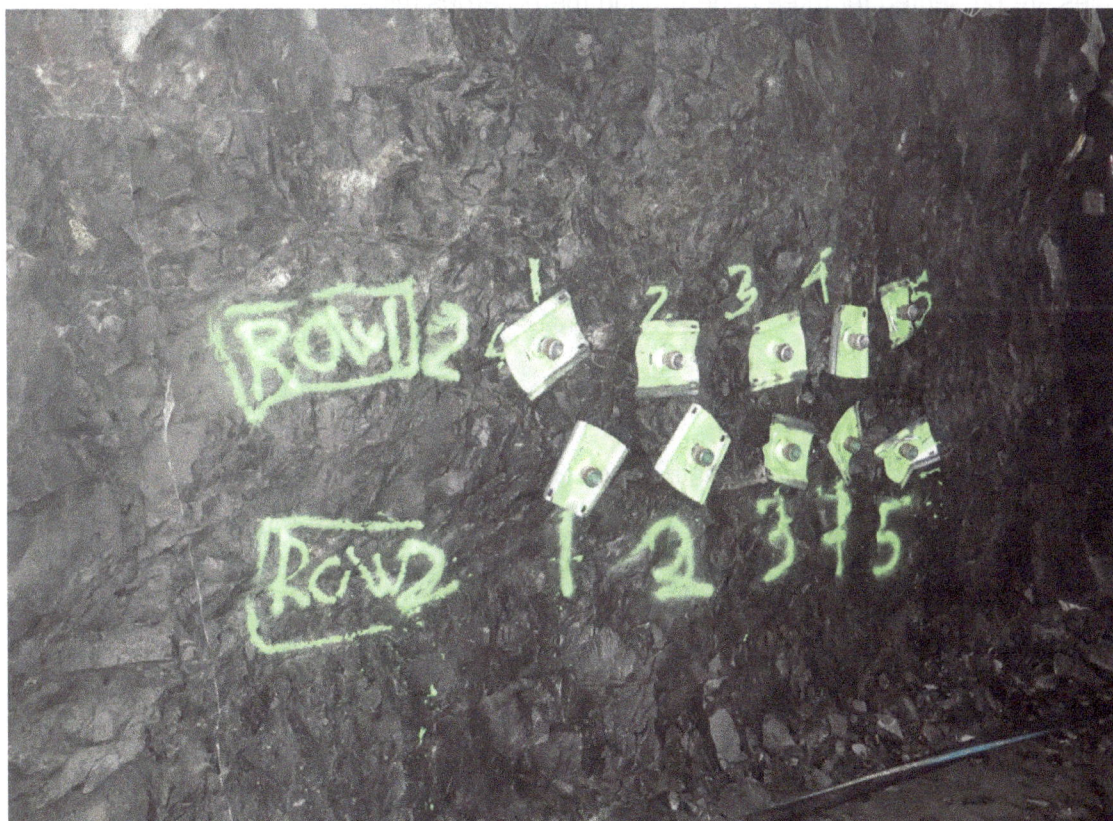

FIG 7 – Example of bolt layout for testing.

Analysis of results

It is important to analyse the results both statistically and in a practical manner. It is often easier to evaluate results shown graphically. Particularly if there were several trials completed for different bolts as shown in Figure 8. Once trials at the specific test scale are completed then larger *in situ* tests can be conducted. Results can then be obtained over a larger data set.

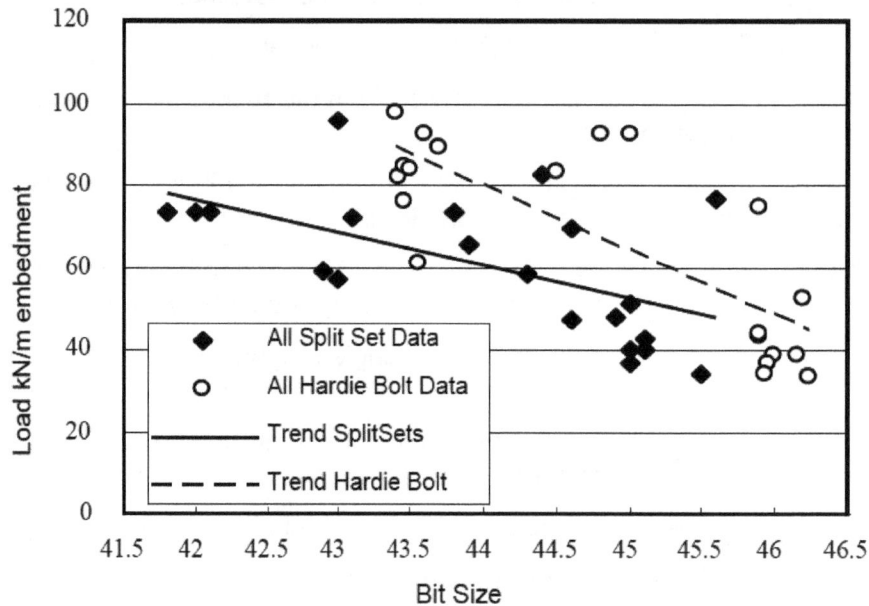

FIG 8 – Effect of bit size on load transfer (Figures 4–5 from Player, 2004).

CASE STUDIES

Examples are presented for three underground metalliferous mines.

Mine A

At this site, the pull tests results went from passing 90 per cent to 50 per cent passing within one month. The immediate reaction for the site was to install bolts a at significantly higher density to ensure the same performance in each excavation. The investigation involved:

- Three suppliers.
- 2.1 m and 2.4 m FRS.
- Two lithologies.
- Bit sizes – new bits 44.5 mm to 45 mm, three sets of re-sharps (43 mm to 44 mm, 42.5 mm to 43 mm and 41 mm to 42 mm).

Trials were conducted over a three month period with the following findings:

- It was discovered that bits supplied at 43.5 mm were in fact 44.5 mm to 45 mm in diameter and this had not been checked by the contractor.
- 42.5 mm to 43.5 mm bit size range produced the best results (Figure 9). This led to only re-sharp bits to be used.
- Operators were encouraged to slow down installation speed due to the data shown in Figure 10.
- Training of bolt installation provided to the operators.

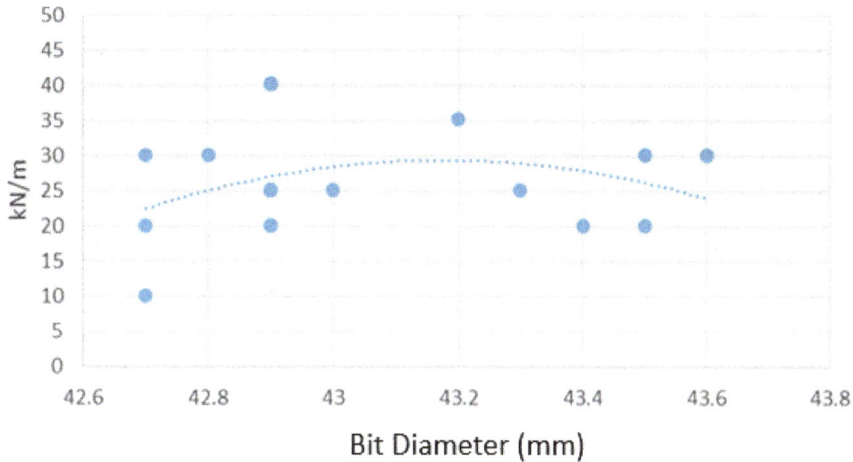

FIG 9 – Load transfer and bit diameter.

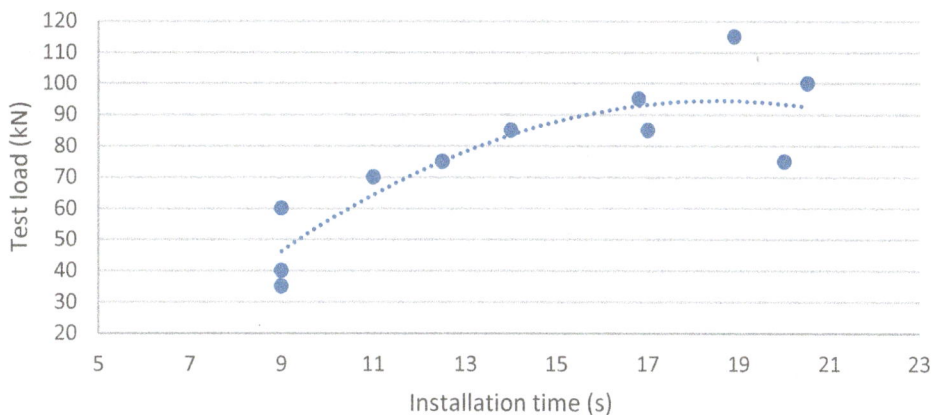

FIG 10 – Pull out load verses installation time.

Mine B

This site has extremely high strength rock mass. Reasons for lower than accepted QAQC rate was not as obvious site A and pass rates have stayed below the 90 per cent QAQC target. The first reaction by site management was to continue development and grout all bolts. This further changed by moving to Spirol bolts, which have a higher capacity, but also a higher cost. However, the change of bolt did lead to the 90 per cent pass target being achieved.

The site undertook a considerable number of trials with different bolt manufacturers. Metallurgical testing showed no material problems with steel quality.

Mine C

This site experienced a drop-in pass rate when they changed contractors. The rate of failure steadily dropped over a period of months. With increased emphasis on the quality of installation by site engineers and additional training the pass rate returned to the required standard. This example demonstrates the need for well-trained operators who understand the need for correct installation practices.

Case study conclusions

The authors experience with the case studies is that solving high FRS pull test failure rates can be complex and time consuming. With the case studies discussed all the sites had started their own investigations but had poor data collection and did not achieve a useful, conclusive outcome. More in-depth investigations were needed to find a solution. Both sites spent considerable time testing and trialling different manufacturers and hole sizes.

A question that the authors were presented with, is how common is pull test failures in the Australian mining industry? This is a tricky question to answer as most of the time this information is confidential, however the authors have worked across several sites that have had recent problems. These problems in some cases have taken months to resolve, in others, the problem was simpler, and a solution found relatively quickly.

When there is insufficient, ongoing analysis of the pull test results, poor installation may not be recognised for some time until a rockfall occurs. An important control is for ground support suppliers to provide feedback to both the technical and operational staff on the routine pull test results prior to leaving site. In addition, these representatives need to give feed back to the crews on installation quality and any improvements that need to be made.

CONCLUSION

Friction rock stabilisers are the most common bolt used in the Australian mining industry, with approximately 15 million units installed per annum. They are a relatively cheap and quick reinforcement system compared to alternatives. However, they are susceptible to poor installation quality and robust QAQC systems are needed to monitor, evaluate, and react to issues that may arise.

This paper highlights that FRS are not necessarily the 'easy installation' bolt it has a reputation for. It is not uncommon for high pull test failure rates to occur with the cause not aways easily identifiable. The primary concern is the safety of personnel working underneath these bolts. But there can also be a significant cost aspect with restricting access to drives, rehabilitation, interruption to production or changes to more expensive reinforcement systems.

Investigating the causes of the high pull test failure rates requires a rigorous and methodical process to ensure that the range of factors that may be contributing are evaluated.

ACKNOWLEDGEMENTS

The authors would like to acknowledge DSI Australia, all Friction Rock Stabiliser supplier site representatives, the jumbo operators that we have talked with over the last three decades for the knowledge flow in both directions, our fellow geotechnical engineers and clients for bring their ground control issues to us.

REFERENCES

Hassell, R C, 2007. Corrosion of rock reinforcement in underground excavations, PhD thesis, Curtin University of Technology, Kalgoorlie.

Hassell, R C, Villaescusa, E and Thomson, A G, 2004. Kinsella, B, Corrosion Assessment of ground support systems, *Proceedings from the Fifth International Symposium on Ground Support.*

Jones, E, Hassell, R and Power, N, 2022. Assessing Corrosion of Reinforcement on a Large Scale, *Proceedings from The Australasian Ground Control Conference, AusRock 2022* (The Australasian Institute of Mining and Metallurgy: Melbourne).

Lindeboom, H, 1978. Friction Rock Stabilizer, United States Patent 4,126,004, 21 Nov 1978, https://patents.google.com/patent/US4312604

Player, J R, 2004. Rock mass damage and excavation response – Big Bell longitudinal Sub level caving operation, Thesis for Masters of Engineering Science, Curtin University of Technology, West Australian School of Mines.

Scott, J, 1975. Friction Rock Stabilizers, United States Patent 3,922,867, 2 Dec 1975, https://patents.google.com/patent/US3922867A

Tilman, M M, Jolly III, A F and Neumeir, L A, 1984. Corrosion of Friction Rock Stabilisers in Selected Uranium and Copper Mine Waters, United States Department of the Interior, Bureau of Mines Report of Investigations RI8904, https://stacks.cdc.gov/view/cdc/10388/cdc_10388_DS1.pdf

Tomory, P B, 1997. Analysis of Splitset performance, Thesis submitted for Master of Applied Science, University of Toronto.

Villaescusa, E and Wright, J, 1997. Excavation reinforcement using cement grouted split set bolts.

Windsor, C and Thompson, A, 1993. Rock Reinforcement – Technology, Testing, Design and Evaluation.

The application of critical chain project management to sublevel open stoping at Olympic Dam

J Light[1], L Alford[2], G Capes[3] and T Woodroffe[4]

1. Senior Mining Engineer, BHP, Adelaide SA 5000. Email: jack.light@bhp.com
2. Superintendent Strategic Infrastructure, BHP, Adelaide SA 5000. Email: liam.alford@bhp.com
3. Manager Mine Planning, BHP, Adelaide SA 5000. Email: geoff.capes@bhp.com
4. Manager Continuous Improvement, BHP, Olympic Dam, SA, 5725.
   Email: tatum.woodroffe1@bhp.com

ABSTRACT

Since 2019, the mine planning team at Olympic Dam has been establishing and maturing a shift from deterministic to probabilistic scheduling techniques. This paper outlines this journey undertaken by the mine planning team, key learnings from implementation and further work required to increase the level of planning maturity.

Olympic Dam (OD) is a highly variable, polymetallic orebody with a vertically integrated value chain from operations to final product. The processing plant has a tight set of quality constraints which are managed through timing of stopes from varying parts of the orebody. Value is eroded for the operation when forecasted quality of the mine feed encounters disruptions from unplanned changes to the timing of stope starts.

As production has moved from the historic Northern Mining Area into the more geologically complex Southern Mining Area, average stope size has reduced, resulting in increased mining fronts required to meet production targets. This has created a problem of increased schedule complexity and difficulties in identifying and communicating resource priorities within a large set of individual stope critical paths.

The problem was approached by considering each stope as an individual project constructed by teams working across multiple projects at any one time. The operation could then be considered a multi-project environment with shared resources and enables the application of Critical Chain Project Management (CCPM) to the mine plan.

Key to CCPM is the addition of time buffers to protect the critical path of a project form upstream or downstream processes. Implementation of time buffers protects the stope start date against compliance to plan variance. The monitoring of each time buffer consumption, as a proportion of total construction activities completed enables proactive prioritisation of resources.

This resulted in an increase in the number of stopes starting in accordance with the mine plan, simplified communication of priorities to all levels of the operation, increased compliance to plan and a clear measure of schedule risk during planning routines.

INTRODUCTION

Olympic Dam (OD) is a polymetallic orebody, 570 km north of Adelaide, South Australia. The mining method for OD is sub level open stoping (SLOS). The economic minerals present in the orebody are copper, uranium, gold, and silver. Ore must be processed to final product on-site due to the presence of uranium within the orebody. This results in a vertically integrated value chain with several quality and quantity stocking points (Figure 1).

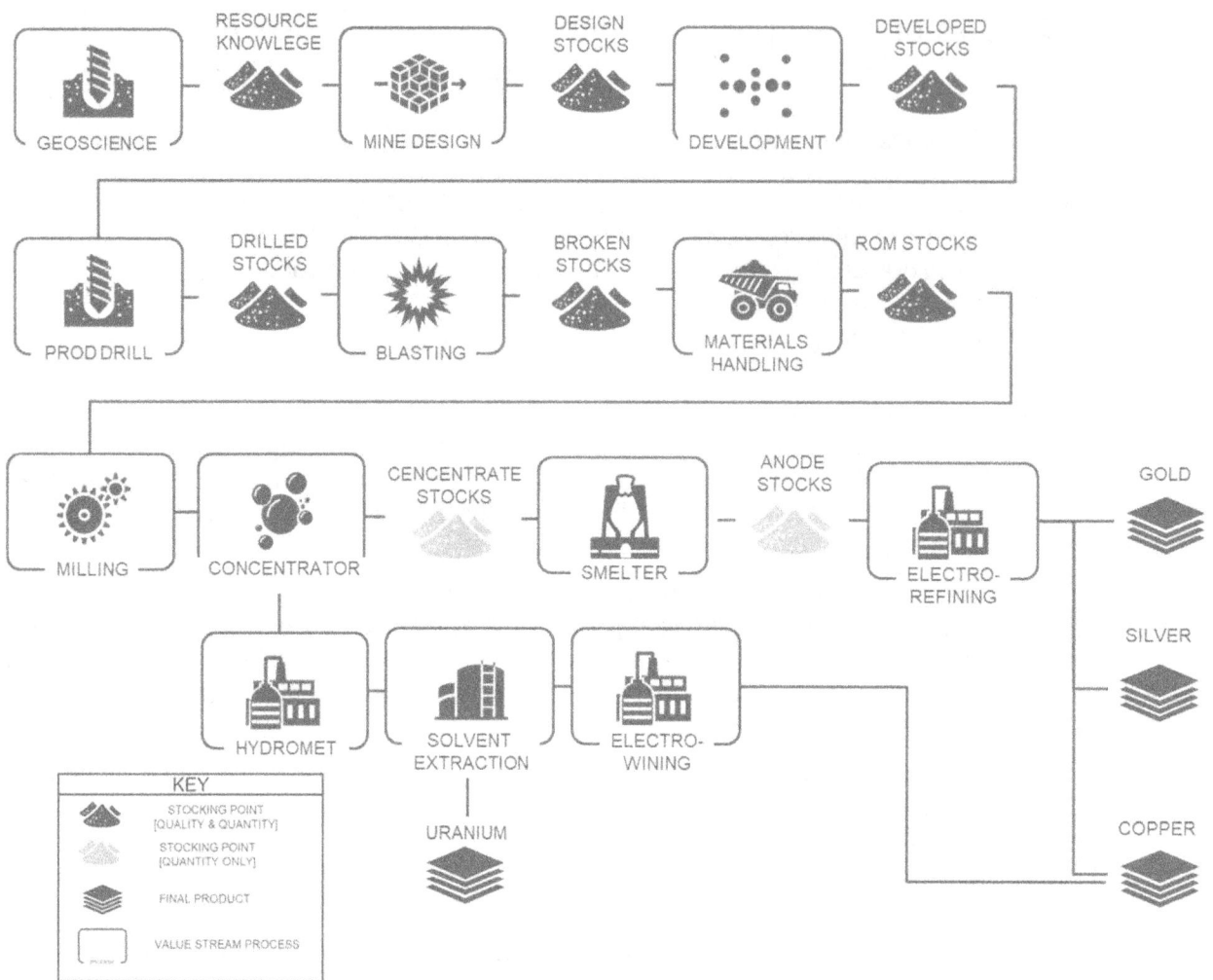

FIG 1 – Olympic Dam value chain.

This integrated value chain from mine to final product results in the mine having a tight set of delivery quality specifications. A lower value outcome for OD will occur if the mine does not deliver on these quality specifications. Along with maximising grade for the economic minerals the main quality constraints are:

- Iron to Silica ratio: This results in lower milling rates.

- Chlorite: If levels are too high gelling of the hydromet circuit can occur (Ehrig, 2013).

- Copper to Sulfur ratio: Lower smelter throughput if below working range.

OD uses the selective nature of SLOS to manage processing quality constraints. This is done through delivering specific stopes across a variety of quality ranges to combine into a single average blend which fits the processing plant requirements. The materials handling system (MHS) has limited ability to limited quality buffers between orepass, run-of-mine stockpile (ROM) and mill feeders.

The stope production plan is made through a mine planning process of life of asset, five-year plan, two-year budget, three month rolling plan, weekly plan and 24-hour shift plan. These plans outline the stope timings and resultant quantity and quality that will be delivered to the processing plant. A review of the quality and quantity is completed at each phase of the planning cycle to ensure it is within specification.

The time taken to deliver a stope to the ROM can take up to 5–10 years from mining block diamond drilling to stope production. Detailed stope design is required up to 24 months prior to execution with typical stope construction taking 6–12 months, resulting in limited ability to change schedule/ sequence within a two-year time frame. This long lead time for stope delivery and inability to bring new stopes into production quickly, results in a high reliance on the system to deliver specific stopes at their scheduled start dates.

Historically, variability in the mine value chain had been mitigated through stocking points (Figure 1). While this is an effective tool for managing production volumes, it fails to protect the mine feed from quality variations. This has suited the historic Northern Mine Area (NMA) resource with consistent ore quality within a limited range. Over the last ten years mining has been moving further south into the Southern Mine Area (SMA) which does not maintain the same properties as the NMA (Table 1).

TABLE 1

Key variations between NMA and SMA.

	Northern mine area	Southern mine area
Footprint	Small	Large
Materials handling	Centralised orepass and rail	Trucking
Stope size	Large	Small (~30–50% reduction)
Deleterious elements	Limited	Greater quantity
Ground conditions	Good	Variable
Quality of processed material	Consistent with limited range	More variable with wide range

The reduction in average stope size has a direct relationship to the number of stopes and mining fronts needed to maintain the target production rate (Equation 1). As production has moved into the SMA the number of required mining fronts has increased due to the relative change in stope size.

$$Mine\ Throughput\ (tpa) = Stope\ Size\ (t) * Mining\ Front\ Cycle\ Time\ \left(\frac{Stopes}{year}\right) * Mining\ Fronts\ (\#) \quad (1)$$

The geological expression of grade and deleterious elements within the SMA is more discreet and has a significantly wider distribution compared to the NMA. These increased concentrations of deleterious elements coupled with a higher number of stopes in production requires multiple stopes to be online simultaneously to balance the mine feed quality.

Bringing a stope into production and closing a stope out requires a large quantity of ancillary tasks and administrative safety controls including:

- set-up of stope ventilation management
- development of stope operating and mine block capital headings
- stope production drilling and loading
- stope tele-remoting and subsequent exclusion zones
- stope void exclusion zones
- backfill surface drilling and filling exclusion zones.

The duration of these stoping tasks is largely independent of stope size and requires multiple handovers between departments where the highest levels of process variability is experienced. An increase in the number of stopes required to meet mine production targets results in a significant increase in ancillary tasks within the schedule and therefore increased complexity and levels of process variability within execution.

These challenges resulted in a high level of required compliance to plan (C2P) for stope construction activities. Volume and Spatial compliance to plan were primary key performance indicators (KPI's) measured by the following definitions:

- Spatial compliance – volumes of physicals completed in the locations defined by the monthly schedule (eg development metres in a heading).

- Volume compliance – total volume of physicals completed within the month compared to the volume planned in the monthly schedule (eg total planned development metres).

Compliance to plan processes and metrics were required to mitigate for individual activity biases by supporting trade-off decisions between competing priorities as they inevitably arise within the complex mining environment. However, this resulted in misalignment where tasks were being completed to maximise compliance KPI's but did not always align to tasks on the critical path.

With widening quality range, higher reliance on planned stope starts, decrease in stope size, and increase in mining fronts, a compounding challenge arose. High complexity within the plan decreased likelihood of success as well as high reliance on meeting scheduled stope start dates to maintain quality constraints. The challenges above combined with the misalignment of KPI's resulted in the following symptoms:

- process variability impacting stope construction resulting in late delivery
- high levels of in-month changes to schedule to deliver material within processing plant constraints
- lack of transparency around mine priorities
- large numbers of priority tasks increasing difficulty to appropriately allocate resourcing
- hard to balance messaging around tasks of most benefit to the schedule versus departmental volume/spatial compliance targets.

An alternate mine planning approach was required to overcome these challenges. Critical Chain Project Management was identified as a potential tool.

Critical Chain Project Management

Critical Chain Project Management is a useful tool which can be deployed within single and multi-project environments with high levels of uncertainty and risks associated with them (Bhan and Whagmare, 2017). Heagney (2016) defines projects as 'Temporary in nature, have definite start and end dates, result in the creation of a unique product or service, and are completed when their goals and objectives have been met and signed off by the stakeholders'. Sublevel open stoping operations can be considered as a multi-project environment with shared resources, fulfilling the criteria above:

- Temporary – Stopes are a useful project for the duration of production.
- Start and end dates – Starting at development and ending at stope backfill.
- Unique Product – Individual ore quality as well as geology/geotechnical variations.
- Stakeholder signoff – Execution teams, processing teams.

The critical path of any project is the longest sequence of activities from start to finish which are required for project completion (Atin and Lubis, 2019). Based on the concept of Theory of Constraints (Goldratt, 1992), Critical Chain Project Management identifies the critical path of a project as the bottleneck and uses time buffers to ensure it is protected against upstream and downstream processes.

Typically, within a project, the duration of each task along the critical path is estimated by the individual departments. As outlined in Goldratt (1997) individual departments almost always overestimate task duration to include their own safety buffer to ensure handover dates are maintained. However due to the 'Student's syndrome' (where people start work just before a deadline), the internal safety buffer is wasted and tasks over run due to process variability. Critical Chain Project Management exploits the internal safety buffer by removing it from the individual tasks and aggregating it to protect the critical path. There are three types of time buffers used in the application of CCPM.

1. Project Buffer – Between the final activity on the critical path and project completion.
2. Feeding Buffer – Placed were the non-critical paths feed onto the critical path.
3. Resource Buffer – Where resources are required for a critical path activity. These are used for specific 'bottle neck' resources which will vary with fleet planning over time (Figure 2).

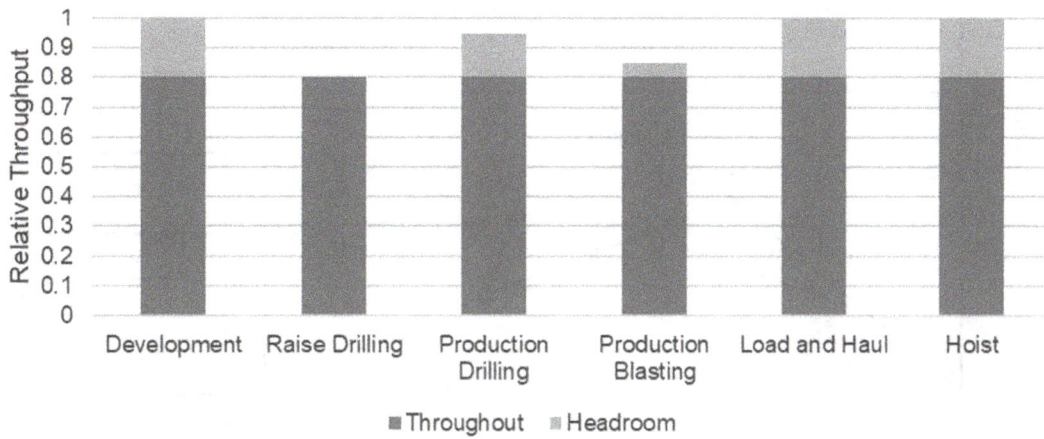

FIG 2 – Overall equipment capacities highlighting bottleneck resource (Raise Drilling).

Unlike a typical project environment, task durations within a mine schedule are created using input rates for equipment. These are often based on average daily actuals which do not include unplanned downtime, such as equipment failure, interactions, or unintended dependencies. This results in planned durations within a schedule having very limited internal safety buffer and leads to mine plans often becoming unachievable. Variance between planned duration and actual duration is measured as compliance to plan (C2P) and is reported and used as a tool to attempt to reduce plan variability in the future. To apply a project buffer, additional time buffers would need to be added to the end of a critical path in a mine plan.

Bloss *et al* (2020) showed that by plotting the amount of time buffer consumed relative to the percentage of stope construction complete along the critical path (Figure 3), categories could be defined to provide a current view on whether stopes are expected to reach the extraction phase late, at-risk (unstable), on-track (stable), or early (exploitable). The graph is referred to as a fever chart as colours are used to visually represent status (yellow = Early (Exploitable), Green = On Time (Stable), Red = At-Risk (unstable) and Black/Grey = Late).

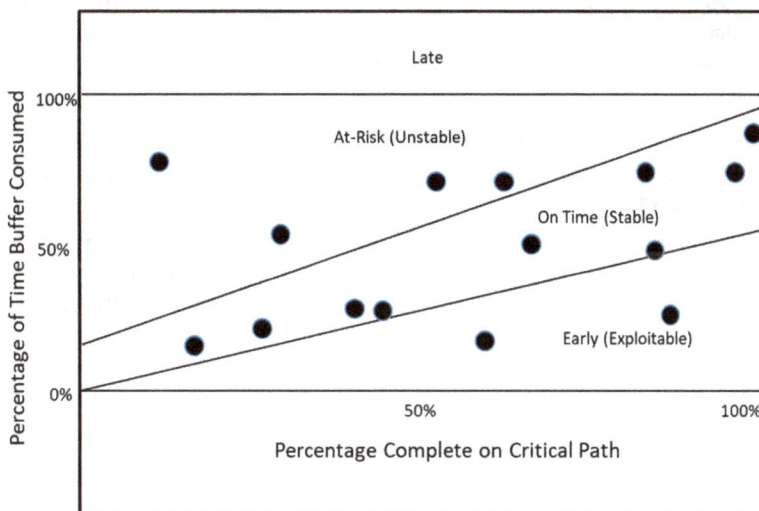

FIG 3 – Stopes plotted on fever chart used to assess overall schedule risk and resourcing priorities (Bloss *et al*, 2020).

On a monthly scale, tracking the path of individual stopes (Figure 4), can be used as a leading indicator of stope completion risk. This enables proactive management of critical path to allow for escalation of works before a stope falls behind. Direction of the trend line indicates rate of buffer consumption relative to critical path completion and determines if time buffer is being consumed faster than the planned buffer, being maintained, or building/recovering time buffer. A stope in the stable category which is trending into the at-risk area can be proactively managed prior to ensure it does not become at-risk.

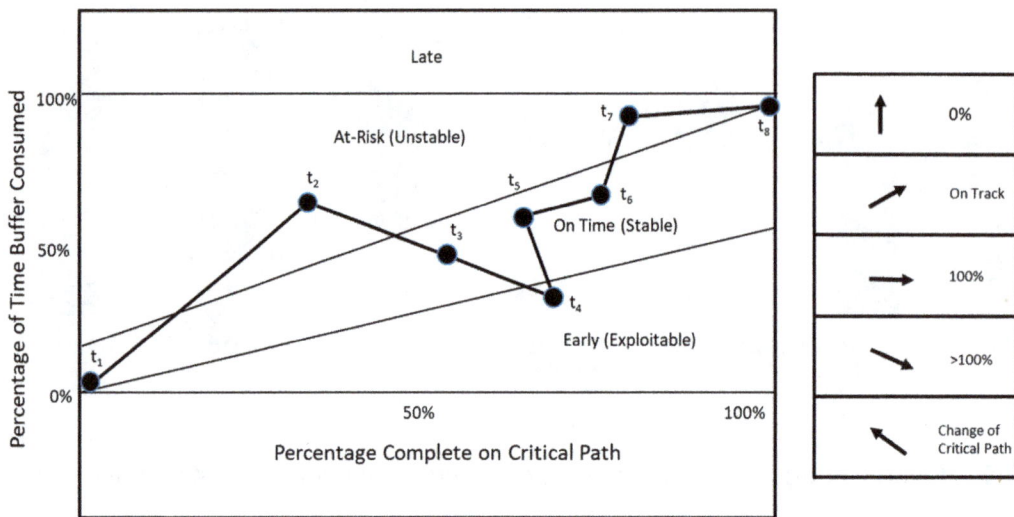

FIG 4 – Individual tracking of stopes – direction of line indicated rate of buffer consumption relative to critical path completion (Bloss *et al*, 2020).

METHODOLOGY

All stopes within the schedule were separated into individual mining fronts (project streams). Each front was defined by a geotechnically linked sequence of stopes where the preceding stope was required to be backfilled before production of the next stope in sequence.

The critical path for each stope was defined with stope extraction being the first critical path task for the next stope in sequence. Additional considerations including access, ventilation, services, and materials handling constraints were then overlaid and results in some cases with dependencies across multiple mining fronts.

Buffer placement

Time buffers were added between the final construction activity and the start of stope production (Stope Start). Schedule milestones (Buffer tasks) were added and set to 'as soon as possible' to enable a buffer to be built between stope start and the end of construction. All stope dependencies were linked to the buffer task. Non-critical paths were required to start sufficiently early to provide a 'Feeding Buffer' (Goldratt, 1997) to protect the critical path (Figure 5). Feeding buffers for non-critical paths were added by setting tasks to as soon as possible within a mining front. Typically, non-critical tasks were upstream constrained by preceding stopes. Unconstrained paths are set 'as late as possible' and use the final task as a 'start no earlier than' to generate a suitable fixed duration lag. To ensure appropriate prioritisation of stope planning and construction activities and protect against process variability, the buffer task date was communicated in lieu of the planned stope extraction start date.

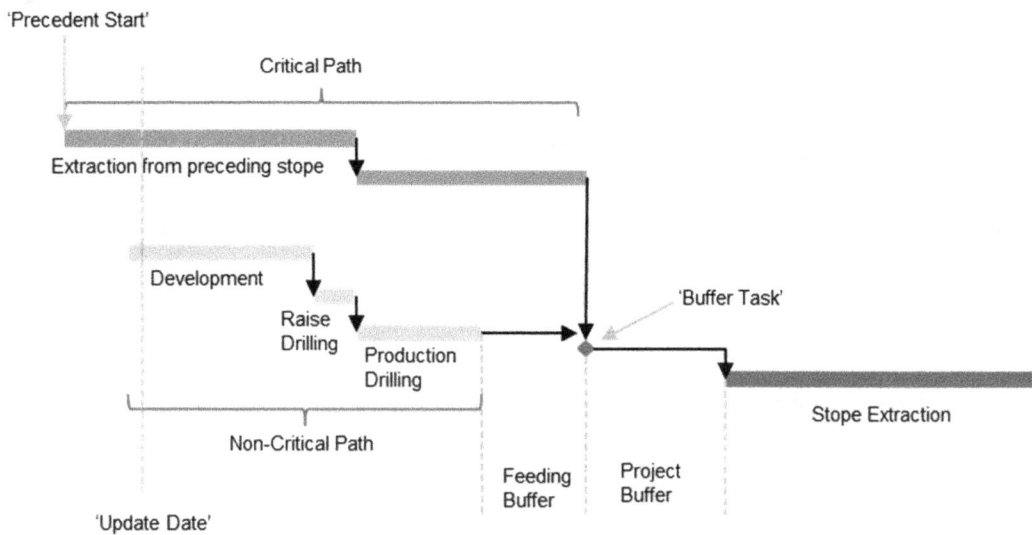

FIG 5 – Schedule structure to support time buffers between stope construction and 'Stope Start' (Bloss *et al*, 2020).

Buffer sizing

Rate-based scheduling results in task durations being driven through assumed rates and total physicals for each task. Input rates were calculated from historical average rates and therefore had limited internal safety buffer to exploit. Additional time was therefore required to be added to each critical path to generate a project buffer. Stope cycle time is directly linked to the mine footprint. It was important to ensure time buffers were correctly sized to minimise the cost associated with additional mining fronts while still ensuring the stope starts were sufficiently protected. Fixed duration buffers were unable to be used due to variations in the length of individual critical paths.

The time taken to complete a task can be broken down into two parts. The planned, deterministic task duration and the additional time taken to complete the task (Time Slippage) due to process variability (Figure 6). Compliance to Plan was used as a proxy for variability within the system. The additional time required to complete a task is proportional to the compliance to plan based on the following relationship:

$$T_{task} = Duration + Time\ slippage$$

$$T_{task} = Duration + [(1/C2P)-1]$$

$$T_{task} = Duration + Duration/C2P - Duration$$

$$T_{task} = Duration/C2P$$

$$T_{task} = Duration * C\ where\ C = 1/C2P$$

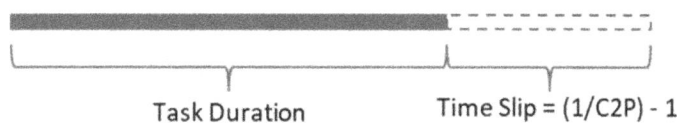

FIG 6 – Time slippage of any task is the additional time spent to complete due to compliance to plan (Bloss *et al*, 2020).

Historical analysis of schedule compliance was undertaken to estimate the size of buffers. The size of buffers has been adjusted over time relative to local factors influencing variability.

Defining stope risk

To be able to plot each stope on the fever chart outlined above, key dates were required to be extracted from the mine schedule.

Definition of a mining front's position was undertaken using the following schedule outputs and metric calculations:

- Precedent Start/Construction Start – Dictates the start of construction of each stope. In active mining fronts, this is the start in production of the previous stope. New areas without a preceding stope as a dependency were given a milestone task to define its construction start.

- Buffer Milestone – Used to define the end of the Critical path and the start of the stope start buffer.

- Update Date – Date on which the data is updated from the schedule. This indicates the start of the remaining activities.

- Stope Start Date – Indicates the end of the buffer and the start of stope production (also the start of construction for the next stope).

Using the input dates from the schedule the percentage of buffer consumed and percentage critical path complete can be calculated using the equations in Table 2.

TABLE 2

Calculation of metrics required to plot percentage complete and % buffer consumed.

Metric	Description	Equation
T_0	Initial duration of critical path	$Buffer - Precedent\ Start$
T_1	Current duration of critical path	$Buffer - Update\ Date$
$\%C_i$	% of Critical path complete	$\dfrac{(T_0 - T_1)}{T_0}$
C	Compliance factor	$\left(\dfrac{1}{Critial\ Path\ C2P}\right) - 1$
B_r	Required Buffer based on duration of initial critical path length	$T_0 * C$
B_a	Actual buffer remaining	$Stope\ Start - Buffer$
$\%B_i$	% Buffer Consumed	$\dfrac{(B_r - B_a)}{B_r}$

Application

As a starting point, buffers were applied based on a C2P of 80 per cent. A time buffer of 25 per cent $((1/C2P) -1)$ was applied to critical path construction activities for all stopes in the annual plan. Stopes were plotted on a fever chart which showed 62 per cent of stopes were in the at-risk category. This indicated the stopes were consuming too much time buffer and were at risk of late delivery. Resourcing was able to be adjusted to ensure 'at-risk' stopes were prioritised and reduce the number of at-risk down to 25 per cent (Figure 7). Critical Path C2P inputs originally estimated as 80 per cent have been modified to include unique area-based conditions such as more difficult ground conditions with higher process variability.

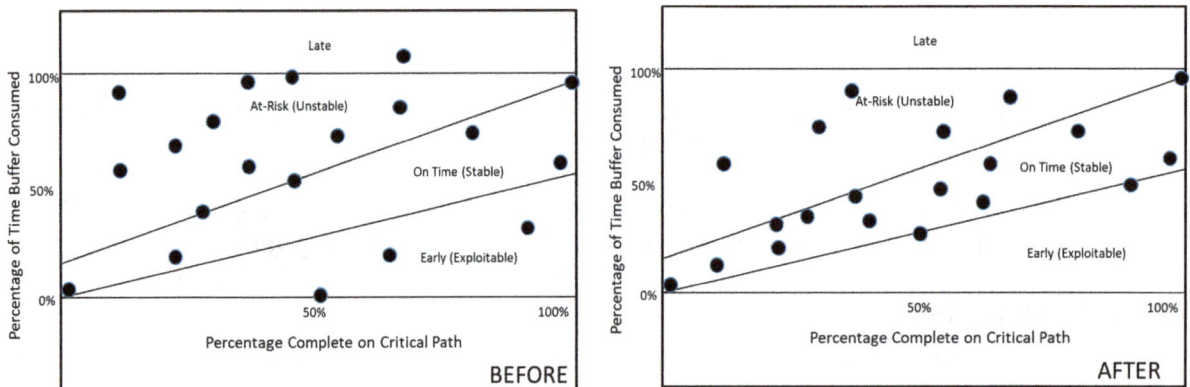

FIG 7 – Use of fever chart to prioritise at-risk critical paths and resulting risk profile (from 62 per cent to 25 per cent at-risk stopes).

OUTCOMES

The outcome of this process was demonstrated over a 12-month period as shown in Figure 8. There was a decrease in overconsumption of the planned time buffer. Key changes were made including a shift from volume and spatial compliance to critical path compliance as a KPI. Critical path compliance considers progress of planned tasks against actual progress along the stope critical path. A large shift in performance was also noted through the advancement of critical path management into shift planning.

Mining Front Monthly Buffer Performance

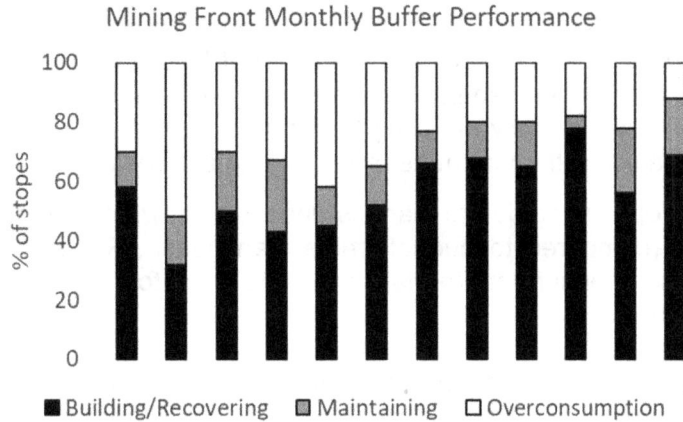

FIG 8 – Sample 12-month period of buffer performance.

This was then supported by a monthly Plan, Do, Check, Act (PDCA) routine on this process, where variance reasons were documented, and corrective actions assigned. A pareto chart was developed looking at factors influencing critical path stope performance. Some of the main factors have included stope overbreak, ground support rehabilitation incomplete, planning site re-prioritised in month, backfill hole blockages, and emerging interactions from balancing vent loading where activities were not completed on time.

Focus on the critical path within each mining front aims to increase the stope turnover rate within each mining front. This can result in less mining fronts required to maintain a constant mine throughput. Investment decisions can then be made to trade-off sprint capacity within resource pools compared to the capital investment required to develop, maintain, service and ventilate additional mining fronts.

The tool has also been expanded to include upstream considerations for diamond drilling, infrastructure planning, and mine design timing aiding in team resourcing and prioritisation of work packages.

LEARNINGS

The application of CCPM has been implemented at OD since 2019. Some key learnings to date in the application of the process have been the following.

Application:

- To track, visualise and report on buffer consumption, the schedule is required to be exported and post processed within Excel to generate dashboards. The additional step makes iterating plans based on project buffer consumption difficult and is prohibitive to expanding the critical chain methodology into tracking feeding buffers and resource buffers within the schedule.

- Non-critical paths with small feeding buffers, need to be managed closely to ensure they do not become the critical path.

Inputs:

- A high confidence in mine design layout, regular schedule hygiene and peer review is required to ensure activities are linked and consider area based geotechnical, ventilation, and other operational considerations to remove mine planning induced variability.

Planning and feedback routines:

- Establishing and maintaining the broader team understanding on value chain excellence (Bloss *et al,* 2020), Theory of Constraints and critical chain processes is required for this process to be successful.

- Strong feedback loops between execution and planning teams are required with continuous improvement through a Plan-Do-Check-Act cycle. Actions are tracked and iterated into the next planning cycle to protect critical path and stope delivery.

- The use of the fever chart in communicating to shorter term planning horizons needs to be considered. Sole use of the fever chart for setting short-term priorities without sufficient context can result in additional unplanned interactions. Plan on Plan routines are required to provide this context for weekly and daily planning to inform trade-off decisions where resources can be redirected from a site with more buffer remaining to a site with more risk of on time delivery.

- Re-prioritising of stope timing due to plan disruptions or downstream quality requirements will result in absolute (compared to budget) mine plan health. Realignment of time buffers is required in this case to ensure appropriate priority is given to right activities.

Further work:

- Work with industry partners to integrate tools into mine scheduling software and reduce post-processing requirements.

- Tracking of resource buffers for critical path activities is yet to be implemented within the schedule. Ongoing work to understand constraint resources, unlocking additional capacity where required and work to ensure bottle neck moves to designed bottleneck unit.

- Currently assuming equal process variability across entire critical path (eg 80 per cent C2P), next iteration will be to breakup into individual activities based on rate probability.

CONCLUSION

Having a standardised process to plan stope performance which provides visibility of stope start risk enables the mine to take corrective action to recover or exploit individual mining fronts with respect to the available buffer across multiple mining fronts. Key planning and feedback routines enable the mine to identify gaps against targets and to develop and manage improvement projects to support improved outcomes. This process will continue to mature with ongoing application and feedback from different stakeholders.

REFERENCES

Atin, S and Lubis, R, 2019, November. Implementation of critical path method in project planning and scheduling, in *IOP Conference Series: Materials Science and Engineering,* 662(2):022031 (IOP Publishing).

Bhan, M A and Whagmare, A P, 2017. Monitoring of Construction Project through Buffer Management, *International Journal of Engineering Sciences and Research Technology.*

Bloss, M L, Capes, G W, Seib, R, Alford, L V, Light, J L, Minniakhmetov, I and Nielsen, C, 2020. Value chain excellence–managing variability to stabilise and exploit the mine value chain, *Mining Technology,* 129(4):187–205.

Ehrig, K, 2013. 'Gelling': An operational challenge due to mineralogy, 10th Annual SA Exploration and Mining Conference.

Goldratt, E M, 1992. *The Goal,* The North River Press.

Goldratt, E M, 1997. *Critical Chain,* The North River Press.

Heagney, J, 2016. *Fundamentals of project management,* Amacom.

Project and execution of the excavation for a haulage shaft with raise boring techniques

M Orunesu Preiata[1], C Seymour[2], I Thin[3] and A Pretorius[4]

1. General Manager Technical Services, IGO Ltd, South Perth WA 6951.
   Email: marco.orunesu@igo.com.au
2. Principal, Dempers and Seymour Pty Ltd, Balcatta WA 6021.
   Email: cseymour@dempersseymour.com.au
3. Principal Geotechnical Engineer, KSCA Geomechanics Pty Ltd, Perth WA 6076.
   Email: iain.thin@kscageomechanics.com.au
4. Group Project Manager, IGO Ltd, South Perth WA 6951. Email: andre.pretorius@igo.com.au

ABSTRACT

The Odysseus Project is part of the high-grade Cosmos Nickel Complex that lies 30 km north of Leinster. The Odysseus definitive feasibility study (DFS) was approved by the Western Areas' board in October 2018. Early works commenced in January 2019 with rehabilitation of the existing underground infrastructure. In June 2022 Western Areas was acquired by IGO Ltd. The project is now well advanced to meet its target of first concentrate.

The project included the construction of a haulage shaft. The design of the hoisting equipment and the shaft barrel evolved from the initial DFS design stage to the final execution, as the data were made available from the initial investigation to the data gathered during execution (although this did not change from its initial concept).

The design philosophy for the haulage shaft incorporated, from the beginning a multidisciplinary approach with inputs from three main areas involved in such projects, namely excavation techniques, geotechnical assessments, and infrastructure construction. This interaction is a fundamental component for the success of such a critical project. In fact, each of these three areas influences each other in a cyclical manner, triggering a continuous review of the project inputs and outputs, while maintaining the original project goals: the design of an infrastructure able to deliver the desired throughput.

The throughput dictates the infrastructure that needs to be built, which in turn controls the size of the shaft. The key assumption applied to this is that the rock mass can accommodate the excavation void, and the excavation technique with its tolerances and limits that ultimately need to be verified back with the initial assumptions used for the infrastructure design.

The aim of the paper is to present a case study of the experience gained with the Odysseus hoisting shaft and to propose a possible logic path for similar projects elsewhere.

INTRODUCTION

The Cosmos Nickel Project is located about 1000 km by road and 790 km by air, north-east of Perth, in the Northern Goldfields District of Western Australia. Western Areas was recently acquired by IGO Ltd in June 2022. Cosmos includes high-grade deposits at Cosmos, Cosmos Deeps, Prospero, Tapinos, AM1, AM2, AM4 and Odysseus North and South massive sulfide, and AM5 and AM6 deposits – where production mining has not yet started in Odysseus North and South and AM6. Previous open pit and underground mining occurred between 2000 and 2012. A DFS on the Odysseus deposit was completed in 2018, and included a hoisting shaft of 1020 m of length.

Site investigation work for the shaft identified the first 14 m of material from surface was unstable, being made up of soft rock. Establishing a boxcut was selected as the preferred option, with the boxcut excavated with a trencher. Although seen as a novel methodology to establish a boxcut, from an overall risk mitigation perspective considering cost, schedule, impact on surrounding assets (the presence of an active haul road approximately 50 m to the west, a rehabilitated waste dump 25 m to the east and foundations for the raker legs 20 m south of the sub-brace area) and ground stability, the use of the trencher was deemed the preferred approach.

The excavation of the shaft barrel with raise boring techniques took place in 19 months, starting with the piloting of Leg 1 in September 2020, and ending with the completion of the reaming of Leg 2 in April 2022. Stripping and concrete lining of the final 140 m of the shaft is planned to commence December 2022.

WHY A SHAFT

At a scoping study level, an initial trade-off study was carried out in 2016 to evaluate which material transport system should be used for Odysseus on a total cost basis, starting from the initial production target of 6.4 Mt (including Odysseus, AM6 and remnants). This production target had been identified by the previous owner, with an assumption to expand the production target to 7.9 Mt of ore and 2.0 Mt of waste. Three scenarios were evaluated:

- Truck 100 per cent. Conventional truck and haul via decline only.

- Truck and hoist. Decline and raise bore barrel, strip and line a shaft.

- Hoist 100 per cent. Blind sink shaft and hoist all waste and ore.

Capital and operating costs were estimated for the study for each scenario using industry databases, also considering the necessity to dewater the mine, sustaining capital, trucking, and owners' costs (no ventilation costs were considered at the time). The final cost estimate is summarised in Table 1.

TABLE 1

Scoping study – summary of capital expenditure and benchmark.

Item	m.u.	Truck 100%	Truck/hoist	Hoist 100%
Tonnage hauled	Mt	9.9	2.5	
Tonnage hoisted	Mt		7.4	9.5
Years of hoisting	Yrs		9.2	10.5
Capital expenditure (M$)	A$ M	69.5	132.0	92.0
Operating expenditure (M$)	A$ M	101.6	59.4	36.1
Total expenditure	A$ M	171.1	191.4	128.1

Item	m.u.	Truck 100%	Truck/hoist	Hoist 100%
Additional capital cost	A$ M	-	62.50	22.50
Additional operating costs	A$ M	-	42.20	65.50
Discount factor for operating only	%	-	0.5	0.5
Additional operating costs (2016 actualised)	A$ M	-	21.1	32.8
Delay in achieving full production	Yrs	-	-	1
Notional revenue delay (2016 data)	A$ M	-	-	20
NPV impact	A$ M	-	41.4	9.8

Considering the truck 100 per cent scenario as a base case, and assuming an average discounting factor of 0.5 for 10 years, the final comparison between the three scenarios is summarised in Table 1.

The pre-feasibility study (PFS) then progressed based on this assessment with the truck only scenario achieving a Reserve of 4.1 Mt at 2.3 per cent Ni, with a production of 750 ktpa, and the shaft as ventilation intake only.

The technical and economic landscape in between the various studies started to change with increased fuel costs, and increased health and safety focus on diesel emissions, leading to a review of the material transport concept to adopt in the DFS at the beginning of 2018. A trade-off study was then conducted based on preliminary schedules showing a target from 700 to 900 ktpa potentially achievable with a raise bored shaft.

The trade-off study was conducted on a net present cost (NPC) basis, using the fuel price at the time, an interest rate of 8 per cent, capital and operating costs estimates from experienced mining contractors in both underground haulage and hoisting, zero CPI, and a tax rate of 30 per cent. For the capital, only the cost of the winder, headframe, shaft equipping and conveyances were used as the excavation cost for the shaft barrel would be common to the two options. The results of the study are summarised in Figure 1, where the net prevalence for the shaft option is evident. Based on these results, the DFS proceeded with the shaft option.

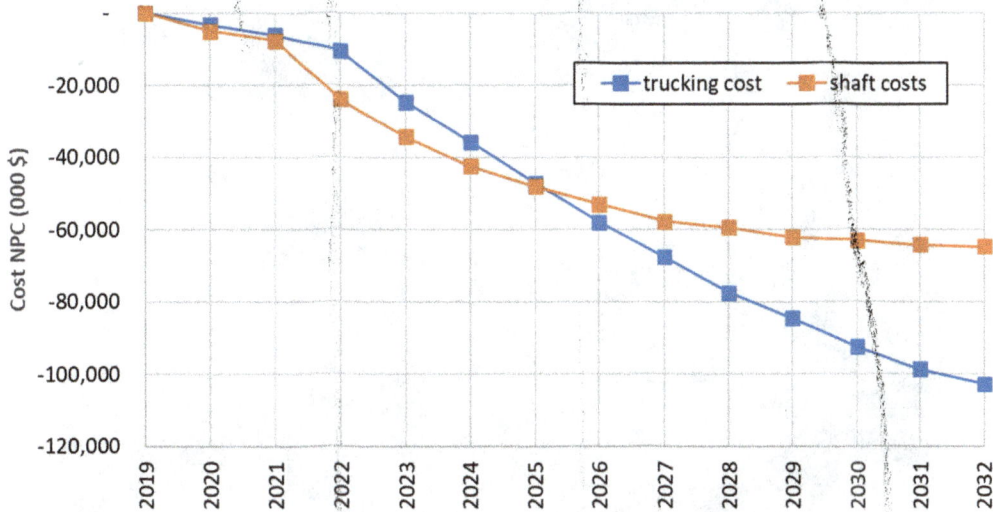

FIG 1 – NPC curves for shaft and trucking options.

SHAFT DESIGN

Shaft length

The length of the shaft was determined on the basis of the underground capital expenditure, and the cash flow expected from the project that required an early revenue to minimise the capital payback period.

The mining method selected for Odysseus North and South was top-down centre out, longhole stoping with paste fill. This requires the upper parts of the orebody were in production sooner than the bottom, not requiring the full development of the underground capital before commencing production; and, if the shaft ended at the top of the orebody, this would allow for an early start for hoisting. If the loading station was positioned at the base of the orebody, all the capital access had to be to developed first, delaying the hoisting and therefore the payback. Based on these considerations, the loading station was located at the top of the orebody at 1000 m below surface (see Figure 2), and the excavation was divided in two legs of 664 m (Leg 1) and 356 m (Leg 2). Leg 1 from surface to the mid-shaft level at 9833 mRL and Leg 2 from the mid-shaft level to shaft bottom at 9468 mRL.

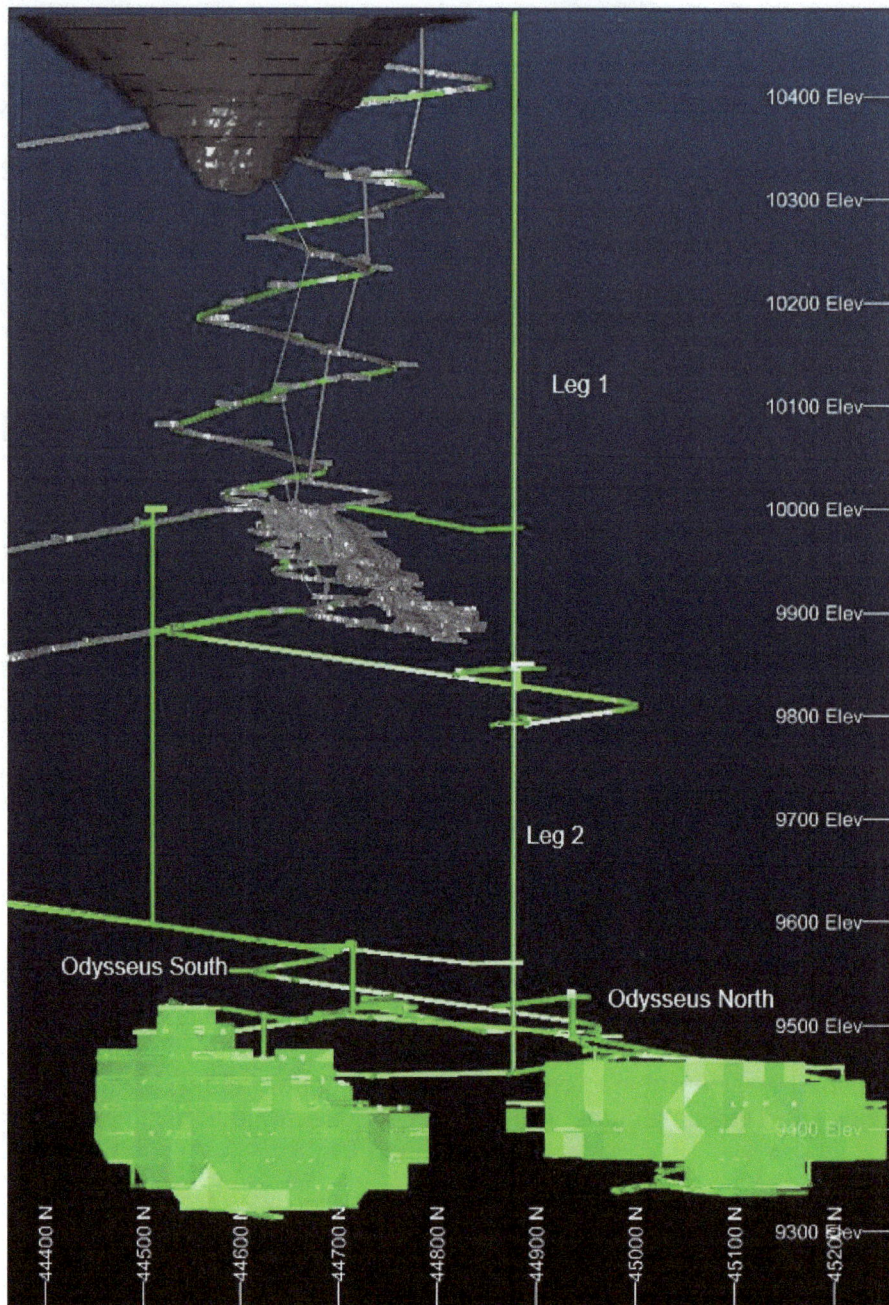

FIG 2 – Final Definitive Feasibility Study design (looking west).

Another advantage of the selected shaft length was related to the depth of the orebody and the knowledge of the geological and geotechnical environment of a deep mine in general at a DFS level. For deep mines, the amount of drilling available can be limited due to its costs and complexity. Therefore, the geological and geotechnical environment is not well understood at the beginning of the project, and, not linking completely the revenue stream to the full capital expenditure, de-risk the project.

Dimensions and tolerances

At the initial preliminary design of the shaft, the excavation was based on a diameter of 5 m, with the assumption to install a new hoisting system. Once established, it was possible to use the Impala 12N infrastructure (see sections below and Figure 3), this diameter was increased to 5.5 m to accommodate this infrastructure.

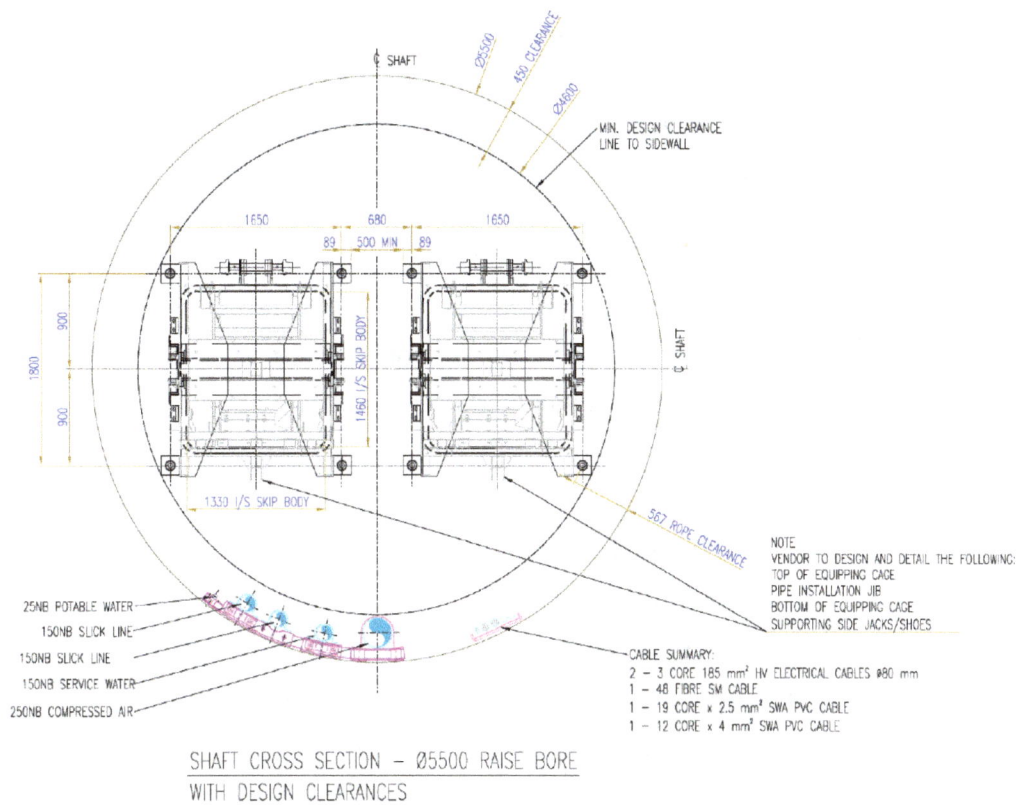

SHAFT CROSS SECTION – Ø5500 RAISE BORE
WITH DESIGN CLEARANCES

FIG 3 – First shaft diameter design.

This initial diameter was evaluated against key parameters that ultimately influence this dimension: excavation tolerances, dynamic behaviour of the conveyances and their clearance against the shaft walls and characteristics of the rock mass.

In consultation with recognised experienced contractors, and after reviewing the available technology at the time of the tender, a tolerance of 300 mm radius on the shaft central axis was adopted for the entire length.

The dynamic behaviour of the conveyances was analysed according to the norm ISO (CD) 19426-7 (still in draft at the time).

All the clearances were verified, except the one against the shaft walls. With the tolerance adopted, the effective diameter available was 4.9 m, and the analysis produced clearances between 63 mm to 147 mm, in relation to the air velocity, with a minimum clearance required by ISO (CD) 19426-7 of 150 mm.

To satisfy the tolerances required, and considering the geotechnical constrains, especially in the last 100 m of Leg 1 (see geotechnical sections), the decision to adopt an initial diameter of 5.7 m was made for the tender. Also considering the raise boring excavation techniques adopted, a final excavation diameter of 5.8 m was selected to account for the cutter changes.

DFS initial design

The DFS design of the shaft and rock hoisting system was based on the following metrics as presented in Table 2.

TABLE 2

DFS hoisting system parameters.

Design life	10 years
Nominal annual production	1.1 Mtpa (Ore and waste @150 mm)
Nominal shaft depth	1000 m
Shaft final diameter	5.5 m – raise bored in two legs, meshed and bolted
Guidance system	Steel wire ropes (four per conveyance)
	Spear guides in headframe and loading plat
Skip loading system	Measuring flasks (same as skips)
Shaft function	Production, refrigerated air intake
Plats	Chip handling (temporary)
	Mid-shaft (temporary)
	Skip loading
	Bottom
Ventilation	Downcast (250 kg/s, 14–16°C mixed) via brace openings and concrete duct between bulk coolers and sub-brace
Production winder	Double drum (3.6 MW)
Conveyances	12.5 t skips
	Maintenance and inspection basket
	Shaft equipping cage
Headframe	45 m high steel structure (Integrated sky-shaft)
Shaft services	Piped compressed air and water services
	Paste delivery line
	Power (11 kV) cables
	Communication and control cables

Lower capital option Impala 12N shaft

During the feasibility study, an opportunity was presented to purchase a winder, winder house and headframe from Impala 12 North (12N) Shaft (Figures 4 and 5) in Rustenburg, South Africa. Detailed evaluation of this opportunity showed that it provided the best economical outcome for the project and was adopted as the final solution. It included:

- Purchase of the equipment from Impala.

- Dismantling of the equipment at Impala 12N Shaft and transporting it to the various business partners.

- Assessment and refurbishment of all the mechanical and structural components to meet Australian Standards.

- Engineering and supply of a new winder control system to Australian Standards and Legislation, utilising existing equipment where possible (motor, transformers etc).

- Shipping of equipment from South Africa to Australia.

- Storing equipment in the lay-down yard on-site.

- Transporting the equipment to the shaft area, and construction and commissioning of the system.

Note: In the early stage of the DFS, a mid-shaft loading station was adopted to enable production hoisting while the shaft development continued. However, this approach did not prove economical and was dismissed during the implementation stage.

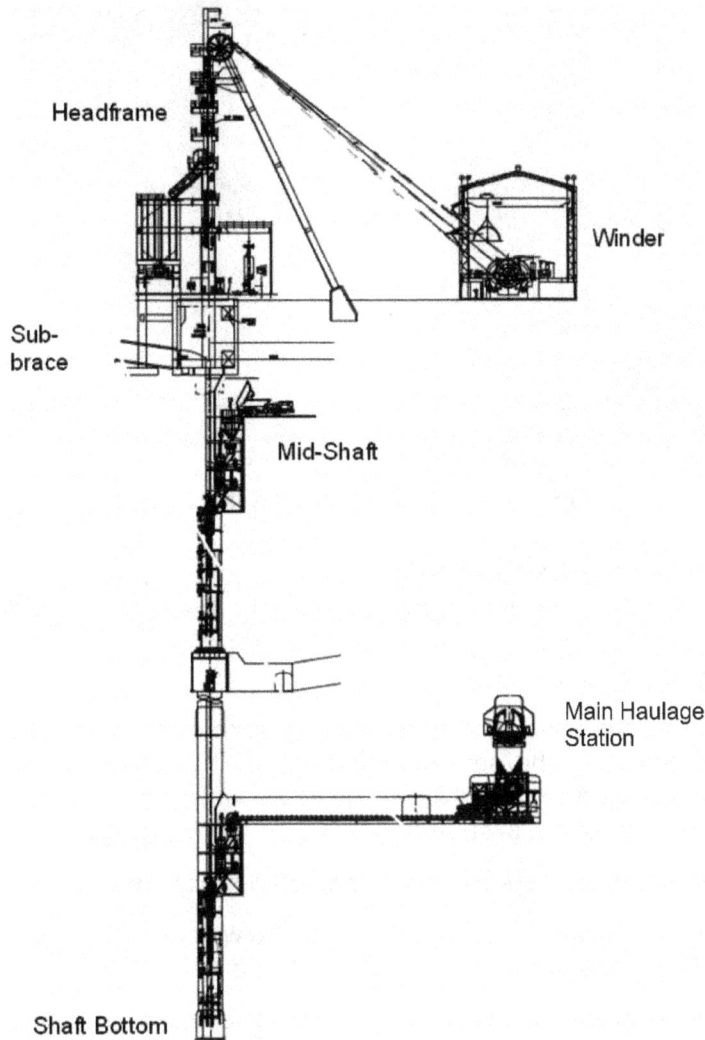

FIG 4 – Impala 12N with mid shaft loading station.

The Impala 12N Shaft was 8.5 m diameter blind sunk and concrete lined with a concrete brattice along the length of the shaft to a depth of 860 m . The shaft went into production in the 2004 and was utilised to hoist platinum ore by means of 15 t skips, and provide ventilation infrastructure in the form of upcast and downcast airways. On surface, the shaft was serviced by a set of ventilation fans and a bulk air cooler supported by a refrigeration plant. The facility had a design life of eight years and was retired more or less at the end of this period. During the life of this facility, the installed equipment operated efficiently at designed capacity with no major incidents recorded.

FIG 5 – Impala 12N side and front views.

Impala 12N shaft adaption to Odysseus

In general, the basis of the Cosmos design was to replicate the Impala 12N design where possible, while ensuring that the facility aligned with Australian Standards and Western Australian regulations. Specifically, the following were addressed:

- Winder house – The steel structure and cladding were dismantled, repaired where required, and erected at Cosmos in the same configuration. The brick walls in lower section of the building were replaced with poured *in situ* concrete walls and minor changes were made to accommodate the new control system and equipment cooling requirements.

- Winder house overhead crane – The crane and crane rails installation were replicated.

- Winder – A new control system was supplied with the winder to align with Australian Standards and Legislation. Rope attachments were checked and re-certified.

- Headframe – The integrated headframe structure was dismantled, refurbished, modified to suit the Cosmos construction methodology and shipped to Cosmos in the largest possible sections. Minor modifications were made to the sheave deck to suit the revised shaft diameter. The sheaves were fitted with new polymer inserts which will be grooved on-site during commissioning.

- Ropes – A set of 51 mm diameter non-spin ropes were supplied with the winder and a full set of eight 42 mm diameter non-spin stranded ropes were supplied for the guidance system. These guide ropes will be replaced with half-lock coil ropes with the first replacement, estimated to be within the first three years to provide a longer lifespan of around ten years.

All supporting infrastructure have been designed to integrate with these facilities.

GEOTECHNICAL ENVIRONMENT

Once the first infrastructure design was finalised, the next challenge was selecting an appropriate shaft location. An understanding of the Odysseus geotechnical environment was further advanced as part of the DFS. The geotechnical work completed specifically for the haulage shaft consisted of reviewing a historically drilled and geotechnically logged drill hole (CGT058) for a proposed fresh air raise that was completed by the previous owners, the drilling and geotechnical logging of a new dedicated drill hole (WAGD010), assessment of the likely inflows to the planned shaft, Acoustic Televiewer (ATV)/Optical Televiewer (OTV) surveys, laboratory strength testing, completion of a raise bore stability analysis, numerical modelling, and an external peer review. The geotechnical environment associated with the haulage shaft can be summarised as a largely competent with exposure to weak/very weak material, within a moderate to high stress regime.

Geotechnical domains

The main rock types associated with the haulage shaft are Felsic porphyry, Felsic Schist, Felsic volcanic and Granite Pegmatite intrusions. While Ultramafic komatiite (referred to as the Western Ultramafic and Cosmos Ultramafic) units exist at Odysseus, predominately associated with the North and South orebodies, the positioning of the shaft was such that it avoids the vertical development. In terms of the main geotechnical domains for haulage shaft, these were simplified as the Felsic and Pegmatite units.

Stress regime

A total of 15 stress measurements were completed within the Cosmos underground mine complex between the PFS and DFS. These measurements encompassed the use of HI Cells (× 7), WASM Acoustic Emission (× 6) and e-Precision Acoustic Emission (× 2) which covered a depth range below surface of 262 m to 1127 m. Of these 15 measurements, five were completed at or below 1000 m below surface (covering the 1000–1127 m depth range). The relationship of principal stress magnitude with depth and orientation is presented in Table 3.

TABLE 3
Cosmos-derived principal stress gradients and orientations.

Component	Magnitude (MPa)	Trend (°)	Plunge (°)
Sigma 1	0.0672 × depth (m)	130	10
Sigma 2	0.0479 × depth (m)	40	0
Sigma 3	0.0335 × depth (m)	310	80

Numerical modelling

The Odysseus underground mine areas represent two sizeable footprints within a rock mass of variable quality and competency, geotechnically significant structures and a moderate to high pre-mining stress state. Collectively, this creates a complex geotechnical environment which presents the potential for many mining-induced stress/strain interactions to exist. While there are many stress/strain mining-induced interactions, their understanding (and subsequent management) can be achieved through appropriate 3D numerical modelling. Numerical modelling completed for the DFS incorporated the use of FLAC3D software, selecting a mine-wide non-linear finite element modelling approach with the work completed externally through Itasca.

The numerical modelling was completed at a global scale to enable the interaction of mine-scale excavations, geological units, rock mass conditions, structures and overall sequence to be examined. Although Itasca numerically assessed the suitability of the mine design (which was progressively modified through four design iterations), excavation stability, and mining-induced seismicity, the work package also encompassed a focus on the stability of the haulage shaft (as well as all the various ventilation related raises).

The FLAC3D modelling approach incorporated a total of four simulations. The four different models evolved as a result of applying variations to the mine design once the subsequent model results had been interrogated and any identified issues addressed – essentially four cycles of geotechnical and planning optimisation. With each of the four simulations, the same key/long-term infrastructure was applied, which included the entire extent of the haulage shaft from surface to full depth.

SHAFT LOCATION AND GEOTECHNICAL INVESTIGATIONS

The review subsequently proposed an alternative position that increased the minimum stand-off distance where the shaft barrel passes the Cosmos Deeps stope mining area. Figure 6 shows the PFS location for the fresh air rise (FAR) and the alternative position for the haulage shaft in relation to Cosmos Deeps. With the original position of FAR, the minimum stand-off distance was 30 m (ie the distance between the shaft barrel and the closest Cosmos Deeps stope – during the DFS, it was not known if the stope voids had been backfilled or not).

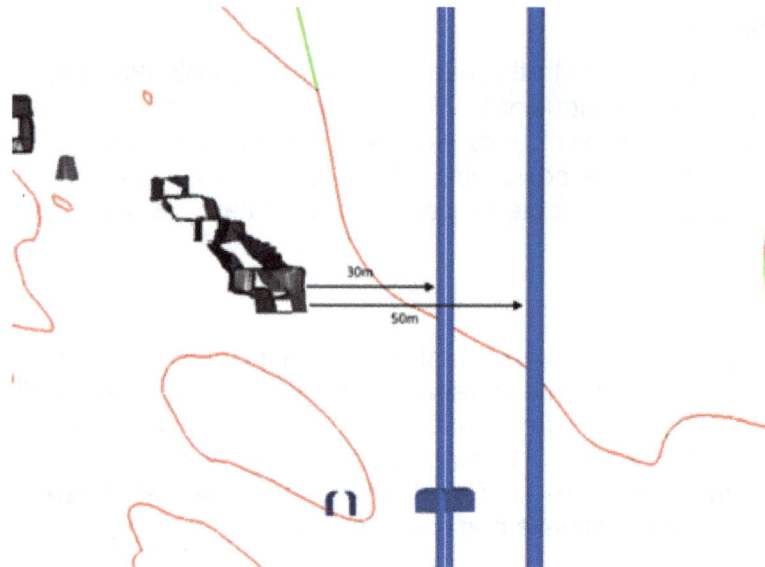

FIG 6 – Cross-section looking east showing the PFS location of the FAR and the proposed alternative location.

With the proposed updated position, the shaft was moved a further 20 m to the north-east, resulting in a 50 m stand-off distance from Cosmos Deeps. However, while increasing the stand-off distance of the shaft from the known stope mining voids, it was important to understand whether this then creates any potential adverse interactions or impacts, such as known structures and/or ground conditions.

The PFS and alternative shaft locations were both positioned within the competent Felsic Porphyry/Felsic Volcanics rock mass. However, it was predicted (based on the understanding at that time) that Pegmatite intrusions would be intersected (Figure 7) – this would be the case for both shaft positions.

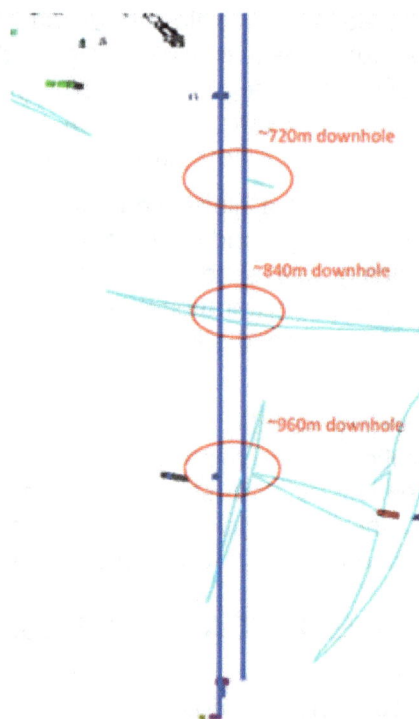

FIG 7 – Cross-section looking east showing the predicted intersections of Pegmatite intrusions for the PFS location and proposed alternative haulage shaft location.

The geotechnical log (from the historical drill hole CGT058) and core photographs were reviewed as part of the raise bore stability assessment for the original FAR location. Nothing was noted from this data to suggest there would be any significant rock mass stability issues associated with the development of the ventilation raise. Some discing was noted in the core (depth interval 832–840 m), which corresponded to the second leg of the haulage shaft. The discing appeared to be associated with a Pegmatite unit intersection (refer to Figure 7).

The quality and condition of the CGT058 drill hole core at the equivalent depth adjacent to the Cosmos Deeps (approximately 460–620 m and corresponds to the first leg of the shaft) did not reveal or suggest that the rock mass had suffered from stope mining-induced effects.

The proposed alternative haulage shaft position was also compared against the known major geological faults. Based on the predicted major geological fault wireframes, it was determined that there would no interactions with the shaft (Figure 8).

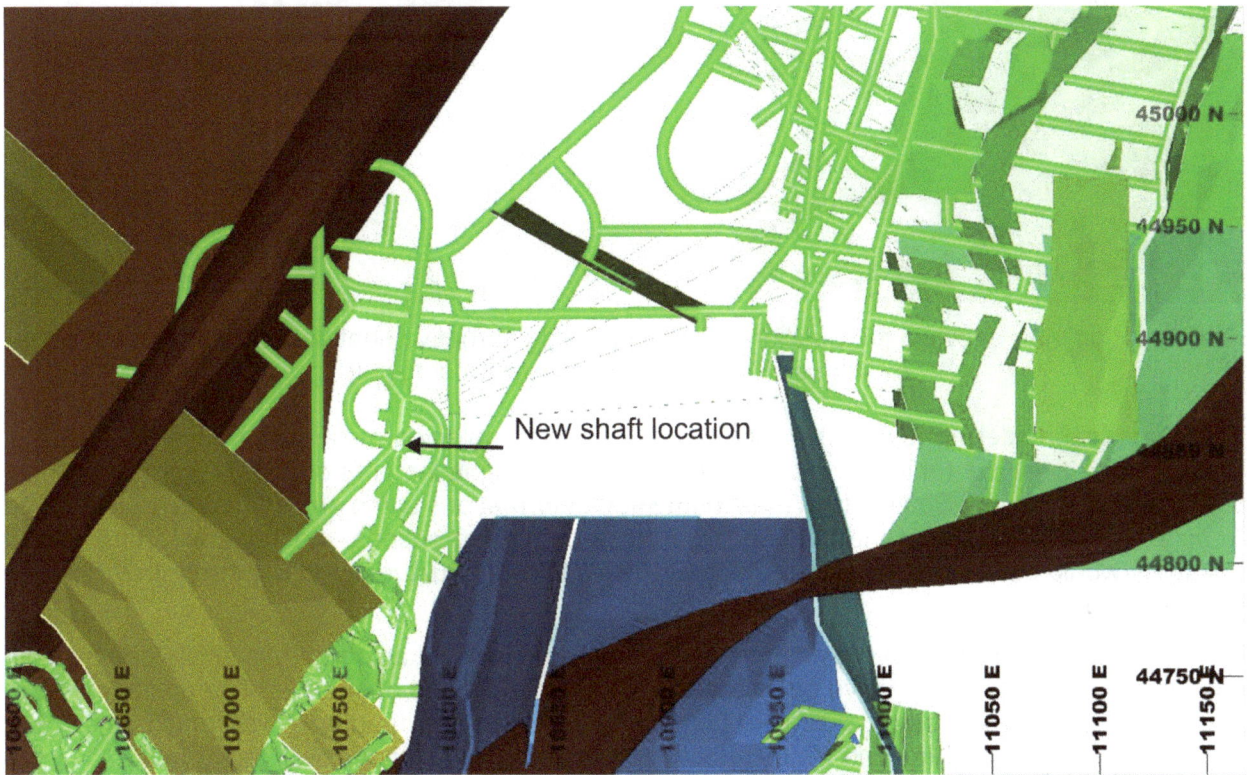

FIG 8 – Plan view comparison between the proposed alternative shaft location and the known major geological faults (note the bright green solids represent lateral development with all coloured solids representing the known faults).

While the focus of the shaft location review was concerned with underground interactions, the alternative position from a surface perspective was also considered. Moving the shaft to the alternative location resulted in the surface position sitting approximately 30 m away from an existing waste dump (Figure 9).

FIG 9 – Plan view showing the position of the proposed alternative shaft location in relation to the existing surface waste dump and Cosmos pit.

The numerical modelling results that incorporated the final selected mining sequence showed that there was no significant mining-induced influence (either displacements or stress) from the mining of Odysseus North and South – total maximum shaft deformations at the end of mining are predicted to be 20 mm at the base of the both shafts. Across both shaft options, the key zones of potential instability would be the presence of geotechnically significant structures and low strength Pegmatite intrusion intercepts.

The modelling confirmed that the alternative shaft position was suitable from a stability perspective, and will not result in any negative geotechnical issues, interactions or impacts. The additional 20 m stand-off distance was seen to place the LOM shaft infrastructure in an improved environment that is outside of any mining-induced stress effect influences that would have resulted from the Cosmos Deeps stope production.

Test work completed

The geotechnical investigation incorporated geotechnical rock mass logging of core from the new WAGD010 diamond drill hole and processing of ATV and OTV survey data to provide rock mass and structural data respectively. Laboratory testing of samples selected from the drill hole were also completed and incorporated into the assessment. The objectives of the work were to provide:

- A geotechnical assessment for raise boring based on the industry standard practice methods established by McCracken and Stacey (1989) and updated by Peck, Coombes and Lee (2011) and Penney, Stephenson and Pascoe (2018).

- Recommendations for excavation and support strategies that would ensure that the final diameter is achieved.

Representative core samples were selected from the drill hole and sent to the Western Australian School of Mines (WASM) for the following tests:

- Uniaxial Compressive Strength (UCS) tests with modulus (24 tests)

- Direct Shear Tests (DST) (13 tests)

- Raise Bore Index (RBI) (19 tests)

- Single Stage Triaxial Tests (three confining pressures, 37 tests in total)
- Slake Tests (three tests with two cycles per test)

Laboratory strength testing – summary results

The UCS tests were conducted on samples from within the proposed raise bore length. The results are presented in Table 4. The main geotechnical domains for haulage shaft (Felsic and Pegmatite) are classified as Very Strong with an average UCS test results of 167 MPa and 145 MPa respectively.

TABLE 4

Summary UCS laboratory test results.

Sample interval (m)		Length (mm)	Diameter (mm)	Mass (g)	Unit weight (kN/m³)	Lithology	Secant Young's modulus (GPa)	Secant Poisson's ratio	UCS (MPa)	Failure mode
From	To									
14.60	14.90	209.5	82.9	2912.1	25.2	Felsic Porphyry	7.0	0.211	46	Quiet
79.60	80.00	210.4	83.0	3047.5	26.3	Volcanics (with Quartz veining)	57.3	0.283	20	Quiet
122.20	122.45	209.4	83.0	2992.1	25.9	Felsic Porphyry	70.0	0.239	161	Violent
157.75	158.00	152.4	60.7	1275.2	28.4	Felsic Porphyry	60.9	0.170	130	Violent
252.35	252.60	153.0	60.9	1238.8	27.3	Conglomerate	74.7	0.232	142	Violent
332.73	332.97	152.9	61.0	1184.8	26.0	Felsic Porphyry	98.1	0.395	70	Quiet
351.55	351.80	152.9	60.9	1181.3	26.1	Felsic Porphyry	78.2	0.274	369	Violent
394.25	394.55	153.0	60.7	1185.2	26.3	Felsic Porphyry	79.1	0.277	178	Violent
449.80	450.00	153.0	61.0	1188.4	26.1	Felsic Porphyry	75.9	0.249	323	Violent
470.10	470.35	152.5	60.9	1161.6	25.6	Granite Pegmatite	79.4	0.287	125	Quiet
525.90	526.15	152.9	61.0	1191.7	26.2	Felsic Porphyry	67.4	0.294	176	Violent
551.10	551.35	154.4	61.0	1201.3	26.1	Felsic Porphyry	84.4	0.142	100	Quiet
565.65	566.00	153.3	61.0	1182.0	25.9	Felsic Porphyry	76.3	0.274	168	Violent
648.10	648.20	153.0	60.8	1195.2	26.4	Felsic Tuff	67.0	0.220	288	Violent
747.46	747.75	154.4	60.6	1238.4	27.3	Felsic Undifferentiated	62.5	0.302	176	Violent
783.63	783.90	154.4	60.7	1264.9	27.8	Felsic Undifferentiated	63.7	0.275	210	Violent
827.72	828.07	153.4	60.9	1182.2	26.0	Granite Pegmatite	74.0	0.525	161	Violent
850.50	850.80	153.0	60.8	1182.7	26.1	Felsic Undifferentiated	61.7	0.216	299	Violent
878.70	879.00	152.3	60.7	1163.8	25.9	Granite Pegmatite	50.7	0.155	150	Quiet
894.90	895.30	153.4	60.7	1230.5	27.2	Orthocumulate (Peridotite)	76.3	0.364	254	Violent
906.60	906.85	152.5	59.7	1189.2	27.4	Orthocumulate (Peridotite)	44.2	0.352	28	Quiet
1040.80	1041.10	153.4	60.8	1191.1	26.2	Felsic Undifferentiated	69.5	0.179	169	Quiet

Slake durability tests

Slake durability index (SDI) tests were completed specifically targeting the upper, near surface section of the shaft location. In the case of the Odysseus shaft, the SDI tests were completed to investigate the influence of the water used during pilot drilling on the superficial rock masses.

Table 5 summarises the SDI results, demonstrating that slaking was not going to be an issue for the Odysseus Shaft (High Durability).

TABLE 5

Slake durability index test results from drill hole WAGD010.

Sample number	Depth (m)	SDI – 1st cycle (%)	SDI – 2nd cycle (%)	Durability
1	20.70–21.05	99.7	99.6	High
2	41.36–41.71	99.3	98.9	High
3	55.85–56.22	99.9	99.9	High

Packer testing

Historically high dewatering rates (around 50 L/s) were needed to maintain dry conditions during mining at the Cosmos Nickel Mine. The hyper-saline groundwater was managed using an extensive mine water disposal system. The disposal system included several mine water receptors, comprising purpose-built ponds and dams, and completed pits. A large proportion of the groundwater inflows to the mine were associated with aquifer intersections in the various vent raises and services holes intersecting the underground operations.

As part of the investigation into the suitability of the proposed site for the haulage shaft, Groundwater Resource Management (GRM) was engaged to assess the likely inflows to the planned shaft. The assessment was based on a review of the recovered core and hydraulic testing using downhole packers, in the dedicated geotechnical drill hole WAGD010.

The drill core review identified 32 transmissive features that could be associated with groundwater inflows to the planned shaft. However, only one feature was considered to have a high inflow potential (ie nominally greater than 5 L/s), based on recorded water circulation losses during drilling. The packer testing program utilised a staged approach. The results indicated hydraulic conductivities ranging from 7.7×10^{-5} to 6.3×10^{-4} m/d. The low hydraulic conductivities were consistent with the low inflow rate during the constant head tests, which ranged from 0.38 to 0.24 L/s. The outcomes from the tests were used to estimate groundwater inflow rates to the haulage shaft. These ranged from 10 L/s at the start of dewatering to between 3 and 5 L/s in the longer term.

Acoustic (ATV) and Optical (OTV) televiewer surveys

Downhole ATV and OTV surveys were conducted in the drill hole from 18 m to the end of hole. The data was analysed creating structural domains based on depth and similar structural characteristics. Set orientations were used in conjunction with the joint shear strength parameters from the site tactile assessment and the laboratory testing to carry out wedge stability analyses for the series of downhole structural domains.

The structural data interpreted from ATV and OTV surveys, was separated into 66 structural domains based on similar structural characteristics such as joint frequency and orientation.

Geotechnical rock mass assessment

Geotechnical logging of the rock mass parameters for drill hole WAGD010 was undertaken using domain geotechnical logging methodology (Dempers, Seymour and Harris, 2010). The logged data was subsequently processed for the raise bore shaft assessment by applying the McCracken and Stacey (1989) method. This method is used to derive the Raise bore Quality Index Q_R and the maximum unsupported raise bore diameter $Span_{max}$.

Crown and short-term stability analysis

The analysis of both the crown and the short-term stability in drill hole WAGD010 showed multiple logged domains that fall below the required $Span_{max}$ for the proposed 5.7 m shaft diameter. While this indicates that these domains will be unstable, most are relatively short domains (<1 m) and are surrounded by very competent rock.

However, there were domains in both the face and short-term analysis that fell below the 5.7 m $Span_{max}$ requirement. These domains were within the 500 m to 600 m and 880 m to 1030 m depth sections in WAGD010 in both the face and short-term stability graphs as highlighted in Figures 10 and 11 respectively. While there are domains of good rock within these two zones, it is the multiple domains of poor rock that dictate the maximum achievable raise bore diameter and which give rise to the recommended three-lift raise boring strategy.

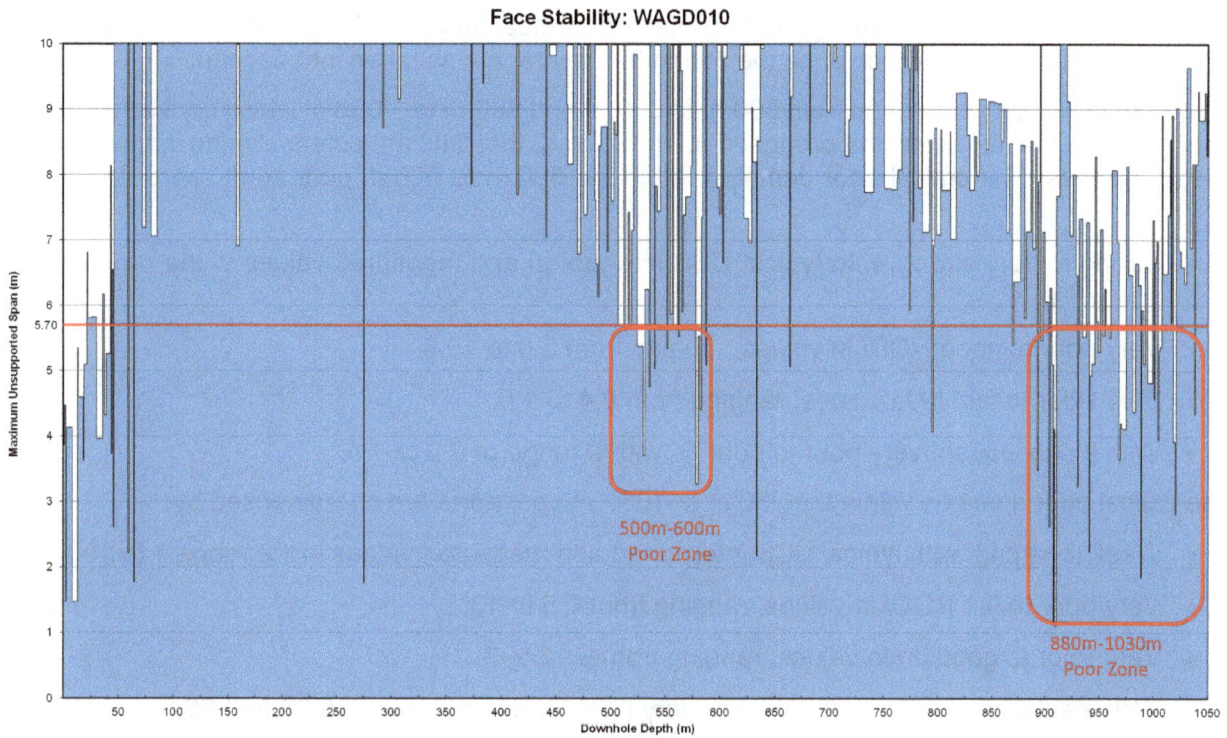

FIG 10 – Maximum unsupported span for face stability.

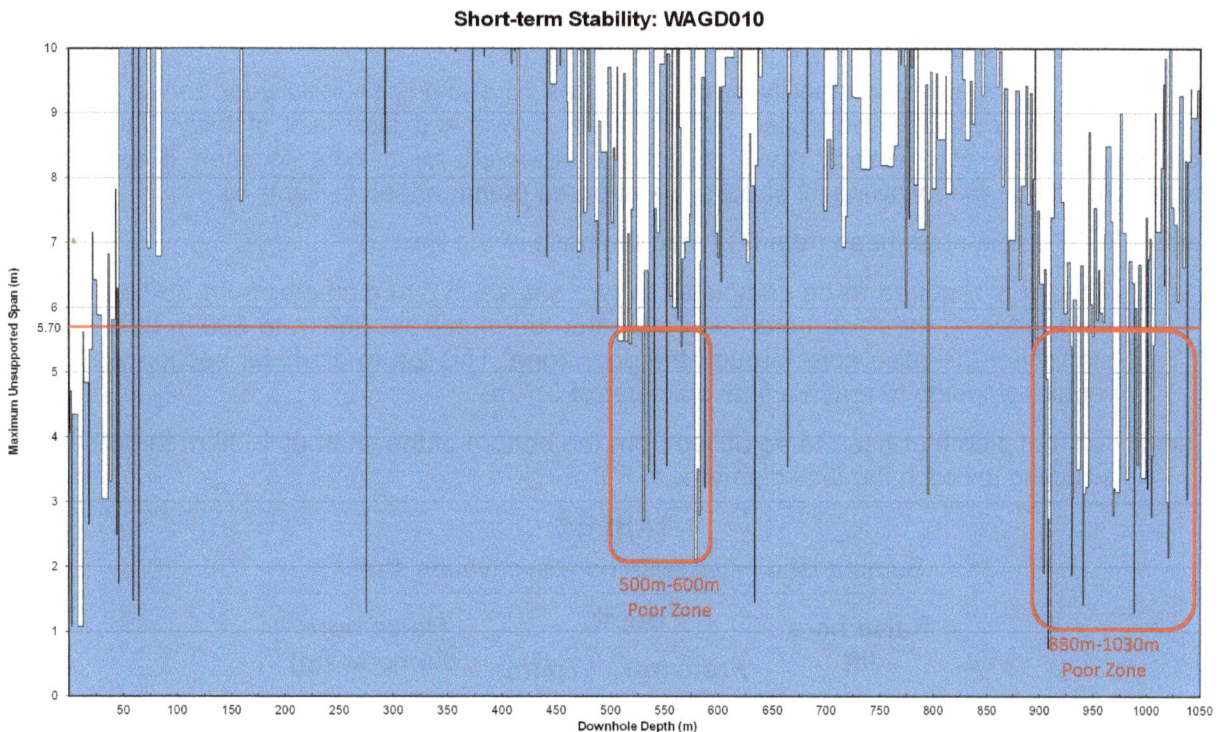

FIG 11 – Maximum unsupported span for short-term sidewall stability.

Typical rock conditions and broken zones

The typical fresh rock mass encountered in drill hole WAGD010 were characterised as follows:

- Moderately strong to extremely strong, with typical UCS interpreted and measured values in the range of 50 to over 200 MPa.

- Fair to good RQD/Jn values (an indication of rock block size), in the range of 9 to 17.

- Very poor to good Jr/Ja values (representing discontinuity shear strength), typically in the range 0.4 to 3.

- A subsequent good to very good rock mass classification with a Q range of 16 to 96, excluding the domains of poor rock once the rolling 3 m average was taken into account.

There are two main poor zones from 500 m to 600 m and 880 m to 1 030 m which consist of multiple poor domains that fall below the proposed 5.7 m $Span_{max}$ that limit the potential raise diameter within the poor zones. The critical poor domains within the 500 m to 600 m poor zone are characterised by:

- Strong to very strong, with typical UCS interpreted and measured values in the range 100 to 154 MPa.

- Very poor to good RQD/Jn values, ranging from 3.6 to 15.8.

- Very Poor to fair Jr/Ja values, ranging from 0.4 to 1.5.

- With a subsequent very poor to poor Q, with a range of 0.2 to 1.5.

The critical poor domains within the 880 m to 1030 m poor zone are characterised by:

- Weak to strong, with typical UCS interpreted and measured values in the range 1 to 100 MPa.

- Very poor to fair RQD/Jn values, ranging from 0.5 to 13.

- Very Poor to good Jr/Ja values, ranging from 0.13 to 3.

- With a subsequent extremely poor to very poor Q, with a range of 0.02 to 0.65.

This poor zone also contains a zone of ultramafic rock that prior to this hole being drilled was thought not to be in the potential raise bore area.

Recommended raise bore lift diameters

Both the face and short-term stability graphs (Figures 10 and 11) showed poor zones in which a 5.7 m raise bore could not be pulled and maintain crown and side wall stability. To meet the geotechnical limits of the rock mass within the selected raise bore location it was initially recommended that the following lift strategy be adopted (summarised in Table 6):

Lift 1: A 5.7 m diameter head from 500 m to the base of the boxcut.

Lift 2: This lift is required to be completed in two stages, 5.7 m diameter from 880 m to 600 m then reducing the raise bore head diameter to 2.4 m from 600 m to 500 m to ensure the stability of the raise bore through the poor zone. This 2.4 m diameter section would then require stripping to reach a final diameter of 5.7 m.

Lift 3: A 2.4 m diameter head to be pulled from the bottom of the shaft at 1030 m to 880 m, then stripping to reach the final diameter of 5.7 m.

TABLE 6
Summarised initial raise bore excavation strategy.

Raise bore lift	Depth		Raise bore diameter (m)
	From (m)	To (m)	
1	45	500	5.7
2	500	600	2.4
	600	880	5.7
3	880	1030	2.4

EXCAVATING THE HAULAGE SHAFT – HOW IT EVOLVED

Based on the completed raise bore stability assessment, it was identified that there were two intervals where it was considered that it would not be possible to raise bore the shaft to the desired full width diameter of 5.7 m. The concern with these two zones (at downhole depths of 500–600 m and 880–1030 m), was driven by the 'extremely poor to poor' rock mass conditions. Without any ground stabilisation work, stable raise boring in these two intervals would only be achieved with a maximum diameter of 2.4 m.

There were two main zones between 500–600 m where the long-term wall stability was assessed to be less than the planned 5.7 m diameter. These zones are between 506.7–511.8 m and 578.2–583.5 m. These weak zones were likely to experience localised wall overbreak with sidewall slabbing from steeply dipping defects. The blocky ground conditions in the above intervals were anticipated to impact on face stability during raise excavation. However, based on the external review of the raise bore stability assessment, the conclusion was that a continuing unravelling type failure would unlikely occur, so there would not be a need to undertake grout stabilisation prior to raise boring from a shaft wall failure point of view in the interval 500 to 600 m downhole depth, and the decision to proceed with a 5.7 m reaming head was made. This then resulted in developing the shaft in two legs rather than the initial approach of three (as presented in Table 6).

The external review in the interval between 880 m to 1030 m was of a greater concern for shaft stability than the interval between 500 m to 600 m. Without a successful form of ground improvement, the review concurred with the McCracken and Stacey (1989) raise bore stability that the interval between 880 m to 1030 m could not be raise bored to the full 5.7 m diameter.

Ground improvement

As part of the external review process, there was an intention to design a conceptual ground improvement grouting program to assist in the costing and feasibility evaluation of this option. However, prior to undertaking this work, there was a review of methods to quantify the extent of improvement that could be achieved with grouting and modelling to assess if this would be sufficient to permit full face raise boring at a 5.7 m diameter. A view was formed that this option would not be viable at the proposed diameters as there would remain a high risk of significant face failures.

The extent of failure in a 5.7 m diameter raise bore in the worst-case ground conditions was still expected to be large even after grouting. This was largely due to the low intact rock strength relative to the magnitude of the *in situ* stresses. The shaft intersection length of the worst-case zones associated with major structures could be double the borehole intersection length. This presents a significant risk of equipment damage and challenges to the safety of personnel working in the vicinity of the shaft brow.

The selected go-forward option – raise bore, strip and line

With the completed geotechnical work package, raise bore strip and line was the selected go-forward option to be applied to the downhole interval 880 m to 1030 m – full face raise boring to the desired 5.7 m diameter would be applied from surface down to 880 m.

The packer testing program was undertaken over broad intervals with a single packer and would not provide the resolution necessary for the design of a ground improvement grouting program. Stress modelling considering ground improvement from grouting suggested that the grout would not provide enough improvement to mitigate face failure (due to the magnitude of the stresses and the low intact rock strength). The extent of face failure is proportional to raise bore diameter. The modelling suggested that the depth of face failure for a 5.7 m diameter raise bore would be significant and would present a high risk of severe damage to reaming equipment including potential for drill string failure. Instances of face failure may also require lowering of the reamer to remove blocks of rock which is both hazardous and time consuming.

With these collective risks, full face reaming in the downhole interval 880 m to 1030 m, even after ground improvement grouting, was not recommended. Raise boring a smaller 1.8 m diameter pilot shaft over the interval 880 m to 1030 m and down bench stripping it out to 5.7 m diameter with progressive support, was subsequently adopted as a lower risk alternative to attempting ground improvement and full diameter reaming.

Ground control

With the selected development approaches confirmed for the haulage shaft, the necessary ground control systems were then established. The vertical development of the shaft was broken into two – Leg 1 (raise bore to full diameter, surface to 880 mbs) and Leg 2 (raise bore strip and line, 880 mbs down to shaft bottom). With the ability to raise bore Leg 1 to full diameter, only permanent support would be installed. However, due to the mixed geotechnical conditions associated with Leg 2, temporary and permanent support would need to be installed.

The permanent ground support system for Leg 1 will consist of Geobrugg galvanised high tensile steel wire mesh MINAX 80/4, installed horizontally around the shaft barrel axis with 1.8 m long galvanised Gewi bolts fully encapsulated with resin in a 1.5 m × 1.5 m bolting pattern. Figure 12 shows a photograph of the start of ground support installation for Leg 1 – note that FRS will be sprayed over the installed mesh and bolts for the first 30 m vertical metres for long-term stability associated with the weathered zone.

FIG 12 – Photograph showing the start of the installation of ground support for Leg 1.

With Leg 2, Tensar mesh and galvanised 1.5 m long split sets on a 1.5 m grid spacing has been adopted as the minimum temporary support requirement. This will be installed prior to the placement of a concrete liner which will form the permanent ground support. For those localised zones in Leg 2 that represent the weakest ground conditions (Exceptionally Poor, Extremely Poor and Very Poor), the temporary ground support will consist of spraying fibrecrete (75 mm thick) as the initial surface support with MINAX 80/4 mesh installed over the fibrecrete, installed with 2.4 m long galvanised Gewi bolts fully encapsulated resin.

Installation of the permanent concrete liner support (250 mm thick with a strength od 40 MPa) will lag the temporary support by 1.5 to 2 shaft diameters (ie approximately 9 m to 12 m). This lag will allow some convergence to occur prior to the installation of the concrete liner, which can help avoid the build-up of high stresses within the liner.

Excavation

The barrel excavation contract was awarded to RUC after a tender process where well established and recognised contractors from Australia and abroad were invited with a fixed/variable schedule of rates contract.

The barrel was excavated with the following equipment (see Figure 13):

- Raise borer – Strata 960 with operating torque of 600 knm, max torque of 1050 knm, rotation speed of 0 to 4 rev/min and hydraulic lift of 960 t with a max of 1180 t.

- Drill rods – 355.6 mm (14 inch) diameter with 279.4 (11 inch) DI 22 connections.

- Rotary Vertical Drilling System (RVDS) – Micon 428.6 mm (12 7/8) inch tool with 406.4 mm (16 inch) pads.

- Pilot Bits – Sandvik P80 tricone 406.4 mm (16 inch) diameter.

- Cutters – Epiroc Mag V × 6 Row hard carbide at 37 mm kerf spacing.

- Reamer – RUC 12 E mod 5800 mm diameter with 32 cutters, increased against the 5.7 m assessed to account for cutter changes.

- Cutter loading was on average 20 000 newtons per button row at 3–3.5 rev/min delivering 0.8–1.2 mm per revolution advance.

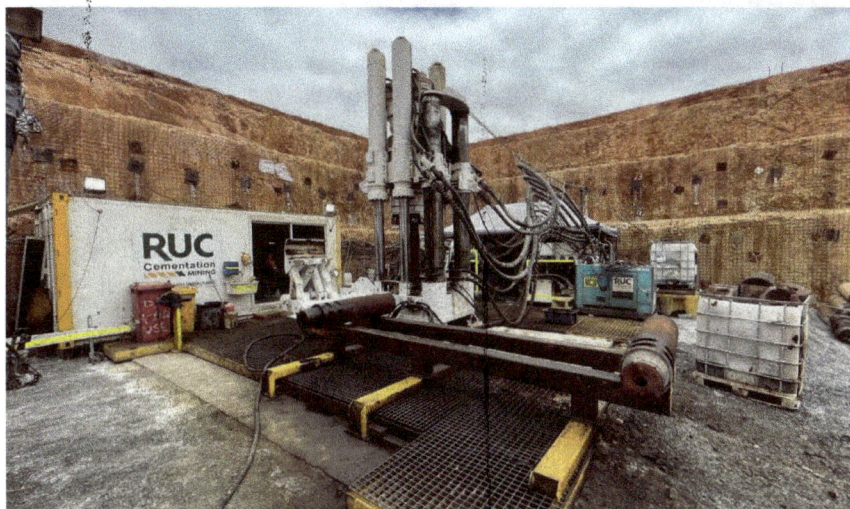

FIG 13 – Strata 960 in the boxcut (note the installed bolts, mesh and cable bolting reinforcement in the vertical boxcut walls).

The solutions to deliver and control the accuracy laid mainly in the collar being located in competent rock, monitoring of the initial deviation through initial frequent surveys, and once the string reached a depth of 4 m, the prompt installation of the RVDS. The collar in hard rock was probably the main control, avoiding the initial deviation that, once induced, cannot be controlled, or adjusted. The survey measures were carried out at the beginning to align the rig, at the installation of the RVDS and at 65 m of depth. After these initial surveys, the accuracy laid entirely in the RVDS tool and the high rock UCS to confine the string, avoiding the deviation.

The initial survey down the hole were conducted initially with gyro technology, but this has been proven not to be effective for the tolerances required. Laser surveys were subsequently used.

Monitoring

The pilot drilling for Leg 1 started in September 2020, and completed at the end of December 2020. After verticality check carried out in the pilot hole, back reaming started January 2021 and completed in September 2021. Leg 2 pilot drilling started in October 2021 and completed in December 2021, back reaming started in January 2022 and completed in April 2022.

The accuracy of the pilot drilling was monitored through daily reports generated interpreting the signals sent back to the surface via the RVDS tool.

Final tolerances achieved/LIDAR radar

After completion of the pilot drilling for Leg 1, a first attempt to measure the deviation was conducted using lasers and piano wires, but both did not produce satisfactory measurements: the laser due to the presence of water (approximately 3 L/s), the piano wire due to the relatively small diameter of the pilot hole and a suspected small deviation that located the wires on one side of the hole. At this stage the deviation was estimated in the order of 250 mm, and considering the fact that the final plan was to re-enter the barrel to support it, the decision to proceed with the reaming, and fix potential verticality defects via the stage, was made.

Once the reaming was completed a final verticality check was conducted via laser again, and LIDAR survey. The LIDAR survey presented some challenges due to the extreme smoothness of the barrel (see Figure 14), that did not give reference points to the LIDAR to orientate it. Final laser and LIDAR interpretations resulted in a final deviation of 103 mm. To correct this deviation, Leg 2 collar was aligned with the surface collar of Leg 1.

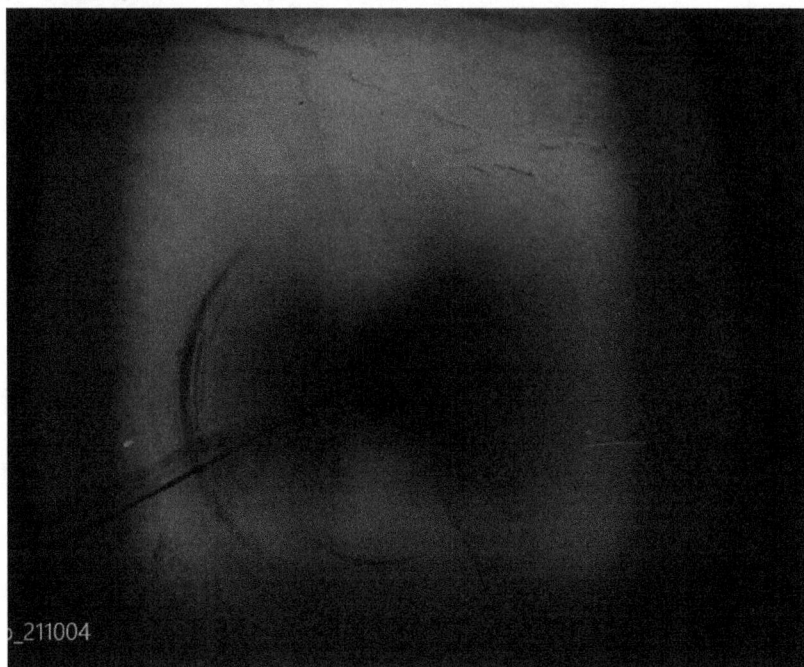

FIG 14 – Camera survey.

Leg 2 pilot drilling broke through within 100 mm of the target position, and final LIDAR survey is scheduled for March-April 2023.

Back analysis of reaming in relation to geotechnical design

During the excavation of the barrel several attempts to correlate the geotechnical data available against the excavation performance have been carried out. Some correlation tentative, like Maximum unsupported span, or RQD, or Raise Boring Index, versus daily reaming rate or cutter duration shown no correlation, while UCS, and Young's modulus from laboratory tests (refer to Table 4) shown some low to mild correlation with daily reaming rate and cutters duration (see Figure 15).

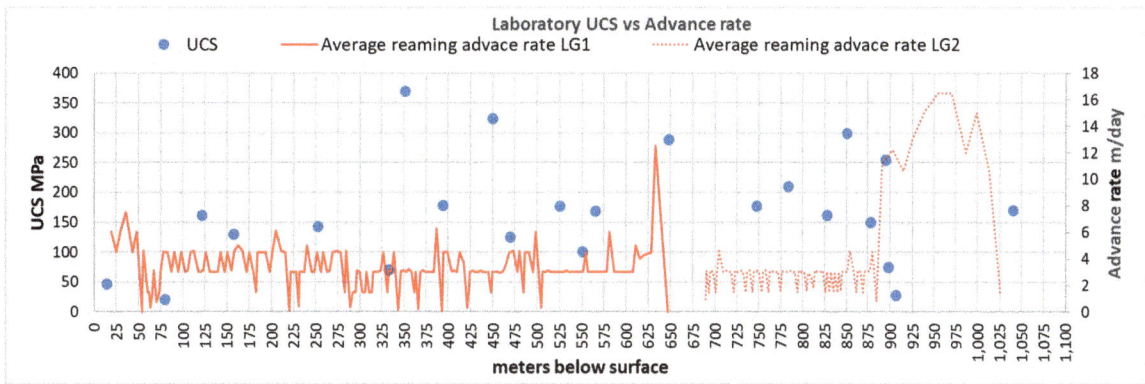

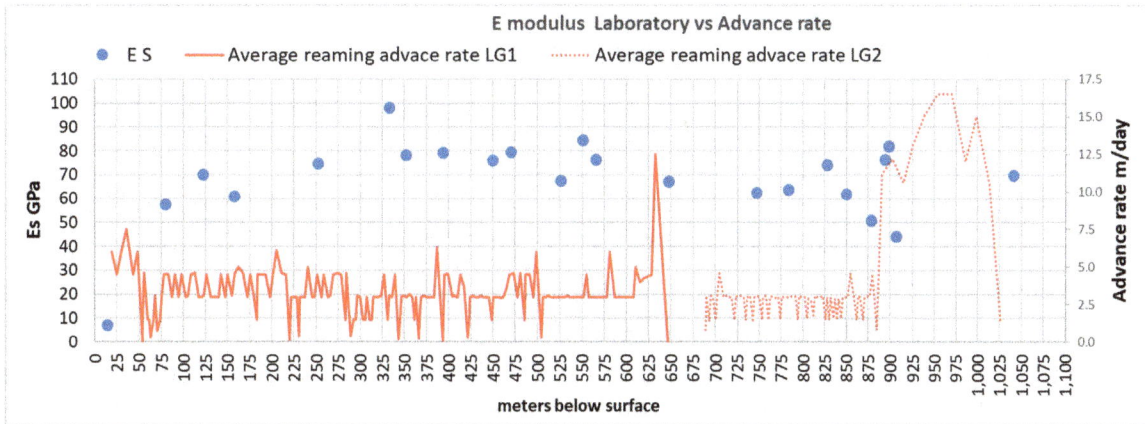

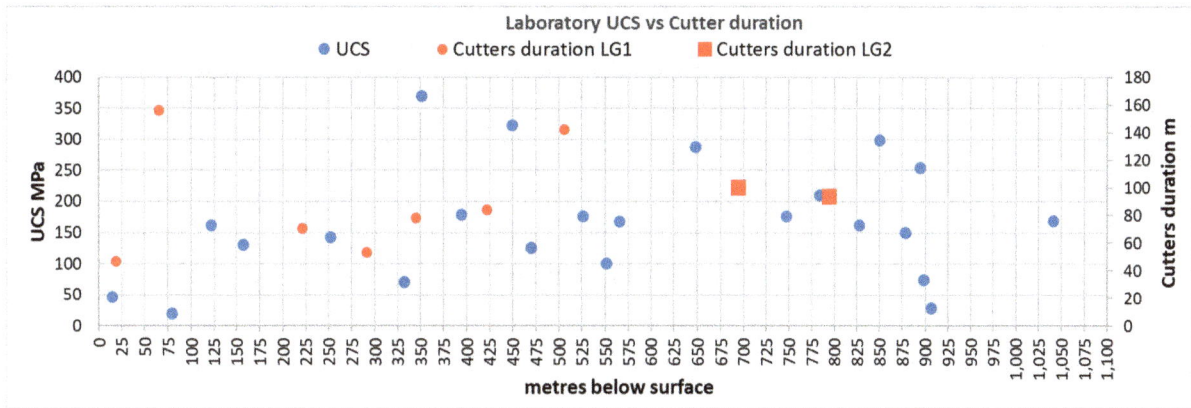

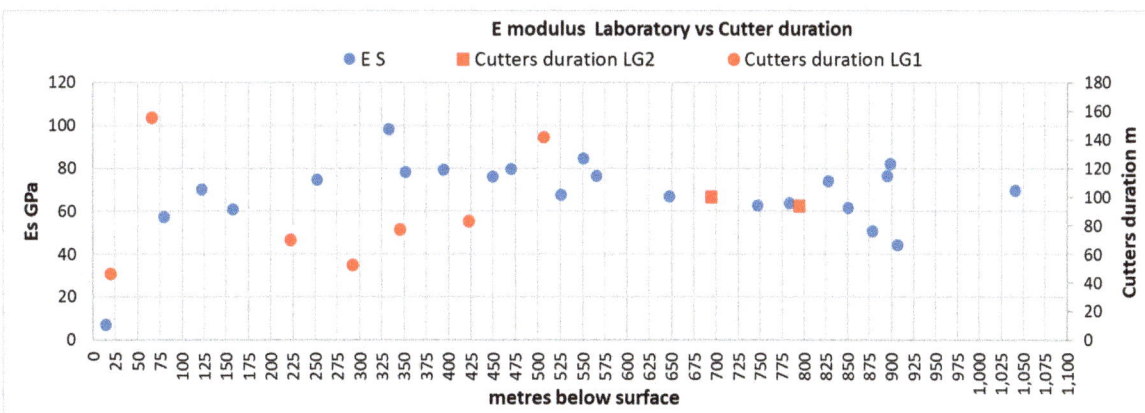

FIG 15 – Excavation parameters versus geotechnical data correlation.

Expansion of shaft capacity to AM6

During the implementation stage the decision to include the AM6 deposit was made, and a PFS completed. The study assumed the currently proposed shaft to hoist AM6 material. This triggered a review of the project choices made.

The Impala 12N winder is capable of hoisting 1.4 Mtpa, running at 8 m/s with 12 t payload skips hoisting rock from 1000 m below surface. However, due to potential rope harmonics, the speed has been adjusted to 9 m/s, resulting in an annual hoisting capacity of approximately 1.55 Mtpa. The entire rock handling system is designed to manage 1.6 Mtpa. This allows for the handling of additional material produced from AM6 as well as the development waste.

LESSONS LEARNED

As it is possible to verify through this paper, the project went through several review and transformations from the initial stages of planning to its execution. Several lessons have been learnt and can be generalised for similar projects:

- The level of changes could not have happened without a strong multidisciplinary interaction and communication approach among the team members, including the value of an external peer review process. Each change triggers a domino effect that needs to be verified with a holistic approach.

- It is essential to use the rock as an ally for the project, collaring in hard rock to prevent initial uncontrolled deviations, use its UCS to control deviation during piloting is paramount. Try to understand the potential of any reaction with the drilling water through slaking tests is also important to prevent unplanned deviation due to lack of confinement of the drill string.

- The selection of a competent contractor with a contract easy to manage is vital to deal with the project itself and unplanned circumstances.

- Unless the magnitude of the deviation is obviously of an order of magnitude higher than expected, it is better to measure the deviation after reaming. The bigger voids allow a better set-up for the measurement devices.

- If the tolerances are smaller than the error of the survey tools, like the GYRO technology, it is better not to waste time trying to measure the deviation, it is better to get organised with better tools from the beginning.

- A bigger number of UCS and Young modulus measurement may help in dealing with the advance rates and ultimately with the contractor.

- McCracken and Stacey (1989) method has to be treated carefully in interpreting the results because it could just be a tool to identify sections that need additional and thorough and different investigations.

ACKNOWLEDGEMENTS

The authors wish to thank IGO for the permission to prepare and present this paper, as well as to all that contributed to the project.

REFERENCES

Dempers, G D, Seymour, C R W and Harris, M B, 2010. Optimising Geotechnical Logging to Accurately Represent the Geotechnical Environment, in *Proceedings 2010 Second Australasian Ground Control in Mining Conference*.

McCracken, A and Stacey, T R, 1989. Geotechnical Risk Assessment for Large-Diameter Raise Bored Shafts, *Trans Instn Min Metall (Sect A: Mining Industry)*, 98(Sept–Dec):A145–A150.

Peck, W A, Coombes, B and Lee, M F, 2011. Fine tuning raise bore stability assessments and risk, in *Proceedings 11th Underground Operators' Conference*, pp 216–225 (The Australasian Institute of Mining and Metallurgy: Melbourne).

Penney, A, Stephenson, R M and Pascoe, M J, 2018. Raise bore stability and risk assessment empirical database update, in *Proceedings The Fourth Australasian Ground Control in Mining Conference (AusRock)*, pp 434–445 (The Australasian Institute of Mining and Metallurgy: Melbourne).

Ground support rationalisation at Golden Grove

R Varden[1], P Brenchley[2], E Jones[3] and J Player[4]

1. MAusIMM(CP), Senior Geotechnical Engineer, MineGeoTech, Perth WA 6000.
   Email: richardvarden@minegeotech.com.au
2. Senior Geotechnical Engineer, 29 Metals, Perth WA 6000.
   Email: paul.brenchly@29metals.com.au
3. Principal Geotechnical Engineering – Projects, MineGeoTech, Perth WA 6000.
   Email: emmajones@minegeotech.com.au
4. MAusIMM(CP), Principal Engineer, MineGeoTech, Perth WA 6000.
   Email: johnplayer@minegeotech.com.au

INTRODUCTION

This paper summarises the process taken to simplify 35 ground support standards accounting for the multiple development profiles, dynamic geotechnical environment and the practical safety implications of a large-scale change in the ground control standards.

This reduction was required after an incident involving incorrect standards being issued. The subsequent investigation identified that:

- Selection of ground support standard for a mining instruction (MI) is not straightforward due to multiple standards, standards not reflecting current reinforcement used and complex geotechnical environment being encountered. Instead, site-specific ground control instructions were being issued.

- Ground support standards were hard to follow and confusing, errors were present and not all standards to scale.

- Ground support standards being used did not reflect what was being installed, ie bolt type incorrect and mesh orientation different.

- No process or tier action response plan (TARP) for operators to follow for ground control standard noncompliance.

- No justification for ground support standards being used in the current geotechnical environment.

- No process for ground support standard selection and thereby no paper trail to document and assess adequacy of performance.

A project was undertaken to reduce and simplify the ground support standards, produce a justification document to support ground support design and develop a process for site practitioners to identify the type of ground support standard required for the mining instruction.

The exercise of rationalising the ground support was met with some challenges, including:

- Technical aspects to determine the level of demand and the capacity for each ground support scheme.

- Training the geotechnical engineers how to determine the level of risk for each area.

- Winning over other technical services staff and the underground mining contractor of the new approach and then implementation.

One of the major successes was the acceptance by the underground operators who found the system simplified their work and gave them increased confidence in management of seismicity. The rationalised ground support standards have now been successfully in place for over two years.

MINE LOCATION AND HISTORY

Golden Grove mine is a deep underground copper and zinc operation located approximately 250 km east of Geraldton, Western Australia. It consists of the Scuddles underground mine and gold pit and the Gossan Hill underground mine and historic copper pit. The Scuddles deposits are located approximately 4 km north-east of Gossan Hill.

The mines have been in operation for just over 30 years with a remaining life of about ten years. The Australian Consolidated Minerals constructed the project in 1990 and sold it to Normandy Mining Ltd in 1991. Normandy was acquired by Newmont Australia Ltd in 2002. Oxiana Ltd purchase the mine from Newmont in 2005, Oxiana merged with Zinifex Ltd in 2008 to form OZ Minerals. OZ Minerals sold the mines to MMG Australia Ltd in 2009. EMR Capital acquired Golden Grove from MMG in 2017. It is now operated by 29Metals Limited after a 2021 divestment by EMR Capital. Multiple changes of ownership has made maintaining the mine history challenging particularly as the operation was run for extensive periods of time with limited geotechnical services.

GEOLOGY

The Golden Grove Volcanic Hosted Massive Sulphide (VHMS) deposits are located in the Murchison Province of the Archean Yilgarn Block within the Yalgoo Greenstone Belt. The regional setting has been described by Watkins and Hickman (1990), Clifford (1992) and Sharpe (1999). The Gossan Hill Group comprises four formations: Gossan Valley (lower), Golden Grove, Scuddles and Cattle Well (upper).

The Golden Grove formation, Figure 1, is the host of the mine and can be described as re-sedimented juvenile tuffaceous debris of rhyolite to andesite composition, minor sedimentary rocks, and volcanics ranging from andesite to rhyodacite. It varies from 75 m to 800 m wide, with economically significant horizons including stringer to massive sulphide and magnetite and hydrothermal alteration associated with these horizons.

The Scuddles formation, consists of dacite volcanics with minor rhyolite and sediments. It exhibits a conformable upper contact and conformable to transgressive (peperitic) lower contact. The formation varies from 730 m to 990 m thick and is commonly in direct contact with underlying GG6 Zn sulphides.

The economic VHMS can be broadly divided into two parts, a stratigraphically lower copper zone hosted by GG4, and a zinc-rich zone, with associated footwall stringer and massive pyrite-copper, hosted by GG6. Separating the two zones is an approximately 50 m wide package of relatively unaltered massive epiclastic of GG5. The GG packages are chronological based, with each a mix of four lithologies.

The sequence is intruded by over 120 dolerite, rhyolite, and dacite intrusives cut across the Golden Grove sequence. The thickness of the inclined dykes ranges from less than 1 m to 25 m, averaging 4 m thick. An example of this at a development drive scale is shown in Figure 2.

The complexity of the geometry has resulted in several geotechnical domains, where the intrusives are considered to be the highest risk where dynamic rock mass response can occur. This response is more prevalent with depth and is expected to be an ongoing risk to address until the end of the mine life.

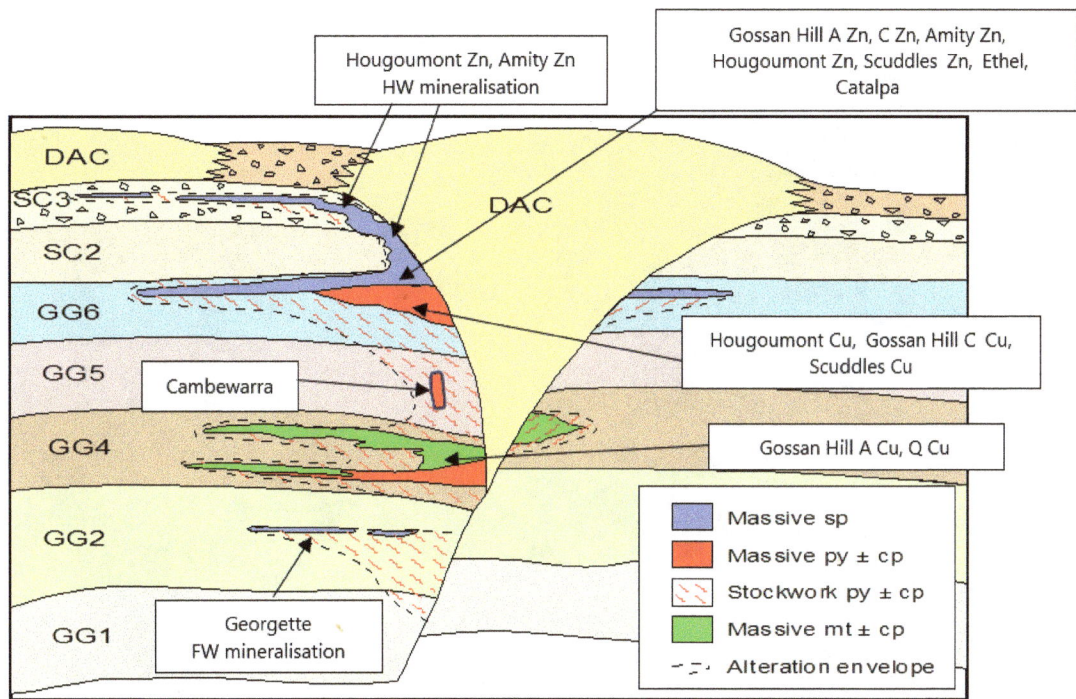

FIG 1 – Generalised geological model of Golden Grove VHMS.

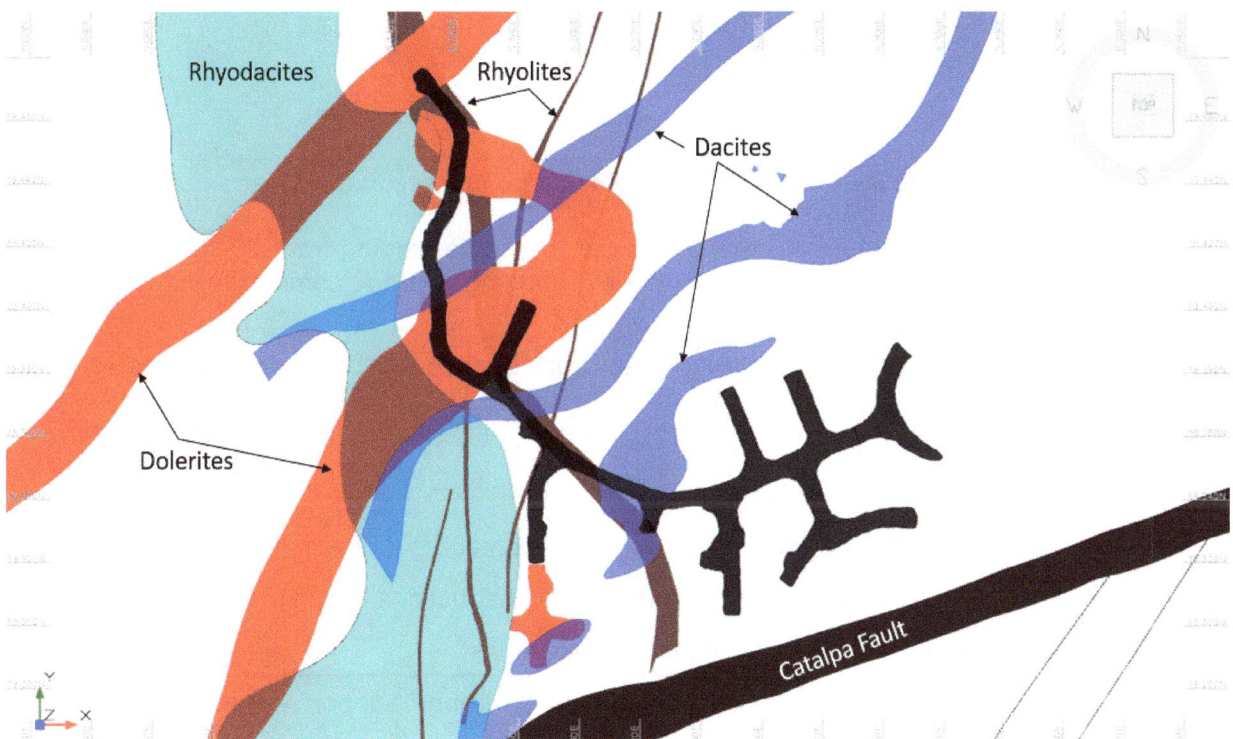

FIG 2 – Level plan showing complexity of intrusives.

MINING METHOD

All ore is mined using diesel loaders and hauled to the surface stockpile in 65 t underground trucks or shaft haulage for a total ore movement of 1.6 Mt per annum.

Gossan Hill (Figure 3) predominantly utilises longhole open stoping with stope widths ranging from 4 m longitudinal stopes to 15 m wide transverse stopes. The Hydraulic Radius (HR) ranges from 6 to 10. Production occurs across a number of orebodies from 300 m to 1400 m below surface. Production sequence for mining blocks varies with depth and size of orebodies and can be top down, bottom up, continuous or primary-secondary stoping. Development is currently ~1550 m below surface and will ultimately reach a depth of ~2000 m.

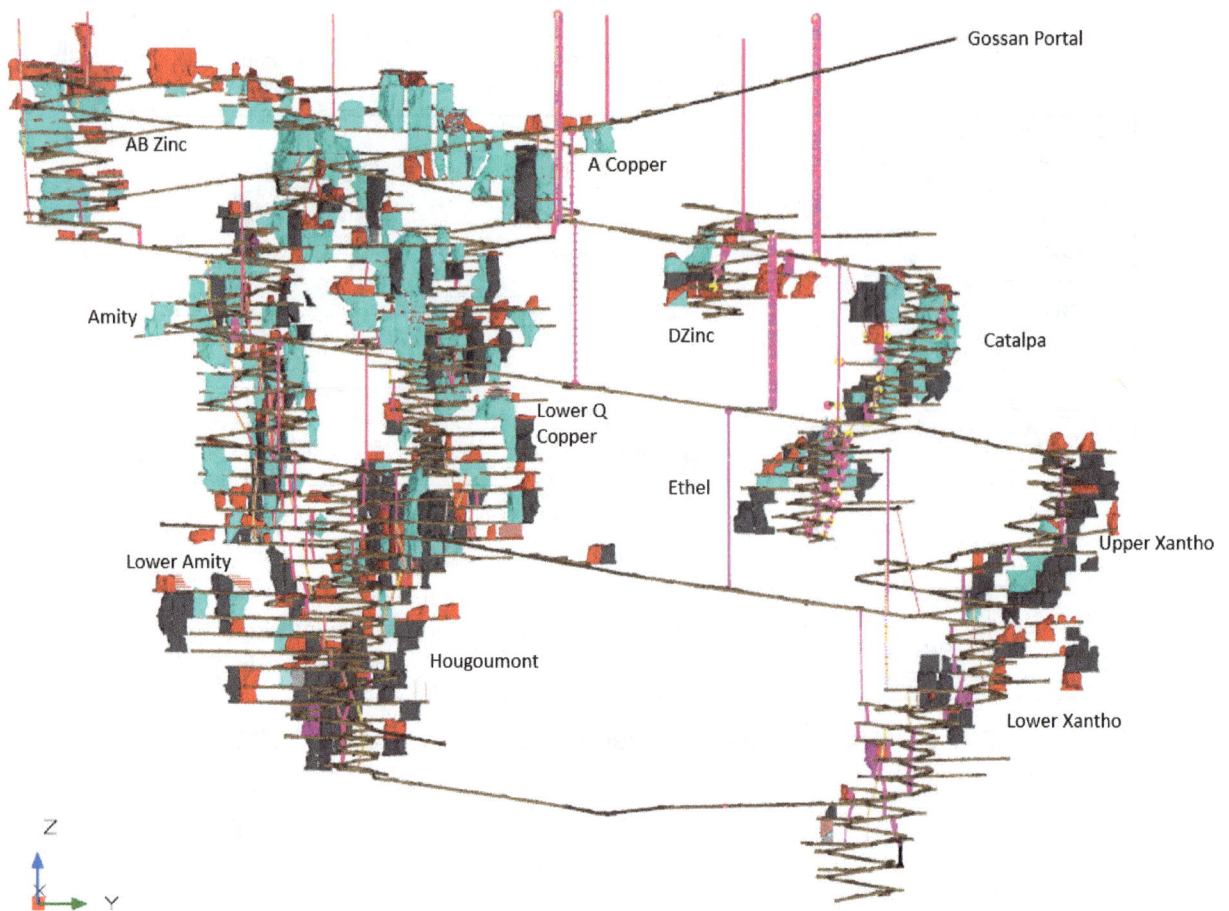

FIG 3 – Gossan Hill Life-of-mine Long Section 2022.

New deep mining areas (>1400 m) have increased the stope heights from 20–25 m to ~45 m to reduce stress damage from level interaction while managing orebody complexity with a maximum individual stope wall HR of 6.

Cemented Hydraulic Backfill (CHF) is reticulated from surface via borehole and pipe to backfill stopes. The cement content is adjusted to suit exposure spans and heights. A paste backfill plant (due for completion Q3 2022) is to be used for the backfill of the future deep orebodies.

BACKGROUND FOR STANDARDS

In July 2020, MineGeoTech were engaged to help the operation:

- Review the incident involving errors in site-specific instructions circulated.
- Review other site-specific instructions for similar issues.
- Investigate reported issues with communication of ground support standards and effective implementation.
- Investigation of rationalising the number of ground support standards and site-specific instructions.
- Complete a ground control review on the geotechnical systems and processes.
- Assist with action items relating to incident investigation outcomes.

This work identified underlying issues with the ground support standards, justification for this standard and suitability for the changing geotechnical environment the mine was heading towards. The conclusions from this initial review work are summarised:

- Ground support standards had not been updated/reviewed with change to an underground mining contractor change of reinforcement elements or changes to development profiles.

- New standards were added for dynamic geotechnical conditions, but these were not scaled and errors identified.

- Site specific instructions were used in place of standards where standards did not cover current reinforcement elements.

- The number of site specific instructions that were required to fill the ground support standard gap had increased the workload of the geotechnical department beyond capacity. This resulted in copying previous instructions resulting in errors being transferred.

- Ground support standards were changed multiple times in one mining instruction resulting in operators having to determine when to start and stop new profiles.

- Operators reported that there was confusion and errors with the issued site-specific instructions.

- No TARP for operators to follow where non-compliance was encountered.

PROPOSED SCOPE OF WORKS

Following on from the initial work, MineGeoTech were then engaged to:

- Reduce the number of standards to a manageable number without loss of effectiveness and improve crew understanding and installation outcomes.

- Reduce the number of ground support elements used across the standards.

- Stop the occurrence of headings being supported incorrectly.

- Reduce the amount of administration across technical services and contractor with improved compliance to ground support standards.

- Establish a standard and methodology for ground support design to account for the change of environment and improvements in understanding of seismic sources.

- Confirm the appropriate level of ground support installed in the appropriate areas of the mine.

- Implement a standard that can be practically and safely installed.

- Reduce the time that geotechnical engineers spend on creating site specific standards, and develop a process map for ground support design that will result in a consistent design and can be documented.

APPROACH

A phased approach was adapted to complete this project.

The operation could not wait for the forecasted six months of work planned to collect data and complete stability analyses to justify the new support standards due to:

- Underground installation of ground support schemes not complying to standards due to changes in elements and equipment.

- Using site-specific instructions instead already resulting in incidents.

- Ground conditions expecting to become more complex as the mining progressed at depth.

To mitigate the current risk on-site but also complete the parallel work to reduce and justifying for a long-term solution, the work was broken down into four phases plus a check on implementation after one month, as summarised below and presented in Figure 4.

Phase 1 – Immediate response: Update Ground Support Standards to reflect current practice and replace the common site-specific standards. Streamline the standards where possible without changing coverage or design. No analysis work completed for this phase.

Phase 2 – Justification and optimisation: Undertake stability analysis for:

- Intersection cable requirements.

- Structural control (kinematic analysis) for multiple drive dimensions.
- Dynamic requirements, this was split into two phases:
  - Phase 1:
    - Assess rockfall database.
    - Map current ground support schemes installed.
    - Determine performance of support to date.
  - Phase 2:
    - QAQC seismic database.
    - Seismic source investigation.
    - Numerical modelling back analysis and performance review.

Outputs of this analysis work would then feed into creating a new ground support standard, TARP, justification document and process map.

Phase 3 – Change of management: Undertaking a Change of Management process to quantify the impact and risk assess the process to eliminate or mitigate potential hazards with this change.

Phase 4 – Implementation: Mining Standards Check – method statements on installation to account for new elements being introduced. Order elements in readiness for implementation. Update of the Ground Control Management Plan (GCMP) and associated processes.

Selection of a two-week period to roll out the new standards targeting two areas – crew rollout and technical services rollout.

- Crew rollout:
  - Present the new standards and required changes to allow the crews to ask questions and become aware of what was going to happen.
  - Updated ground awareness presentation.
  - Trial standards.
- Technical services rollout:
  - Present the new standards with explanation to reasoning, how they will be used and how they can be applied to the multiple drive dimensions.
  - Updated ground awareness presentation.
  - System change presentation – how is this going to work.
  - Update of mine instructions documents.

Performance review and monitoring: After one month, check in with crews and technical services to identify any required tweaks or issues with the new standard and processes.

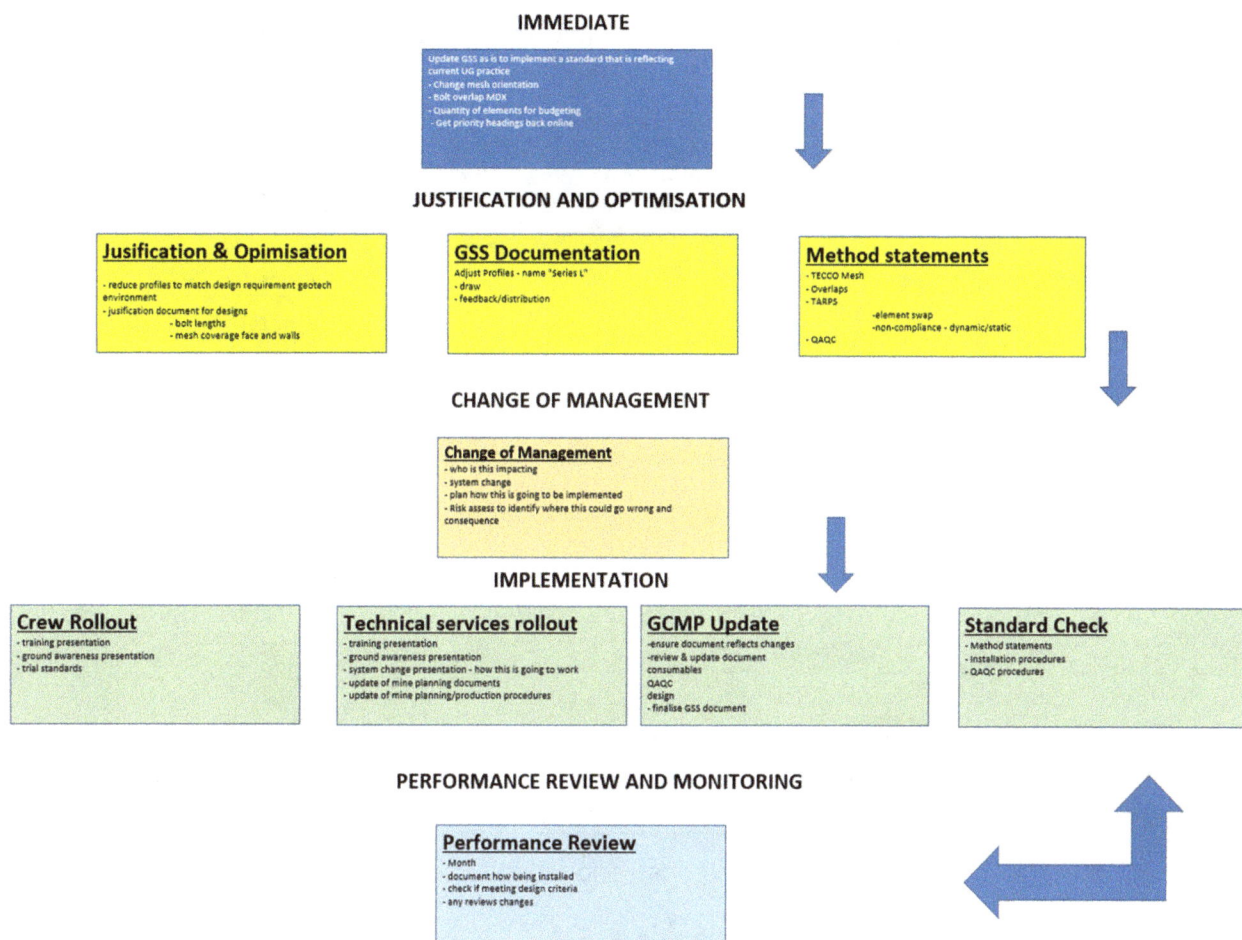

FIG 4 – Flow diagram of five phases to roll out new standards.

The phases are discussed further in detail.

Phase 1 – Immediate response

The 35 standards were reduced to 21 in this phase. The number remaining reflects the multiple geotechnical domains and drive profiles that are present in the mine.

Three new profiles were created to reduce the number of site-specific instructions and to introduce a new face-standard as required by the underground mining contractor.

All standards were adapted to represent what was being installed underground and remove any errors found.

Standards were rebranded to differentiate to old standards. These were labelled as 'Series K' to anticipate any future corrections and/or changes required depending on outcomes of the implementation. The naming system was selected for traceability underground to ensure operators had the correct copy.

The rollout was planned as per the following list and stylised in Figure 5 to reduce the risk of incorrect standards being issued and confusion to operators and supervisors:

- All Jumbo operators received a copy of new standard.

- The selected standard for the mining instruction would be photocopied on the back.

- One mining instruction per ground support standard. A change in standard would require a new mining instruction (in the past there were multiple standards on one mining instruction).

- All new Plan Of Intent (POI), will use the new standards. All existing POI's will be updated with new standards and then reissued.

- All updated MI's will be released at the same time – swap out old for new on an allocated day that would best suit the underground contractor.

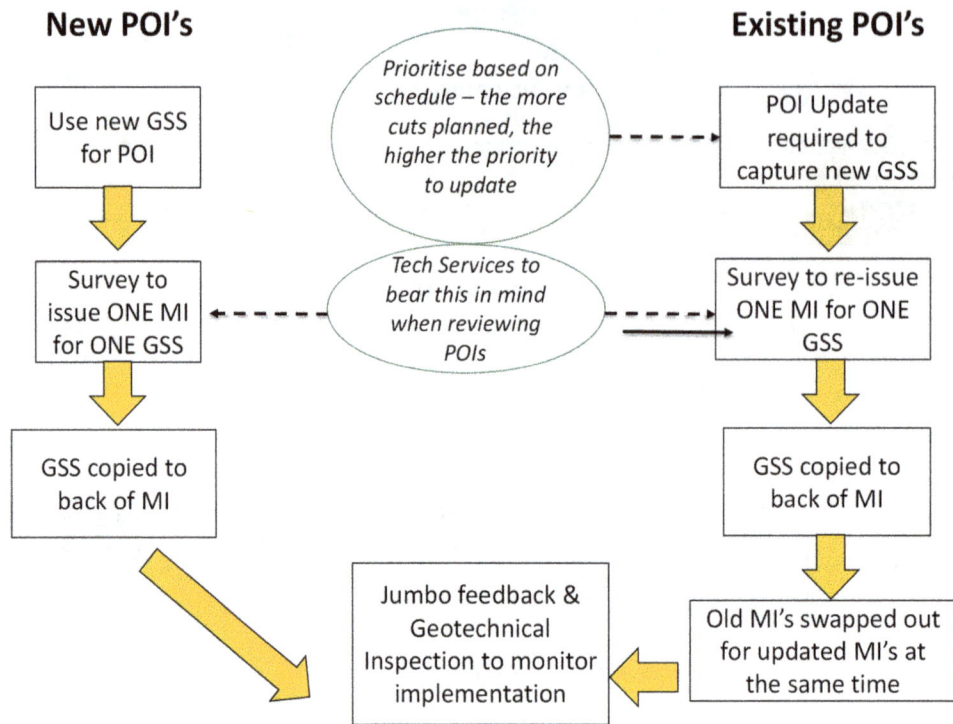

FIG 5 – Immediate response and improvement (Phase 1).

Phase 2 – Justification and optimisation

This phase comprised of undertaken stability analysis work to justify new standards and the development of these standards.

Stability analysis methodology

The approach undertaken to form the basis for ground support scheme selection required kinematic and dynamic studies following the updating of the mine rock properties:

- Kinematic:
  - Structural setting review processing historic Sirovision images and new SR3 lidar mapping data for development headings. Only structures with a >2 m trace length were shown to have reliable set orientations and be significant for kinematic analysis.
  - Comparison of mining area structural setting which showed similarities in the formational units and additional structure and complexity within the intrusives.
  - Static and probabilistic analysis of the structural assessment and undertake ground support scheme design and validation.
- Dynamic:
  - Seismic Hazard assessment using clustering (last two years) and assign a ranking.
  - Energy Demand review – event magnitude, depth of damage, and loading velocity for a large event that encompass the development heading.
  - Ground support scheme of appropriate capacity for the forecast demand.

This was followed by the development of five process maps to show how the effective ground support scheme design is done at Golden Grove. This is summarised by Figure 6 and Table 1.

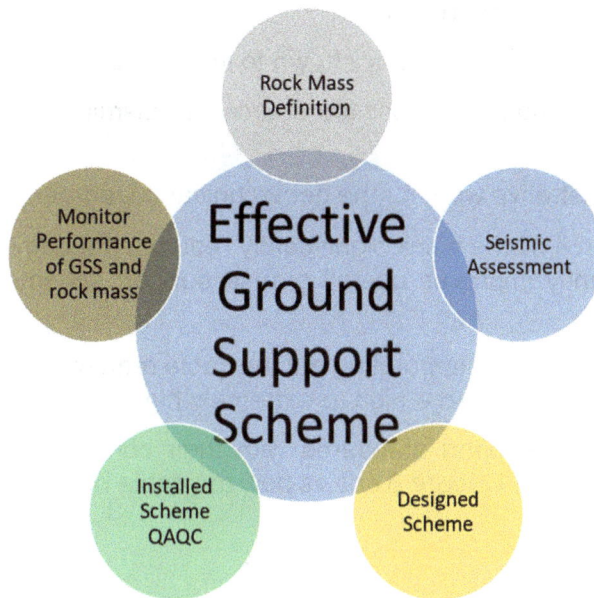

FIG 6 – Effective ground support scheme.

TABLE 1

Effective ground support scheme.

Rock mass	Seismic assessment	Designed scheme	Installed scheme QAQC	Monitoring performance
Rock strength testing	Cluster analysis	Empirical design (Q-system)	Pull testing	Seismic system suitability
Geotechnical Logging	Hazard analysis magnitude re-occurrence	Kinematic design (wedges)	Grout testing	Extensometers
Structural setting from mapping/ scanning	Activity rate and acceleration	Dynamic design energy capacity	Bolt spacing	Observation holes
Overstress rock mass damage from modelling and observations	Energy Demand		Mesh height and overlaps	Compiled Documentation of Ground Support scheme performance and rock mass observations Seismic database
Stress state			Installation observations	Scanning, (both visual and CMS) of appropriate excavations

Ground support scheme standards changes

The following are the main changes made to the ground support scheme:

- Starts with a TARP addressing the most likely reasons for non-compliance the required action dependent on the seismic hazard.

- Standards are non-dimensional (except for the specific CHF standard). This approach allows for operators to be within acceptable coverage requirements and substantially reduces the number of standards and site-specific standards.

- Flexibility of having five designs that cover a range of seismic hazards and energy demand which is calculated to select the appropriate Ground Support Standard (GSS) standard.

- The GS4.0 standard (refer to Table 2), can be upgraded easily with one, two or three rows of cable bolts significantly increasing the energy and deformation capacity.

- Identifying zones/areas of hazard will:

- o Simplify Plan of Intent – MI process.
- o Allow for MI's to have hazard clearly shown to improve operator awareness.
- o Optimise GSS in response to possible source of seismic activity rather than applying a global estimate.
- o Reduce the administrative work for the geotechnical department to allow more engineering.

- TECCO MESH as a roll of chain link significantly reduces the number of overlaps in high-risk areas and consequently improves overall ground support scheme performance to dynamic loading.

- Flexibility within the standard – separating out the cable requirements allows GS4.0 to be used for three levels of seismic hazard – high to extremely high.

Table 2 lists the new ground support standards with their corresponding elements and energy capacity. Figure 7 shows GS4 and Figure 8 shows how the cable bolt variations (drive cables: DC 1, 2, 3) can be applied.

TABLE 2

New ground support scheme.

GSS Standard	Ground Support Scheme Element used in Standard							Energy Capacity (kJ/m2) (approx)	Demand Category	Comments
	Split Sets	MDX	DeBonded Posimix	17.8 Cable Bolt	Weld Mesh	Fibrecrete	TECCO			
GS1	✓				✓			n/a	Aseismic Profile	
GS2		✓			✓			15	Medium Hazard	
GS3		✓			✓	✓		17	Medium Hazard	Also can be used to reduce dilation of foliation/bedding in north-south drives
GS4.0			✓			✓	✓	30 to 32	High Hazard	Minimum for Dacites at Depth
GS4.1 (GS4 + DC1)			✓	✓		✓	✓	30 to 35	High Hazard	(similar KJ/m² with MDX)
GS4.2 (GS4 + DC2)			✓	✓		✓	✓	35 to 40	Very High Hazard	(similar KJ/m² with MDX)
GS4.3 (GS4 + DC3)			✓	✓		✓	✓	40 to 45	Very to Extremely High Hazard	This configuration to be used if MDX bolts need to be replace debonded posimix due to ground conditions
FM	✓				✓			n/a	Aseismic Profile	For Limited use at Face
CHF	✓					✓		n/a	CHF only	3m cut – 4.5mW*5.0mH

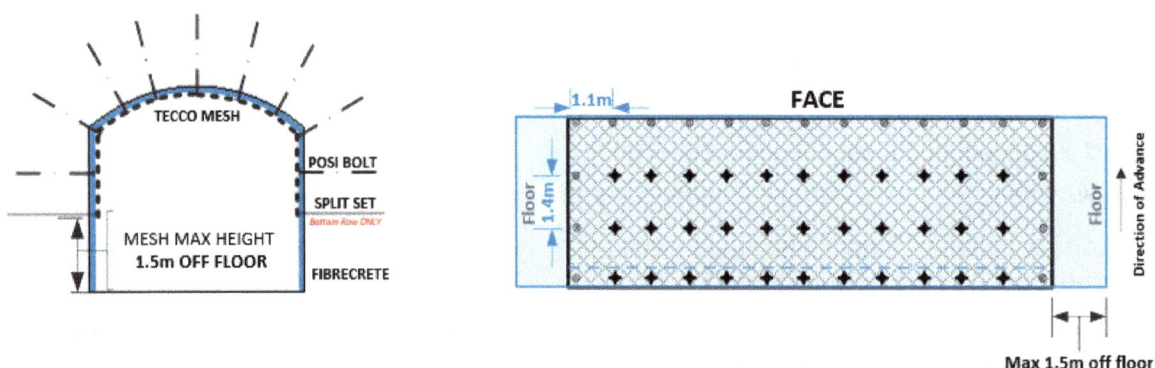

FIG 7 – GS4 section and plan view.

DRIVE CABLES

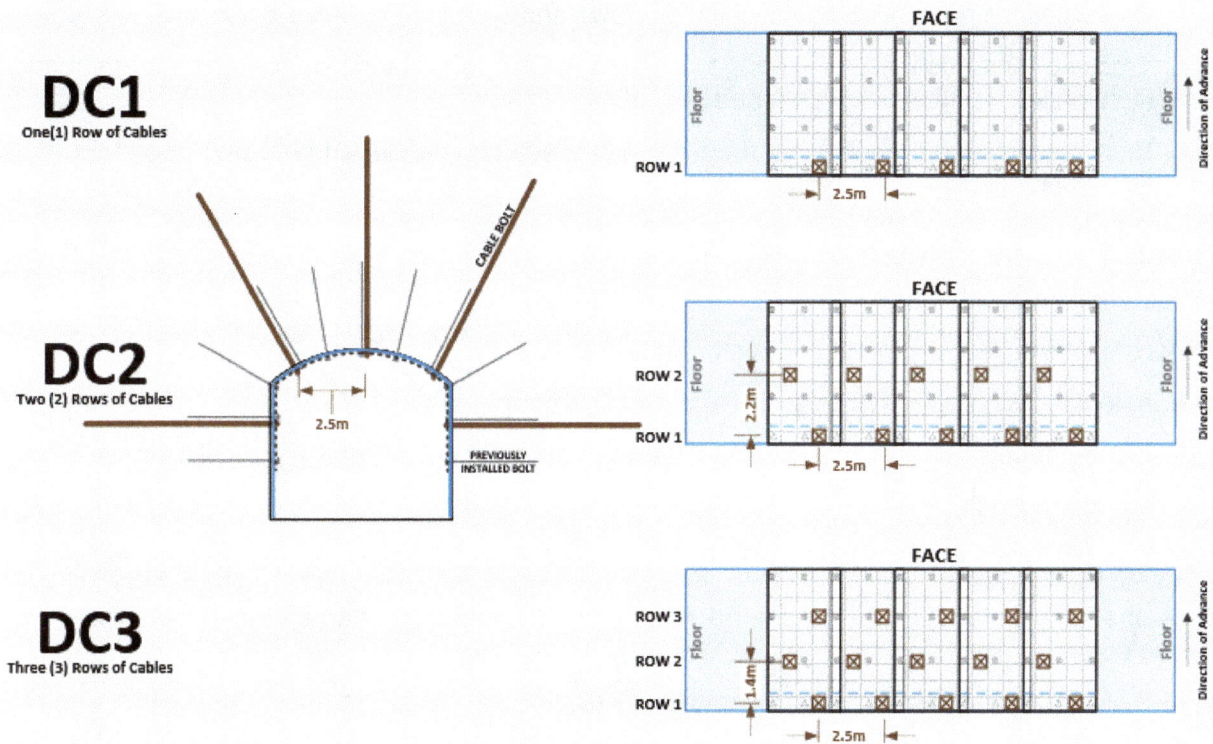

FIG 8 – GS4 cable bolt layouts DC1, DC2 and DC3.

Supporting documentation

In addition to updated standards, two support documents were developed to aid in implementation and ongoing use of a new mesh, Geobrugg G80–4, for dynamic ground conditions and a TARP to reduce delays where operators and supervisors could carry on with instructions when faced with a ground control issue that could be resolved at the operations level.

Geobrugg G80–4

G80–4 mesh is a very high-tensile chain link mesh and is ordered to a length and width to suite profile of excavation and development advance. This style of mesh had not been used at the mine before and challenges that were identified with implementation that required further discussion and planning included:

- Overlap on development heading advance.
- Rehabilitation or extending mesh.
- Intersection stripping.
- Production blasting.

TARP

Tier action response plan (TARP) was developed to respond to a non-compliance. This being dependent to the level of seismic hazard as identified on the mining instruction.

This TARP was reviewed by the operators and supervisors to ensure understanding and coverage of issues they face and can then manage.

Phase 3 – Change of management

A change of management review was undertaken in the week leading up to the roll-out of the new standards. This involved a high-impact risk assessment to identify stakeholders, risks and consequences relating to the change out of the new standards.

Outcomes of this work included:

- Additional senior geotechnical engineer site support for the month after implementation.

- Presentations developed specifically for crews and technical services personnel.

- Tracking spreadsheet for change of ground support standards on mining instructions highlighting geotechnical risk for each.

- High communication levels required between geotechnical department and operators during implementation.

Phase 4 – Implementation

Implementation consisted of:

- rollout to the underground contractor and operators

- rollout to technical services group and management

- GCMP update

- standard check.

Underground operators

Once the new ground support scheme had been accepted by management the system was presented to the workforce as:

- Seismic awareness and overview of new GSS.

- Additional details of new GSS, TARP and quality of installation were provided to Jumbo operators, nippers, and UG supervisors.

Presentation to the crews included the following:

- reason for change

- key changes

- new standards – each standard talked through

- TARP

- flexibility of the new system

- feedback – operators given the chance to give feedback during implementation.

Operator acceptance of the new system has been very positive with many giving good feedback to the geotechnical department. Operators found the new range of standards far simpler to manage. The quantities per profile are not specified, however crews quickly determined the requirements.

Technical services – mining instruction process

Two aspects were addressed to implement the new system with the Technical Services team:

- Explanation and details regarding naming convention and seismic hazard level. Template for mining instructions changed to identify the hazard level. This was covered in a specific presentation.

- How the changeover of the standards was practically going to be undertaken. A process map was developed by senior engineers and comprised of:

  o **One week leading up to changeover** – begin to update mining instructions with new standards, save in a specific folder. Sign-off process to follow but these are not to be issued. If any new instructions were required that week, two versions would be made; one with the new (to be saved in specific folder) the other with the current instruction is issued.

  o **Day of changeover** – all those instructions saved and signed in the specific folder will be used to supersede existing instructions. All instructions will be swapped out. Survey and all technical services personnel will follow up with the underground contractor to ensure the switch is complete.

Geotechnical department – ground support design process flow

The primary method of assigning ground support schemes is through the POI system managed by the planning department.

To aid the geotechnical department to identify the seismic hazard level and thereby appropriate ground support standard, a process was developed to identify the hazard and document the decision process.

A template was developed to follow and complete as an excel spreadsheet with separate tabs detailed as follows:

- **Tab 1: Information** – summary of mining area, level, PIO name, the engineer and date.

- **Tab 2: Data** – the engineer can collect all supporting evidence and paste into this tab, eg sections of geology, seismic data etc.

- **Tab 3: Checklist** – this is a list of questions that the engineer answers and then attaches to the POI and to be signed off by the engineer, includes:

  o results of inspection of the area and including rehabilitation instructions

  o static rock mass conditions

  o seismicity including level of risk and stress conditions

  o development intersections, pillars – sizes and interactions

  o ground support – recommended support regime for development and intersections

  o backfill – is the development in the vicinity of backfill and what is the strength of the fill

  o probing – is probing for voids required.

- **Tab 4**: Energy demand calculations.

- **Tab 5/6**: Seismic clustering for Gossan and Scuddles.

- **Tab 6**: Library of reference material.

Training of this process was completed during the implementation phase. Support was provided by senior geotechnical engineers throughout this phase and beyond to ensure the system was implemented in full.

This new template enabled documentation of ground support design decisions, a process that could be followed consistently and thereby reduce overall workload for ground support design for mining instructions.

Performance review and monitoring

The implementation and performance of the standards was allocated to run for one month. During this time, the geotechnical department in particular were monitoring, documenting, and communicating to operators how the ground support standards were being received, interpreted, and followed.

An inspection sheet was developed specifically to document installation compliance to design with a focus on how the new G80–4 mesh was installed.

Outcomes of the one-month review and monitoring did capture small errors and slight tweaks to the ground support standards at the request of operators. The standard was re-issued with these minor updates after which the site decided that the standard was implemented successfully.

CONCLUSIONS

The introduction of a ground support scheme based on seismic risk has resulted in:

- Number of standards reduced from 35 to five.

- Acceptance by the operators, followed up with additional one-on-one training and feedback.

- Increased flexibility within a specific risk category – allows a standard to be allocated to a range of different type of drive profiles in the same area.

- A process that provides consistent recommendations.

- An ongoing structural analysis program of scanning heading and analysing data per mining area.

- One of the major successes was the acceptance by the operators, who found the system simplified their work and gave them increased confidence in management of seismicity. The rationalised ground support standards have now been successfully in place for over two years.

ACKNOWLEDGEMENTS

The Authors would like to thank Golden Grove, 29Metals management for the opportunity to undertake this important work and present this paper, the site technical staff and contractors for supporting the process, and the operators for applying the outcomes.

REFERENCES

Clifford, B A, 1992. Facies and Palaeoenvironment Analysis of the Archaean Volcanic-Sedimentary Succession Hosting the Golden Grove Cu-Zn Massive Sulphide Deposits, WA.

Sharpe, R, 1999. The Archean Cu-Zn magnetite-rich Gossan Hill VHMS deposit, Western Australia: evidence of a structurally-focussed, exhalative and sub-seafloor replacement mineralising system, PhD thesis, University of Tasmania.

Watkins, K P and Hickman, A H, 1990. Geological Evolution and Mineralization of the Murchison Province Western Australia, Bulletin (Geological Survey of Western Australia), vol 137, 267 p.

Operational readiness – what is it and how to get it done?

A G L Pratt[1]

1. FAusIMM(CP), Principal, Adrian Pratt Mining Consultancy, Hampton Vic 3188.
   Email: adriangpratt@bigpond.com

ABSTRACT

Operational readiness is a commonly used term in mining and many other industry settings. However, translating this term into a coherent scope of work that is practical and meaningful is a challenge faced by many mining projects. This inhibits the allocation of resources to the work and its completion.

The problem is with understanding how the terminology applies to establishing a mining operation. Resolving this problem requires clarity around what operational readiness means for a mining project's transition to operations and who owns the tasks of completing operational readiness. The motivation for resolution of this problem is in appreciating the consequences arising from any lack of clarity in definition of this task, the timing of the work, and ownership for its completion. In broad terms, these consequences amount to an erosion of value at a critical time for both the new mine and the company and shareholders.

This situation often arises where operational readiness is not adequately addressed in project's study phases. As a result, neither time, resources nor money are adequately provided to complete the work required. In other cases, operational readiness is sometimes assimilated with the delivery of the project's physical assets by the project's construction team, rather than aligned with the establishment of the operating team to deliver the desired operational outputs.

This paper argues that operational readiness starts in a project's study phases and flows through into the start of operations. The definition of operational readiness is discussed as a task, and a model for its planning and completion is presented. This discussion includes assigning ownership for operational readiness planning and delivery to the General Manager of a new operation and the operations leadership team, who are central to the success of this work. After all, the reputations of this leadership team depend on the performance of the operation in its first few years.

The author's work over the last 20 years, leading operational readiness programs and the preparation of feasibility studies for mining projects, has provided the insights and evolution of ideas that are central to this paper. Operational readiness is fundamentally directed at ensuring a new mining operation gets away to a successful start and builds a solid foundation for the future of the mine/operation/province. It requires clarity of purpose, ownership and coordination by people in the operating team, and their preparation of the requisite process definition and establishment of the significant relationships required for operations. Without its successful completion, it is unlikely the new operation will meet and sustain its start-up targets. This paper discusses a practical approach to operational readiness for mining projects.

INTRODUCTION

Large amounts of time and effort in the study phases and money in the execution stage of mining projects is directed at selecting and establishing the physical assets that define the project, quantifying its production, and estimating its cash flow. Figure 1 illustrates the key principles defining the basis for this work.

Key principles

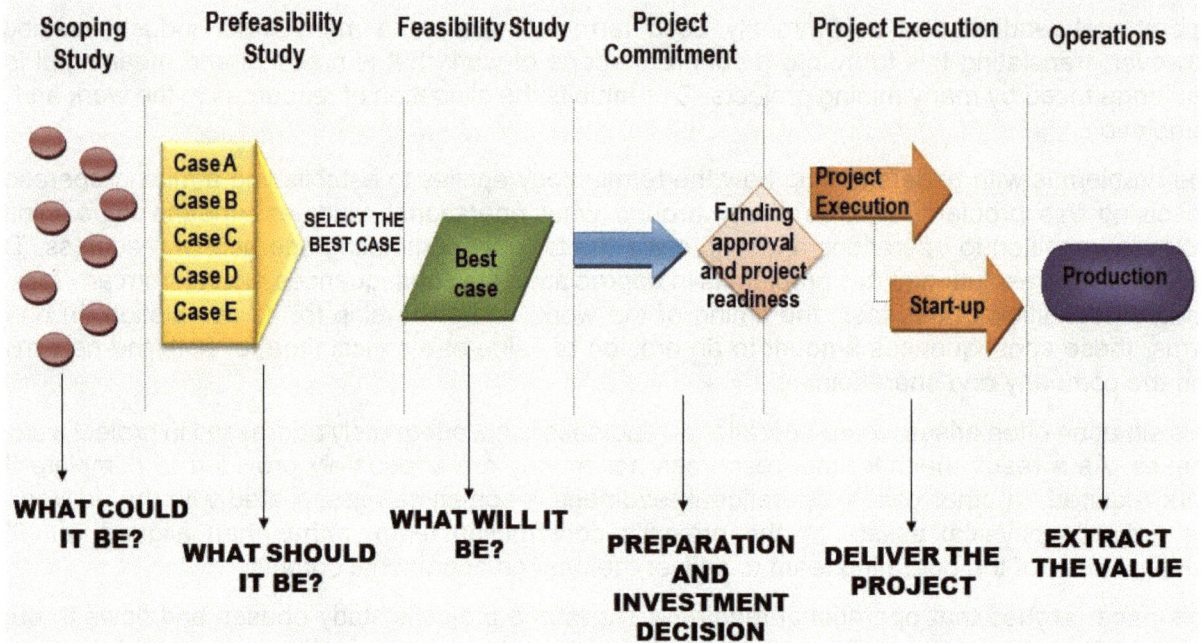

FIG 1 – Key principles that underpin the basis of studies (source: N Cusworth, pers comm, 2022).

Working through this process provides the project's proponent with tangible signs of progress for the definition and delivery of the physical components of a project that create a large part of its identity. A Feasibility Study (FS) describes these physicals and their delivery model in some depth and is succeeded by a Project Execution Plan (PEP). The PEP describes the delivery detail for the project's physical components and managing its construction, including:

- Scoping, engineering, design, and construction of infrastructure and facilities.

- Specification, procurement, delivery, and commissioning of infrastructure, facilities, and equipment required for mining and processing operations.

What can be, and is often, overlooked as part of the evolution from study phases to implementation are those components of a project's future identity as an operation that are less tangible than the physical assets. It is these less tangible components that determine how the physical components operate and how the operating team interact both internally and externally. This area of work, typically, does not fit with the skill sets of a project (execution) team. At the heart of this work is the question of the operating team's operational readiness.

For almost all new operations, getting away to 'a good start' is universally assumed to be true in the base case of the FS. In this context, 'good' means achieving all the targets set for its first 12 month to two years. Achieving these targets takes place against the background of the owner's operating team coming to grips with:

- New equipment and how they are going to use it.

- The processes for running operations to the satisfaction of the owner.

- The project location.

- The deposit (possibly the first time it has been mined and processed).

- Building the relationships within and outside their team that are essential for its performance.

- Inheriting the relationships, good and bad, that the project team established during the construction period – that is, the project's identity and reputation with its stakeholders.

The complexity and challenges of this task is influenced by the familiarity of the business behind the project with the scale and type of operation. Key to this is the operating knowledge they possess and the access they have to an appropriate business management framework. Central to success is the operating team's commitment to the company's mission and vision statements and supporting set of values (after Kaplan and Norton, 2001):

- Mission – provides a starting point for a business and defines why it exists for its owners.

- Vision – provides a tangible picture of the future of business or operation, which shows people what needs to be achieved in a way that they can imagine sharing in this achievement.

- Values – provides support for establishing the business' cultural environment, including how they aspire to get their work done and measure performance ie what is expected of people and their behaviour.

This author's experience with operational readiness covers several different perspectives at the commencement of mining operations. These perspectives inform the operational readiness model and approach outlined in this paper, including:

- As a member and leader of the operating team tasked with bringing the mine into production.

- As a member of the head office technical support team tasked with assessing progress and operational readiness performance of the operating team.

- As a member of the construction team observing the efforts of the operating team attempting to bring their new operation up to speed.

This paper provides a description of operational readiness for mining projects and a model for getting this work done using a consistent and coordinated approach to preparations for operations.

WHY IS OPERATIONAL READINESS IMPORTANT?

For the purposes of this paper, the term 'operational readiness' describes the work required to ensure the owner or operator of a project or asset under construction is equipped to take delivery of the asset at handover from the construction phase. Implicit in this definition is that, as the operator, they can accept responsibility for the asset and operating it in a way that is consistent with their business needs and chosen risk appetite.

To match this definition requires:

- Accessing and developing the skills and capabilities (knowledge) required to support the operation, including its maintenance – that is, the recruitment of operations and maintenance personnel and their initial training.

- Labour resources planning, allocating accountability, and defining communication within the organisation to ensure an operation can achieve its operational and business objectives, including transition of personnel from project to operations, inclusive of contractors, while optimising local employment.

- Defining the production ramp-up program and setting performance targets.

- Identifying and developing the protocols and processes required to commence operations and equipment maintenance, including risk management, and gaining timely access to project documents and vendor information.

- Consultation with stakeholders, community, and regulatory authorities to build effective relationships and provide appropriate information to establish and maintain their support for the operation.

- Introduction of equipment to the mine and the ramp-up program.

By and large, this work delivers nothing visible that can compete with the delivery of the physical components associated with the project's assets. However, if these requirements for a new operation are not addressed, the consequence can include some or all the following (Deloitte Touche Tomhatsu, 2012):

- poor safety, health, and environmental performance
- slow production ramp-up
- failure to meet planned ore tonnage and grade targets from the mine
- failure to meet design throughput
- failure to achieve design metal recovery
- excessive downtime due to equipment failures and time to repair
- unprepared workforce
- excessive waste of resources
- energy inefficiency
- elevated operational costs.

Project value is destroyed as the delivery of a best-case scenario is eroded, as illustrated in Figure 2.

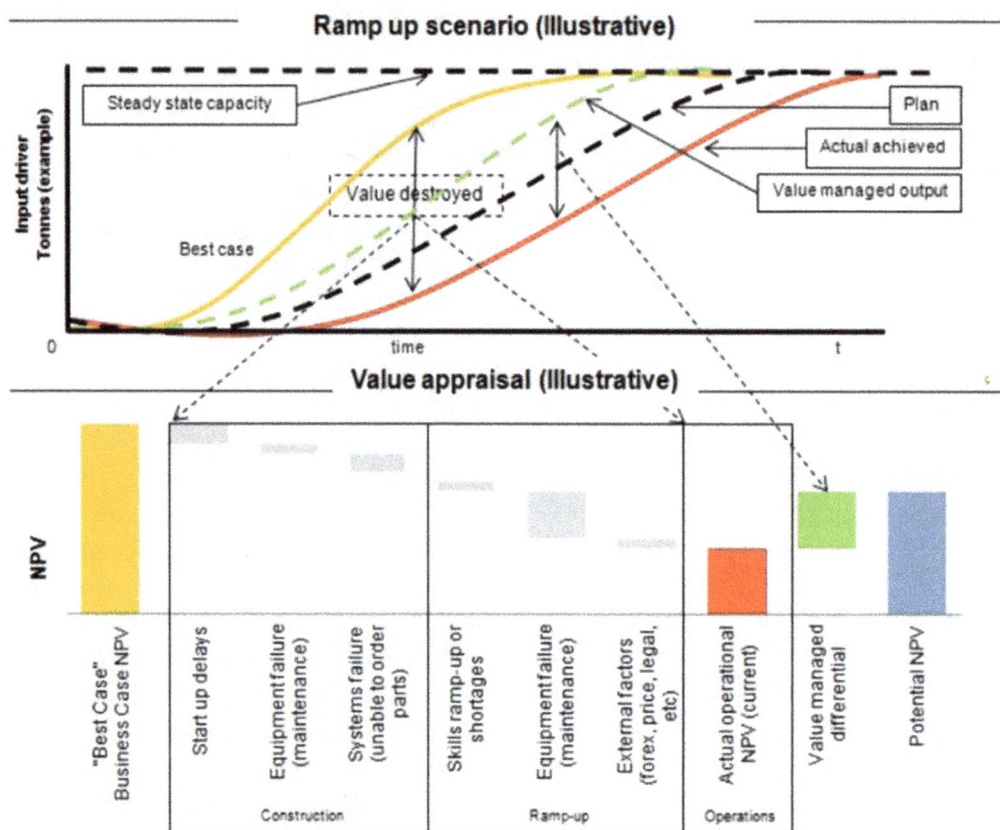

FIG 2 – Capital project value destruction (source: Deloitte Touche Tomhatsu, 2012).

This is not a theoretical point. Numerous authors have reported on mining project performance against their declared operability, costs, and schedule expectations, and it is not a pretty picture (Mackenzie and Cusworth, 2016, 2007; Ward and McCarthy, 1999). Ward and McCarthy also make the point that feasibility studies generally underestimate the effect of the 'learning curve' on production ramp-up. This author contends that operational readiness is a key factor in this learning curve phenomenon.

Members of the future operational team are sometimes involved during a FS to ensure that it is operationally sound. However, accomplishing this effectively requires thought around how their involvement is structured (Shillabeer, 2001; Kear, 2004). Similarly, project personnel are sometimes brought in to help with the development of FS schedules and cost estimates. All of this is to strengthen links between phases to make sure the operating logic that forms the basis of the FS is fit-for-purpose, soundly conceived, well-engineered, properly estimated (time and costs), and effectively delivered. The integration of study and project teams improves the chances of predicable outcome for projects (Mackenzie and Cusworth, 2016). A project where these aspects are adequately addressed has the best possible chance of meeting its performance targets if its operational readiness planning is effectively addressed. However, a poorly conceived, engineered, estimated (time and costs), or delivered project will unlikely be saved by its operational readiness. To prevent or mitigate the potential impacts of the consequences cited earlier requires preparation and coordinated completion of a specific program of work: the Operational Readiness Plan (ORP).

Preparation of an ORP entails:

- Understanding the business's capacity to support the new operation against the background of its business management framework.

- Evolving the business management systems used to support the project's construction phase and adding those systems needed for the operating phase.

This work is required to ensure a smooth transition from the project phase to initial operation and to sustain the operation's future performance. The ORP must address what the operation needs to get right from the start of operations to meet its best-case scenario (Deloitte, 2013). The key questions for an OPR to address include:

- What objectives does it need to pursue to achieve the critical outcomes targeted?

- What are the tasks and process areas that support these objectives?

- Who are the operation's business partners?

- What are their roles and the nature of their relationship?

- How is ORP progress measured and communicated?

The owner's business management framework is a key part of the foundations for operations readiness, as it is for their approach to 'project readiness', their capacity to function as the 'owner' of their project under the PEP, and their relationship with the Project Management Office (PMO) model chosen for the project.

The business management framework anchors the development of the PEP and ORP. As the project transitions from construction into operations, the PEP and ORP need a level of integration to enable them to seamlessly support establishing and maintaining the project's fundamental performance criteria of:

- social and legal licence to operate

- business performance.

These two fundamentals 'bookend' (Figure 3) a business's ability to establish a modern, sustainable, mining business and rely on:

- People – their organisation, engagement, capacity, competence, and deployment.

- Process – what the work is, how it gets done and by whom.

- External relationships – with regulators, community, business partners, etc.

These outcomes need support of a Sustainable Mining Business Model

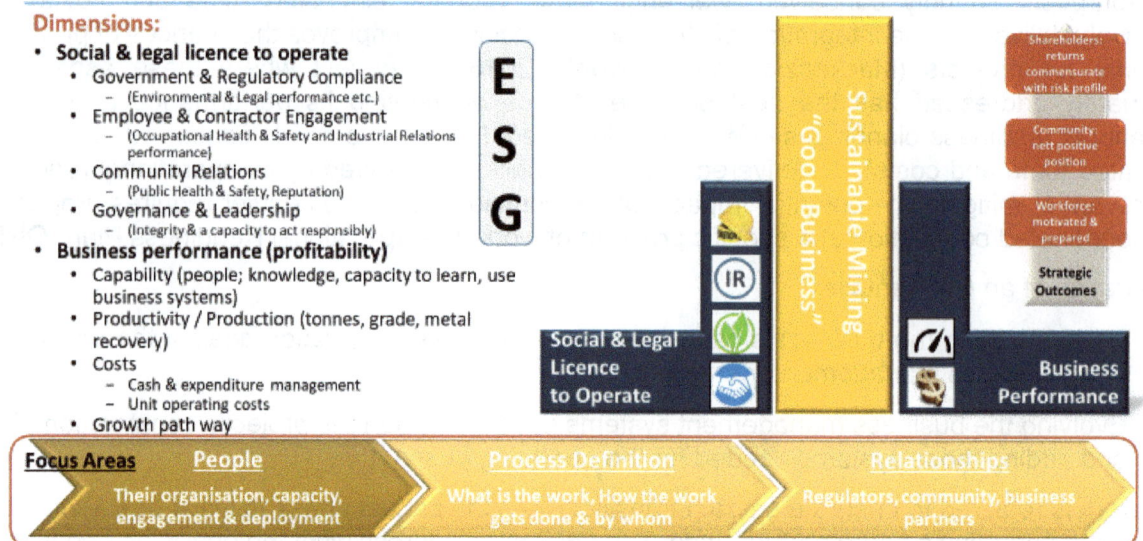

FIG 3 – Sustainability for a mining business.

Delivery of this type of sustainable operating model requires a business management framework that enables and maintains this focus (Figure 3). A typical business management framework used by the author is shown in Table 1. It contains five business management systems, which are divided into 24 system elements. These systems enable its people to get the work of the business done, ideally with no duplication, no overlaps, and no conflicts. Their purpose is to:

- Establish a seamless connection between project delivery and commencement of operations, and its interaction with its stakeholders.

- Provide alignment of business processes for effective planning and conduct of its activities.

- Enable a shared understanding of business risks and opportunities, and how they are managed.

TABLE 1

Typical business management framework.

Focus areas	No.	Business management systems	System elements
People – Their organisation, capacity, engagement and deployment	1	**Human Capital**	1.1 Organisation 1.2 Recruitment 1.3 Training and Development 1.4 Working Conditions and Benefits
Process – What is the work, How the work gets done and by whom	2	**Enterprise**	2.1 Governance, Risk and Compliance 2.2 Strategy and Planning 2.3 Managerial Leadership 2.4 Workplace and Work Environment 2.5 Response and Resolution of Incidents 2.6 Assurance and Business Reporting
	3	**Value Chain**	3.1 Exploration and Resources Definition 3.2 Project Development 3.3 Mining and Ore Reserves 3.4 Metallurgy and Processing 3.5 Gold Sales
	4	**Business Support**	4.1 Commercial Management 4.2 Asset Management 4.3 Infrastructure 4.4 Information Management 4.5 Administration 4.6 Environment
External Relationships – Regulators, community, business partners	5	**Stakeholder**	5.1 Community 5.2 Government and Regulator 5.3 Business Partners

The 24 elements divide into management plans with their supporting procedures. An owner will need many of these management plans in place to support its role in the PEP – that is, to be project-ready as an organisation. How many management plans depends on the owner's role in the project delivery. In the authors experience, as many as three-quarters of the operational management plans will be required for project execution, Figure 4.

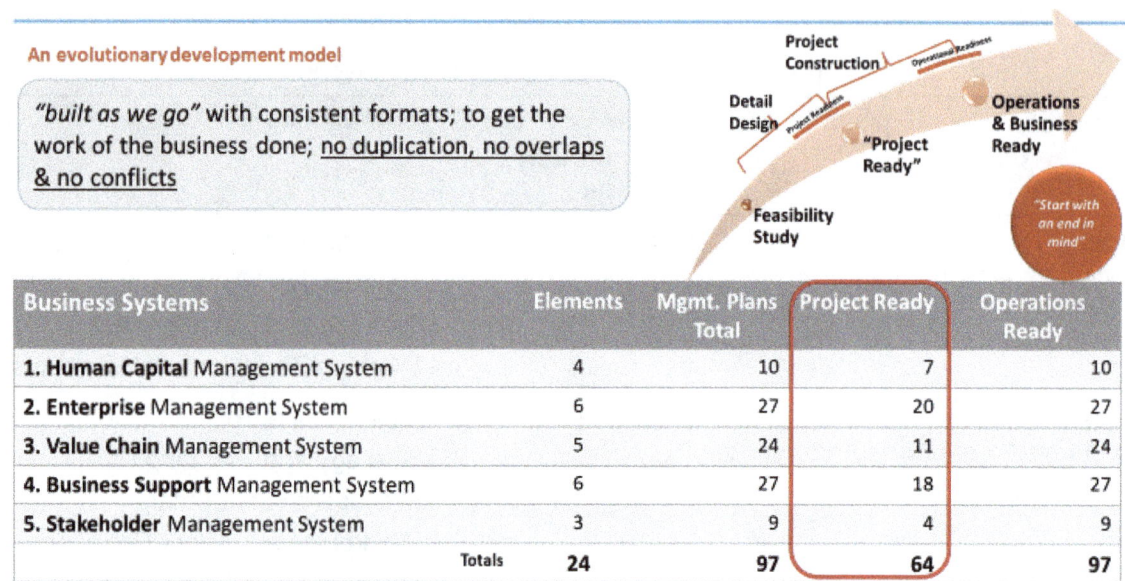

FIG 4 – Management System Evolution example.

An evolutionary development model

"built as we go" with consistent formats; to get the work of the business done; no duplication, no overlaps & no conflicts

Business Systems	Elements	Mgmt. Plans Total	Project Ready	Operations Ready
1. **Human Capital** Management System	4	10	7	10
2. **Enterprise** Management System	6	27	20	27
3. **Value Chain** Management System	5	24	11	24
4. **Business Support** Management System	6	27	18	27
5. **Stakeholder** Management System	3	9	4	9
Totals	24	97	64	97

Consistency in the structure and approach taken throughout this framework is important so that, where reasonable and practical, the documentation has the same look-and-feel across the business. A document hierarchy is provided in Figure 5. This structure assists and supports personnel with understanding and using this framework in a way that is consistent with the owner's vision and its values.

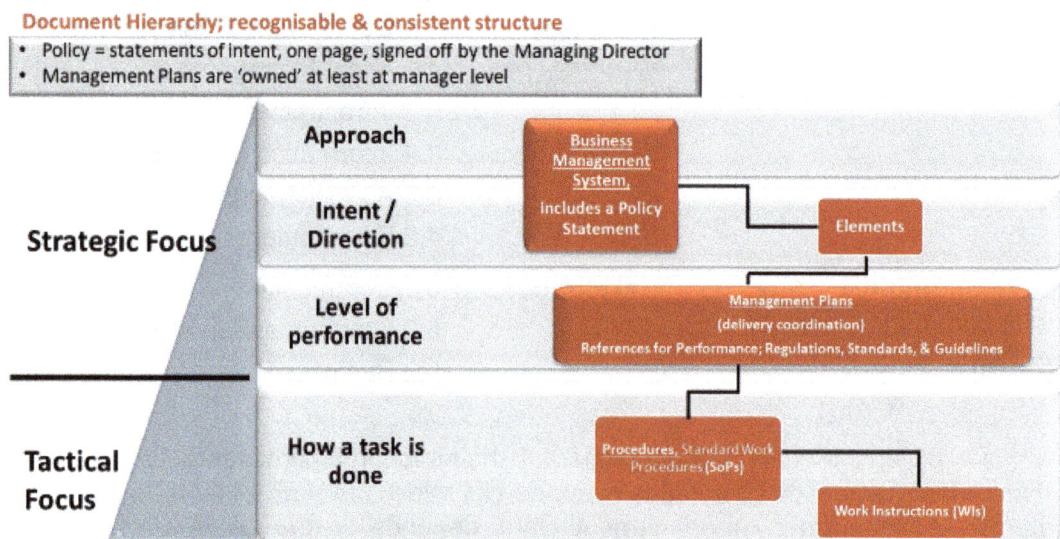

Document Hierarchy; recognisable & consistent structure
- Policy = statements of intent, one page, signed off by the Managing Director
- Management Plans are 'owned' at least at manager level

FIG 5 – Business management system document hierarchy.

The structural concepts for the documentation of this framework are:

- Business management systems have a clear association with a focus area (eg people), supported by a policy statement.

- System elements breakdown work within a business management system.

- Management plans address specific aspects of elements, coordinating the application of performance standards and definition of how work is done by referencing levels of performance and procedures.

- Government directives and regulations and industry standards and guidelines inform levels of performance.

- Procedures and work instructions describe how individual tasks are performed. These documents must be appropriate for communication with the personnel expected to use them.

Although the structure and development of a business management framework is a foundational input to operational readiness, it is not discussed further in this paper because it is not central to the main topic of the paper. The key point is that there needs to be a workable business management framework in place for project readiness, so the operational readiness work has something to build on.

The description of operation readiness at the beginning of this section is theoretical. In practice, a future operation needs to come up with a description for operational readiness that is meaningful to its circumstances and its people. This is important to building their commitment to operational readiness preparations. The intent of operational readiness is to avoid an operation falling victim to the fallacy that the personnel who operate and maintain the business will build the processes they need as they go (DiStefano, Gotetz and Storino, 2015).

WHO OWNS OPERATIONAL READINESS?

The handover process for a project from study to project stage is done at, or around, the time of the decision to proceed with the project. Those who lead the study phase step back to make way for those focusing on leading construction. The construction and execution tasks require a different skill set. The problem space is all about delivering the project's physicals: the scope, cost, and time to complete this task. It is this problem space that is at the heart of the PEP, the delivery of the project's physical components, and managing its construction, including:

- Scoping, engineering, design, and construction of infrastructure and facilities.

- Specification, procurement, delivery of these items and the commissioning of equipment and facilities required for mining and processing operations.

This is the realm of the Project Manager and the project team, whose focus is very different from the General Manager and the operating team. The project team has a short-term involvement, which continues until plant commissioning is complete and project custody transitions to the General Manager. Whereas the General Manager and operating team are there for the long-term, focused on the operation of the project's assets and maximising their operating life. That is, the operation's production. The operation's mining team likely lead this process off through their work in pre (plant) production pit and/or mine development.

The ORP is the operating team's equivalent to the project team's PEP. It is designed to deliver the people and establish the management processes and external relationships required to start operations and support production ramp. This work establishing management systems, their management plans and procedures, provides the basis for the operation's ESG performance.

The success of the operation's start-up relies heavily on the operating team's commitment and ownership of the ORP. Attracting and retaining people with the capability and capacity to complete operational readiness work is central to this plan. The selection of personnel for operational roles is typically based on the candidates' experience and performance in similar operating roles. Ideally, this would include experience gained during the establishment phase of an operation. This will equip the successful candidate with the awareness that, at least initially, their role in a new operation will be quite different to their role in an existing operation. The former involves significant additional work to establish new teams of people, define and embed the processes needed for operations, and to establish and maintain relationships. In most cases, simply recycling what they did in their previous job will not suffice.

The General Manager's ownership (accountability) of operating readiness does not mean they are solely responsible for facilitating and managing the process. Nor does it mean they can simply delegate establishing the ORP to their managers and functional area leaders. Instead, the executive leadership team need to consider supporting the General Manager's role to address this task. They can do this by establishing a specialist or champion engaged solely to lead and coordinate the ORP task, on behalf of the General Manager. The ORP Leader needs sufficient experience with mine start-ups and the skills to engage with the operating team leaders, managers, and superintendents.

The ORP Leader works with operating team leaders to prepare ORPs for each area of the business using a consistent approach and processes.

GETTING THE WORK DONE

The definition of ORP needs to start in the study phase. This builds in the roles, time, and money for the ORP work. Ideally, the FS should address the future operating model for the project and outline the components of an ORP, its resourcing requirements, and the implications for the project's capital costs.

The inclusion of an ORP outline as part of the FS provides the initial operating team with a core component of a scope of work prior to commissioning and the start of operations. The operating team's role is to address operational readiness and adjust its delivery based on insights offered by project's construction experience and their own experience. The timing of this work and its relationship with the delivery milestones for a mining project is important, as illustrated in Figure 6. Operational readiness has a clearly defined window for each project. If it is too late its relevance and impact is lost as the operation ramps up to its nameplate.

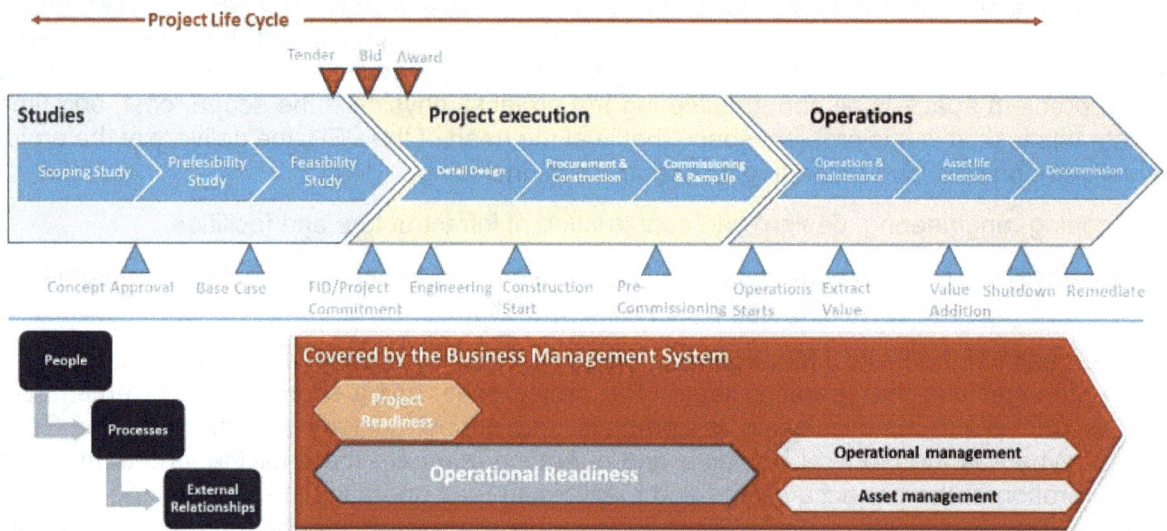

FIG 6 – Operational readiness and its place in the project timeline (after: Jahnig and Agoston, 2016).

An ORP provides a framework for the delivery of a set of objectives to establish an operation. The ORP focuses on the non-physical requirements of the project implementation – that is, the people, their organisation, the operating processes (how work is done), and the relationships with the community, regulators, and suppliers. These points of focus for the ORP are the same as those at the foundation of the owner's business management framework outlined earlier.

The creation of an ORP with a focus on these requirements at start-up gives the project the best possible chance of avoiding the consequences associated with failure to meet its planned ramp-up targets, as highlighted in Figure 2.

The project team supports the preparation of the ORP by providing relevant project documentation, equipment vendor information, as-bult drawings etc in a timely manner.

ORP objectives

To address the outcomes targeted by the ORP, it needs to incorporate the following objectives:

- establish effective labour resources planning for all the operation's functional areas
- select the right people for the right roles and appropriate training support
- develop a collaborative and effective working culture
- establish effective operating management systems, their plans (processes and procedures) and controls

- establish and maintain the external relationships needed by the operation with community, government, regulatory bodies, and supplies of good and services used by the operation

- ensure the operation has the tools it needs to commence and sustain production, eg minor equipment, software applications, and consumables.

The first three of the points above highlight the role played by the owner's/operation's human resources (HR) team. Establishing and resourcing the HR team and the processes they implement sets the tone for the personnel attracted to the operations. Underperformance by the HR team will make it difficult for the operation to get the people they need, when it needs them. Access to the people, at the right time, is a consistently observable key factor in the completion of all the operational readiness assignments the author has undertaken. These people need to be competent and confident in their roles at start-up for an operation to 'hit the ground running, safely' (C Moorhead, pers comm, 2016), and can rarely be employed too early (within reason…).

A second success factor in operational readiness is the structure and coordination of operational readiness work across the operations. This can be approached by building and applying a common workflow structure across the operation. To support this process, the operation is divided into its functional areas, with each area aligned with how costs are commonly consolidated in a mining operation: mining, processing, and general and administration (G&A). Functional areas are typically led by manager or superintendent level roles. The structure is based more on the type of work performed rather the organisational structure or who reports to whom.

Operations and operational readiness are about teamwork. Excellence in one or just a few areas will not achieve the goal. In this respect, it is like many areas of human endeavour, as said by basketballer Michael Jordan, 'talent wins games, but teamwork and intelligence win championships'.

ORP workflow structure

The ORP workflow illustrated in Figure 7 can be applied to each of the functional areas. The purpose of this structure is to link the ORP's objectives to tasks undertaken by each of the operation's functional areas to support commencement of the operations in line with its targeted outcomes. Importantly, this common workflow for functional areas ties the approach of each functional area together, which supports their capacity to learn from each other and provides cohesion in reporting progress. This cohesion improves the chances of achieving a meaningful understanding of an operation's actual state of readiness for operation. The alternative is for each functional area to 'run its own show' and tell their story their own way. This approach makes it much harder to gauge the holistic assessment of an operation's true state of operational readiness.

The workflow structure follows the framework used for the business management system's framework. This aids integration of business management systems with the ORP and its delivery, particularly around the management plans and their associated procedures.

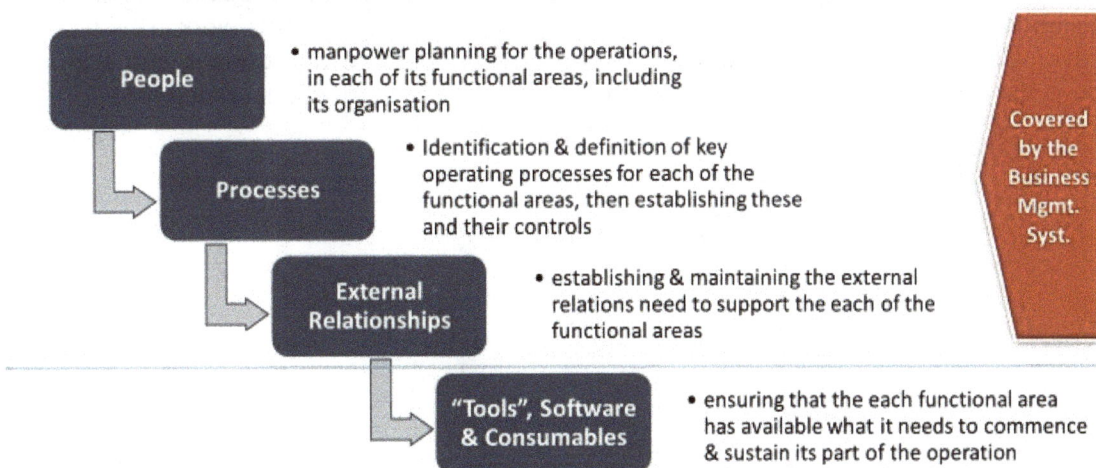

FIG 7 – ORP workflow structure.

Links with the operating risk register

The ORP links with the risk register prepared for the operation phase as part of the FS. The register identifies events and circumstances that have the potential to adversely impact the mine's operating performance – that is, the register identifies what could go wrong. It complements the ORP, which focuses on what the mine's operating team needs to get right from the start for the operation to meet its targeted operating outcomes. This is not simply about addressing high-risk events, which do need the requisite attention. It is also important to build the core capability for a successful operation (Deloitte, 2013). To achieve this, the ORP focuses on those human capabilities, work processes, and external relationships identified for the operation to meet its operating outcomes.

The development of a risk register for the operation and an ORP are both important and required to support any new mining and processing operation. The ORP is part of the transition process from construction project to an operation, whereas the risk register for operations is an enduring document and one the operation maintains and monitors throughout its life. The ORP supports functional area leaders prepare for the establishment of the operating systems and processes required to prevent or mitigate operating risks. The initial operational risk register, developed as part of the FS, is the source from which operational readiness tasks are identified and prioritised in the ORP.

ORP roles and responsibilities

The roles and responsibilities associated with a model for preparing and managing the work of an ORP are outlined in Table 2. Specifically, the ORP roles entail:

- Managing Director or Chief Operating Officer: ORP endorsement and support for operating team's leadership to enable it to deliver against the ORP's outcomes and objectives. Operational readiness is as much a mind-set as it is an approach or methodology. Only with the right direction from senior leadership will programs be able to bridge the gap between those who design and build, and those who will eventually operate (Deloitte, 2013).

- Project Manager: contributes to and supports the ORP process by ensuring the timely availability of project information, cooperation and coordination with the project team when required.

- General Manager Operations: responsible for managing the compilation of the mine's ORP and the coordination between operations managers to ensure the effective delivery of ORP outcomes and objectives.

- ORP Leader: facilitates the ORP, mentoring, monitoring, audit and review of progress against the ORP and distribution of this information to all required levels of the business.

- Operating team management: functional area leaders for operational and support focused work groups. Responsible for leadership and coordination of their teams to deliver the ORP requirements as they relate to their areas of the operation. That is, building their teams, the preparation of the operating systems and their associated procedures of work, and collaboration with other site leaders to support them to complete their work. This also includes developing their community and regulatory relationships as appropriate and ensuring they have access to the equipment they require to perform their functions effectively.

TABLE 2

RACI table for operational readiness planning.

ORP processes roles	Managing Director/COO	Project Manager	GM of Operations	ORP Leader	Functional Area Leaders	ORP teams
Plan endorsement	C	I	A	R	C	I
Preparation of the ORP document	C	C	A	R	C	I
Monitor and tracking progress against the ORP	I	I	A	R	C	I
Tasking and tracking of the work teams at functional area level to complete ORP task group work and tasks.	I	I	A	R	C	I
Tasking work teams at section level to complete ORP tasks	I	I	C	A	R	I
Capture of ORP intellectual property	C	C	A	R	C	I
Integration with and support of wet commissioning plans	I	C	A	C	R	C

R: **responsible** – owns the plan.
A: **accountable** – must sign-off/approve the plan.
C: **consulted** – needs to be consulted in plan preparation.
I: **informed** – informed of the results but does need consultation for its preparation.

Establishment of ORP activities

The implementation process for ORP begins with each functional area leader taking responsibility for a gap analysis in their area:

- Reviewing the business management systems, management plans and procedures that relate to their functional area prepared to date as part of the project phase. This identifies those aspects that need to evolve to meet the needs of an operation or need development for operations. These ORP tasks need to be raised with the General Manager Operations and ORP Leader and confirmed for inclusion in the ORP.

- Linking or associating each ORP task identified in their area either with those identified as part of the business management system and operations risk register in some cases.

- Reviewing the equipment and consumables needed and aligning these with either a business management plan or procedures, whichever has the best logic. The procurement arrangements for equipment and consumables becomes a component of the completion of ORP tasks rather than a separate standalone item.

- Allocating the responsibility for completion of ORP tasks amongst their team's members.

- Requiring each of their team members to develop realistic schedules for completion of their allocated ORP tasks, including the nomination of any necessary preceding work/tasks and those they believe are dependent on their work's completion.

A simplified flow chart for the preparation and communication of the ORP is shown as Figure 8.

ORP Preparation; planning, the work itself, & reporting

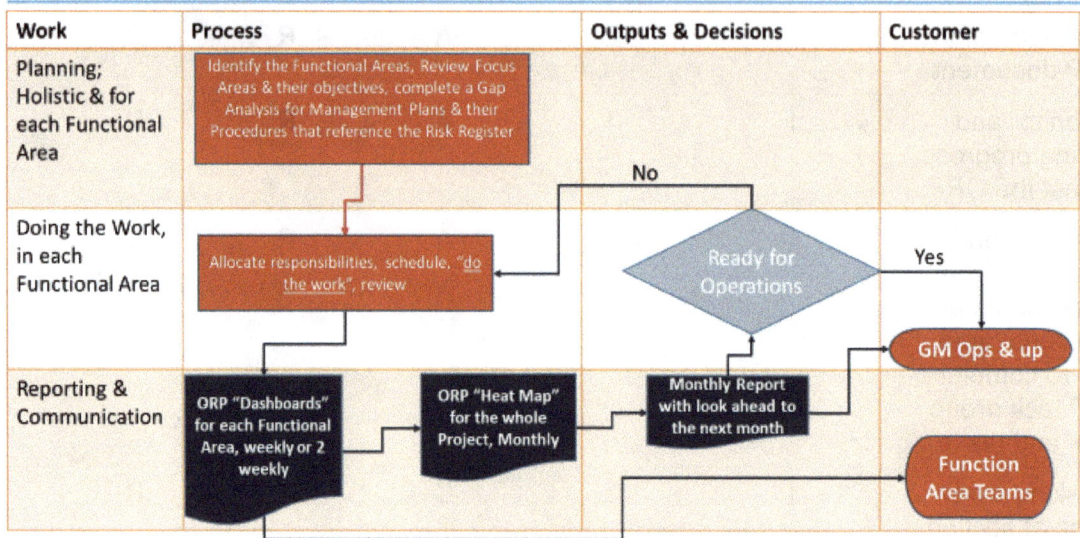

Work	Process	Outputs & Decisions	Customer
Planning; Holistic & for each Functional Area	Identify the Functional Areas, Review Focus Areas & their objectives, complete a Gap Analysis for Management Plans & their Procedures that reference the Risk Register	No	
Doing the Work, in each Functional Area	Allocate responsibilities, schedule, "do the work", review	Ready for Operations — Yes	GM Ops & up
Reporting & Communication	ORP "Dashboards" for each Functional Area, weekly or 2 weekly — ORP "Heat Map" for the whole Project, Monthly	Monthly Report with look ahead to the next month	Function Area Teams

FIG 8 – ORP – Simplified flow chart.

What is important in this process is the ownership by individual managers, superintendents, and their teams of the ORP work they need to complete to provide them with an acceptable foundation to commence operations. The capacity of functional area leaders to effectively communicate and manage the work within their own functional area and their interactions with the mine's other functional areas supports a successful ORP implementation. A component of good communication is clarity around the areas of responsibility for the ORP work. Functional areas, leaders and their potential scope are shown in Table 3, based on an Indonesian Gold project.

The objective is to complete the establishment phase of the ORP so that all ORP tasks are defined, responsibility for their completion allocated, and their timelines for completion estimated. The sequencing of tasks for completion requires logical consistency with the operations risk register. That is, those ORP task associated with, or linked to, higher risk ranked items get done first.

During this establishment phase, the ORP Leader meets with each of the functional area leaders and then with their teams. The purpose of these meetings is to explain the ORP process, its intent, and to support functional area teams in the definition of their ORP work, preparation of a schedule for their work, and assigning responsibilities for task completion. It is important they understand it is their work, they own it, and are responsible for getting it done. The ORP Leader is there to facilitate the process: not to do a functional area team's work.

Op. Cost Ass.	Functional area	Leaders	Potential scope
Mining	Geology and mineral resources	Geology Superintendent	Review and refresh existing processes plus the establish the additional requirements associated with grade control and reconciliation for gold mining
Mining	Mining, geotechnical and ore reserves (includes survey and mobile plant)	Mining Manager	Review and refresh with minimal change to existing processes
Processing	Metallurgy mineral processing (includes fixed plant)	Processing Manager	Review and refresh existing processes plus the establish the additional requirements associated with process control and reconciliation for gold processing; oxide and primary ore
Processing	Tailings storage facilities	Senior Engineer / Responsible Tailings Facility Manager* – TSF, Water and Sediment Control	Review and refresh existing processes plus the establish the additional requirements associated with gold processing
G&A	Human resources (HR)	HR Manager	Review and refresh with minimal change to existing processes
G&A	Occupational health and safety (OH&S)	OH&S Manager	Review and refresh with minimal change to existing processes
G&A	Site services and accommodation management	Manager Site services	Review and refresh with minimal change to existing processes
G&A	Site access and security	Manager Security	Review and refresh existing processes plus the establish the additional requirements associated with gold security
G&A	Supply – purchasing, logistics and warehouse	Commercial Manager Logistic and Warehouse Superintendent	Review and refresh with minimal change to existing processes
G&A	Information and communications technology (ICT)	ICT Superintendent	Review and refresh with minimal change to existing processes
G&A	External (stakeholder) relations	External Relations Manager	Review and refresh with minimal change to existing processes

Op. Cost Ass.	Functional area	Leaders	Potential scope
G&A	General and business management	General Manager	Review and refresh with minimal change to existing processes
G&A	Water and sediment management	Senior Engineer – TSF, Water and Sediment Control	Review and refresh with minimal change to existing processes
G&A	Non-Processing Infrastructure, facilities, engineering and maintenance	Maintenance Manager	Review and refresh existing processes plus the establish the additional requirements associated with gold mining and processing
G&A	Community engagement	Community Relations Manager	Review and refresh with minimal change to existing processes
G&A	Environment	Environment Manager	Review and refresh with minimal change to existing processes

* As defined by the Global Industry Standard for Tailings Management (GISTM, 2020).

Tracking, monitoring, and communicating progress

The tracking and monitoring of progress for ORP management and communication is outlined in Figures 9 and 10. Table 4 outlines key participants in meetings, their frequency, the structure of this communication, areas of interest, information horizons and supporting documents. The purpose of these meetings is to drive ORP performance, to identify and to commit to corrective actions when required. They provide a forum to discuss progress against the ORP workflows and to address issues that may arise regarding priorities in relation to ORP activities and minimum levels of achievement to commence operations.

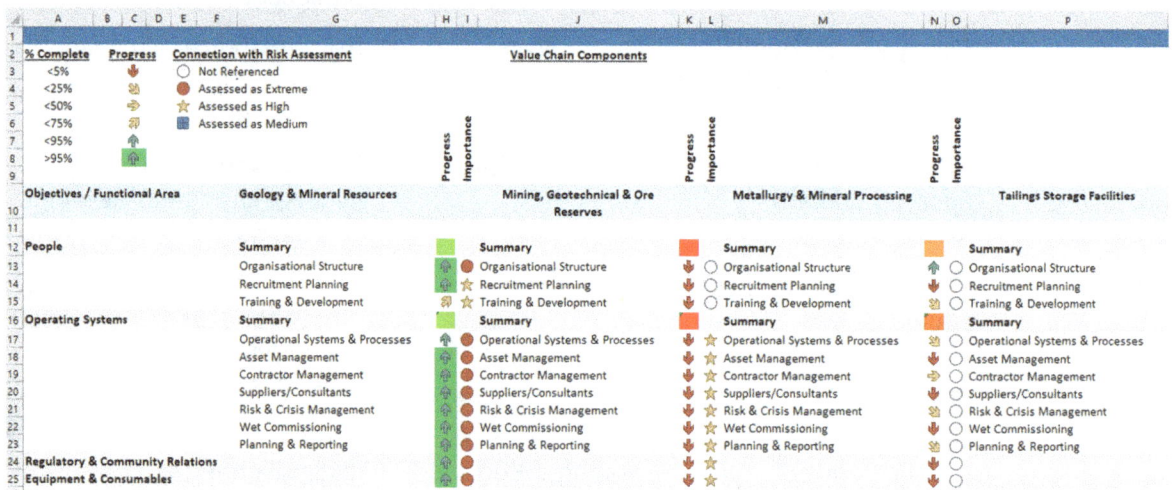

FIG 9 – ORP 'Dashboard'.

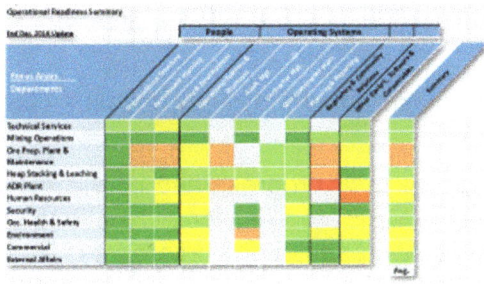

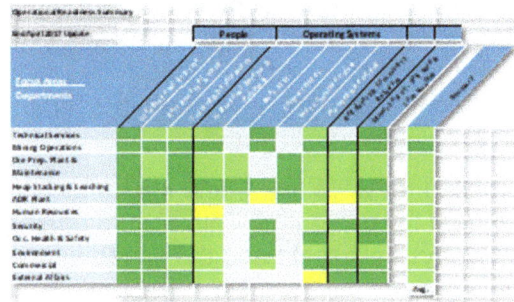

FIG 10 – Example ORP heat map, greenfield gold project, months at 5 and 9 (commissioning).

TABLE 4

ORP reporting and scheduled monitoring, example.

Frequency	Participants	Order of business	Supporting documentation
Monthly	Managing Director/Chief Operating Officer General Manager Operations Project Manager ORP Leader	Overall progress Priorities Deferrals outside plan	Heat map Notes addressing progress from the previous month and identified issues
Fortnightly (to start with)	General Manager Operations ORP Leader Functional area leaders	Overall progress Priorities Area progress against plan	Heat map Dashboards
Weekly	Functional area leaders Functional area teams	Area progress against schedule (complete, in progress and forecast to complete, deferrals within schedule, conflicts, priorities)	Heat map Dashboard Forecast completion of all task groups associated with a risk ranking of extreme or high Rolled up Microsoft Project schedule to task group level
Daily	Functional area teams	Progress at an area level	Area Microsoft Project schedule

In addition to using a schedule to manage ORP task status tracking and activity forecasting by each functional area, an operational readiness dashboard is useful for the ORP Leader and functional area's leaders to maintain focus on important ORP tasks. The dashboard covers each functional areas tasks, consolidating them and referencing the business management systems and their management plans and procedures.

These are typically constructed as a Microsoft Excel spreadsheet, similar to the dashboard shown in Figure 9. The dashboard format is configured to highlight:

- Progress status – based on percent-complete, with a range from no material progress (not started), to work in progress, to complete.

- Importance – this is linked to mine's risk matrix rankings, extreme, high, medium, and not referenced in the risk assessment.

The dashboard assists functional area teams to communicate their progress and connection from their schedule of tasks through to overall 'heat map' shows the ORP progress for all functional areas. The heat map is best suited to external reporting and high-level presentations on progress. Both are constructed from progress tracking in each functional area's ORP schedule. An example ORP heat map is shown in Figure 10.

The process for developing and implementing an ORP involves staging with a gradually increasing requirement for personnel. This occurs against the background of the construction and delivery of a project's physical assets and the scheduled start of wet commissioning. The foundation to this work is the capacity of the business to recruit a leadership team for the operation with the experience and capacity to build the business management systems, management plans and procedures needed for the operations phase in a time frame that enables the business to commence operations in a way that is consistent with the owner's risk appetite.

Once the focus moves beyond the initial focus on the recruitment of the operation leadership team, this team drives the evolution of the operational aspects of the business management framework and supports recruitment of leaders within their teams. These are the people needed to define operational processes and to establish the relationships that sustain the operation's performance. What this means in practice is the heat map (Figure 10) starts to change from red to green, initially on the left side, spreading to the right side.

Recruitment takes time, especially if you a looking for personnel who can provide what is needed for some of the key roles in operational readiness and in the early days of an operation. Often this takes longer than anticipated and involves convincing people to leave secure roles in established organisations where they are successful and have a future.

Relationship building in some key areas takes time and the operations staff need to be in place early. To achieve this, they could potentially be seconded to the project team to allow the opportunities to establish these relationships, which then roll through into the operation.

Implication of project capital costs

The requirement for operational readiness has clear implications for the preparation of the project's capital costs. The work requires people, takes time, and needs money for its completion.

Many guidelines and standards in mining project studies require statement of an operation's operating and maintenance philosophy, organisational structures, mobilisation schedule for operations personnel etc. If this is merely a purposeful description of requirements resolved in spreadsheets of costs, it is easy to overlook the resources, time demands, and scale of the actual process of recruitment. It also does not address questions of what personnel are required to do when they arrive, the time and resources they will need for operational readiness, and their estimated costs. The development of some processes may require input from outside providers, and some will require supporting software applications. The costs of establishing and maintaining software for a new mining operation are often underestimated or overlooked in the study phase.

Within the owner's cost area of a project's Work Breakdown Structure (WBS), it is important to separate the costs for assembly of the operating team and their completion of operational readiness work from the project management costs, construction and commissioning. This is to ensure their visibility and ownership by the General Manager Operations.

Lessons for next time

The final task of the ORP Leader is to ensure lessons are learned from the success or otherwise of the operational readiness program, by recording them, typically 6–12 months after start-up. Capturing these is definitely easier and more likely when a new operation gets away to a good start, but it is even more important to learn the lessons from a poor start. In the latter case, some form of review and adjustments will likely be made to mitigate the impact of a poor start. In either case, recording the lessons learnt is valuable.

CONCLUSIONS

Operational readiness is the program of work required to ensure the operating team, the operators of the project, are ready. It needs to engage them and be meaningful to them to gain their commitment to undertake this work. Operational readiness is more than just a focus on testing and commissioning equipment, although it might be part of it. It is more about establishing teams of people and work processes with sound foundations in the vision and values the company has set for itself. What is needed is more than a review process built around checklists prior to commissioning and commencement of operations. At this late stage, it more likely be reactive and focusing on what is not in place.

Implementation of operational readiness needs to be considered against the operating organisation's current capabilities and its capacity to:

- Undertake the leadership and definition of the work required with the people available to establish an operating team.

- Capture the commitment of the operation's managers and superintendents and senior professionals, for them to embrace it.

- Give the process meaning and identity at their mine – to get a good start for their operations.

Irrespective of the preferred implementation model for operational readiness, the insight gained by the author across several projects, and information reviewed in the preparation of this paper, stresses the importance of:

- Starting early, with the feasibility study, to provide a scope for operational readiness and its assignment to the operating team, capture their time and costs for its preparation and delivery, which is the basis for an operational readiness plan.

- Ensuring senior sponsorship, with executive leadership and commitment to a pragmatic approach to operational readiness. This leadership supports an operating team to drive and manage the implementation of operational readiness tasks and gives it the best possible chance delivering the targeted results for the operation as it transitions into operation.

- Clear definition of what operational readiness is, against the background of the Vision for the new operation that is meaningful to its circumstances and its people, so it matters to them.

- Development of an overarching approach to operational readiness using a set of objectives to establish an operation that aligns with the Owner's Business Management Framework and has a clear connection to the business's Values and Vision for the operation.

- An integrated plan for operational readiness, with a transparent and consistent structure across the functional areas for the operation. This plan needs a realistic assessment of where it is starting from for operational readiness to understanding how to move forward and from there it is all about knowing what the objectives are and how you are going to get there holistically as an operation. Otherwise, 'If you don't know where you are going, any road will get you there' (this is essentially a paraphrase of an exchange between Alice and the Cheshire Cat in ch 6 of Lewis Carroll's Alice's Adventures in Wonderland.).

- Establishing clear roles, covering the definition of the project's components and allocation of responsibility for their delivery, ie all significant physical elements assigned to the Project Team and those which focus on establishing the people, the processes they require, and the relationships they need to sustain the operation assigned to the Operating Team. The ownership and leadership by the General Manager of the operation and their operating team is key – theirs are the reputations made or damaged against the background of the operation's start-up.

- Measuring and communicating progress towards operational readiness, using quantitative and qualitative assessment tools to enable appropriated communication for different stakeholders in the delivery of operational readiness.

- Focus on what needs 'to go right' at the start of operation, building the core capability for a successful operation, focusing too much on what might go wrong can be a distraction.

Operational readiness if not addressed, or poorly done, will likely contribute to the destruction of project value in that first year of operation frequently associated with the start of new mining operations.

Operational readiness matters to the operators, to this point they only need to be reminded:

- 'Operations that start well, tend to be the ones that go well' (C Moorhead, pers comm, 2016).

- 'It takes 20 years to build a reputation and five minutes to ruin it. If you think about that, you'll do things differently' – Warren Buffet.

ACKNOWLEDGEMENTS

The author thanks the many colleagues and clients for their time suffering through various discussions around this topic over the past ten years. This is where the ideas present in this paper come from. The support and encouragement of Colin Moorhead and the team at PT Merdeka Copper Gold's Tujuh Bukit Gold Mine was especially important in the development of this work.

REFERENCES

Deloitte Touche Tomhatsu, 2012. Effective Operational Readiness of Large Mining Capital Projects Avoiding value leakage in the transition from project execution into operations [online]. Available from: <http://deloitteblog.co.za/files/icp/Effective_Operational_Readiness.pdf>

Deloitte, 2013. Operations go-live! Mastering the people side of operational readiness [online]. Available from: <https://www2.deloitte.com/content/dam/Deloitte/global/documents/gx-icp-operation-go-live.pdf>

DiStefano, B, Gotetz, W and Storino, B, 2015. Operational Readiness: Bridging the Gap Between Construction and Operations for New Capital Assets [online]. Available from: <http://reliabilityweb.com/index.php/articles/bridging_the_gap_between_construction_and_operations/> (accessed August 2015).

Global Industry Standard for Tailings Management (GISTM), August 2020. Global Tailings Review.org.

Jahnig, D and Agoston, J, 2016. Risk management – is your capital project ready to deliver business value?, A White Paper Issued by The Navigant Construction Forum™, accessed 5/10/2016.

Kaplan, R S and Norton, D P, 2001. The Strategy Focused Organization: how balanced score card companies thrive the new business environment, pp 72–73 (Harvard Business School Press).

Kear, R M, 2004. Mine project life cycle, in *Proceedings MassMin 2004* (eds: A Karzulovic and M A Alfaro), pp 117–120 (Instituto de Ingenieros de Chile: Santiago).

Mackenzie, W and Cusworth, N, 2007. The use and abuse of feasibility studies, in *Proceedings Project Evaluation 2007*, pp 65–76 (The Australasian Institute of Mining and Metallurgy: Melbourne).

Mackenzie, W R and Cusworth, N, 2016. The Use and Abuse of Feasibility Studies – Has Anything Changed? in *Proceedings Project Evaluation 2016*, pp 133–148 (The Australasian Institute of Mining and Metallurgy: Melbourne).

Shillabeer, J H, 2001. Lessons Learned Preparing Mining Feasibility Studies, in *Mineral Resources and Ore Reserves Estimation – The AusIMM Guide to Good Practice* (ed: AC Edwards), pp 435–440 (The Australasian Institute of Mining and Metallurgy: Melbourne).

Ward, D J and McCarthy, P L, 1999. Start up performance of new base metals projects, in Adding Value to the Carpentaria Mineral Province, Mount Isa.

Thin spray liners for face support at Kanmantoo

B Roache[1], J Jardine[2] and B Sainsbury[3]

1. Principal Geotechnical Engineer, Neboro, Melbourne Vic 3000. Email: broache@neboro.com.au
2. MAusIMM, Mining Manager – Kanmantoo Copper, Hillgrove Resources Limited, Kanmantoo, SA 5252. Email: jol.jardine@hillgroveresources.com.au
3. Professor (Geomechanics), Deakin University, Waurn Ponds Vic 3216. Email: breanne.sainsbury@deakin.edu.au

ABSTRACT

Ground support installation into the face of development headings has become an accepted practice at many Australian hard rock mines. Ground support is installed in the development face to reduce personnel exposure to rockfall while conducting work tasks such as marking up, installing lifter tubes, charging, and face sampling.

The ground support system installed in the face is typically friction bolts and mesh. This support system can be effective at reducing rockfall occurrence, but it also increases the cost of development per metre, in terms of capital for the ground support, jumbo and operator capital allocation and the installation time slowing each drill and blast cycle. Other issues associated with meshing and bolting the development face include management of rock bolts and mesh sheets in the fired dirt of every cut, damage to services such as vent duct, geometry issues when installing ground support onto irregular development profiles and operator error.

Fibrecrete installation onto the development face for support is an option considered by many mines, but it can be difficult to manage due to the removal of visual indications of misfires and slowing the mining cycle due to the wait time for fibrecrete curing.

The use of Thin Spray Liner (TSL) products in underground hard rock mines has been relatively infrequent in the past, mainly due to issues with cost, health and safety, and liner strength. Kanmantoo Mine investigated the use of TSLs for the purpose of development face support, while collecting data on the costs and practicalities associated with their use. Several TSL products were trialled and a preferred TSL product was selected for use as face support. Cost savings, drill and blast cycle timing improvements and other benefits were achieved while meeting the requirement to reduce rockfall risk.

INTRODUCTION

Development faces are formed during each hard rock mining drill and blast cycle. Historically at most mines each development face was unsupported and not reinforced with rock bolts and mesh or surface liners due to their temporary nature. This has gradually become a less widely accepted practice over the last ten years with a switch to installing ground support into the face at many underground mines (Kolapo et al, 2021; Chang et al, 2017; Kanda and Stacey, 2019). This change is mainly due to a desire to reduce the exposure of personnel working at the face to small falling rocks dislodging from higher on the development face.

Installing rock bolts and mesh to support the development face increases the development cost per metre, in terms of capital cost for the ground support, jumbo drill, operator capital allocation and the ground support installation time slows each drill and blast cycle. There is other flow on issues to be considered when rock bolting and meshing development faces, such as the removal of rock bolts and mesh sheets from the fired dirt, geometry issues when installing ground support into irregular face profiles and operator error.

Hillgrove Resources owns Kanmantoo Mine which is located approximately 55 km from Adelaide in South Australia. Kanmantoo Mine investigated the use of TSL products for the purpose of development face support, while collecting data on the costs associated with its use. Several different TSL products were trialled and a preferred TSL product was selected for use. Cost savings and drill and blast cycle timing improvements and other benefits have been achieved while meeting the requirement to retain the face and reduce rockfall risk.

BACKGROUND

Face support background

Ground support installed into development heading faces in static conditions, not seismic, became more common at Australian hard rock mines in the last 10 years. The change to supported development faces occurred in response to ongoing injuries and fatalities experienced by personnel conducting work activities at the face. Mining regulators, such as the Queensland Mining Inspectorate (2016) issued detailed guidance material for mine operators managing development faces and concluded:

> *Current approaches, largely managed by procedural controls have been shown to be ineffective in protecting workers against rockfalls in a dependable manner. Meshing of the face as done by some mines would add a considerably greater level of protection to workers at the face from falling rocks.*

After a fatality in South Australia during 2015 in which an operator was struck by a rockfall at an unsupported development face (South Australian Employment Tribunal, 2018), SafeWork SA took compliance action issuing an improvement notice to all mines in the state to implement adequate controls to manage the risk associated with rockfalls at active development faces.

It is important to acknowledge that not all Australian hard rock mines support development faces, although most development faces will be mechanically scaled by a jumbo. The most common ground support installed to support and retain the development face are friction bolts and mesh, with variations on this being the friction bolt length, mesh sheet size and gauge, and face coverage extent. In rare occurrences, shotcrete/fibrecrete is also used. Most mine operators accept that for static conditions the face support is not required below the grade line. Two examples of friction bolts and mesh sheets installed to support the face are shown in Figure 1. The example on the left provides more complete mesh coverage to the margins of the face, while the example on the right focuses on supporting the higher face region.

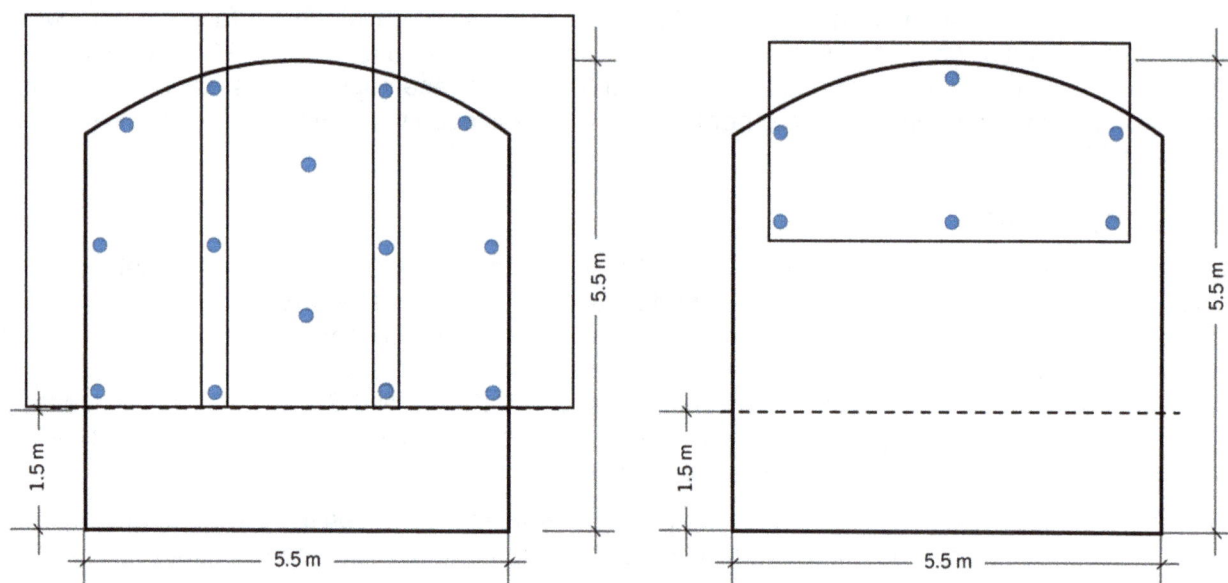

FIG 1 – Development face support examples using friction bolts and mesh.

TSL use in hard rock mines

TSLs have a long history of being trialled for use as surface support in Australian hard rock mines. TSL use in underground hard rock mines and issues related to their usage are offered by Potvin and Hadjigeorgiou (2020). They note the technology has been around for a long time with mine trials in the 1990s, and although commercial products became available that offered the potential for efficient material use, lower mining costs, fast curing and ease of spraying, they were not adopted for use as regular ground support mainly due to a perceived lack of tensile strength when compared to 50 mm thickness fibrecrete.

TSLs with fast curing characteristics, high tensile strength and some ductility will usually contain isocyanates. Isocyanates are highly reactive chemicals that are used to make flexible foams, resins and TSLs. Depending on the TSL product, fast curing products may contain methyl disphenyl isocyanate (MDI) or a variation. MDIs are known to be hazardous to workers, to both skin exposure and inhalation and these aspects need to be considered when selecting a TSL product for use in a mine. Inhalation of the TSL product should be avoided and managing this aspect is difficult when the TSL application process could include creation of an aerosol, vapour or mist.

The reasons for moving away from meshing and bolting the face

There are several ongoing issues related to the use of mesh and bolts in the development face. The time taken to mesh and bolt the face and the associated cost of this activity are key issues, but the less discussed issues relate to the impact on mining efficiency and safety.

When used as face support, the mesh and bolts end up in the fired dirt following the next cut being blasted. This has significant impacts for material handling in non-truck haulage mines, with steel removal processes and infrastructure required to remove the mesh and bolts from the dirt. In truck haulage mines, ventilation ducting is commonly damaged due to protruding steel from loaded trucks. Steel separation from ore usually takes place on the ROM which is labour intensive, hazardous, time consuming and costly.

Geometry issues are often experienced when mesh positioning on faces leads to ineffective ground support installations, particularly at intersections and turns, or where ore drives follow ore on a cut by cut basis, with wall stripping often required. The reason that faces are supported is to reduce rockfall risk when personnel are working at the face, but when overbreak occurs there will often be gaps in the ground support coverage. This may mean the installed ground support is not compliant with the design as shown in Figure 2. To address gaps in ground support coverage caused by overbreak, some mines will install additional mesh strips cut to size to fill all gaps in the ground support coverage. Other issues such as incorrect bolting angles can cause rock bolts to be speared into the backs of the next cut. This delays bogging and face meshing post blast when protruding rock bolts and mesh need to be cut-off by a service crew progressively interrupting bogging and ground support steps in the mining cycle.

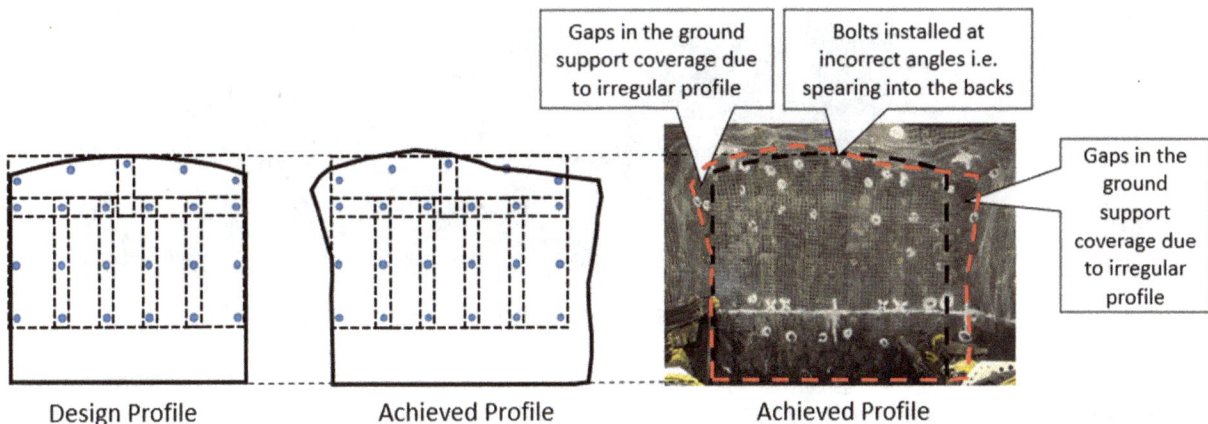

FIG 2 – Design profile verses achieved profile issues.

TSL USE FOR DEVELOPMENT FACE SUPPORT

Site trials of TSLs

Kanmantoo Mine trialled five TSL products for potential use as development face support. The selection of a preferred product was achieved by conducting trial sprays at the mine, with the operational crews involved in applying each TSL product. The operational crews gave their opinion on the usability aspects of each TSL product. The overall selection criteria were broadly based around the following:

- Ease of use, including storage requirements, transport and product temperature for spraying.

- The resources required to spray the TSL, such as the number of operators and machines.

- The material safety data information with consideration to the amount of direct exposure to personnel and whether the product formed a dust, aerosol, mist or vapour when sprayed.

- Spray performance such as ease and speed of TSL spray application, adhesion, rate of hardening and material loss.

- Usage cost.

The selection criteria enabled the eventual choice of a preferred TSL product (Silcrete™ TSL from Polymer Group Ltd) which was supplied as a two-component liquid that is mixed while spraying. This product demonstrated several positive attributes, such as ease of spraying, simple spray set-up, minimal reliance on mine resources such as personnel and machinery, no production of a vapour or mist while spraying and fast strength gain. Some of the less favourable aspects were the cost of the TSL, the requirement to maintain the TSL liquid components at or above 17°C for spraying and the need to keep one component agitated prior to decanting. A development face supported during the trial with the preferred TSL is shown in Figure 3.

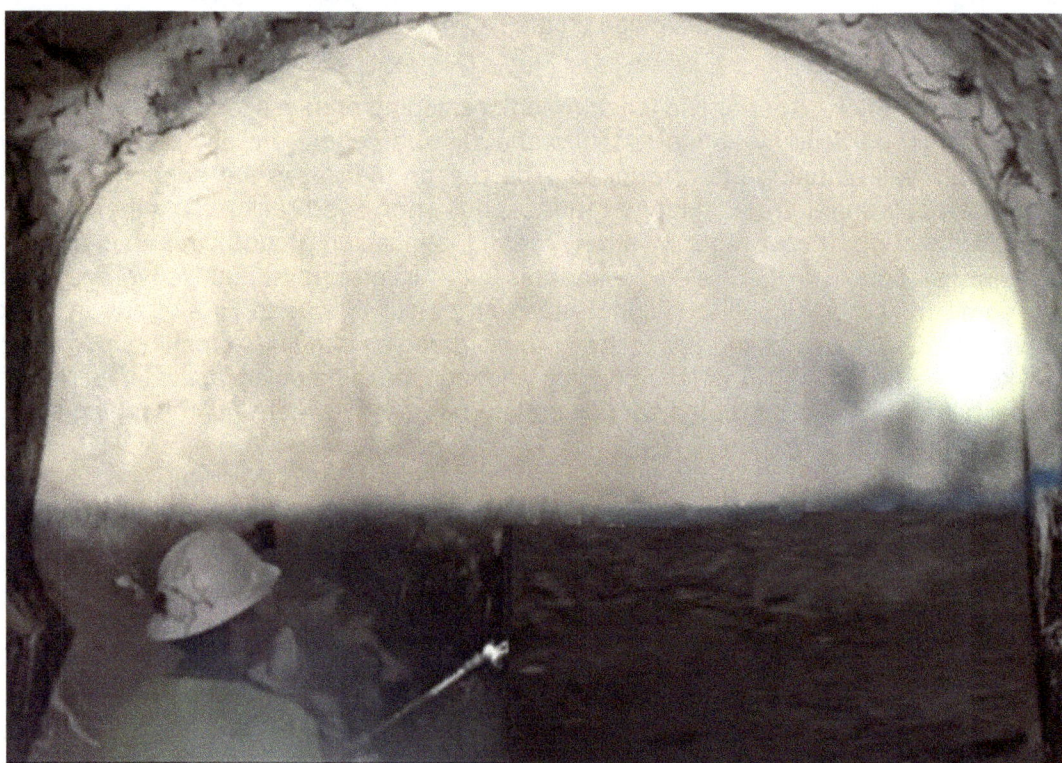

FIG 3 – Kanmantoo development face supported with Silcrete™ TSL.

Personnel exposure to TSL liquid via splashes or spills was controlled by using full arm clothing, rubber gloves and goggles. A face respirator was used while spraying underground, with a 10 m restricted area established when spraying at the face. Periodic atmospheric monitoring to ensure underground mine areas are complaint with workplace exposure standards for airborne isocyanate is an ongoing occupational health and safety requirement.

An important part in reducing the occurrence of airborne isocyanate in vapours, mists or aerosols while spraying is the pump selection. At Kanmantoo a gear pump was used that provides smooth pulse free flow and could run effectively for spraying the preferred TSL at low pressures. These aspects reduce the generation of airborne isocyanate while spraying.

TSL sample selection and testing

A series of laboratory tests were conducted to understand the performance of the preferred TSL product. In each case the samples were prepared and cured on-site and shipped to the laboratory for testing. A summary of some of the tests conducted are provided.

Laboratory adhesion testing

Direct adhesion testing was completed on small-scale (20 mm diameter) TSL cylinders that had been cast onto high-strength cement cores (40 MPa indicative strength) as shown in Figure 4. The samples were tensioned at a rate of 0.01 mm/min until failure.

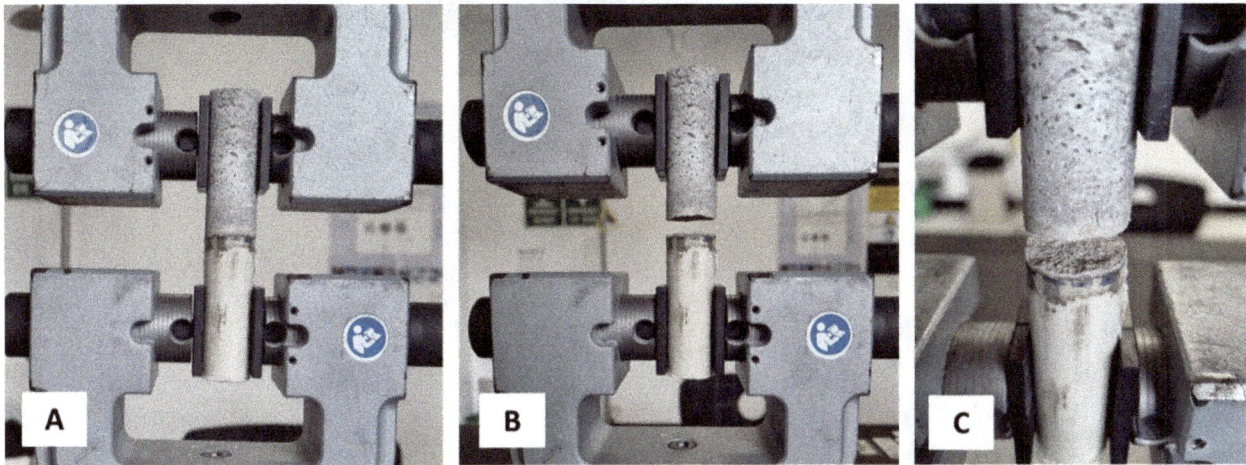

FIG 4 – Adhesion testing samples of Silcrete™ TSL.

In each instance the tension failure occurred in the cement cylinder and not in the TSL or at the cement/TSL bond surface. The force at failure was >315 N in all cases (eg >1 MPa) that suggests the bond strength of the TSL is sufficiently high, eg will not result in slabbing of the product off the surface after placement (EFNARC, 2008).

Laboratory linear block support test

The block push through test (TSL linear block support test) using the EFNARC (2008) process was chosen as the most relevant for understanding how the TSL would behave in loading situations at the development face.

The linear block support test was conducted on a 50 kN servo-controlled universal testing machine, with load applied at 0.03 mm/sec. Examples of the specimen seating and failure mode are presented in Figure 5.

FIG 5 – (a) Test set-up (b) measuring TSL thickness at failure points (c) failed TSL sample.

The load and displacement of the specimen during testing are recorded and the linear load capacity (LLC) is calculated based on the Equation 1:

$$LLC = \frac{Fmax}{2e}$$

(1)

where:

LLC	is the Linear Load Capacity
F_{max}	is the maximum load at TSL failure
e	is the TSL gap length ie the width of the test sample

The results are presented in Figure 6 along with a general relationship to predict LLC based on TSL thickness. The TSL thickness was measured on the failure planes at the completion of testing and is considered an average value +/- 0.25 mm.

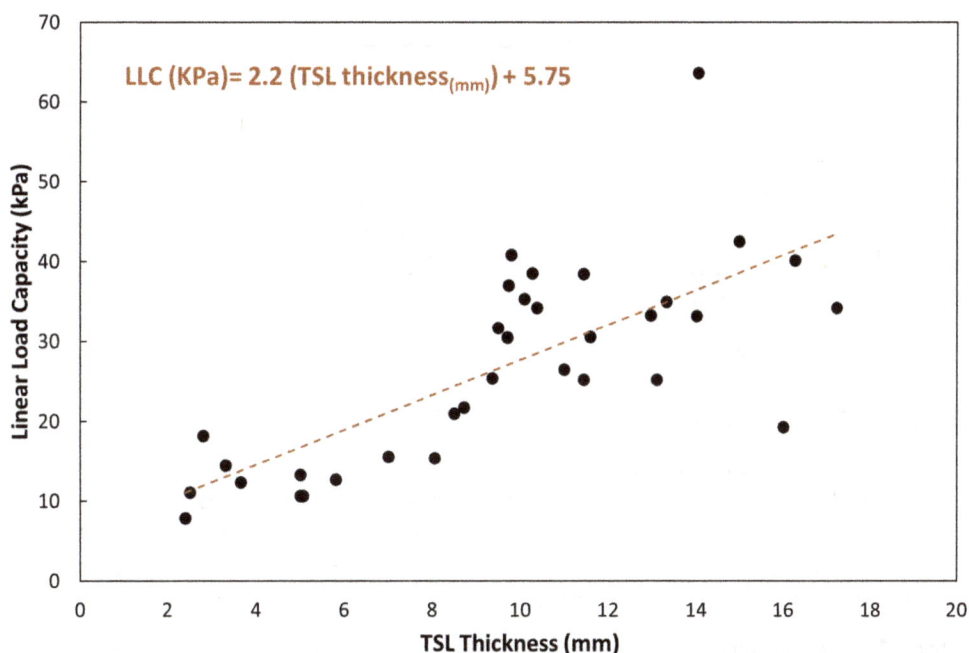

$$LLC\ (KPa)= 2.2\ (TSL\ thickness_{(mm)}) + 5.75$$

FIG 6 – Measured Linear Load Capacity of Silcrete™ TSL based on thickness.

The LLC provides an indication of the bearing capacity of a TSL *in situ* by multiplying the LLC by the perimeter of the maximum potential loose block. For a range of block sizes and TSL thicknesses the bearing capacity is calculated and provided in Table 1.

TABLE 1

Bearing capacity of Silcrete™ TSL based on block edge length and TSL thickness.

| | | Bearing capacity (kPa) | | |
| | | TSL thickness (mm) | | |
GSI	Indicative block edge length (cm)[1]	2	4	6
90	100	41	58	76
80	30	12	17	23
70	21	9	12	16
67.5	15	6	9	11
62.5	10	4	6	8
57.5	5	2	3	4
52.5	3	1	2	2
Kanmantoo	70 × 30	20	29	38

1 – based on Cai *et al* (2007).

In situ strength

Early strength gain is difficult to test due to the very fast strength gain of the TSL. Laboratory testing by Polymer Group indicates 90 per cent of total compressive strength is developed in about 20 minutes, as shown in Figure 7. Operationally this has been verified by observing the TSL react after spraying and generate heat, up to about 85°C and then observe the heat dissipate as the chemical response slows and maximum strength is reached in 30 minutes. At this stage Kanmantoo Mine has not invested in performing field trials of early strength gain using a non-destructive field-based technique but may do so in the future.

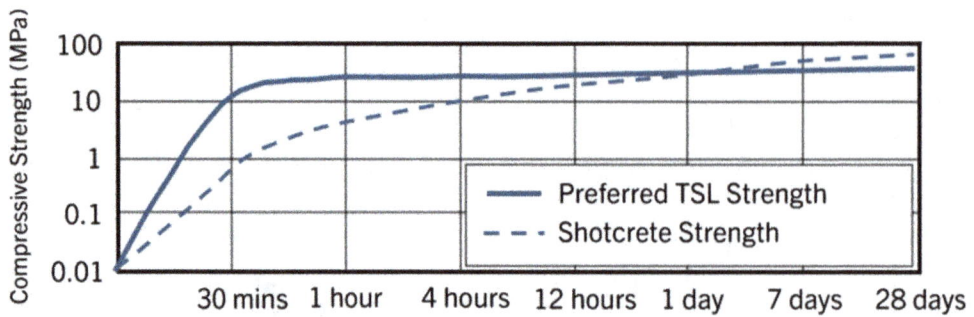

FIG 7 – Silcrete™ TSL and shotcrete early strength gain.

DEVELOPMENT FACE SUPPORT REQUIREMENTS

Kanmantoo's selection of a development face ground support system considered the need for a TSL to adequately support the development face and reduce rockfall exposure to personnel. It was always the intention for the ground support system to be a TSL, with no rock bolts in the face. The TSL had to be capable of retaining the expected failure mechanism, specific for the ground conditions and mining environment at Kanmantoo Mine.

The Kanmantoo Mine is developed within the GABS rock unit (Garnet andalusite biotite schist). The GABS has high intact strength and is blocky but often considered a massive rock as the spacing of natural structures is wide. Underground development within the GABS rock unit is considered favourable for mining, and all development faces encountered to date have been considered suitable to support with TSL.

The predicted failure mechanism of development faces at Kanmantoo was based on experience and the site structural model. Potential wedges were calculated to occur at some drive orientations. The expected size and geometry of the face wedges were considered likely to be scaled out during mechanised scaling of the face with a jumbo. The most likely failure mechanism for a development face within GABS was concluded to be a falling rock, originating from high on the development face or shoulder, formed through blast damage and an occurrence of inadequate mechanical scaling on a portion of the development face. The least favourable location from a loading perspective was found to be a downward gravitational load on the lining, rather than the more favourable situation of supporting a rock on the face with some component of resisting shear strength due to rock on rock contact. Establishing a volume for a falling rock was taken as the maximum spacing between blastholes, or 0.7 m (perimeter blasthole spacing) × 0.3 m (damage zone around a blasthole) × 0.4 m (depth of damage zone) or 0.084 m^3 and a weight of 260 kg. This equates to a 2.55 kN force over an area of 0.21 m^2 (~12 kPa). Since the TSL will hold the rock in place in its original position, there was no requirement to consider a dynamic load.

Calculating the holding capacity of the TSL was based primarily on the results from the TSL linear block support test as shown in Table 1. The TSL linear block support test was found to be a good fit for load bearing assessment of the liner, as the observed failure model fit well with a direct shear failure, or punching shear failure, as described by Barrett and McCreath (1995) and shown in Figure 8. Adhesive failure and flexural failure models that rely on a ductile TSL pulling away from the rock substrate were not applicable due to the very high adhesion strength between the TSL and the rock.

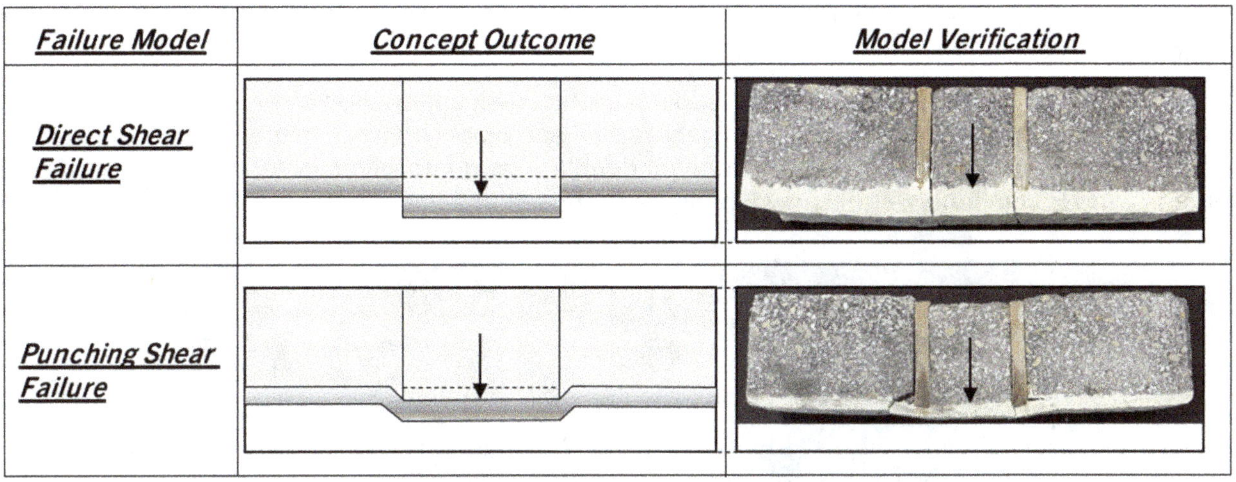

Failure Model	Concept Outcome	Model Verification
Direct Shear Failure		
Punching Shear Failure		

FIG 8 – Verification of TSL failure models at Kanmantoo.

Based on the expected block size from Kanmantoo, the bearing capacity of a 4 mm TSL can be predicted at 29 kPa (Table 1). This far exceeds the ~12 kPa required and provides a factor of safety greater than 2. A thinner liner could be specified (eg 2 mm), but is operationally difficult to achieve.

The dynamic capacity of the TSL for use in seismically active areas was not investigated for Kanmantoo's current mine plan, which is expected to be low stress conditions. TSLs have previously been shown to be potentially useful in mitigating rock bursts (Archibald and Dirige, 2006; Baidoe, 2003; Komurlu, 2018). Further work would need to be conducted to determine the effectiveness of a 2–6 mm thickness liner for this application.

QUALITY ASSURANCE AND CONTROL (QA/QC)

The adopted QA/QC approach for ongoing TSL use at Kanmantoo Mine involves the following processes:

- The surface coverage standard is checked by Shift Supervisors and audited by Technical Services.
- Spray thickness is measured during audits by Technical Services, who measure thickness at several face blasthole positions.
- Grout cubes filled with TSL are used for compressive strength testing.
- Linear Block Support Testing is used to assess the flexure response.

MINING SEQUENCE

The development mining sequence with application of TSL as face support involves the following work tasks:

1. Bog out the fired cut. Resources required: Bogger and operator.
2. Rattle and install ground support to the backs and walls then rattle the face. Once completed the jumbo pulls out of the heading. Resources required: Jumbo and operator.
3. Spray the face with TSL. Resources required: Ute, pump module and operator.
4. Bog the rattle dirt. Resources required: Bogger and operator.
5. Mark up, if required and bore the face. Resources required: Jumbo and operator.
6. Charge and fire. Resources required: Charge up rig and operators ×2.

SET-UP FOR TSL SPRAYING

Strata Consolidation Pty Ltd assisted site personnel in training and procedures and facilitated use of their piston and gear pumps for the TSL trials. At Kanmantoo the A and B liquid components were stored in a sea container with a reverse cycle heating unit for temperature control of the Silcrete™

TSL product. As the underground mine is developed deeper and the ambient mine temperature reaches or exceeds 21°C, TSL storage will be moved underground.

When preparing to spray a development heading, the first task undertaken by the TSL spray operator is to ensure the light vehicle has the removeable spray equipment secured to the tray. The equipment includes a pneumatically driven two component double gear pump with two sealable storage drums for the A and B components (as shown in Figure 9), hoses and the spray gun with nozzle.

FIG 9 – Strata Consolidation's Modular LV Tray set-up and decanting TSL components from the climate-controlled storage.

The operator decants the required quantity of TSL from the intermediate bulk containers (IBCs) at the TSL storage area as shown in Figure 9 and transfers both components to the light vehicle. Two important aspects at this stage of the work process include managing the temperature of the decanted TSL components and ensuring the IBC storages are mixed well prior to decanting, using an air lance or a mechanical IBC mixer.

Following arrival at the development heading the face is sprayed with TSL. This work process is a one-person task, with all spraying conducted on foot. Resin mixing occurs at the static mixer which is part of the spray gun held by the operator. The jumbo can remain parked at the development face during spraying, with a hose run from the light vehicle and past the jumbo to the face. The spray activity can be conducted within 30 minutes.

The spray operator visually checks the applied TSL thickness by spraying until the rock is no longer visible through the applied white coloured TSL layer. Current experiences with Silcrete™ TSL indicate that this method reliably achieves a 4 mm minimum layer TSL thickness.

When the face is sprayed, the operator packs up by running a small amount of TSL Part A to flush the hose and spray gun, packs up the hose and departs. The hoses are capped, and the Part A and Part B components can be left in the lines for an extended period for subsequent applications. Any leftover TSL product is returned and decanted back into the TSL IBC storages for future use.

The jumbo can commence boring immediately after departure of the spray operator. Kanmantoo allows for 20 minutes from completion of spraying before commencing any personnel activity at the face, such as digging out the lifters.

FACE SUPPORT COST COMPARISON

The costing of TSL as face support compared to mesh and bolting the face is summarised in Figure 10. A scenario of 4 mm thick TSL was compared to mesh and bolts with nine friction bolts (0.9 m long) and two sheets of mesh (4.5 m × 2.4 m). Kanmantoo have found that at the current purchase price of TSL product, the application of TSL as face support delivers a reduced development cost per metre, compared to supporting the face with mesh and bolts. The costing comparison conducted by Kanmantoo does not consider the hidden cost advantages, such as the potential to speed up the mining cycle, or the benefit of eliminating steel from the fired dirt.

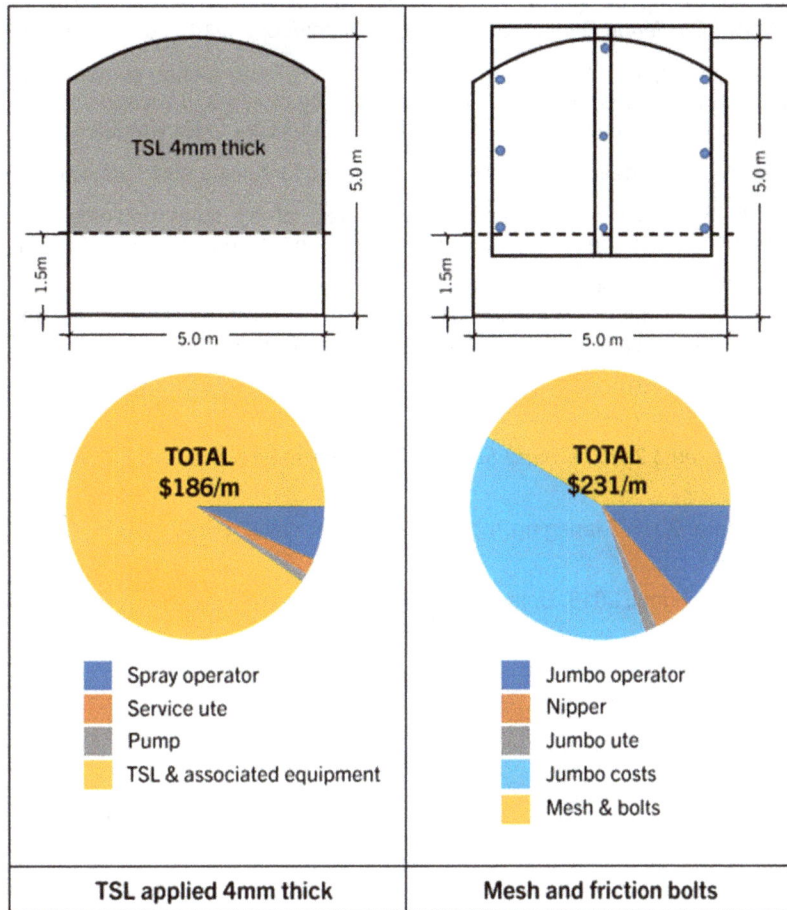

FIG 10 – Cost comparison of Silcrete™ TSL usage verses mesh and bolts, per m of advance.

Other cost implications for TSL use such as the initial TSL spray training, ongoing atmospheric monitoring and PPE use are considered minor costs and have not been accounted for. Initial test work shows no adverse effects of the TSL in the ore stream through the processing plant, particularly given the very small quantitative ratios to ore provided by the 4 mm thick TSL and inert nature of the cured product.

A total cost of $186/m for Silcrete™ TSL used as face support is lower than the $231/m for mesh and bolts.

The cost of TSL usage per metre is very sensitive to the purchase price of the TSL, and the thickness of applied TSL. Purchase price of TSL products is likely to reduce over time if their usage increases in underground hard rock mines.

CONCLUSIONS

Laboratory and field investigations associated with the feasibility of using TSL in place of mesh and bolts on development faces at Kanmantoo have been completed. Investigations suggest that the preferred TSL product at 4 mm thickness will provide sufficient face support and provide a reduced development cost per metre.

REFERENCES

Archibald, J F and Dirige, P A, 2006. Development of thin spray-on liner and composite superliner area supports for damage mitigation in blast-and rockburst-induced rock failure events, Liverpool, UK, in *9th International Conference on Structures Under Shock and Impact (SUSI)*.

Baidoe, J, 2003. Assessment of rockburst-mitigating effects of the area liners, MSc Thesis. Queen's University at Kingston, Ontario, Canada.

Barrett, S V L and McCreath, D R, 1995. Shotcrete support design in blocky ground: towards a deterministic approach, *Tunnelling and Underground Space Technology*, 10(1):79–89.

Cai, M, Kaiser, P K, Tasaka, Y and Minami, M, 2007. Peak and residual strengths of jointed rock masses and their determination for engineering design, in *Proceedings 1st Canada-US Rock Mechanics Symposium – Rock Mechanics Meet Social Challenge Demands,* 1:259–267. https://doi.org/10.1201/noe0415444019-c32.

Chang, S, Choi, S-W, Lee C, Kang, T-H, Kim, J and Choi, M-S, 2017. Performance comparison between thin spray-on liners with different compositions, in M Hudyma and Y Potvin (eds), *Proceedings of the First International Conference on Underground Mining Technology,* Australian Centre for Geomechanics, Perth, pp 79–85.

EFNARC, 2008. Specification and Guidelines on Thin Spray-on Liners for Mining and Tunnelling, www.efnarc.org.

Kanda, M and Stacey, T, 2019. Review of the practical effectiveness of thin spray-on liners based on information from suppliers and observations from the mining industry, 443–458, https://doi.org/10.36487/acg_rep/1905_27_kanda.

Kolapo, P, Onifade, M, Said, K O, Amwaama, M, Aladejare, A E, Lawal, A I and Akinseye, P O, 2021. On the Application of the Novel Thin Spray-on Liner (TSL): A Progress Report in Mining Operations, *Geotechnical and Geological Engineering,* 39:5445–5477. https://doi.org/10.1007/s10706–021–01861–5.

Komurlu, E, 2018. Support performances of elastomer thin spray-on liners (TSLS) against the rock burst: a case study, in *12th Regional Rock Mechanics Symposium*, Traban, Turkey.

Potvin, Y and Hadjigeorgiou, J, (eds) 2020. *Ground Support for Underground Mines* (Australian Centre for Geomechanics: Perth).

Queensland Mines Inspectorate, 2016. Managing rockfall hazards at development headings, *Mines Safety Bulletin No. 159,* 15 December, ver 1.

South Australian Employment Tribunal, 2018. Boland v BHP Billiton Olympic Dam Corporation Pty Ltd [2018] SAET 167.

Improvement strategies for accurate underground drilling and blasting effectiveness at the Barrick Kibali Gold Mine

F M Senda[1], J P S Kayumba[2], I Traore[3] and K Musumali[4]

1. Production and Short Term Planning Superintendent, Kibali Gold Mine, Wtsa, DRC. Email: felly.senda@kibaligold.com
2. Mine Manager, Myrafalls Mine, Quebec, Canada. Email: jean-paul.kayumba@myrafallsmine.com
3. Group Technical Services Manager, Barrick-Africa. Email: ismail.traore@barrick.com
4. Principle UG Mine Manager, Anglo American, Johannesbourg, SA. Email: kayombo.musumali@angloamerican.com

ABSTRACT

This paper highlights strategies used at the Barrick operated Kibali Gold Mine (KGM) to improve the drilling accuracy and blasting effectiveness for longhole open stoping. The mine currently uses Solo DL421 for the required stope drilling and has experienced an average deviation of 10 per cent. Strategies are raised to eliminate or at least mitigate the adverse effects of blasthole deviation and the impact of blasting induced effect at the stope faces including the stope crown. The integration of deviation trends in the design would allow the actual blastholes to be established within their required positions to then facilitate a proper energy distribution with the eventual production blasting. The implementation of stope undercutting reduces the hole length from 35 m to 30 m which in return decreases blasthole deviation by 3 per cent. Another strategy adopted to improve the drilling accuracy is to slow down the penetration rate with the use of several operating approaches, including 135 MPa percussion pressure, 65 rev/sec rotation speed and 17 MPa feed pressure. To this penetration rate, a constant k factor is applied to increase the rate when the bedding inclination is less than 45° from the horizontal or when the hole length is under 10 m. This reduces deviation by a further 4 per cent. Where the rock mass is highly jointed it is advised to use the average penetration rate over the entire length of the hole. Furthermore, the paper shows the effect of deviation on the blasting results through a prediction of fragmentation blasting model optimised with reduced deviation and highlights the effect of presplit mining and timing delay on the blasting effectiveness. Although the size pattern increases with deviation, a strategy pertaining to timing delay with blasting would be to align to the work of Grant (1990) where the optimum hole to hole delay is found to be between 7–10 ms/m of spacing, while the optimum delay between adjacent ring varied between 110–160 ms. Stabilisation of crowns and wings will perform with the implementation of presplit blasting.

INTRODUCTION

Profitability and safety remain the driving force for mining shareholders. Blasthole deviation and blasting inefficiencies have adverse economic impacts on mining operations. The adverse impacts are exhibited as ore loss, underbreak, overbreak, ore dilution, lower mining recovery, and, poor fragmentation.

Blasthole deviation is defined as the difference between the actual and planned blasthole coordinates. The difference is manifested in terms of hole location, orientation, collaring, and alignment. Jones (2018) and Adebayo and Akande (2015) link the mining profitability to hole deviation and notice that for a mineable stope shape, dilution of ore due to overbreak can be significantly decreased with the improvement of drilling accuracy. A note pertaining to general mine design is that the drilling azimuth should be located parallel to the stope wall. Morrison (1995) discovers that excessive damages and unsuccessful blastings can be observed if blastholes are not located on the azimuth parallel to the stope wall. This means that there is an intrinsic link between blasthole deviation and blasting results.

Drilling and blasting parameters have the largest impact on fragmentation. They must be designed in such a way that each blasthole achieves its planned energy distribution in the rock mass. Accurate drilling will allow the planned energy distribution in the rock mass and the blast to generate the expected fragmentation. In other words, if there is no hole deviation, the blasting parameters will not be altered, and the blasting effectiveness will be improved. Over the blasing parameters, Gleen Heard, a chief minng director for Barrick across the world, correlates the stabilisation of slope in opencast mining through presplit blasting to the stabilisation of crowns and wings for open stopes in underground mining. One of the objectives of presplit blasting in underground is to control the stope wings and crown in order to avoid any kind of fall of ground over mining given the inevitability of negative side of blasting induced effects. In the design pattern, presplit holes need to be drilled at half budern time half spacing using same hole diameter as the main production holes.

Since the quality of blasting results is not only controlled by the burden and spacing, the column charge and delay timing will require specific attention. Hettinger (2015) investigates the effect of timing on rock fragmentation. It is found from the investigation that there is a link between fragmentation quality and a well-defined timing delay. The author further determined the best range of timing delays for an optimised rock fragmentation. Raskidinov (2006) reveals that blastholes and hole charge constructions are the main factors leading the outcomes of drilling and blasting operations. Cunningham (2011) explained in his paper, '*The Use of Blasting to Improve Slope Stability*', the effect of detonation beyond the burden and spacing. Wherein, he found that strain waves and expending gases lead to a development of cracks for a continuous rock mass. He therefore, raised the necessity of firing first the presplit holes first prior to the main productions holes in order to create a buffer zone where the energy will dissipate to avoid backbreak. Applied to underground mines, this theory will play a major role in preventing fall of ground from the stope crown while improving the stope's stability over mining. From the abovementioned it is an established fact that efficiencies with blasting are a process that results from accurate drilling of blastholes that are correctly charged, timed, and well initiated.

This paper discusses strategies used at the Barrick operated Kibali Gold Mine (KGM) to improve the drilling accuracy and blasting efficiency for the underground operation.

KIBALI UNDERGROUND GOLD MINE

The mine is operated by a joint venture between Barrick, Sokimo, and Anglo Gold Ashanti. It is situated in the Moto Goldfields, ~560 km north-east of Kisangani in the Oriental Province of the Republic Democratic of Congo. The deposit occurs in the Kibalian sequence of deformed and metamorphosed volcano sediments where the gold mineralisation is contained within the volcaniclastics sediments and banded ferruginous cherts (Beck Engineering, 2018). The gold mineralisation is in most cases disseminated in style and associated with the brittle formation of pre-existing structures and alteration zones flooded with albite, silica, and carbonates. The orebody is mined using sublevel open stoping where ore is recovered by open stopes and paste filled after mining. The stopes are 20 and 30 m wide and mined from top to bottom due to the stability of the orebody's hanging and footwall. Blasthole drilling lengths are constrained by the mine sublevel height which varies between 35 m to 40 m.

For a better understanding of drilling and blasting parameters, the geological and geotechnical features should be well estimated as the accurate estimation of the rock factor (as defined in the Kuz-Ram fragmentation model) depends on the mapping of structural features (joints, foliations, bedding planes and significant structures). Strelec (2010), emphasised how the blasting design was significantly influenced by the joint orientation and spacing. Findings from the geotechnical assessment reveal that joint spacing varies from 0.1 to 2 m while the rock factor is calibrated at 9.7 using the predicted cumulative distribution of Kuz-Ram. This implies to classify the rock mass in a fair to good range where the UCS varies between 117 MPa to 131 MPa. Due to the hardness and the competency of the rock mass, drilling should be done with more caution, especially with regard to the management of the rig's operating variables.

PRODUCTION DRILLING

Due to the mining method used and the size of stopes, percussive longhole drilling machines such as the Solo DL421 and ITH Aries are being used at the mine. Designed for large scale production drilling, the rigs offer the ability to be run in automatic mode which in turn, offers the advantage of drilling over shift change and blast re-entry. Utilising auto drilling allows for a potential increase of 40 m per rig per day leading to a potential gain of 500 t extra per day.

Drop raising techniques are practiced to create a free face for an effective advance per blast. This technique makes use of electro hydraulics with an ITH for drilling holes of large diameter. In comparison to a Solo, an ITH offers more accuracy with minimal power loss and steady penetration rates (Bruce, Lyon and Swartling, 2014).

Production blastholes are drilled using a Solo DL421 with a top hammer. Drilling is done using ST 68 tube rods of 1.8 m and small buttom bit with four flushing holes for a proper cleaning while drilling. In contrast to an ITH, drilling of blastholes using a top hammer may result in significant deviations. Blasthole deviations are measured using a bore track system. The adverse effects of these deviations in mining operations have been argued by several researchers including Singh (1998) who stated that hole deviations have significant economic impacts on mining operations. Adebayo and Akande (2015) and White (1984) described poor consequences of blasthole deviations on the mining chain value, including poor fragmentation, extra drilling, loss of drill strings, ore loss, ore dilution, extra explosive consumption, and increased mining development costs. Almgren (1981) found that the mine planning and operating costs could be adversely affected by the lack of blasthole accuracy. Such problems become inevitable for bulk mining methods which rely on longhole drilling in their basic design.

As such, it comes to understand that reducing blasthole deviation allows to improve the overall mining profitability. As emphasised by Sinkala and Granholm (1987), economic savings of up to 20 per cent of yearly mine operating costs could be generated from reducing blasthole deviations.

Deviation survey and analysis

Blastholes are surveyed using a bore track system. KGM makes use of Renishaw PLC bore track which operates a probe containing a series of electronic inclinometers. Data is collected by lowering the probe into the blasthole using a number of connected rods. The probe detects several points throughout the hole length while showing the actual blasthole trajectory as presented in Figure 1.

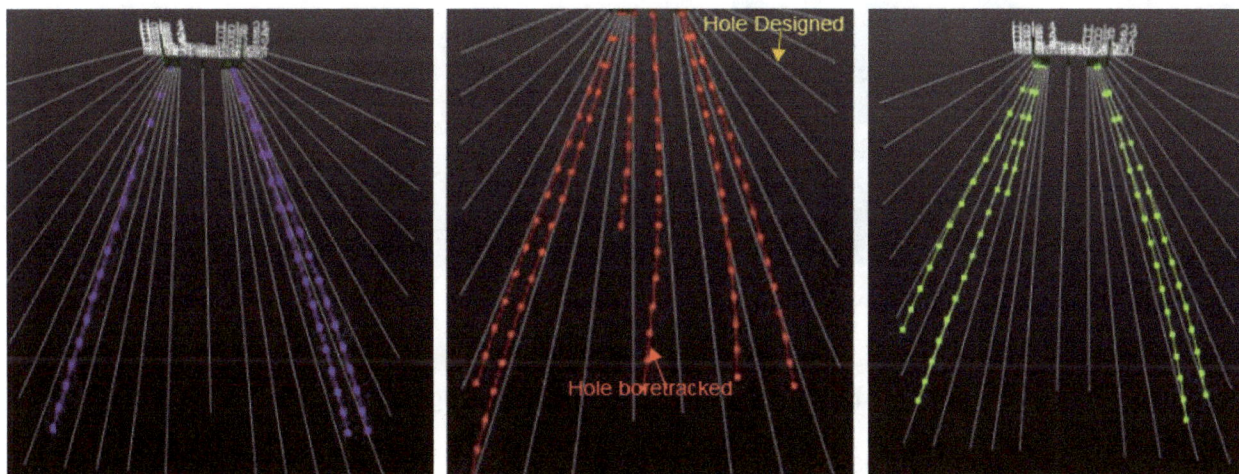

FIG 1 – Blastholes bore tracked.

Conceptually, drill hole deviation occurs from various sources where components of deviation may arise from an offset in the initial set-up position, or from high operating variables (percussion pressure, rotation speed, and feed pressure), or from the deviation of the drill bit (Chabiot, 1995). In addition to this, blasthole deviation is also influenced by the rock type, presence of significant structures, rock mass joints, hole length, and the blasthole trajectory. Chieniany et al (2012) link all the factors influencing blasthole deviation to the penetration rate and find that among those parameters, only the operating variables are the more manageable. On the other hand, the former operating variables are the driving force of the penetration rate. With regards to this, it is felt that hole deviation can be controlled with an improved management of penetration rates.

Many investigations have been undertaken to correlate the penetration rate to the operating variables and rock factors. Kivade, Murthy and Vardhan (2015) investigated the penetration rate of a percussive drill rig and determined how it was influenced by the rock properties. They further found that air pressure and thrust had significant impacts on the penetration rate of the drill rig. Nguyen *et al* (2017), predicted the penetration rate using both rock and operational factors. Podney (2002) estimated the main factors influencing rotary blasthole drilling.

Considering the rock mass as a non-manageable constraint, the penetration rate of a top hammer such as Solo DL421 is driven by the rotation speed, percussion pressure, and the feed pressure which are managed by the operator accordingly with the rock mass. The harder the rock mass is, the more efficient should be the management of penetration rates. Hamarin (1995) confirmed that blasthole deviations may arise from using a higher rate of operating variables. In the same perspective, Worden (1978) noticed that blasthole deviation could be controlled by maintaining a constant penetration rate. This supports the idea that meant to improve the drilling accuracy with a proper management of penetration rates.

Figure 2 presents deviation trends per hole length for diverse stopes at KGM. This is a case study where four operating variables for the Solo DL421 were constantly monitored while drilling. Each hole drilled was bore tracked and surveyed to accurately define deviation trends while the rock mass strength and rod strength were considered non-manageable constraints.

Deviation	Rotation Speed	Percussion Pressure	Feed Pressure
Deviation 1	70	135	16
Deviation 2	80	150	20
Deviation3	85	150	20
Deviation 4	65	135	17

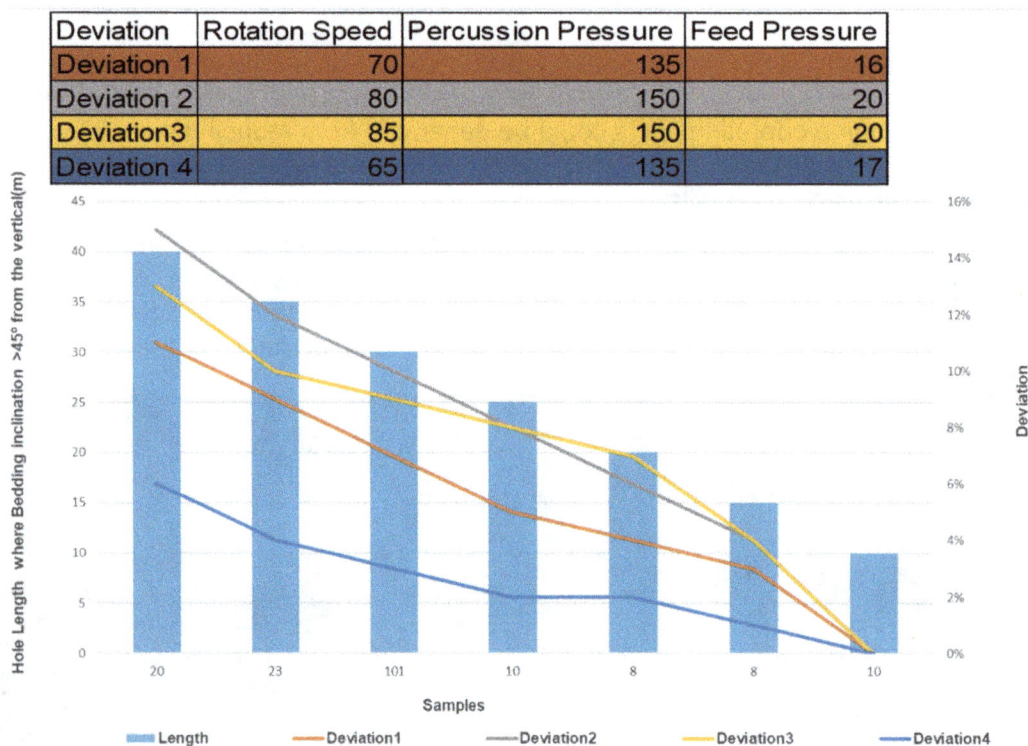

FIG 2 – Deviation versus hole length.

Figure 1, shows that most holes tend to deviate in the same direction where the bedding inclination is less than 45° from the vertical. A strategy pertaining to improving drilling would be to integrate deviation trends in the design. This allows the actual holes to be established within their required positions to facilitate a proper work of each hole while avoiding any risk of underbreak and overbreak at both walls. The effect of the integration on the design is manifested in terms of hole length and angle of rotation. Further observations revealed that holes start deviating beyond 10 m (Figure 2). This arises from drill string deflection due to the interaction between the rock mass, increased number of rods, and the use of higher rates of operating variables. Another strategy would be to increase the rigidity of the drill string to avoid any deflection issues. However, increasing the rigidity of drill rods makes them heavier and difficult to handle by operators. This makes the strategy difficult to implement. As all the strategies raised above are in some ways linked to the penetration rate, it goes that the most manageable factors for controlling blasthole deviation remain the management of the operating variables.

In addition to the management of penetration rates, the use of sufficient water and air pressure and the application of rig stingers play a significant role in the drilling accuracy of blastholes. Stingers stabilise the feed for accurate collaring and drilling while water and air are used for cooling and cleaning purpose. For proper hole cleaning while drilling, the expected water and air pressure should be approximately 7 Bar. Since this was not considered as an issue at the mine, it was not raised among the improvement strategies.

The results from Figure 2 show that blasthole deviation decreases with the use of a lower rate of operating variables where penetration rate is associated with the rotation speed, percussion pressure, and feed pressure. The problem would be, how far the rate should be lowered so as the planned drill stock is achieved with minimum deviations. The monitoring has proven that the following operating variables generate minimum deviation when drilling longholes where the bedding inclination is less than 45° from the vertical:

- rotation speed: 65 rev/sec
- percussion pressure: 135 Ba
- feed pressure: 17 Ba
- average penetration rate: 0.45 m/min.

These parameters are not applied to the blastholes where the bedding inclination is less than 45° from the horizontal. Such holes were drilled with penetration rates beyond 1.1 m/min and the monitoring from the bore track system revealed minimum deviation. For an efficient and optimum drilling rate, a rule pertaining to the drilling is to apply a correction k factor that increases the penetration rate to the best ranges for such holes. This goes the same for the drilling of each blasthole when the length is under 10 m. The correction factor is estimated in the range of [1, 2] and should be managed by the operator according to the rock mass. As illustrated in Figure 2, using the above operating variables decreases deviation by 4 per cent.

Where the ground is totally jointed or full of significant geological structures and discontinuities, it is suggested to use constantly the estimated average penetration rate over the length of the blasthole. This will avoid the bit from following the path of less resistance created by the existence of structures and joints. It should be noted that where joints and structures exist, blasthole deviations are not easily managed.

Stope undercuting

Another strategy in improving drilling accuracy is through the reduction of the blasthole length. This can be possible with the implementation of stope undercutting. It is evident that short holes experience minimum deviation compared to longholes. Given that deviation is linked to the length of the hole, the more the blasthole length will be decreased, the more deviation will be as well. Undercutting a stope means to decrease the size of the stope by drilling from the top and bottom levels. The drilling from the bottom level is exclusively done with short holes accurately drilled to ensure that the area has been totally covered with blasting while the drilling lengths from the top level have decreased to a certain range when compared to the initial lengths. Znibe *et al* (2009) went further to investigate the effect of stope undercutting on its wall overbreak and linked the mining profitability to the stope undercutting. This means, except for the drilling accuracy, the stope undercutting will positively influence the blasting results, mining recovery, and profitability.

KGM has undertaken the undercutting of its longhole stopes which has reduced the stope height from 35 m to 30 m and reduced potential deviation by a further 3 per cent. The undercut is totally drilled with short up-side holes ensuring minimum and negligible deviation.

Previously, site drilled only downholes into the stope floor to ensure a flat bogging level. Due to deviation, the floor was not always fully mined and underbreak was experienced. Current methods drill from the extraction level with short holes to ensure consistent explosive distribution (see Figure 3). This has improved the drilling accuracy and mining recovery, although a minimum underbreak was still experienced at the side and footwall. To mitigate the underbreak at the side and footwalls, it is advised that all side holes to be sub drilled by 1.5 m outside the stope shape with the last ring being drilled on the footwall shape outline.

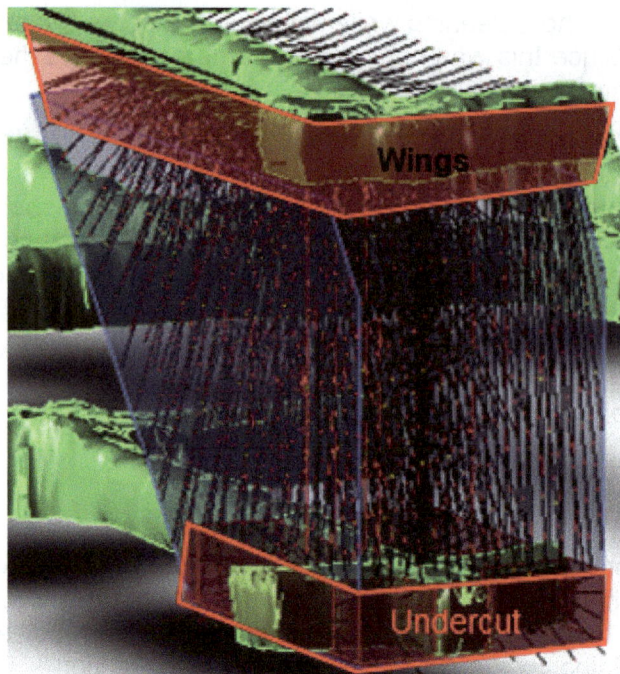

FIG 3 – Stope undercutting at 5 m from the floor.

An improvement to the drilling would be to use a more appropriate drilling rig. Solo DL421 uses pivot points from the centre of the drive. A drill rig with a boom drifter would allow more horizontal holes to the floor ensuring good explosive distribution at the floor.

PRODUCTION BLASTING

Blasting remains the main technical process in the extraction of ore from the underground. New technologies that improve the blasting effectiveness have been embraced by mining industries which approve the effectiveness of electronic blasting systems. Kortnik and Bratun (2010) depicts the advantages of electronic blasting systems including better ground vibration, better control of the rock movement and muck pile, better fragmentation, and improved productivity. He further looked at the disadvantages of the former system and finds that electronic blasting is expensive and requires more skills than Nonel detonators. An overview of the use of electronic blasting systems is presented by Mui *et al*, (2012).

KGM makes use of electronic blasting systems for both up and downholes, with all downholes being drilled at 102 mm diameter, while it is advised that all upholes to be drilled with a small diameter such as 89 mm or less to avoid fall of explosive with gravity. Charging is done with emulsion due to the high presence of water and rock moisture content. ANFO is as well used where totally dry conditions are observed. It is recommended for the first opening (Raise) to be charged with emulsion to avoid any risk related to dilution of ANFO. Priming is done with I-Kon electronic detonators used in conjunction with:

- 250 g pentolite boosters or 400 g Trojan boosters to initiate the emulsion.
- I-Kon Logger for timing attribution per detonator.
- I-Kon Centralised Electronic Blasting System (CEBS).

Several researchers have proven that blasting is the cheapest method of rock fragmentation. The cost of the mining chain decreases with an optimum fragmentation which is evident with the use of smaller drill patterns and accurate blasthole drilling. However, the use of smaller patterns increases the explosive and drilling costs which in turn affect the operating costs. For an optimum operating cost, fragment size from blasting should be linked to the plant performance through the mining chain process. Fortunately, there exists further methods for the prediction and estimation of optimal fragmentation.

Blasting control

The evaluation of blast designs was approached through the prediction of fragmentation models based on the Rosin Rammler's Kuz-Ram model used in parallel with photographic methods for the estimation of fragmentation. Using this approach, a rock factor describing the nature of the geology, and the uniformity index characterising the loading and the pattern type are calculated. The fragment size distribution is determined using Cunningham (1987) theories developed from the Rosin Rammler (1933) formula presented in Equation 1.

$$\% passing = 1 - R_x = 1 - \exp\left[-0.693\left(\frac{X}{X_m}\right)^n\right] \tag{1}$$

Where X_m is the characteristic value for the fragments, n is the uniformity index calculated from the Rossin Rammler expression, and R_x is the the fraction retained on screen opening x.

The mean particule particule X_m is calculated from the udapted equation of Kuznestov.

$$X_m = AK^{-0.8}Q^{\frac{1}{6}}\left(\frac{115}{RWS}\right)^{\frac{19}{20}} \tag{2}$$

Where A is the calibrated rock factor calculated from the rock mass description; k is a powder factor in kg explosive per cubic metre of rock; Q is the the explosive quantity per hole; and RWS is the weight strength relative to Anfo. The calculated drilling pattern integrating deviation trends is presented in Table 1.

TABLE 1

Calculated drilling pattern linked to deviation.

Burden, B (m)	Spacing, S (m)=	n	X_c	k (kg/m³)	Xmean	S / B	Cumulative Fraction Participation										
							0.01	0.05	0.09	0.30	0.60	0.90	1.40	1.60	1.90	2.40	2.80
2.7	3.33	1.22	0.432	1.079	0.320	0.90	1	7	14	47	78	91.4	98	99	100	99.97	99.99
2.8	3.42	1.23	0.453	1.150	0.336	0.90	1	6	13	45	76	90.2	98	99	100	99.96	99.99
2.7	3.36	1.22	0.434	1.088	0.322	0.91	1	7	14	47	77	91.3	98	99	100	99.97	99.99
2.6	3.24	1.21	0.411	1.011	0.304	0.90	1	8	15	50	79	92.4	99	99	100	99.98	100.00
2.8	3.45	1.23	0.456	1.159	0.339	0.91	1	6	13	45	75	90.1	98	99	100	99.96	99.99
2.7	3.38	1.23	0.437	1.097	0.324	0.92	1	7	13	47	77	91.2	98	99	100	99.97	99.99
2.6	3.27	1.21	0.413	1.019	0.306	0.91	1	7	15	49	79	92.3	99	99	100	99.98	100.00
2.8	3.48	1.23	0.459	1.168	0.341	0.92	1	6	13	45	75	90.0	98	99	100	99.96	99.99
2.6	3.29	1.21	0.416	1.027	0.307	0.92	1	7	14	49	79	92.2	99	99	100	99.98	100.00
2.7	3.41	1.23	0.439	1.106	0.326	0.93	1	7	13	47	77	91.0	98	99	100	99.97	99.99
2.9	3.51	1.24	0.475	1.222	0.353	0.90	1	6	12	43	74	89.0	98	99	100	99.94	99.99
2.6	3.32	1.22	0.418	1.036	0.309	0.93	1	7	14	49	79	92.1	99	99	100	99.98	100.00
2.7	3.44	1.23	0.442	1.114	0.328	0.94	1	7	13	46	77	90.9	98	99	100	99.97	99.99
2.9	3.54	1.24	0.478	1.232	0.356	0.91	1	6	12	43	73	88.8	98	99	100	99.94	99.99
2.5	3.15	1.20	0.391	0.945	0.288	0.90	1	8	16	52	81	93.4	99	100	100	99.98	100.00
2.6	3.34	1.22	0.420	1.044	0.311	0.94	1	7	14	48	79	92.0	99	99	100	99.98	100.00

The cumulative distribution is shown in green (see Table 1). The fragment distribution presented in Figure 4 is determined by changing the burden and spacing with spacing being linked to burden and deviation. It should be noted that the pattern will change with different deviation values. The more deviation will increase, the bigger will be the spacing and the drilling pattern. To optimise the process, blasthole deviation needs to decrease. This paper points out several strategies that tend to eliminate or at least mitigate deviation and among which the use of a lower rate of operating variables which remains a crucial strategy. The effect of this may slow down the drilling rates but the daily drill stock target will be still achievable. The cost savings of these strategies may be seen across the mining chain.

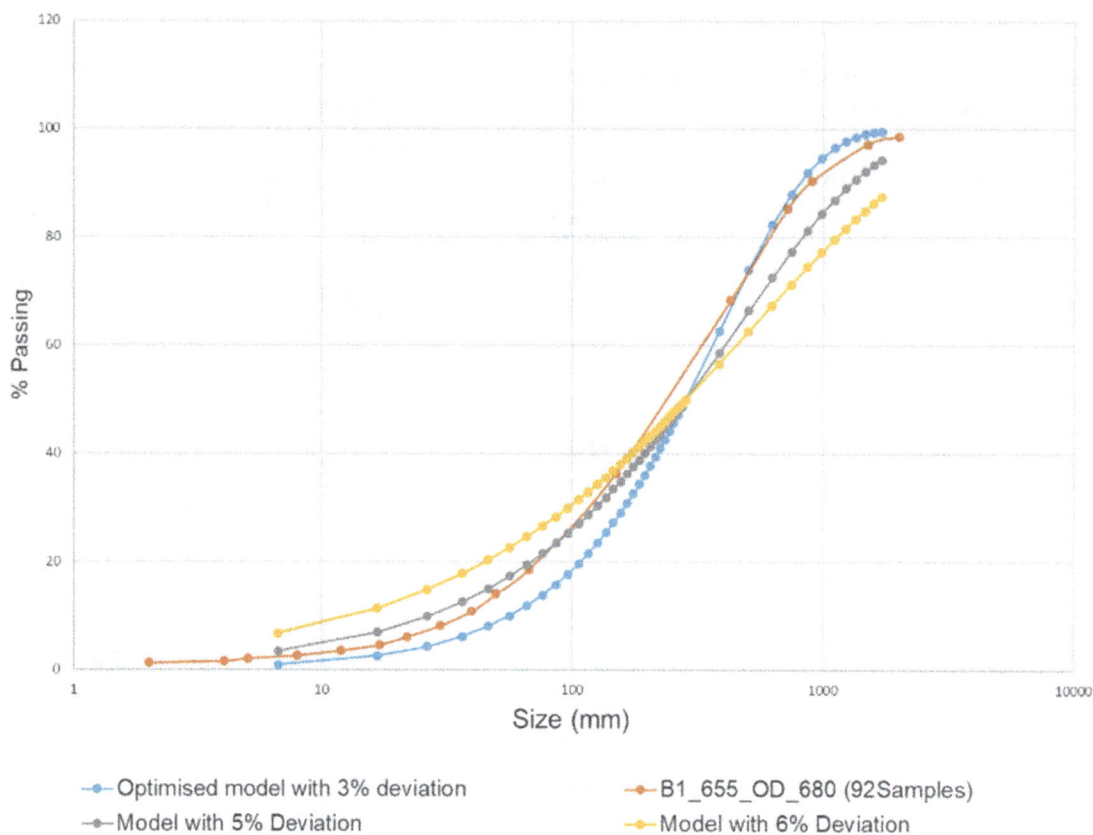

FIG 4 – Fragmentation distribution linked to deviation.

Presplit blasting in underground mining

The main objective of presplit mining in opencast mining is to control backbreak, excessive ground vibrations and filter the effects of explosive gases from production blasting while being performed in a row of blasting holes drilled behind the main blast pattern (Dindarloo, 2015). Calder and Tuomi (1980), emphasised that these holes are to be blasted before or at the same time as the main production holes to create a preferential fracture plane behind which energy will dissipate while preventing this energy from backbreaking over the designed blasting surface. To the presplit blasting, buffer holes are to be drilled and partially loaded to achieve the first control of the released energy.

As stated above, similar practices were applied to the underground mines, and in particular to the Kibali underground mine. The former practices were raised as a potential method to mitigate the propagation of energy released from the main production holes to the stope crown. Blasting induced effects will likely cause the following:

- backbreak (overbreak)
- loose rock
- ore dilution where the upside of the stope is sitting in waste material.

The impact of the above may unfavourably affect safety and production over mining. In this experiment, presplit holes were drilled at the same time as the main production holes. While production holes were drilled at a specific budern (B) and Spacing (S), presplitng rows were drilled at half burden (B/2) and half spacing (S/2) from the buffer holes which are the first and last production holes constrained by a minimum charge length due to their designed lengths which are in return constrained by the stope shape.

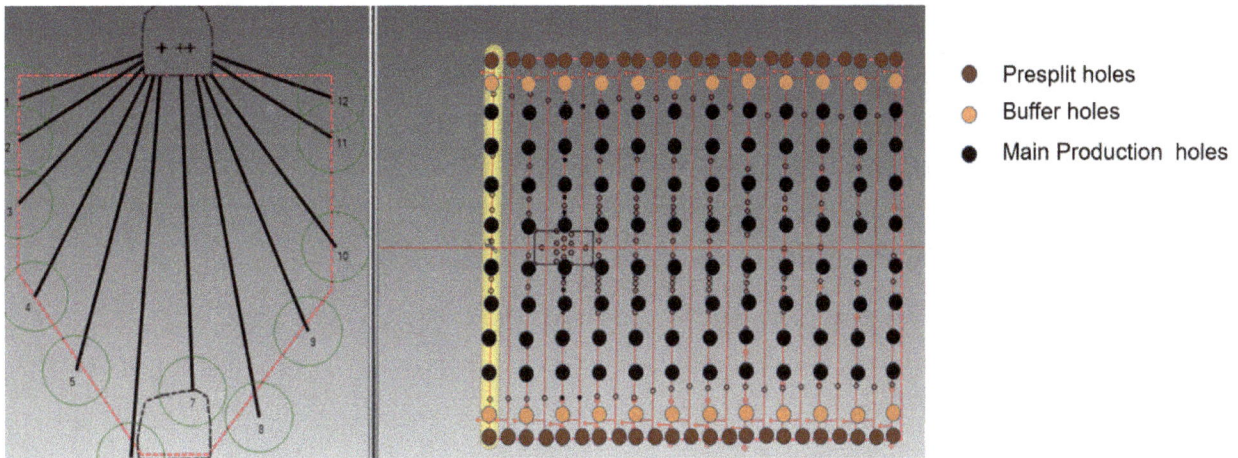

FIG 5 – Drilling pattern with presplit holes included.

Since the prespliting method is used for separation purpose rather than fragmentation, presplit holes are therefore loaded with a law density explosive. Several alternatives may be offered for such a purpose, including air decking, reduction of charge diameter, and the addition of salt to Anfo in a specific ratio. The efficiencies of each of these methods are explained by Dindarloo (2015) in his paper '*Design of control Blasting (pre-spliting) in Goleghor Iron Ore Mine, Iran*' wherein the author emphasised the fact that where the hole diameter is smaller, the reduction of charge diameter is not an option. With the bigger hole diameter used at Kibali Gold Mine, the reduction of charge diameter was the method used to control the powder factor of presplit holes. Explosive cateridges were tighthened through an electronic detonator to a distance that relates to the hole length and in a way far from hole confinement to reduce the strength of the energy released from detonation and to promote the stability of the stope crown. Cases achieved on-site revealed that the ELOS were very minimum in such a way that all the crowns were sitting in stable zone (see Figure 6).

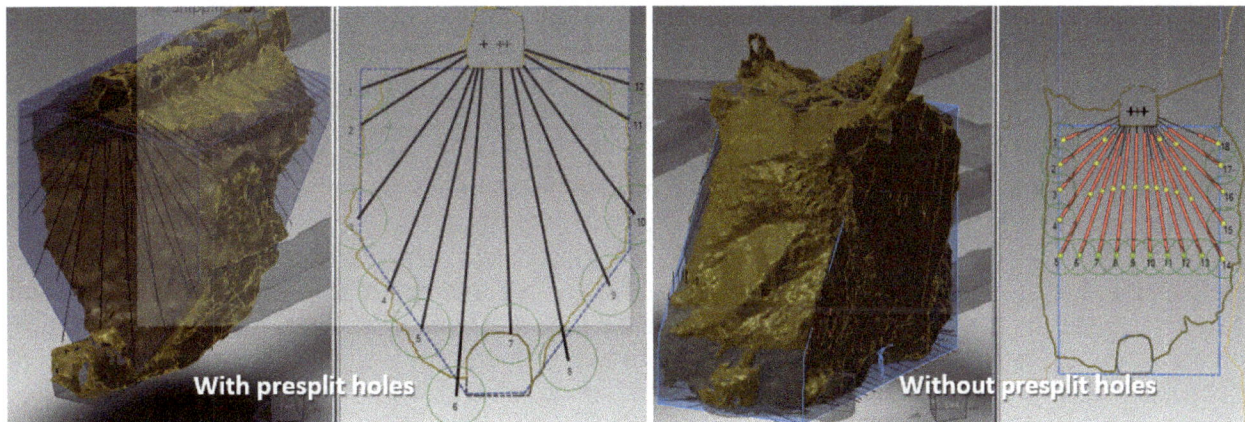

FIG 6 – Comparison between outcomes from blasting with and without presplit holes.

Figure 6 clearly shows that presplit blasting improves the crown shape and stability after blasting. However, This method is not subjected to maximum perfection as damage may happen when buffers are fired at the same time as the first and last production holes. Loose rocks may be left after basting causing production loss therefore blast timing plays a key role in addressing these issues. The following discussion assumes that the necessary in terms of drilling and blasting patterns including the prespliting to limit damage have been already addressed.

Blasting delay influence

In addition to the blasting parameters, delay timing significantly influences the blasting results and the throw of the muck pile. The degree of blast confinement is related to the time it will take for the rock mass to respond (Hettinger, 2015). According to Raskildinov (2001), blasting efficiency improves by increasing the time of the explosion in the rock mass to be blasted and strengthening the interaction of the shock wave. This means that charging designs should be guided by the initial pressure of the explosive product over the charged blasthole length, the application of several groups of detonation waves, and the timing influence in the rock mass. Figure 7 shows the timing influence on the energy released to dislocate a specific rock mass. Two or more holes timed with the same delay will release more energy than when these holes are timed differently. The frequency of duplicate timing delays should be managed with consideration of the vibration effects on the environment.

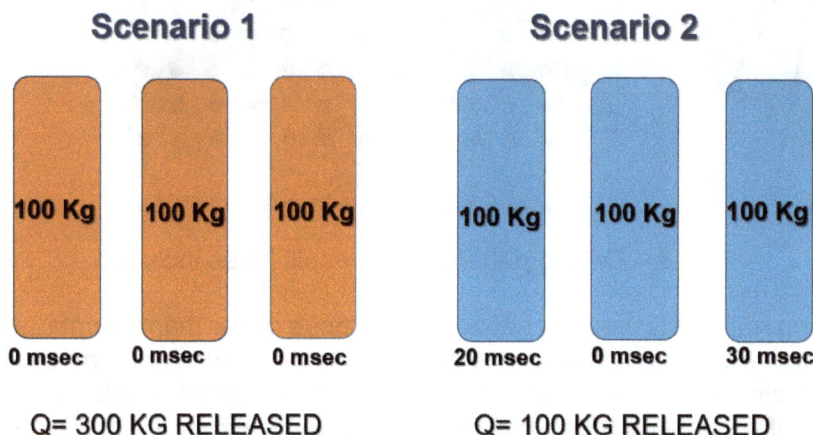

Scenario 1   **Scenario 2**

100 Kg   100 Kg   100 Kg   100 Kg   100 Kg   100 Kg

0 msec   0 msec   0 msec   20 msec   0 msec   30 msec

Q= 300 KG RELEASED   Q= 100 KG RELEASED

FIG 7 – Timing influence of the energy released on the rock mass.

The utilisation of a proper timing delay is necessary for a muck pile formation and the maintenance of a proper balance between required confinement and the creation of crack extension and face movement. ISEE (2011) highlights that timing delay is a means that directs the rock displacement to create desired muck pile and location. In light of this, it should be noted that blasting efficiencies are not only limited to the quality of fragments but involve as well the muck pile location.

A number of typical timing ranges have been suggested by several researchers. Floyd (2013), proposes that hole to hole delay should be at least 10 milliseconds per metre (ms/m) of spacing with an inter ring row delay of 1.5 ms/m of burden. In contrary to Floyd's suggestiions, Grant (1990) earlier found that the timing delay for an optimised fragmentation was between 3–8 ms/m of spacing. Katsabanis and Lui (1996) made a study of the delay effects on the granite bench using manual digitisation of high speed films. They found that zero delays between holes resulted in boulders while the optimum delay time was estimated at around 8 ms/m of burden. Changping Yi (2013) investigated the effect of short delay time on the underground blasting results using a series of numerical simulation of rock blasting through LS Dyna Software. The investigation was made to test Rossmanith's idea that meant to improve the stress wave interaction by controlling the initiation times. The results from the investigation revealed that fragmentation for simultaneous timing is coarser than for delay timing. When initiating with two or more detonators, one should ensure that each detonator has been assigned with a different timing delay.

To the production hole delays, Dindarloo (2015) revealed that the simple approach to get the best preferential plane is to breakup the split into groups of holes that are fired simultaneously with each group being separated by a short delay that is just long enough for the strain waves from the previous group to disperse. Where air deck charging is introduced it is preferable to use the inter-timing delay between decks although this practice came sometimes with issues such as prevention of decks from dislodging or desensitising around them. Since the development of the inertial thrust of a split blast is substantial, the arrangement of blastholes is of paramount importance. The length of timing delay relies on the prevailed conditions. In underground mines, these conditions are not limited to the status of the free face. Where the free face is available, it is preferable to fire at first the presplit holes with a 0 ms before the main production holes while in the absence of the free face the ideal will be to start by creating a free face, before the presplit blasting and main production holes blasting. In both cases, the timing delays between these groups of blasting will vary between 5–10 ms.

At KGM, the inter hole delays of 50 ms resulted in poor fragmentation. Such delays could not allow a good interaction of energy released by blasting in the rock mass. In order to improve such interaction, a strategy was made to reduce the inter hole delay to the interval between 20 ms to

30 ms while maintaining an inter ring timing delays in the range of 110 ms to 160 ms. The delays were estimated using traditional timing design principles. Although findings from Changping Yi's (2013) investigation were not tested, the strategy came with successful results where the thinking behind aligns with Grant's suggestions as the optimum hole to hole delay for best fragmentation is estimated at 7–10 ms/m of spacing with an internal row of 110 ms to 160 ms. The strategy is currently used in most of the stopes at KGM where the sequential blasting machine is used to achieve delay times from 0 to 10 000 ms. This timing has been proven to be effective in terms of stope recovery as since its implementation the overall under and overbreak have been consistently reduced.

CONCLUSION

The main challenge with underground mining using longhole stoping remains the improvement of drilling accuracy and mining profitability. It is found that profitability improves with accurate drilling and blasting effectiveness. Fortunately, several strategies are available to eliminate or at least mitigate deviation. The implementation of stope undercutting reduces the stope length from 35 m to 30 m which in turn reduces deviation by 3 per cent (from 10 per cent to 7 per cent). Another strategy pertaining to drilling is the inclusion of deviation trends a in the hole design and the implementation of presplit blasting within the drilling and blasting pattern. Whilst presplit mining promotes the stability of the stope crown, the key strategy for accurate drilling would be to drill with the estimated operating variables (135 MPa percussion pressure, 65 rev/sec rotation speed and 17 MPa feed pressure) while applying a correction k factor that increases the rate when the bedding inclination is less than 45° from the horizontal or when the hole is under 10 m length. This reduces deviation by a further 4 per cent (from 7 per cent to 3 per cent).

Furthermore, blasting has a significant impact across the mining chain, and should be designed in such ways that fragmentation optimises the downstream results. The blasting layout can be calibrated using Kuz Ram madel which provides a site specific tool for blasting optimisation. This has been used at the KGM to produce an optimum blasting layout that integrate deviation trends (see Table 1). In addition to the optimised blasting layout, timing delay influence significantly on fragmentation. Findings from several trials of small and mass blastings at KGM led to conclude that the blasting effectiveness improves with the use of the optimal range of hole to hole time delays (7–10 ms/m of spacing) and row delays (110–160 ms),and a timing delay length varying in between 5–10 ms to move from one group of blasting detonation to the presplit blasting.

REFERENCES

Adebayo, B and Akande, J M, 2015. Effects of Blast-Hole Deviation on Drilling and Muck-Pile Loading Cost, *International Journal of Scientific Research and Innovative Technology*, pp 64–73.

Beck Engineering, 2018. Life of Mine Deformation and stability Asssessment for Kibali, Report.

Bruce, A D, Lyon, R and Swartling, S, 2014, February 05. Retrieved from The History of the down the hole and the use of water powered: ww.w.geosystemsbruce.com/.../288%20-%20History%20of%20DTH%20Drilling%20

Chabiot, L, 1995. Modelling and Simulation of Longhole Drill Deviation, Montreal, Canada: Department of Mining and Metallurgy Engineering McGill University.

Cunningham, C V B, 1987. Fragmentation Estimations and the Kuz-Ram Model-For Years, in *Proceedings of the Second International Symposium on Rock Fragmentation in Blasting*, pp 475–487.

Floyd, J, 2013. Efficient Blasting Techniques, Blast Dynamics, Reno, NV.

Grant, J R, 1990. Initiation systems-What does the future Hold?, in *Proceedings of the Third International Symposium on Rock Fragmentation by Blasting* (The Australasian Institute of Mining and Metallurgy: Melbourne).

Hettinger, M R, 2015. The Effects of Short Delay Times on Rock Fragmentation In Bench Blasts, Missouri University of Science and Technology.

Jones, P, 2018. Improved drilling accuracy results in reduced ore dilution, Cracow, Technical Services Superintendent, Cracow Gold Mine, Evolution Mining and Minnovare.

Kivade, S B, Murthy, C S N and Vardhan, H, 2015. Experimental Investigations on Penetration Rate of Percussive Drill, in *Global Challenges, Policy Framework and Sustainable Development for Mining of Mineral and Fossil Energy Resources* (GCPF2015), pp 92–98.

Kortnik, J and Bratun, J, 2010. Use of Electronic Initiation System in Mining Industry, *RAMZ-Material and Geoenvironment*, 57(3):403–414.

Mui, S W B, *et al*, 2012. The use of electronic Detonators in Vibration control for Blasting, retrieved from https://www.dsd.gov.hk/EN/Files/Technical.../HATS1207.pdf

Nguyen, V H, *et al*, 2017, May 11. Penetration rate prediction for Percussive Drilling with Rotary in very Hard Rock, retrieved from HAL Id: hal-01521017: https://hal.achive-ouvertes.fr/hal-01521017

Rosin, J and Rammler, E, 1933. Laws governing the fineness of the Powered coal, *J Institute of Fuels*, pp 29–36.

Singh, S P, 1998, January. The effects of rock mass characteristics on Blasthole deviation, *CIM Bulletin*, pp 90–95.

Sinkala, T and Granholm, S, 1987. A bug step towards Cost Saving mining, Atlas Copco Seminar, p 14, Zambia.

White, L, 1984, September. In-The-Hole drills: Making the most of them underground, *E&MJ*, pp 52–57.

Worden, E P, 1978. How to successfully drill the raise pilot holes and minimise deviation, *World Mining*, pp 42–45; 47–51.

Znibe, H, *et al*, 2009. Effect of Stope Undercutting on its Wall overbreak, ReseachGate.

An initial review of conceptual alternative layouts for production areas in Kiirunavaara mine

F K Altuntov[1], B Skawina[2], J Greberg[3], H Engberg[4] and I Niia[5]

1. PhD Student, Luleå University of Technology (LTU), Luleå 971 87, Sweden. Email: firdevs.kubra.altuntov@ltu.se
2. Senior Research Engineer, Luleå University of Technology (LTU) and Luossavaara-Kiirunavaara Aktiebolag (LKAB), Luleå Norrbotten 971 87, Sweden. Email: bart.skawina@ltu.se or bartlomiej.skawina@lkab.com
3. Associate Professor, Luleå University of Technology (LTU), Luleå Norrbotten 971 87, Sweden. Email: jenny.greberg@ltu.se
4. General Manager Mining Technology, Luossavaara-Kiirunavaara Aktiebolag (LKAB), Kiruna Norrbotten 981 86, Sweden. Email: hans.engberg@lkab.com
5. Mine Planner, Luossavaara-Kiirunavaara Aktiebolag (LKAB), Kiruna Norrbotten 981 86, Sweden. Email: ida.niia@lkab.com

ABSTRACT

The study presented in this paper is based on the Kiirunavaara underground mine located in the northern part of Sweden. The deposit is mined using the sublevel caving (SLC) mining method. This paper briefly reviews the alternative SLC layouts, as well as presents challenges for the application. In Kiirunavaara currently, the traditional SLC design is employed, however, continuous improvement of material transportation systems used in mines generates a need for investigation of the current and future production area layouts in sublevel caving mines. When identifying the potential benefits of the alternative production area layouts, the production targets and effects must be further investigated to understand the benefits and potential effects. In this study, several different layouts have been investigated and evaluated. The designs include different development configurations and orepass infrastructure moved further away from the orebody and still require further investigation.

KIIRUNAVAARA MINE

Luossavaara-Kiirunavaara Aktiebolag (LKAB) is an iron ore mining company based in Sweden. LKAB is operating both surface and underground mining operations. Three surface mines are located in the Svappavaara field from which one is currently in production (5.1 Mtpa). The underground operations are Kiirunavaara (26.9 Mtpa) and Malmberget (15.5 Mtpa). There is an ongoing process of reopening the old Mertainen mine and mining study of Per Geijer located north of the Kiruna township. The Kiirunavaara deposit is an orebody with a strike length of ~4 km with an average width of 80 m. The current operating depth varies between 1051 m and 1137 m depending on the area with currently planned mining extending to 1365 m below the surface (Figure 1).

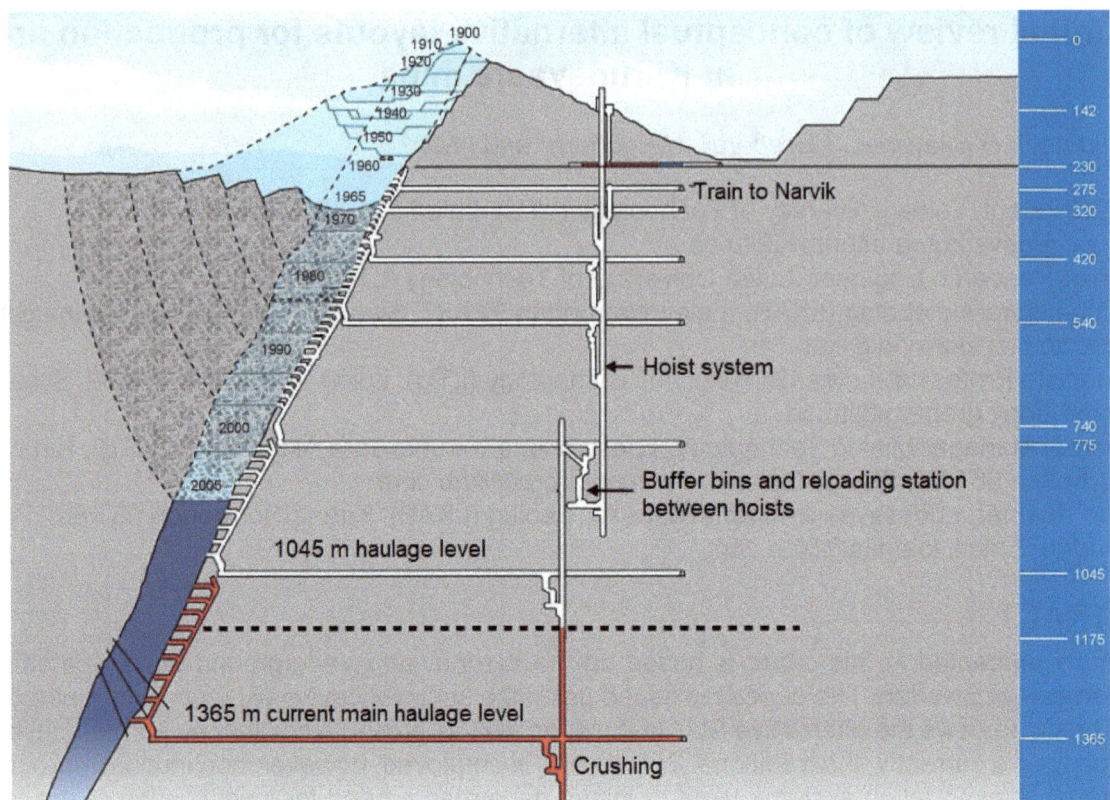

FIG 1 – Kiirunavaara mine (Courtesy of LKAB).

With depth, the orebody narrows, however, it is still open at depth. Orebody strike is north–south and dips at approximately 60° to the east. Kiirunavaara uses the sublevel caving (SLC) mining method in which the ore is extracted via sublevels from top to bottom and the hanging wall caves as the ore is extracted from various production sublevels (Salama *et al*, 2015). The ore consists of igneous and metamorphic rocks and the most common rock formation is syenite porphyry (Skawina *et al*, 2022). The width of the orebody varies from several metres to 200 m and the mine is separated into ten major production zones which are named 'blocks'. There are ten sublevels in every extracted block with a typical length of 400 m to 500 m and orepasses are placed relatively close to the production areas (Skawina *et al*, 2022). Blasted ore is loaded from drawpoints by load-haul-dump machines (LHD) and transferred into orepasses. The passes transfer the ore to the main haulage level where the ore is hauled by trains to the crushers and then hoisted up via skips (Figure 1).

CURRENT PRODUCTION LAYOUTS

The traditional SLC level layout used in the Kiirunvaara mine is presented in Figure 2. The footwall drift is located between orepasses and production cross-cuts. The production areas are connected to the ramps (or declines) and media drifts via access points.

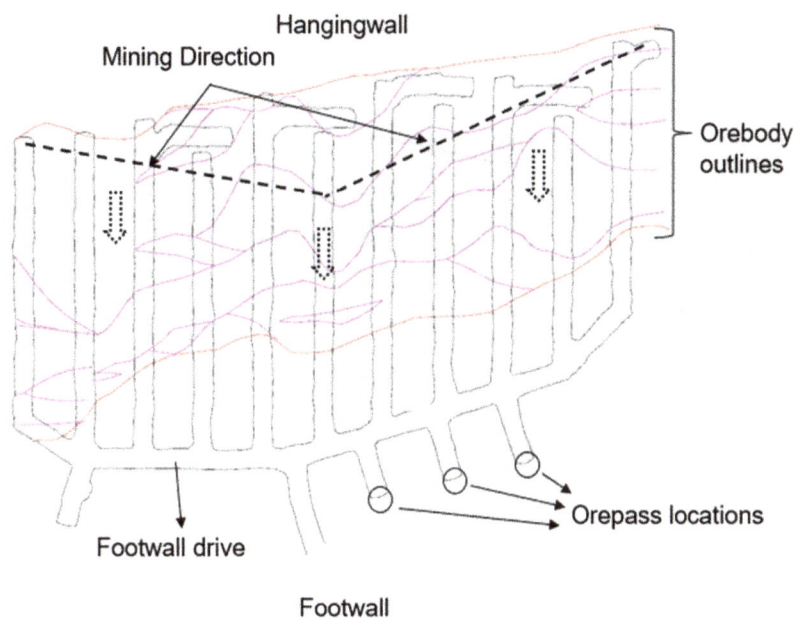

FIG 2 – Currently used production layout in Kiirunavaara mine for a single block/panel. The mine currently uses ten blocks/panels.

The mining sequence is governed by operational limitations related to geotechnical conditions, operational capacities, seismicity, and ore geometry. In the first phase of mining in the Kiirunavaara mine the cave front is kept flat (Shekhar, Gustafson and Schunnesson, 2017). As the production advances depending on the factors such as nearby mining blocks, seismicity, and structural stability of the drifts the pattern of the production area is changed to a chevron front (sometimes also called V-shape) where it converges towards the entrance cross-cut (Shekhar, Gustafson and Schunnesson, 2017). The cave front is maintained in such a way that the production rings in neighbouring cross-cuts are always within two to three blast rings of each other. This makes the cave front relatively flat at ~15° to 20°, and this is common practice in most SLCs. The advantages of this layout are the connection to multiple production drifts with relatively short tramming distances and the ability to adjust the lengths of the drifts depending on the length of the orebody.

In 2018, Quintero proposed a modified traditional layout called fork layout (Figure 3). In this fork layout, the distance of development drifts and tramming of LHDs is intended to be decreased (Quinteiro, 2018). Footwall drift and orepasses were located further away from the caving zone. In this way, it is anticipated to have increased stability. The fork layout is currently being trialled in the Kiirunvaara-Konsuln test mine. In this layout, the loading path of the truck consists of four intersections intended for automation and productivity. In the Kiirunavaara-Konsuln test mine, the material is currently transported by trucks.

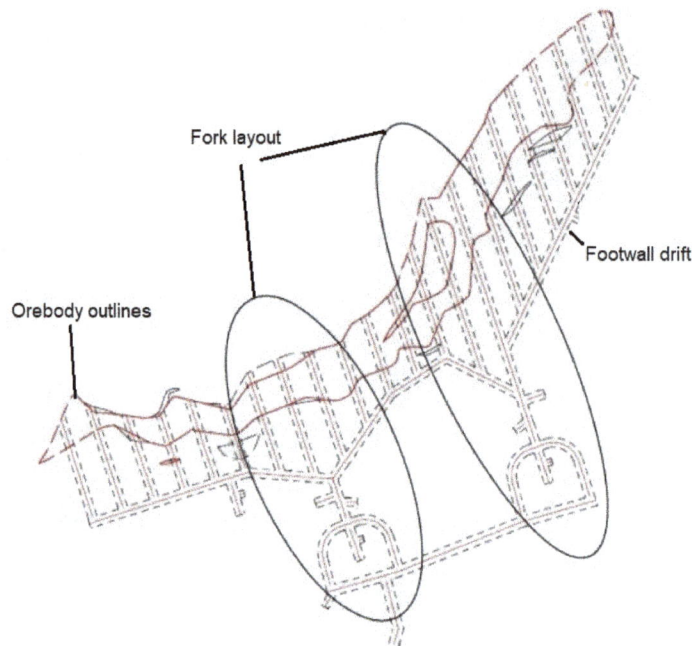

FIG 3 – Fork layout (modified after Quinteiro, 2018).

In this fork layout, the distance of development drifts and tramming of LHDs is intended to be decreased (Quinteiro, 2018). Footwall drift and orepasses were located further away from the caving zone. In this way, it is anticipated to have increased stability. The fork layout is currently being trialled in the Kiirunvaara-Konsuln test mine. In this layout, the loading path of the truck consists of four intersections intended for automation and productivity. In the Kiirunavaara-Konsuln test mine, the material is currently transported by trucks.

ALTERNATIVE PRODUCTION LAYOUTS

In this study, different production-level layouts for the SLC are initially conceptualised and more studies for potentially viable layouts are planned in the next phases of the project.

The heart layout was conceptualised for deposits where orepasses should be placed further away from the orebody (Figure 4).

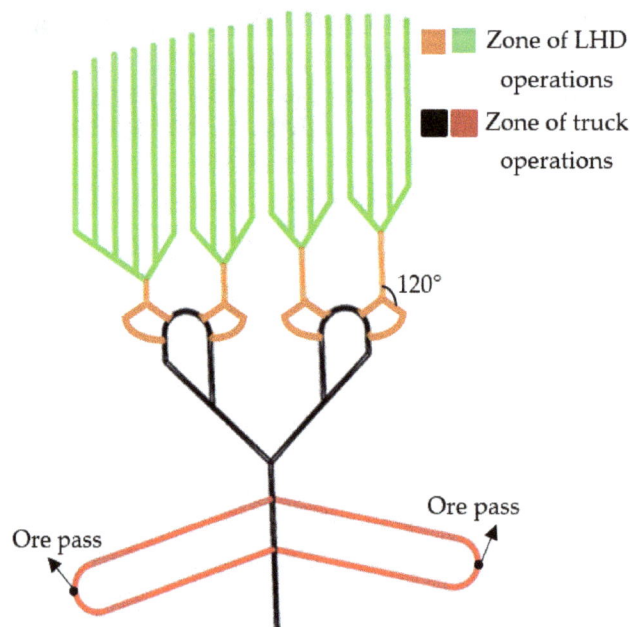

FIG 4 – Heart layout.

In this case, trucks are used to transfer ore from the SLC production drifts to the orepasses. The trucks can use haulage loops for loading and unloading material and are intended to maintain productivity on longer haul distances. The truck travels, after dumping at the orepass, back towards the loading areas and can serve four different forks (four different LHDs). Each side of the 'heart' has two possible loading stations from the two separate forks and a dispatch system could be used to give directions to the truck to either wait for being filled by one LHD at one point or to get its total cargo from both forks. The layout is also independent of whether none, one of, or both the LHD and/or the trucks are manually driven or automated since it is possible to separate traffic between the truck loop and the loading area. The distance from the heart to the orepasses can be varied due to the desired distance wanted and this is to avoid the expected high-stress environment. Like the fork layout (Quinteiro, 2018) and crow's foot layout (Mooney *et al*, 2021), the cross-cuts are divided into groups of production drifts which divide the areas into separate loading zones. Current rock mechanical investigation and modelling in Kiirunavaara mine suggest that the distance between orebody and orepasses should exceed, at least, 400 m at 1400 m depth and exceed, at least, 600 m at 1900 m depth. Stress analysis has suggested these increasing distances with depth to keep the differential stresses below failure levels of the rock mass. Geological information on the deeper part of the Kiirunavaara orebody indicates that the strike length of the orebody decreases. Today's production blocks with 400 m strike length allow two LHDs to work simultaneously in each block. That gives the possibility to operate one LHD per 200 m. The fork layout in a heart-shaped mining system would allow one LHD per 100 m and thus enable to increase in the number of LHDs to four in total and fulfill the production capacity requirements in combination with horizontal truck haulage to the orepasses. The design should be optimised if the heart layout is built and tested. For instance, the turning angle for the LHD when coming out from the loading area (the fork) should be as small as possible to get into position for filling the truck. Also, the orepass dumping configuration should allow for a straight LHD position for stability, safety, and maintenance purposes. Ideally, the three-way crossing should be symmetrical with a 120° angle (Mercedes star-shaped) and the loading point should be at least 15 m long to allow the loading machine to be articulated in a straight position.

There are also investigations of modifying the existing layouts specifically the fork layout and crow's foot layout, as shown in Figures 5 and 6 respectively.

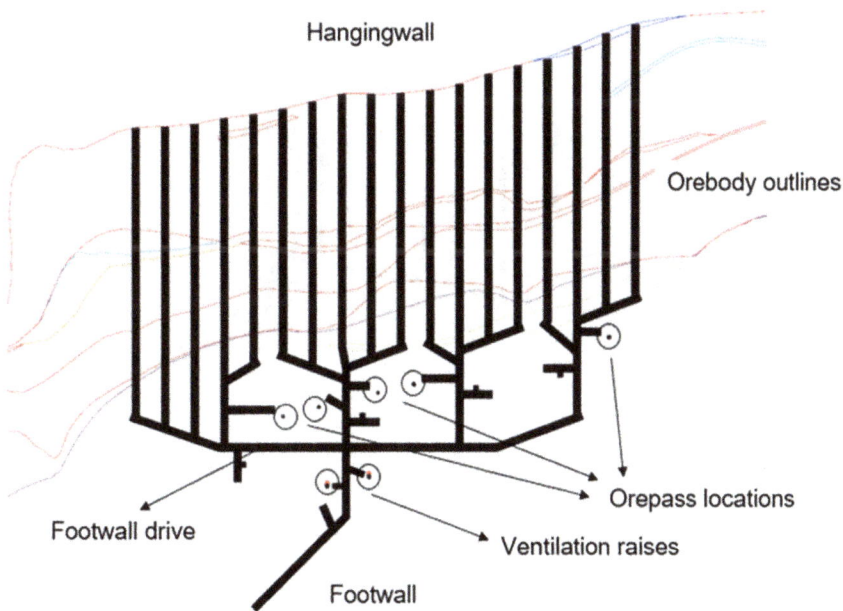

FIG 5 – Modified crow's foot layout.

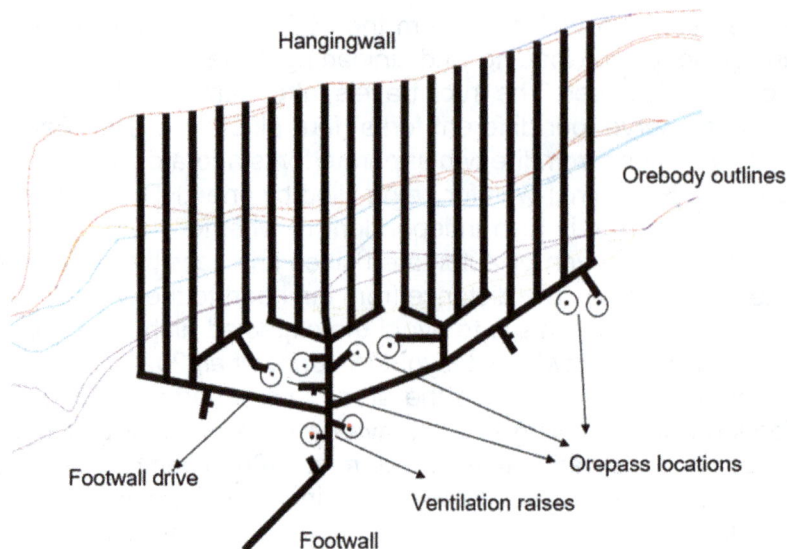

FIG 6 – Fork layout.

The crow's foot layout was presented by OZ Minerals to eliminate loader and vehicle interaction of the footwall drives so that safety, productivity, and automation could be increased (Mooney *et al*, 2021). This layout is a modified version of the traditional layout, and the main difference is that orepasses in the crow's foot layout are placed between the perimeter drive and the cross-cuts. Geotechnical analysis of each layout identified that layouts 5 and 6 were subject to the least adverse conditions in terms of high-stress concentration in working areas, and potential rock mass damage and deformation in the drives and infrastructure areas. For these layouts, the advantages are lower stress around footwall drift and the least rock mass damage around footwall drift compared to the traditional layout.

The double footwall drift layout (Figure 7) consists of one footwall drift that follows the ore boundary and one that is developed later to recover the remaining ore near the footwall.

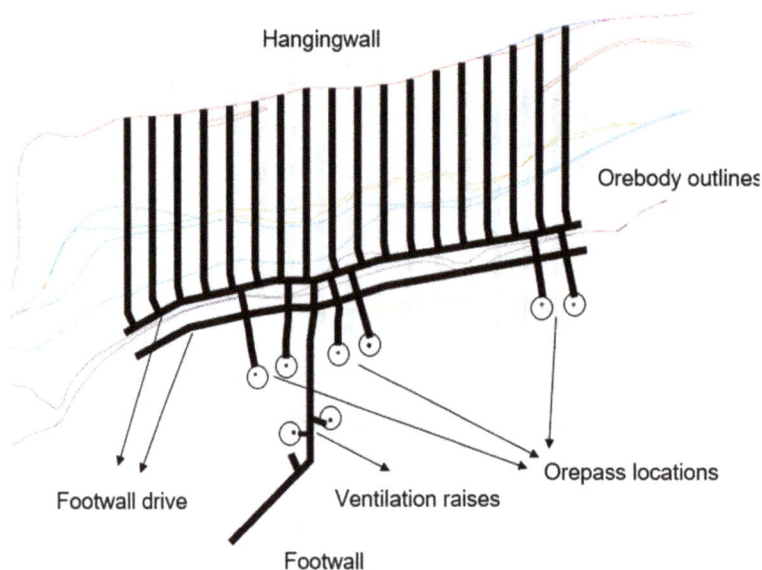

FIG 7 – Double footwall drift layout.

The second footwall drift is mined as a cross-cut to mine the rest of the ore at the level above. The reason for this is related to seismicity where during the first stages of mining the seismicity is mostly concentrated in the waste rock, thus footwall drift is developed inside the orebody. Once reaching closer to the footwall drift most of the energy is released during rock mass yielding, which includes stress redistribution and seismic activity thus the second drive located in the waste rock can be developed and the rest of the ore scheduled for that level can be extracted. This can be much more difficult in the case where mine has only one area to mine. In the case of Kiirunavaara, there are ten blocks on each level where the mining can still progress during the development of the second footwall drive. However, the downsides are high stress and potential deformation to layouts with footwall drift located in the stress abutments, as well as difficult loading conditions for the LHDs as the machines are required to muck drawpoints of the final production rings while articulated due to the short length of the remaining drift.

The layout inside the ore reduces development metres since the footwall is made inside the orebody. In some levels where the orebody extends beyond the existing perimeter, there is a need for an extra footwall drive to not leave ore behind (Figure 8). The downsides are similar to the double footwall drift layout.

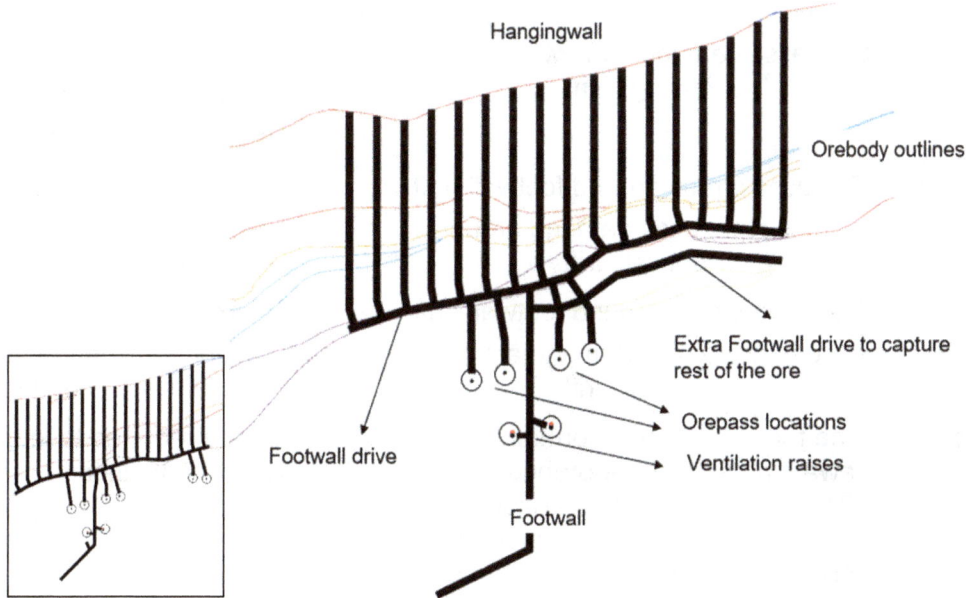

FIG 8 – Inside the orebody: Small left figure represents every level and the right figure levels where the orebody outlines extend beyond the existing footwall drive.

The less-development idea (Figure 9) considers the footwall drift located in the ore boundary and cross-cuts running in two directions.

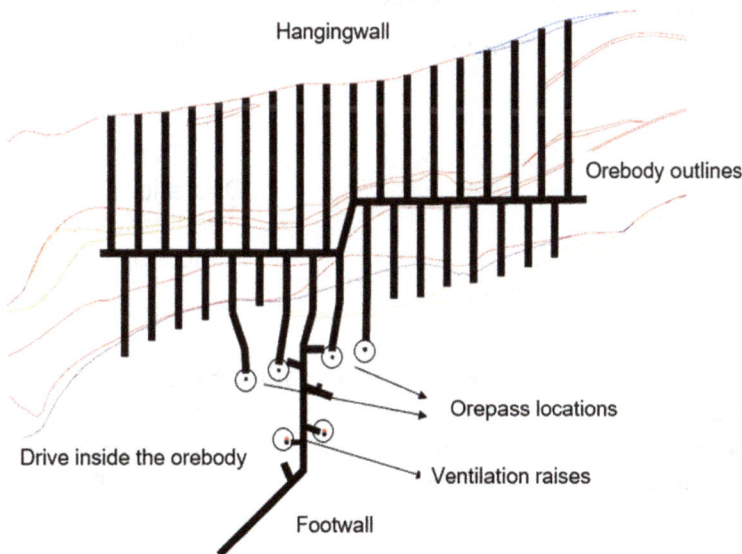

FIG 9 – Less-development concept.

This enables several accesses to the orepasses and requires less reinforcement which results in faster development of the area. However, the downsides are the diminishing pillar formed between the mining fronts on each level, cave interaction between levels, too many drifts that need extra fans for ventilation, and increased production drilling metres.

A summary of the advantages, disadvantages, and status of each concept design is presented in Table 1.

TABLE 1

Comparison of layouts.

Layout	Main advantages	Main disadvantages
Traditional layout	Short tramming distances. Ability to adjust the lengths of the drifts depending on the length of the orebody.	Footwall drift is more exposed to mining induced stresses.
Heart	Orepasses are located away from the mining front.	Long tramming distances. Increased development. Reduced level productivity
Modified fork	Reduced stress around footwall drift. Less rock mass damage around footwall drift.	Increased development. Less redundancy in the number of orepasses for each fork.
Modified crow's foot	Reduced stress around footwall drift. Less rock mass damage around footwall drift.	Increased development. Less redundancy on the number of orepasses for each fork.
Double footwall	The footwall drive is normally further away, outside of the orebody zone. Double footwall drifts are inside and closer to the orebody outline zone which gives less development metres in total.	LHDs are required to load in the straight articulated position. High loading on the liner.
Inside the orebody	Similar case as in double footwall - less development.	LHDs are required to load in the straight articulated position. High loading on the liner for extra footwall drive.
Less development	Similar case as in double footwall - less development.	High support is required due to the diminishing central pillar formed on each level. LHDs are required to load in the straight articulated position. Operational issues with extra fans for ventilation and extra openings for production drilling due to cross-cuts in two directions.

This summarises which concept designs would be taken forward for further review and analysis and which concepts have been discarded. The next step of the layout evaluation project is to consider future requirements and constraints including new technology and transitioning technology such as electrification and automation. The accommodation of these factors is important for productive and safe mining operations.

CONCLUSIONS

Early concept designs for alternative sublevel caving production level layouts have been developed. A preliminary review of these concept designs identified the following:

- The traditional layout is most favourable overall due to the comparatively short haulage distance from production rings to the orepasses. However, the orepasses and footwall drift are close to regions of high mining-induced stress which has the potential to impact the stability and longevity of nearby orepasses.

- The concept of the Heart layout has the potential to help mitigate stress-induced instability in the orepasses and the footwall drift for deeper levels as these are placed significantly further away from the SLC. Since the orepasses are located far away from the production area, trucking is required to transfer ore from the production drifts to the orepasses. The required development per level is also significantly increased.

- The modified crow's foot layout has the advantage of shorter hauling distance in production drifts and eliminates loaders tramming across the footwall drive and potentially interacting with other vehicles or equipment. Production capacity is enhanced in this layout due to the isolated work areas for each panel and operating loader. The disadvantage of this layout is the increased haulage distance if the orepasses servicing each crow's foot become unserviceable due to damage or wear.

- The modified fork layout has the potential advantage of lower mining-induced stress at the location of the footwall drift. However, depending on different versions of the fork layout the development requirement may be higher than in other layouts.

- The double footwall drift, inside the orebody and less development layouts, have the advantage of reduced development. On the other hand, numerical analysis shows high stresses, yielding, and high load on the liner due to the placement of footwall drift for these alternatives.

This paper discussed initial conceptual designs of different production-level layouts for SLC mines. Some of the design concepts have the potential to provide some benefits compared to contemporary SLC layouts. Further assessment to determine the potential advantages and drawbacks is required, and some of the concepts discussed in this paper are not economically, technically, or operationally viable. Increasing the understanding and capabilities of more flexible re-arrangement of the production areas may enable further improvements and modifications that will increase the safety and effectiveness of future SLC operations. Additionally, new layouts should accommodate newly developed equipment such as further electrification and automation of the mines.

ACKNOWLEDGEMENTS

The authors acknowledge the LKAB for funding and support towards the completion of this work.

REFERENCES

Mooney, A, Grosser, H, Marsden, J and Dunstan, G, 2021. Carrapateena crows foot level layout – Safety in design to support sublevel cave production efficiency and automation, in *Proceedings of the Underground Operators Conference* (The Australasian Institute of Mining and Metallurgy: Melbourne).

Quinteiro, C, 2018. Design of a new layout for sublevel caving at depth, in *Caving 2018: Proceedings of the Fourth International Symposium on Block and Sublevel Caving* (eds: Y Potvin and J Jakubec), pp 433–442 (Australian Centre for Geomechanics: Perth). Available from: <https://doi.org/10.36487/ACG_rep/1815_33_Quinteiro>

Salama, A, Greberg, J, Skawina, B and Gustafson, A, 2015. Analyzing energy consumption and gas emissions of loading equipment in underground mining, *CIM Journal*, 6(4):179–188, Available from: <https://doi.org/10.15834/cimj.2015.24>

Shekhar, G, Gustafson, A and Schunnesson, H, 2017. Loading Procedure and Draw Control in LKAB's Sublevel Caving Mines: Baseline Mapping Report, Research report, ISBN 978-91-7583-807-6 (electronic), Luleå University of Technology, Sweden, 60 p. Available from: <http://urn.kb.se/resolve?urn=urn:nbn:se:ltu:diva-61938>

Skawina, B, Salama, A, Gunillasson, J, Strömsten, M and Wettainen, T, 2022. Comparison of Productivity When Running Filled, Near-Empty, or Flow-Through Orepass Using Discrete Event Simulation, *Mining*, 2:186–196. Available from: <https://doi.org/10.3390/mining2020011>

Construction and monitoring of development access into highly stressed rock masses

R Talebi[1], E Villaescusa[2], N Bustos[3] and C Drover[4]

1. Senior Geotechnical Engineer, Northern Star Resources, Kalgoorlie WA 6430. Email: rtalebi@nsrltd.com
2. Chair of Mining Rock Mechanics, WASM, Kalgoorlie WA 6430. Email: e.villaescusa@curtin.edu.au
3. Senior Research Engineer, WASM, Kalgoorlie WA 6430. Email: n.bustossalgado@curtin.edu.au
4. Principal Engineer, Beck Engineering, NSW 2067. Email: cdrover@beck.engineering

ABSTRACT

A research project to better understand the rock mass behaviour and the seismic response has been conducted during the design and construction of the 9215 level at the E-Block sector of the Northern Star Kanowna Belle Mine. The research program was undertaken at a tunnel located 1100 m below the surface with the results applicable to feasibility studies on underground mining projects in high-stress conditions, where failure can occur very soon after tunnel construction.

Semi-circular excavation shapes coupled with the use of de-stress blasting techniques were designed to monitor the excavation in conditions of very high stress. The ground support was designed considering high energy dissipation ground support schemes (HED) as described by Arcaro et al (2021). The results indicate a decreased rock mass demand due to the change of excavation shape. Also, localised reduction in rock mass stiffness was achieved due to the implementation of de-stress blasting. Finally, increased ground support capacity was gained by the implementation of the high energy dissipation ground support schemes. The Kanowna Belle Mine has significant ore reserves at depths exceeding 1000 m below the surface. The results of this research will be used to optimise the extraction of the remaining E Stoping Block at depths exceeding 1350 m.

INTRODUCTION

As hard rock mining progresses deeper, the likelihood of hazardous rock mass behaviour escalates (Villaescusa et al, 2022). Deep, hard rock, underground mines are extracted in high-stress environments and, when combined with excavations and geotechnical features of the rock mass, can result in dynamic rock mass rupture and falls of ground (Heal, 2010).

In recent years several underground mines in Australia have stopped operating at great depth. This is partly due to the geotechnical concerns posed by dynamic rock mass rupture in a high-stress environment and partly due to conventional excavation design and construction procedures not being sufficient in managing the risk of dynamic rock mass rupture. Some of these operations have been compelled to abandon workings at such depths. This is concerning as many other mines are approaching depths where dynamic rock mass rupture is becoming more common and intense.

Figure 1 shows a long section view of the Kanowna Belle Mine, where the deepest sector being mined is the E-Block (highlighted in the figure). Figure 2 shows a zoom-in of the E-Block, displaying the area where large seismic events have been recorded resulting in significant damage. The observed damage has shown a strong correlation with geological features, high strength rock mass, high mining rate or a combination of these factors. Figure 3 shows the distribution of recorded damage identified in the lateral development and classified according to Table 1. Mapped damage in the area extends to 1800 m of tunnels, with one-third of mapped areas being severe damage. It is worth noting that the cost of re-building tunnels after seismic damage is often more expensive than the cost of conventional development. Rehabilitation often involves hindrance to production, lost opportunities and reconstruction and installation of ground support in a broken rock mass.

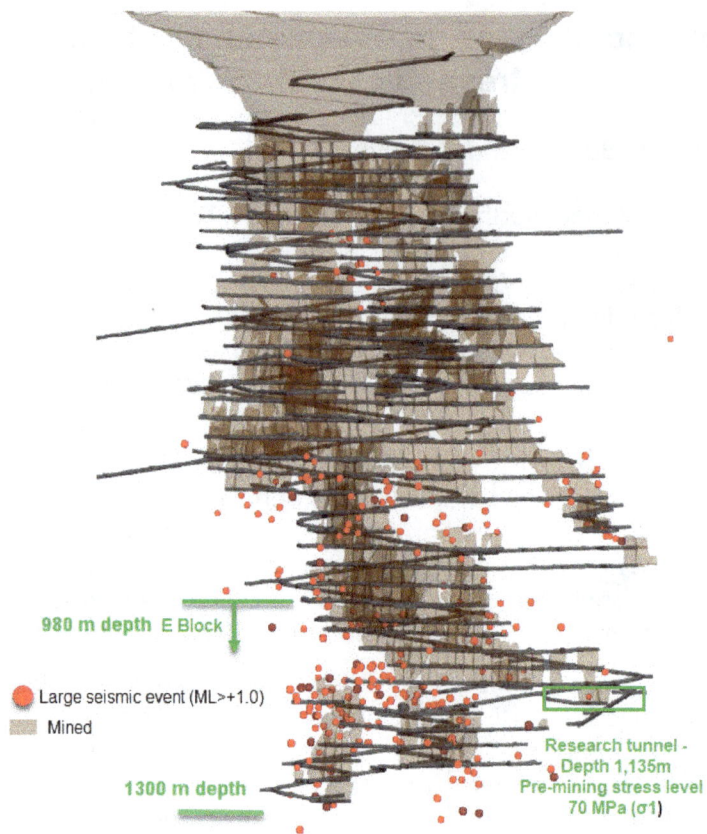

FIG 1 – Vertical view of the Kanowna Belle mine.

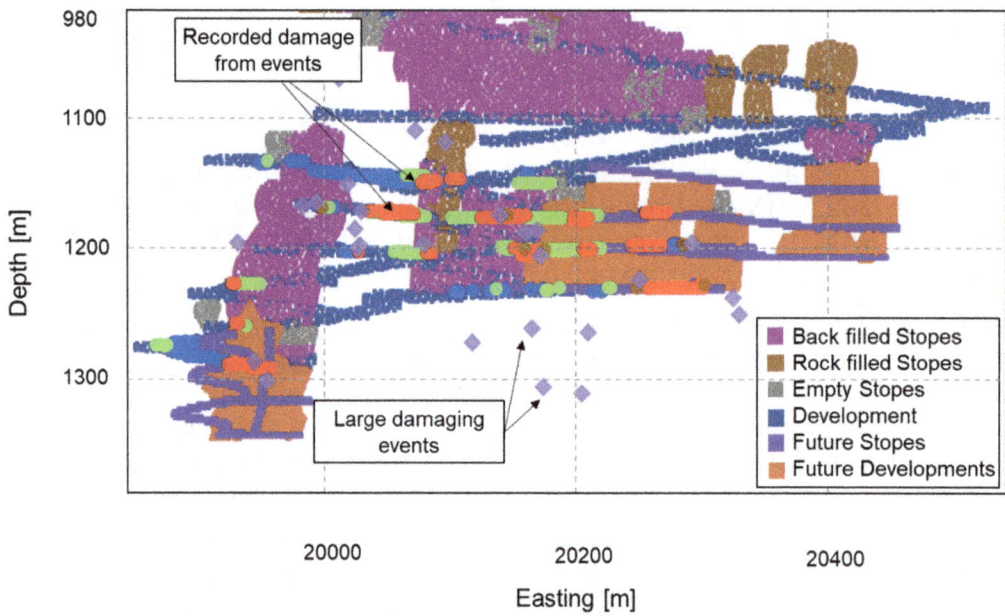

FIG 2 – Large damaging events location and recorder damage at E-Block.

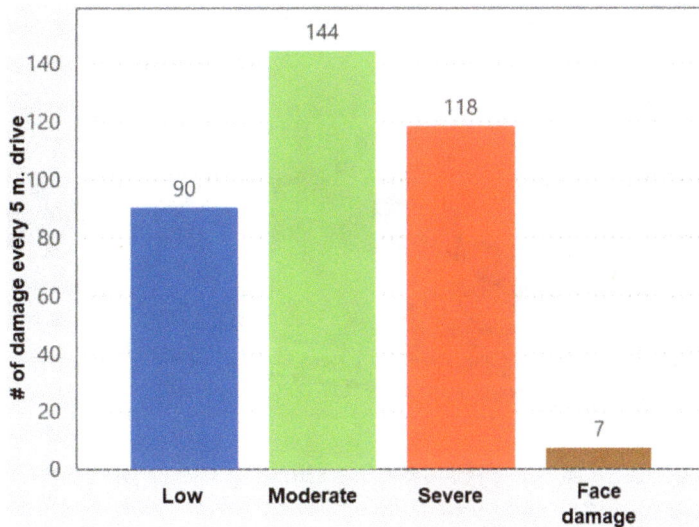

FIG 3 – Damage type distribution in the E-Block.

TABLE 1

Reinforcement and surface support damage scale. Modified from Drover and Villaescusa (2015).

Damage	Reinforcement damage	Surface support damage
Low	Minor loading and mostly deformation of reinforcement surface fixtures. 50 to 100 mm of displacement.	Fibrecrete cracking and possibly small blocks dislodged, minor mesh bagging. The retention function is still effective.
Moderate	Significant loading and deformation of reinforcement surface fixtures, resulting in some broken bolts. Isolated loss of reinforcement function. 100 to 200 mm of displacement.	Fibrecrete fractured and small blocks de-bonded from rock, moderate bagging of mesh with some strand failures and separation of the overlap. Isolated loss of retention.
Severe	Many broken rock bolts with rock unravelling around the elements. May experience floor heave. Significant rehabilitation is required. Greater than 300 mm of displacement.	Fibrecrete and mesh are heavily deformed or damaged. Rock ejection where mesh is torn. Significant ejection of any external layer of fibrecrete. Significant rehabilitation is required.

Large damaging seismic events at Kanowna Belle

Prior to the acquisition of the mine by Northern Star Resources in 2014, one of the largest seismic events occurred at the E-Block and detrimentally affected the mining strategy of the lower, deeper levels of the mine. The seismic event was an ML +2.2 ± 0.3, which was recorded in June 2013, 10 minutes after the 9115 development heading was blasted. No personnel were exposed to the event as the area was still under exclusion at the time. The damage caused by the event was significant and affected three levels in the mine, including the 9115, 9145 and 9175 eastern ore drives. The active structures in this area were only known from their reaction to the development firings. Their location and orientation were derived from seismic spatial trends. The structures in this area are parallel and angled at 70–80° to the orebody structures (see Figure 4). The 9115 and 9145 eastern drives are located close to the Fitzroy shear zone (~5 m). This shear zone consists of a soft rock that should deform easily under high levels of stress.

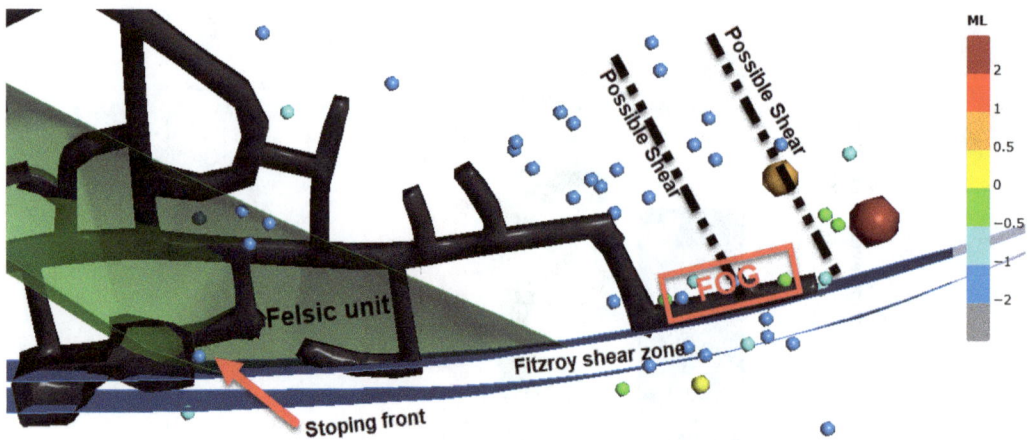

FIG 4 – Geology scheme at the E-Block zone showing the large seismic events (FOG).

Damage – 9115 level (1245 m below surface)

The main damage occurred in the ore drive's right-hand sidewall (southern wall). Figure 5 shows a picture of the main seismic damage on 9115 level. The damage in the right-hand corner looks to have originated from a source somewhere above. It is not clear in the figure, but the right-hand damage breaks up to about 3–4 m into the hanging wall. The shear zone containing the Fitzroy Fault is located in the hanging wall of this drive. Damage here was severe and completely closed the drive making it inaccessible.

FIG 5 – Photo from the seismic damage in 9115 OD25E.

Damage – 9145 level (1215 m below surface)

Significant sidewall movement in the 9145 level is evident in Figure 6. The area where the largest bulking occurred coincided with areas having a lack of sufficient dynamic support elements. The bending of the straps indicates that the excavation was compressed, resulting in floor heave and closure at the vertical walls.

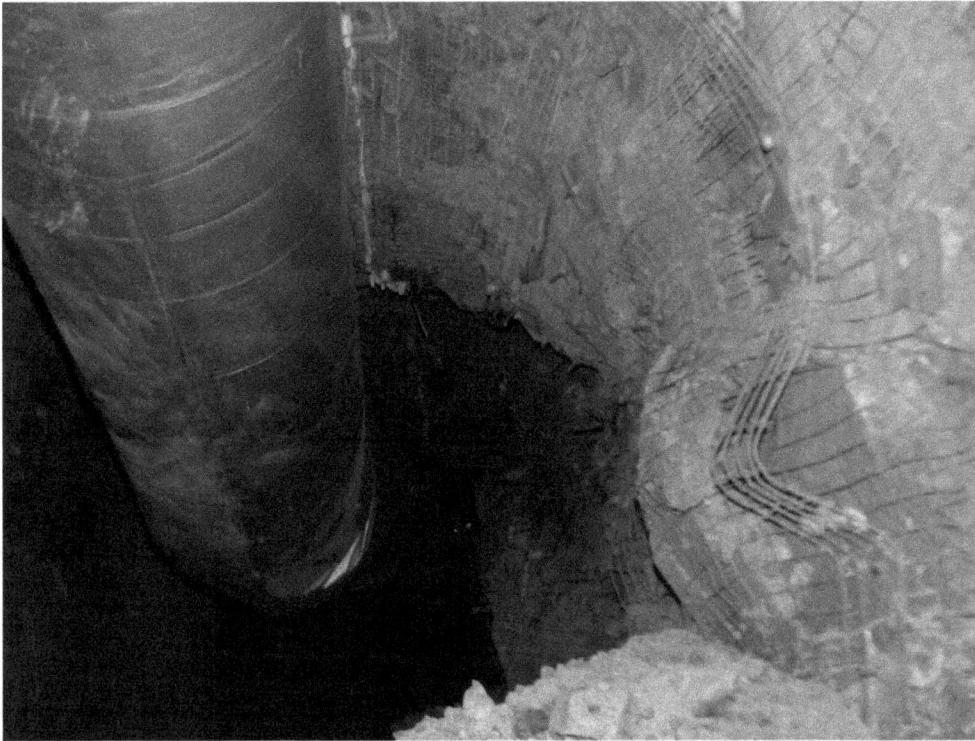

FIG 6 – Photo from the seismic damage in 9145 OD22E.

Damage – 9175 level (1185 m below surface)

A region of major floor heave extended for approximately 15 m along the drive. The maximum floor heave was measured at approximately 1.5 m. The rocks from the floor heave were fragmented into relatively small pieces, as shown in Figure 7.

FIG 7 – Photo from the seismic damage in 9175 OD22E.

Failure mechanism

The rock mass within the Fitzroy shear zone consists of a soft rock mass likely to deform under a high level of stress, but the area immediately outside this shear zone contains a felsic contact which behaves as a brittle rock. This causes a high stiffness contrast between both rock mass units under high stress. Additionally, a degree of freedom for strain occurs near the shear zone in the vicinity of the Fitzroy Fault. This is illustrated in Figure 8, where the magnitude and the orientation of stress are consistent with slippage along the structure.

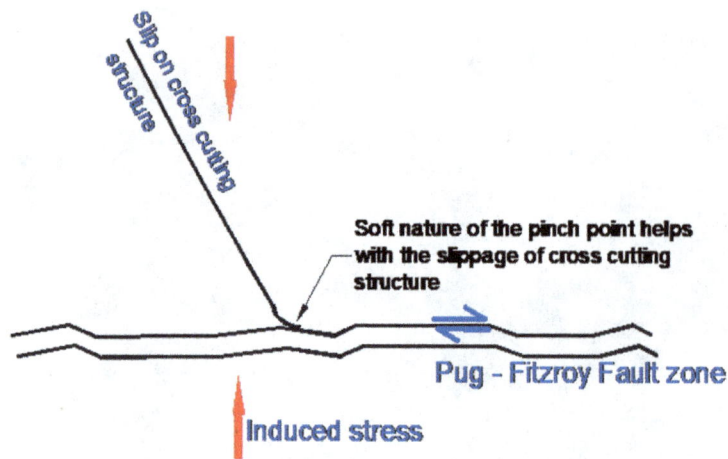

FIG 8 – Schematic view of the failure mechanism.

The risk of shear failure on geological structures can increase with depth due to stress concentration along the discontinuity planes. In addition, the formation of areas having a low confinement stress also contributes to failure. The review of historical seismic damage at Kanowna Belle mine clearly indicates evidence of shear failure, which is observed in most of the affected locations.

The concept of shear failure is further explained in Figure 9. Due to mining over time, stress changes occur in the vicinity of the geological features. When the magnitude and the orientation of the induced stress reaches a level that overcomes the shear strength of the structure, slippage along the structure occurs. The rupture results in stress change around the structure, which can result in dynamic failure of the rock mass. This means that the stress concentration reaches a level in which ejection of rock may be initiated. It is noted that ejection is usually linked to stress concentration locally. Rupture is usually the instability along a structure.

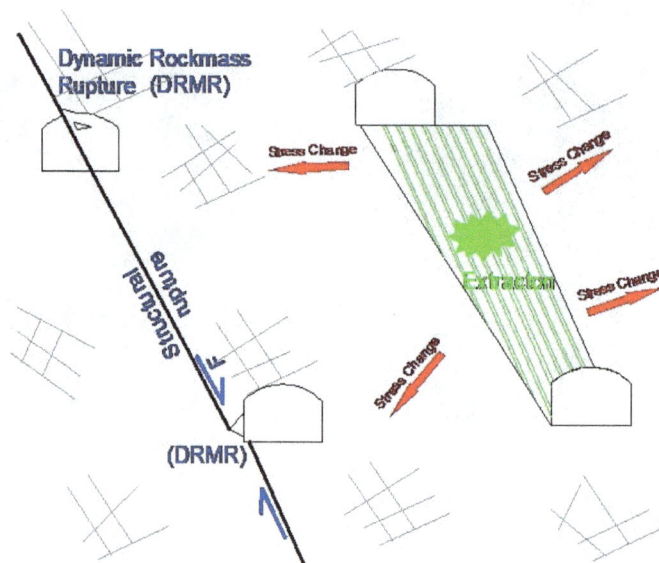

FIG 9 – Concept of shear failure on geological features.

RESEARCH TUNNEL

Layout design

Several steps were followed to design the trial mining area, including a review of historical seismic damage drive locations, the mining direction, and numerical modelling of the proposed area. Historical damage in the lower E block has shown significant excavation damage in areas where footwall development intersects with oblique mineralised structures and in ore development subparallel to the Fitzroy Fault. Pinch points between the Fitzroy Fault and the cross-cutting felsic contact have been the source of significant seismic activity.

Based on the available geological information at the time, the development was designed as shown in Figure 10. The stoping extraction direction was chosen from East to West moving away from the major structures. The cross-cuts are designed subparallel to the *in situ* major principal stress orientation. This was confirmed by the results from a numerical model exercise undertaken in Abaqus. The model suggested that instability occurs primarily at the western abutment of the stoping front and the seismogenic zone affects the nearby cross-cuts and access drift. As illustrated in Figure 11, a high seismic potential was forecast on an oblique shear structure bisecting the first two 9215 stopes during the production stage, as indicated by the modelled rate of energy release (RER) (Levkovitch, Beck and Reusch, 2013).

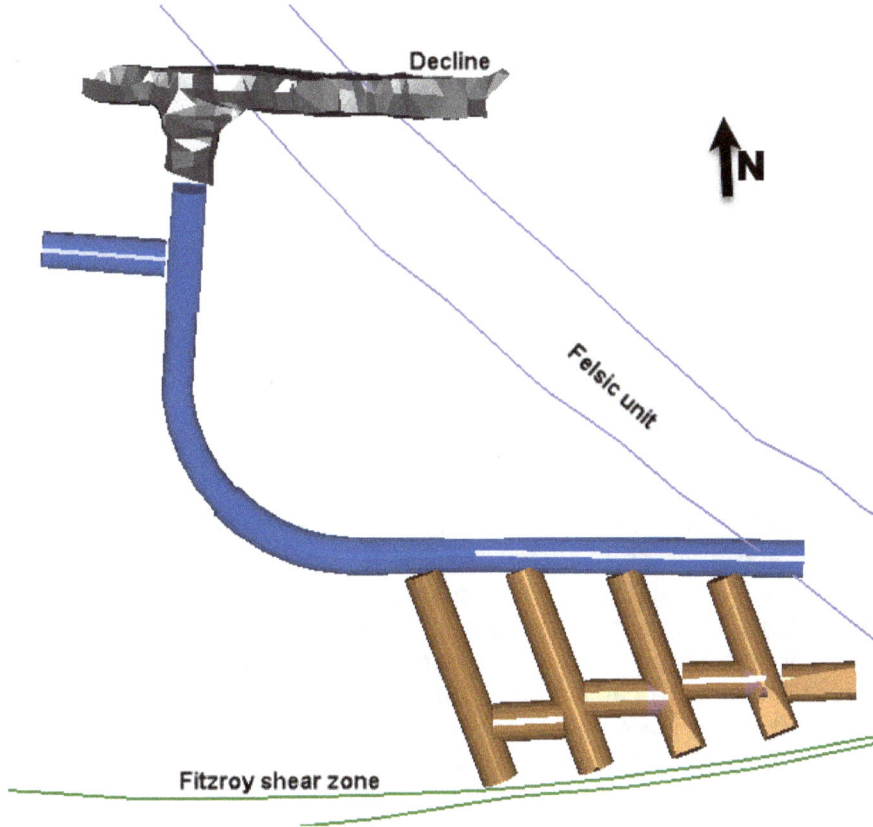

FIG 10 – Plan view showing initial layout design of 9215 level.

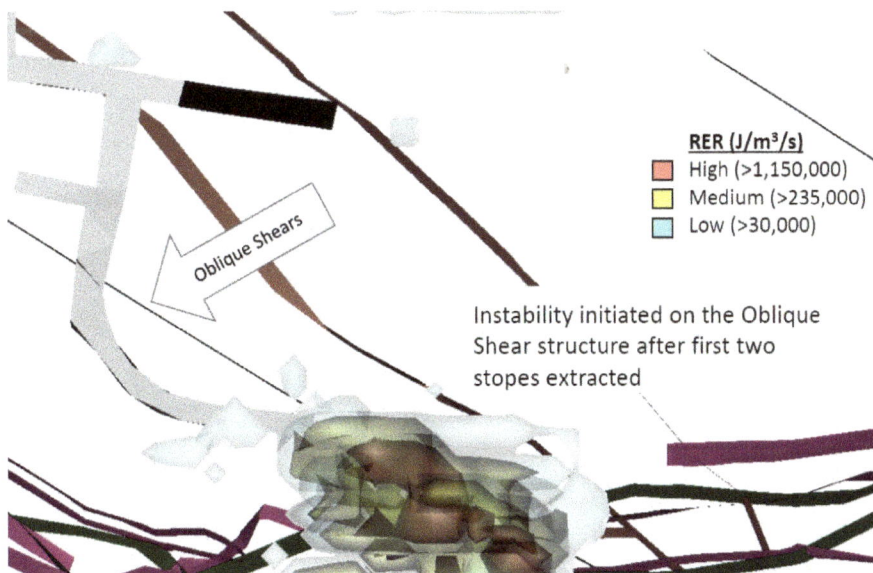

FIG 11 – Modelled seismic potential during production. From Drover and Villaescusa (2020).

Instrumentation

Prior to the commencement of development in 9215, a high-resolution array of eight seismic sensors (3A25K type) was installed, as shown in Figure 12. The objective was to use the seismic system for the analysis of face de-stress blasting, including induced seismicity. This also included loading from production stoping. An additional objective was to use the seismic data for the confirmation of medium and large scale geological structures in the area. The accuracy of the seismic system in the area was 0.3 m with the ability to record an event larger than -2.5 ML.

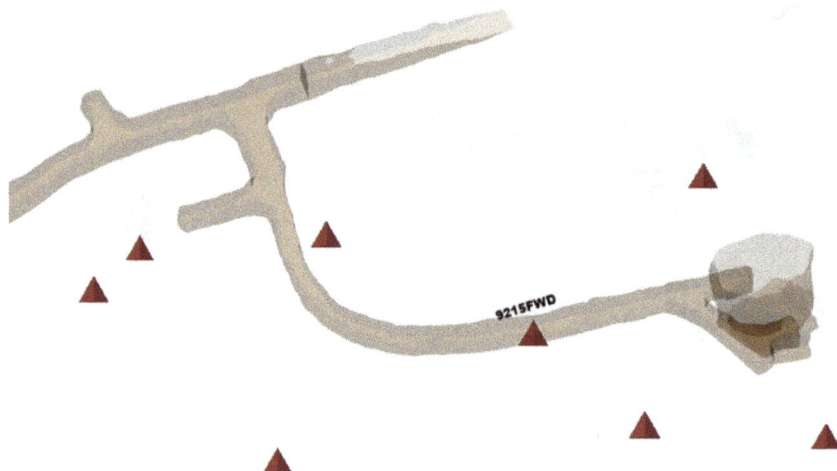

FIG 12 – View of the Seismic array installed around the 9215 research tunnel.

During the calibration of the seismic system in this area, two diamond drill holes (9245 and 9232 levels) were selected and a series of electronic detonators were placed in the holes at a known positions and initiated sequentially. The initiation time of each explosion was accurately registered by the seismic system using fibre-optic cables connected to the detonators. The seismic system measured the artificial seismic waveforms generated by the explosion and the P/S wave seismic velocities of the rock mass were calculated. A velocity model was then generated within the IMS seismic processing software allowing great accuracy in the calculated seismic event locations. The outcome of the calibration blast is shown in Figure 13.

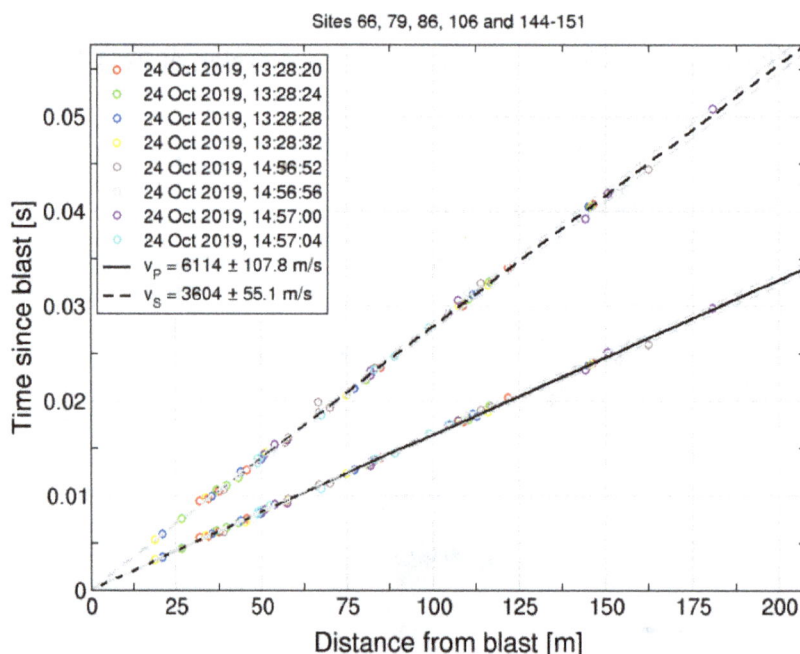

FIG 13 – Example fit of travel times using a homogeneous velocity model for 9215 research tunnel.

Construction

Construction was undertaken using a high energy dissipation (HED) ground support strategy, as described by Arcaro *et al* (2021). The stabilisation is provided in sequential layers to limit the combined rock reinforcement and support displacement at the boundaries of the excavations.

The primary support consisted of an initial layer of 50 mm thick shotcrete and G80/4.6 tensile mesh installed with 2.4 and 3.0 m long decoupled Posimix bolts with a central debonding sleeve. The support scheme was completed by applying a 25 mm thick second spray of shotcrete over the initial layer. The shotcrete overspray (fill-in) was designed to fully encapsulate the high-tensile mesh. Experience has shown that a mesh-reinforced shotcrete has significantly greater energy dissipation capacity than shotcrete with an exposed (external) mesh layer. Mesh-reinforced shotcrete also has superior load transfer characteristics. Since both surface support components were rigidly connected, they deformed at the same time and rate. This allowed the ultimate capacity of both components to be achieved over a compatible range of displacement. An important requirement of the second shotcrete layer is to minimise the thickness of the spray over the G80/4.6 mesh, such that slabbing, delamination or spalling of the second shotcrete layer does not occur.

An innovative set-up that allows the installation of rolls of chain link mesh using a conventional twin boom jumbo was utilised. The methodology involves the introduction of an 'L Pin' device that allows mesh installation without losing the functionality of one of the booms (Arcaro *et al*, 2021). While one boom was fitted with drilling and bolting equipment, the mesh could be installed using the other boom, which later can quickly and easily return to normal functionality by simply removing the L Pins following the installation of the chain link mesh.

De-stress blasting

De-stress blasting is a tunnel construction technique implemented to fracture the rock in such a way that strain energy is dissipated within the rock mass while causing minimal deformation (Drover and Villaescusa, 2019). The technique is implemented to reduce the intensity or frequency of strain bursting or spalling events in high-stress mining environments. It relies on small explosive charges which are used to propagate or mobilise geological structures present within the rock mass environment at the immediate vicinity of the development headings. Thus, the ideal outcome of the de-stress blasting is that small-scale damage allows for the induced stresses and accumulated strain energy at the face to be relieved. This can occur through slip of discontinuities or fracture propagation near the excavation boundary immediately following the blasting event.

The de-stress blasting design implemented at Kanowna Belle consisted of ten holes, 55 mm in diameter, 6.1 m long designed for a 3.7 m cut length, as shown on the left side of Figure 14. The scheme for the de-stressing charge arrangement is shown on the right side of Figure 14. A minimum 0.5 m uncharged buffer zone was maintained between the toe of the development and collar of all de-stressing charges to avoid sympathetic detonations. The holes were charged with ANFO at the last 1.5 m (see Figure 14) and blasted simultaneously.

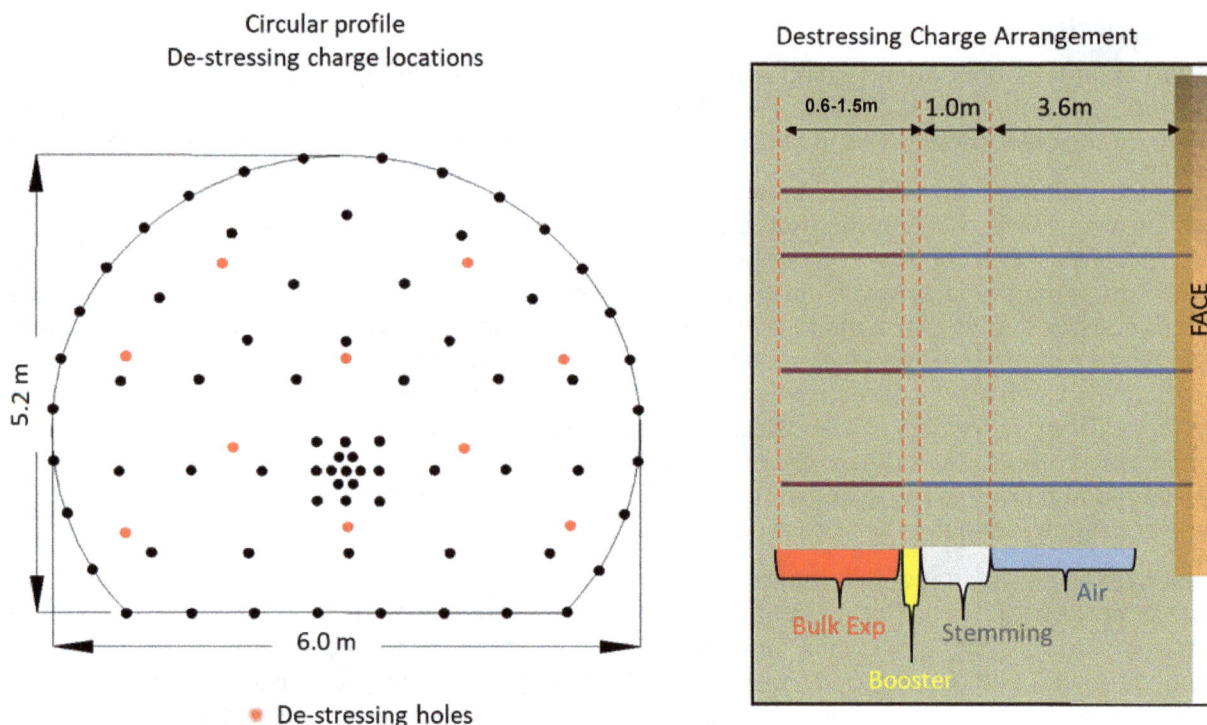

FIG 14 – De-stress blasting charging plan.

All de-stressing charges were initiated simultaneously on the first delay. The design was applied to alternate faces with conventional blasting to compare the results of de-stressed faces with conventionally blasted faces.

Drive profile design

Based on the level of seismic risk, different excavation profiles and ground support schemes were designed for this development, as per Figure 15. In order to minimise stress concentrations at the excavation corners, the excavation drives were modified from having an arched profile in the backs to a semi-circular excavation shape. These shapes are recommended for use for excavations located at great depth where stress-driven failures are common.

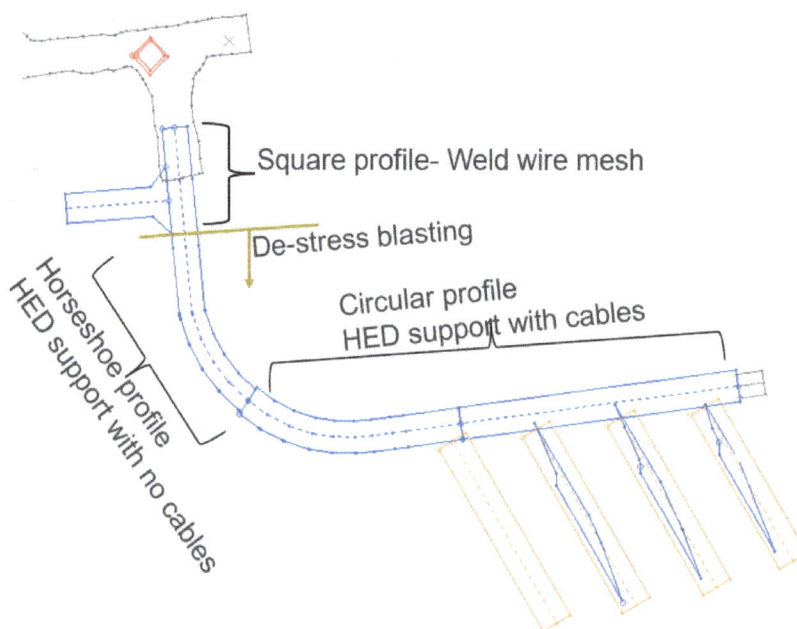

FIG 15 – Ground support regime and profile for 9215.

Compliance with the designed circular profile varied, as shown in Figure 16, was dependent on jumbo operator experience, charging practices and stress orientation. During the development phase, the concept of mining with this profile was proven. The main contributing factors for the successful execution of the circular profile were identified as crew awareness, operator training and constant monitoring of the profile performance.

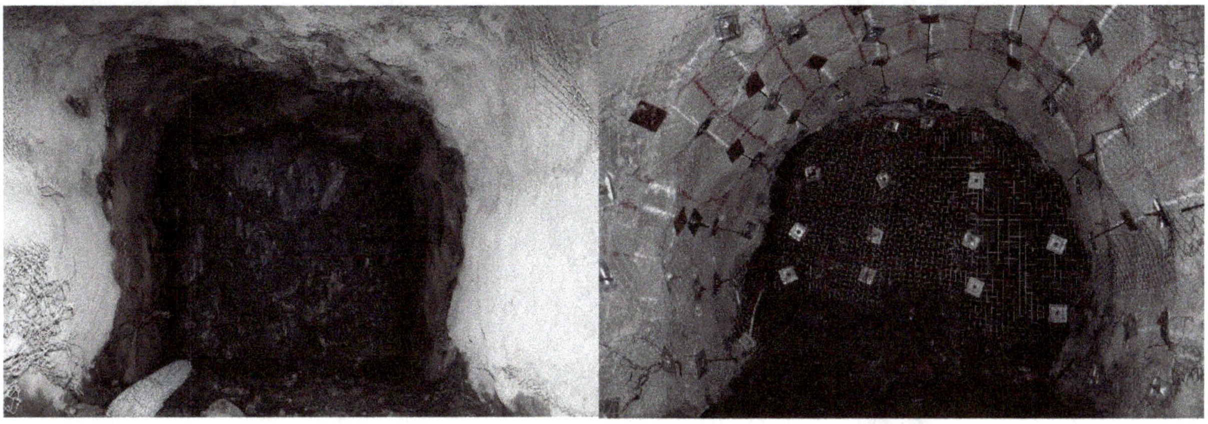

FIG 16 – Examples of semi-circular excavation profile performance in 9215 from poor to as per design.

RESULTS

Tunnel performance during construction and production stages

Ideally, any open stope requires that access infrastructure remain stable and undergo minimum to no ground support rehabilitation (Villaescusa *et al*, 2022). During the development phase of 9215, a number of small and medium-size events were recorded. However, none of the events resulted in any significant rock mass damage. Figure 17 shows that the seismicity (and potential for damage) occurred after the stress change resulting from the creation of the sublevel open stoping geometry (Villaescusa *et al*, 2022). As expected, the number of seismic events significantly increased with the start of the production phase.

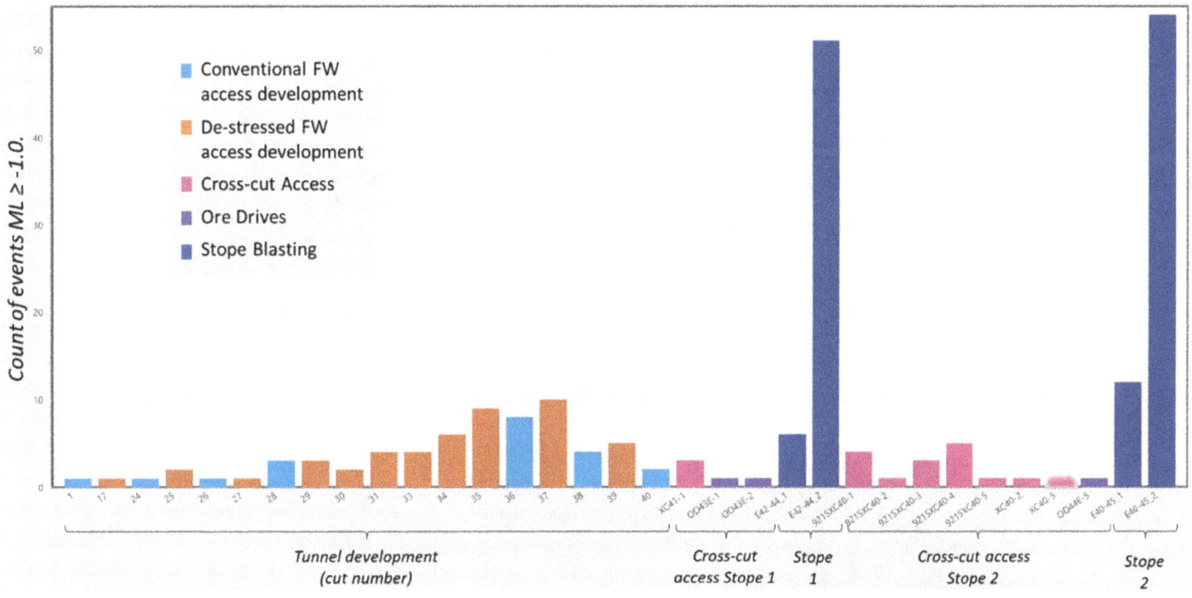

FIG 17 – Number of events with ML ≥ -1.0. From Villaescusa *et al* (2022).

During the production phase, seven seismic events with ML > 0 were recorded in this area. One seismic event ML +1.0 occurred on May 2021, close to the intersection of XC41, following the second firing (F#2) in stope 1. The event was recorded at the same time as firing in the proximity of Wilson lithological contact and the HX06 cross-cutting structure. Also, a cluster of small size events were recorded following the main event.

Figure 18 shows the location of the seismic event and the extent of seismic damage following the +1.0 ML event. No rock mass ejection or ground support system failure occurred following this event, however, two decoupled Posimix bolts were found pulled/failed in the footwall drive access. It is worth noting that the bolts were connected by G80/4.6 mesh with load transferred to other bolts outside the failure areas. Bagged mesh due to excessive deformation of rock mass in FWD and the cross-cut was the main consequence of this event. The depth of the broken rock mass varied from 0.6 m to less than 1.0 m. It is important to consider that the second layer of shotcrete in these locations was not completed at the time. The second layer of shotcrete was aimed to fill the gap between the first layer of shotcrete and provides stiffness to the Chain-link mesh, reducing the mesh's ability to move away from the profile when under load.

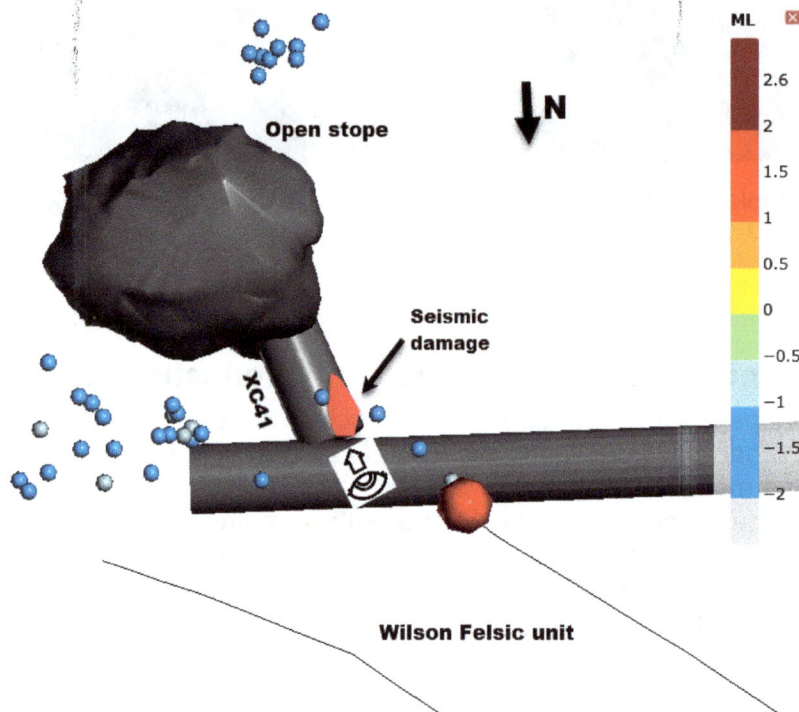

FIG 18 – Location and the extent of seismic damage following the +1.0 ML event.

Another large seismic event (ML +1.5) was recorded on the 8 June 2021, as shown in Figure 19. The event occurred several days after firing stope 1 and was felt on the surface and reported by underground personnel across the mine. The seismicity recorded just before the event is shown on the left side of Figure 20. No event decay could be seen after blasting and the seismic activity led to a large seismic event.

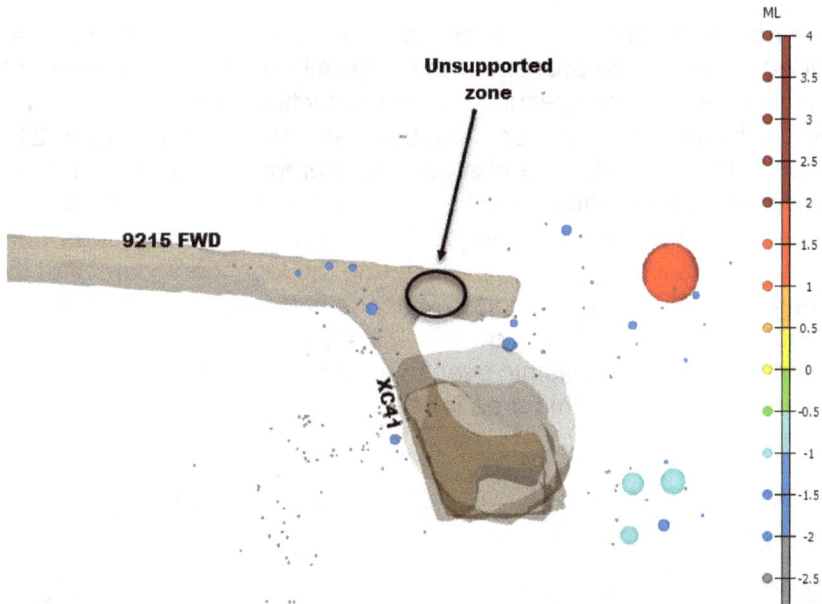

FIG 19 – Location of the +1.5 ML event and the location of the unsupported zone.

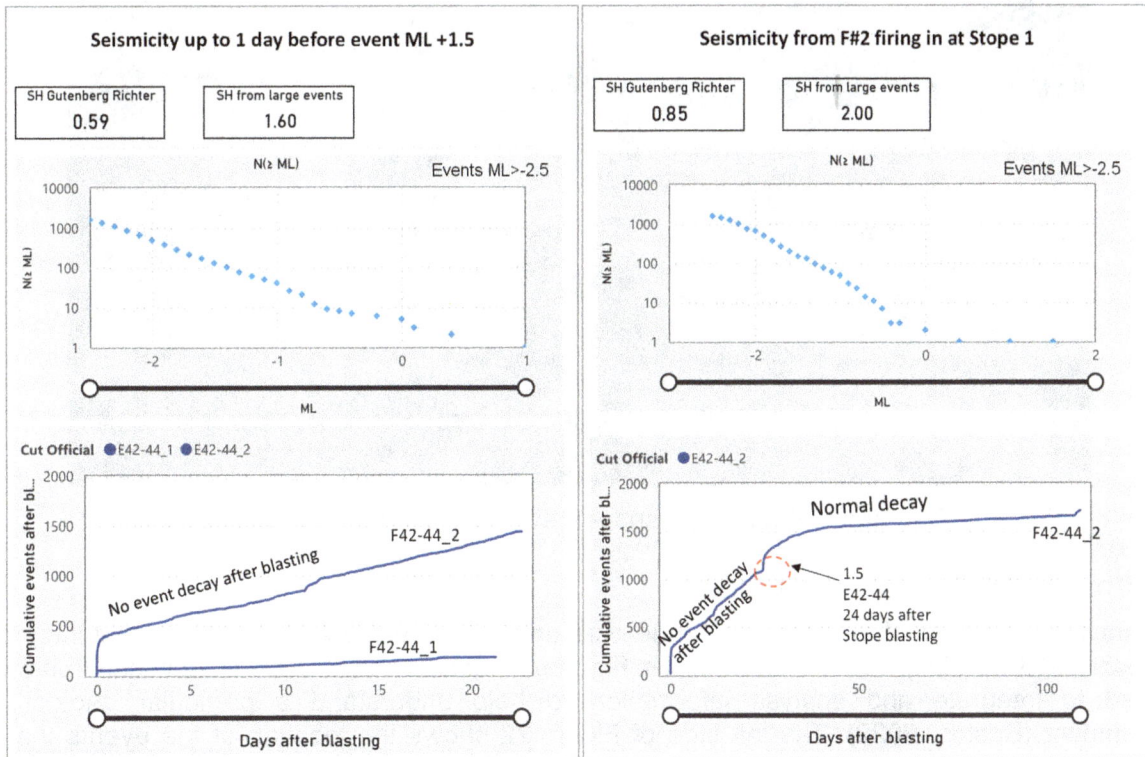

FIG 20 – Seismic hazard and seismic response with time.

There are several methods to calculate seismic hazard. The approach used was that 'SH from large events' is a seismic hazard calculated by adding to the largest event registered (ML+1) the delta value from the second largest event (ML + 0.4). This calculation indicated a seismic hazard of ML+1.6 was expected to occur. On the right side of Figure 20, the subsequent seismic response after the event shows a decrease in the rate of events.

Observations of the area post-event suggested there was no further damage to the ground support, and no bolt/cable failed. The existing bagged mesh in the 9215 XC41 showed a negligible increase in the broken rock mass depth (low damage according to Table 1), which was originally associated with the seismic event ML +1.0 in May. In the 9245 Exploration drive, one failed split set plate was found with a high level of corrosion. Also, a fresh crack on the shotcrete was observed on the southern wall of 9245 Exp, below the grade line.

Following the ML +1.5 event, one observation was the performance of rock mass where several zones of unsupported ground were present in the proximity of the intersection of FWD and the XC access. These unsupported patches resulted from production hole rifling as a result of ineffective confinement of the production blasthole, opening the mesh as shown in Figure 21. This is issue has been improved upon with more effective charge hole stemming practices. Personnel access was restricted to this area. The size of these patches was in the order of 1.0 × 1.0 m. After the ML +1.5 event, the area was inspected, and no new damage to the rock mass was observed where the unsupported ground patches were located. Considering the location accuracy of the seismic system in this area was very reliable, the unsupported patches were located no more than 25 m from the seismic event. Hence, no sign of shakedown or visible evidence indicating that the ground was affected by ground motion from the seismic event was observed. This observation again implies that the nature of rock mass instability in this area is related to shear failure on geological features other than factors like ground motion.

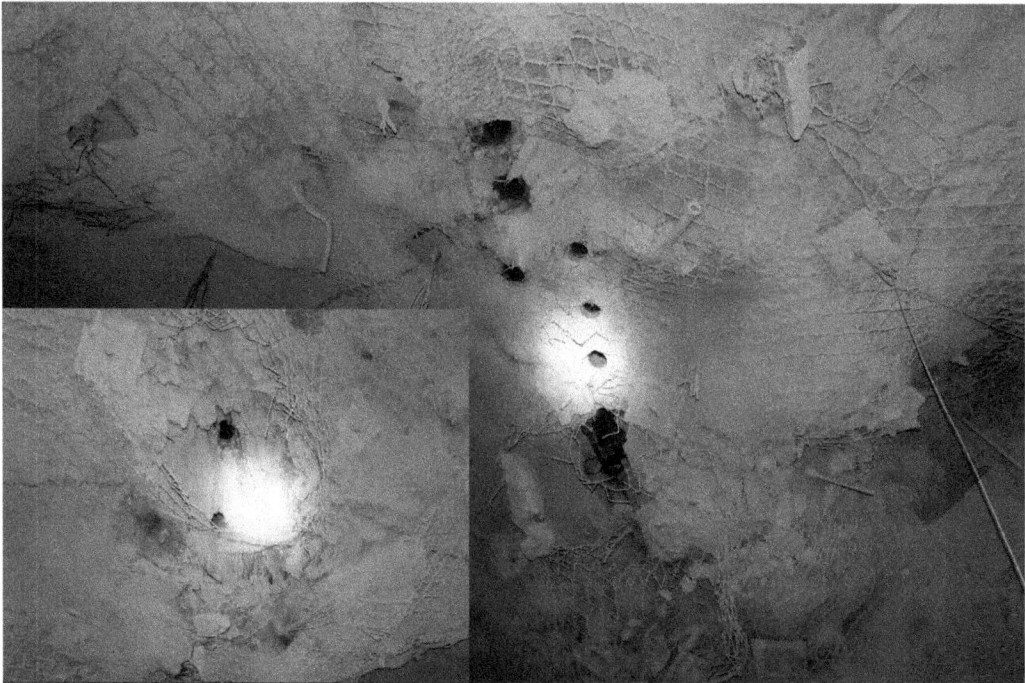

FIG 21 – Damage to the ground support system during the production drilling.

De-stress blasting performance

A quantitative analysis of the research tunnel was undertaken using a newly developed integrated geotechnical data (IGD) platform, as shown in Figures 17 and 20. An IGD is a customised framework created to integrate and analyse information to help understand a particular geotechnical environment (Bustos, 2022). The left side of Figure 22 shows the location of the events the first seven days after blasting, coloured by the blasting technique. From the Easting 20380, most of the cuts included de-stressing holes. The events were located closer to the tunnel up to cut #29, which was 18 m before the Wilson's contact. The additional subsequent events were measured around the tunnel and the Wilson contacts.

The right graph in Figure 22 compares the average cumulative number of events recorded for each blasting technique. It suggested that the de-stressing seismic response had a lower decay rate than traditional development blasting. The lower decay rate obtained by using de-stressing was consistent with the subsequent activation of structures, which were not necessarily mobilised immediately after blasting. Furthermore, the creation of fractures across the de-stressing charges was promoted by pre-existing discontinuities to form the desired planes of weakness (Drover, Villaescusa and

Onederra, 2018) that, once formed, could keep activating existing structures. De-stress blasting pattern and the quantity of charge were slightly modified, based on the quality of the rock mass in different zones.

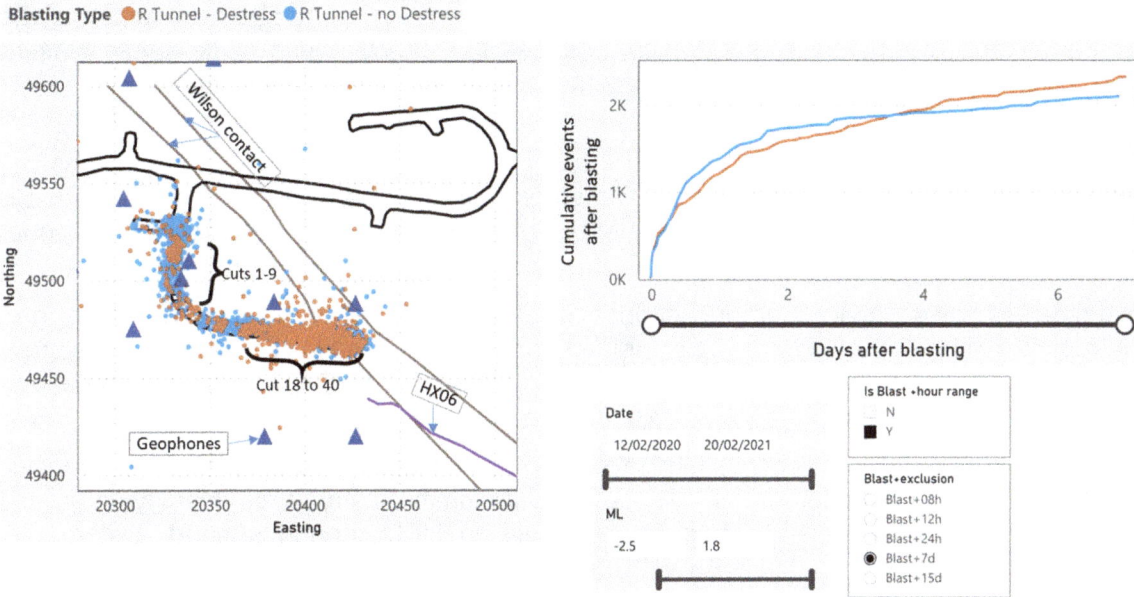

FIG 22 – Left: Plan view showing the location of seismic events. Right: cumulative average number of events for blasting using de-stress and conventional face techniques.

Figure 23 shows the statistical result of 41 cuts undertaken in the tunnel. 24 cuts were developed using traditional blasting for tunnel development, with 86 events on average per blasting (7 days) recorded. A total of 17 cuts used de-stress blasting, recording 136 events on average per blasting (7 days), which represented a 60 per cent increase in the number of events. From cut #29 it was decided to apply de-stressing at all faces after an increased rate of local seismic response, which included events along the Wilson contact.

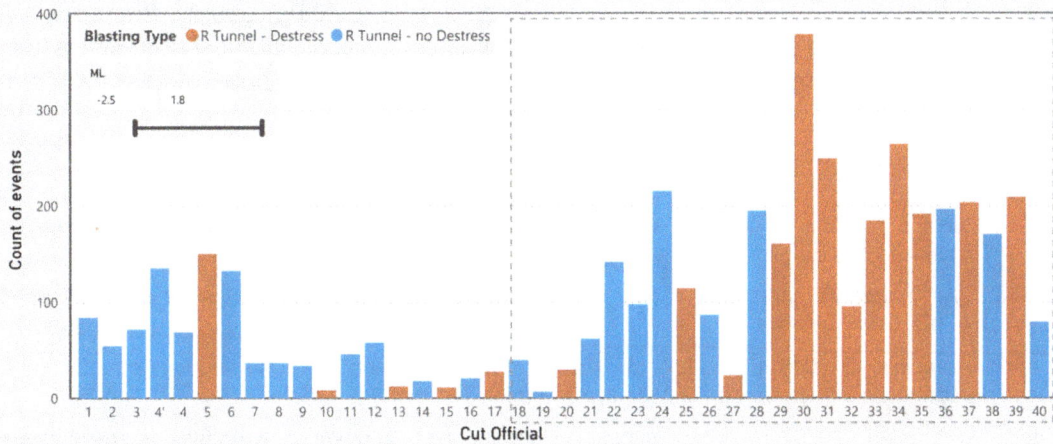

FIG 23 – Number of events per cut blasting up to seven days before the blast.

The statistical comparison of seismicity between the traditional blasting and destressed blasting is was biased because most of the de-stress blasting cuts were located in a more seismically active location close to the contact between the conglomerate and Wilson lithologies, when compared with the majority of conventional blasting cuts. However, this was deemed necessary in order to mitigate dynamic instability potential at the face.

Another interesting result was the review of cuts 18–40 (direction of the tunnel is E-W). The analysis indicated that the average number of events per blasting (considering 7 days) was 75 per cent greater when de-stress was implemented. The distributional nature of the magnitude was similar for both blasting strategies. If a rupture radius of 1 m is considered, the de-stressing cuts generated a larger number of events. Also, if the events filtered for energy greater than 10 J, an average of 59 events per conventional blasting and 113 for de-stress blasting cuts were found. The same analysis for 24 hrs indicated a similar seismic response regarding the average of events. However, events having a source radius smaller than 1 m are 36 per cent in conventional and 55 per cent in de-stressing blasted cuts. This indicated that a portion of the increasing number of events included a greater percentage of smaller events, which indicates the effectiveness of de-stress blasting in decreasing the seismic hazard at the development face during tunnel construction.

CONCLUDING REMARKS

The 9215 research tunnel is a development and stoping area in the Kanowna Belle Mine E-Block, the first mining activity completed in that part of the mine after several years following a period of concerning seismic damages in the past.

Kanowna Belle Mine's previous ground support strategies had relied on a conventional ground support regimen with relatively low capacity for load transfer between the components. This previous strategy was not able to provide the energy dissipation demanded by the rock masses at the current levels of the mine site (at a depth greater than 1000 m).

The 9215 research tunnel was successfully developed by applying a few design principles including the direction of mining for better stress management, proximity to structures, identifying areas that were more susceptible to seismic damage and utilising HED ground support system. These strategies helped with mining this area with no major requirement for rehabilitation during the production stages. The implementation of the de-stress blasting decreased the seismic hazard at the face of the developments with no strain bursting or spalling events affecting the face advance during the development of the tunnel.

As part of the 9215 mining strategy, a comprehensive seismic hazard analysis has been continuously undertaken to limit personnel exposure. The areas more vulnerable to seismic damage were those near the pinch point of cross-cutting structures and the Fitzroy Fault.

ACKNOWLEDGEMENTS

The authors are grateful for Northern Star Resources' permission to publish this work. In particular, we wish to thank Corin Arcaro and Vic Simpson for their ongoing support. Also, we acknowledge the numerous technical and operation staff who assisted in providing quality feedback. We also acknowledge the invaluable assistance and support of research and technical staff at the WA School of Mines. Also, we would like to recognise the time and effort Simon Thomas put into this paper for review and comments.

REFERENCES

Arcaro, C, Villaescusa, E, Hassell, R, Talebi, R and Kusui, A, 2021. Implementation of a High Energy Dissipation Ground Support Scheme.

Bustos, N, 2022. *An integrated approach for design and construction of drawbells in Block Cave Mines,* Doctor of Philosophy PhD, Curtin University.

Drover, C and Villaescusa, E, 2015. Estimation of dynamic load demand on a ground support scheme due to a large structurally controlled violent failure – a case study, *Mining Technology,* 125:1–14.

Drover, C and Villaescusa, E, 2019. A comparison of seismic response to conventional and face de-stress blasting during deep tunnel development, *Journal of Rock Mechanics and Geotechnical Engineering,* 11.

Drover, C and Villaescusa, E, 2020. Global geotechnical mine modelling, *Kanowna Belle Mine, Summary Report,* WASM – Mining 3.

Drover, C, Villaescusa, E and Onederra, I, 2018. Face de-stressing blast design for hard rock tunnelling at great depth, *Tunnelling and Underground Space Technology,* 80:257–268.

Heal, D, 2010. *Observations and Analysis of Incidences of Rock burst Damage in Underground Mines,* PhD, The University of Western Australia.

Levkovitch, V, Beck, D and Reusch, F, 2013. Numerical simulation of the released energy in strain softening rock materials and its application in estimating seismic hazards in mines.

Villaescusa, E, Thompson, A, Windsor, C and Player, J, 2022. *Ground Support Technology,* CRC Press (in print).

Management of dilution and mining recovery to maximise value

R Urie[1]

1. Principal Consultant, SRK Consulting, West Perth WA 6005. Email: rurie@srk.com.au

ABSTRACT

Dilution and loss of ore during the mining process invariably occur in all underground mining operations. This dilution and ore loss can have a significant effect on the performance and value generated from a mining operation. Further, excessive dilution or greater than anticipated ore loss during mining is often one of the key reasons for the underperformance of a mining project when compared to a feasibility study or annual budget.

Mining professionals often go to great lengths preparing detailed mine plans, schedules and economic models only to then apply high-level estimates or rule of thumb values for dilution and mining recovery that can ultimately lead to inaccurate forecasts for the performance of a mine with the resultant loss of value due to the mining of uneconomic material or the loss of potentially valuable ore.

This paper discusses processes and practices that can be practically used to manage dilution and mining recovery in an underground mining operation with a focus on understanding and then maximising value. The processes cover items such as the definition and measurement of dilution and mining recovery, techniques for the estimation of dilution and mining recovery, consideration of the economic implications, management of dilution and mining recovery through the stages of the planning cycle, and practices to reduce dilution and improve mining recovery. Common misconceptions and omissions are also discussed.

INTRODUCTION

Dilution and mining recovery are two of the key modifying factors used in developing a mine plan and estimating the future ore production from both planned and operating underground mines. The actual dilution and mining recovery values experienced during mining can have a significant influence on the value and the overall profitability generated from a mine. In addition, the estimation of dilution and mining recovery for different mining options or scenarios can often strongly influence the selection of the preferred mining method for the orebody.

Although dilution and mining recovery can strongly influence the performance and profitability of a mining operation, the estimation and management of these factors often receives relatively limited attention during both the planning and operations stages compared to other factors such as scheduling, productivity rates and metallurgical recovery.

For example, it is not uncommon for a feasibility study investigating a planned mining operation to spend considerable time and money completing an extensive metallurgical test work program or developing numerous detailed mining schedule scenarios only to complete a single high-level estimate of dilution and mining recovery. In practice it is often changes in dilution and mining recovery that have the most significant influence on the success of the operation.

DEFINING DILUTION AND MINING RECOVERY

The use of a consistent and logical definition for dilution and mining recovery at all stages of the mining cycle is an important foundation for the effective management of these factors. It is common to see several different definitions of dilution and mining recovery in use at the same mining operation which can undermine the effective management of these factors.

Dilution and mining recovery can be defined in several different ways and these definitions are discussed in McCarthy (2014) and Tatman (2001).

Dilution is typically dived into two types – internal or planned dilution and external or unplanned dilution. Internal dilution refers to below cut-off grade material contained within design excavation shapes. The internal dilution is often a function of the mining method or minimum mining width that the equipment dictates.

External dilution is the perimeter or boundary dilution from outside the design excavation shape that becomes part of the ore stream. The majority of the reporting in relation to dilution typically focuses on external dilution with the potentially incorrect assumption that internal dilution is only governed by the already selected mining method. Some of the common definitions that are regularly used in practice for calculating external dilution are shown in Equations 1–3.

$$\text{External dilution (\%)} = \frac{(\text{mass of external waste mined}) \times 100}{(\text{mass of design shape})} \tag{1}$$

$$\text{External dilution (\%)} = \frac{(\text{mass of external waste mined}) \times 100}{(\text{mass of design shape} + \text{mass of external waste mined})} \tag{2}$$

$$\text{External dilution (\%)} = ((\text{undiluted in situ grade} / \text{recovered grade}) - 1) \times 100\% \tag{3}$$

Where below cut-off grade material is encountered in the diluting material Equation 3 is often used to express dilution.

Mining recovery is a factor that measures the recovery of the design ore tonnes during the mining process. Mining recovery recognises that all of the material designed to be mined will not always be extracted successfully due to factors such as imperfect blasting and incomplete bogging. Some of the common definitions for mining recovery in stoping scenarios are shown in Equations 4 and 5. The term mining losses is often used in place of mining recovery, and this represents the portion of the designed ore not recovered rather than recovered during the mining process. This is shown in Equation 6.

$$\text{Mining recovery (\%)} = \frac{(\text{mass extracted from the design stope shape}) \times 100}{(\text{mass of the design stope shape})} \tag{4}$$

$$\text{Mining recovery (\%)} = \frac{(\text{mass of ore and waste extracted}) \times 100}{(\text{mass of the design shape} + \text{mass of waste dilution})} \tag{5}$$

$$\text{Mining losses (\%)} = \frac{(\text{mass of the design stope shape not successfully extracted}) \times 100}{(\text{mass of the design stope shape})} \tag{6}$$

The different definitions can generate quite different values for dilution and mining recovery.

For example, consider a design stope shape containing 1000 t of ore at 4 per cent grade. During mining 900 t of the design shape is successfully extracted (ie fired and bogged out) and the stope also experiences 150 t of dilution at 1 per cent grade. As a result, the stope produces 1050 t at a grade of 3.57 per cent.

Equation 1 gives a dilution value of 150 / 1000 = 15%

Equation 2 gives a dilution value of 150 / 1150 = 13%

Equation 3 gives a dilution value of (4% / 3.57%) - 1 = 12%

Equation 4 gives a mining recovery value of 90%

Equation 5 gives a mining recovery value of 91%

While the methods discussed above are all valid measures, this example does illustrate that quite different values of dilution and mining recovery can be generated based on subtly different calculation methods. The use of a consistent definition of dilution and mining recovery through the mining cycle is important to effectively managing these factors.

It is not uncommon to see different definitions of dilution and mining recovery used at different stages of the mining cycle within the same mining operation. For example, Equation 1 is generally used to define dilution and production forecasts in the planning stage while Equations 2 or 3 are regularly used during the stope reconciliation stage. Inaccurate production estimates and loss of value can arise when actual dilution and mining recovery factors that have been calculated with one method are applied in the planning stage where a different definition of dilution and mining recovery is used.

The calculation methods shown in Equation 1 for dilution and Equation 4 for mining recovery are generally the most commonly used definitions in underground mining and will be used for the remainder of this paper when defining dilution and mining recovery.

MEASUREMENT AND ANALYSIS OF DILUTION AND MINING RECOVERY

The measurement and analysis of the actual dilution and mining recovery values from mining and the comparison of these with the planned values can be a powerful tool in managing dilution and mining recovery and maximising value.

The use of cavity monitoring survey (CMS) equipment and modern mine planning software allow for accurate measurements to be made of the dilution and mining recovery performance of mined stopes and development.

The pictures shown in Figure 1 are from a mine in Australia that has begun separating the quantity of external dilution between the hanging wall and footwall when measuring actual stope performance. This allows for more detailed analysis of the factors causing dilution in the various areas of the stopes.

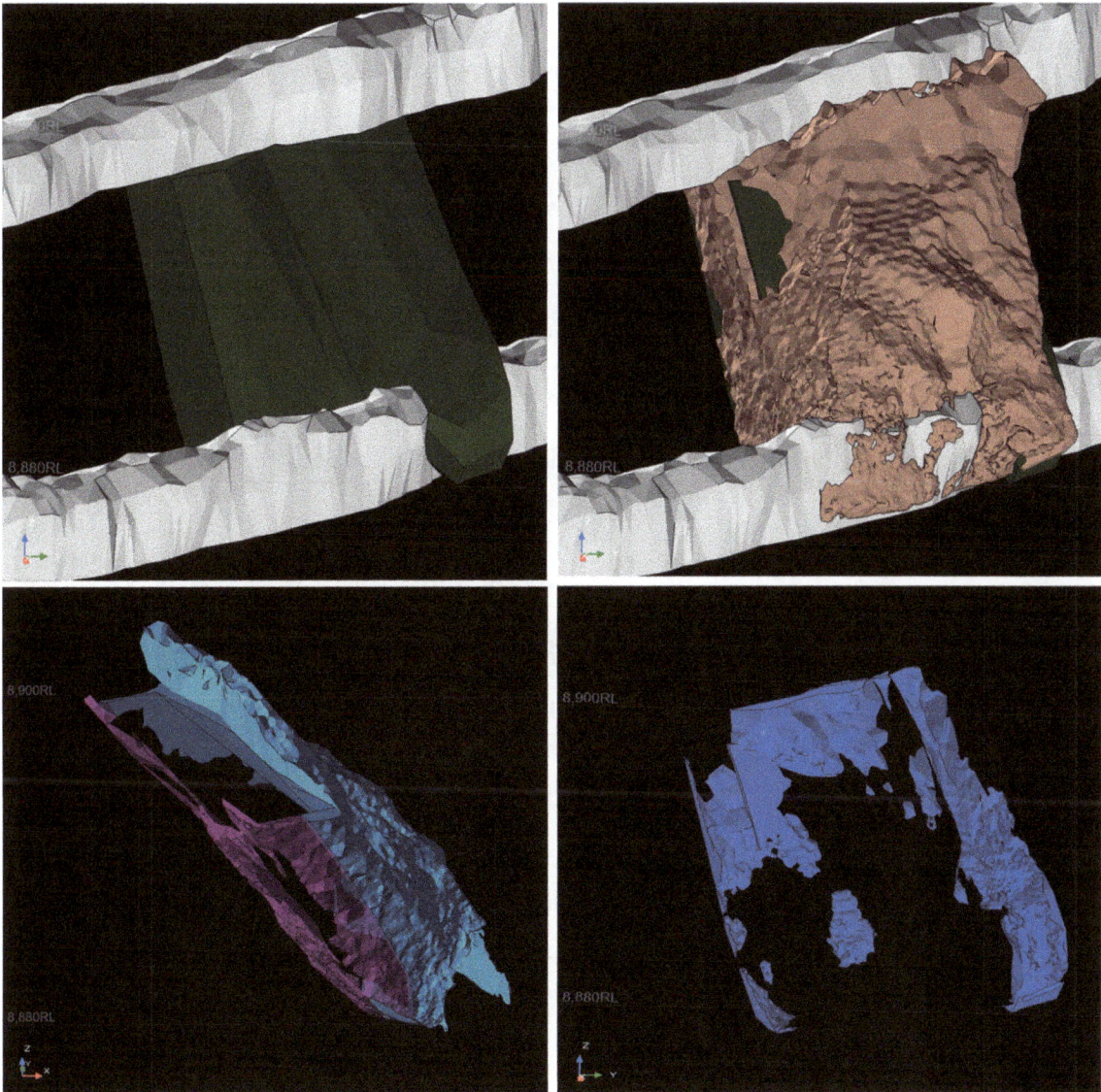

FIG 1 – Pictures from a stope performance analysis showing in clockwise order from top left – the design stope shape, the actual stope shape overlaying the design shape, the areas of underbreak, the areas of overbreak coloured separately between the hanging wall and footwall.

Analysis of actual stope performance data can be an effective method to identify the factors influencing dilution and mining recovery. Where this exercise has been completed, the factors found to be the most significant contributors to dilution are often not the ones that were initially expected to be influencing dilution.

The analysis of stope performance is more effective if all the fields considered as potential influences of dilution and mining recovery are converted to numerical values to allow more efficient analysis of the data. Further, evaluation of each particular face of a stope separately with the corresponding influencing factors has been shown to be more effective in identifying the factors controlling dilution than considering the stope as a whole. Some of the parameters that are commonly included in dilution and mining recovery analysis are:

- Stope design factors such as:
  - stope strike length (m)
  - stope hydraulic radius (m)
  - hanging wall dip (deg)
  - stope depth below surface (m)
  - stope width (m)
  - presence of cable bolts
  - distance of stope wall undercutting by development drives.
- Geology and geotechnical factors:
  - rock quality designation (RQD)
  - modified stability number ('N')
  - lithology unit of stope sidewalls
  - distance of geological structures from stope walls (m).
- Drill and blast parameters:
  - blasthole burden and spacing (m)
  - blasthole stand-off distance from stope sidewalls (m)
  - explosive type.

Dilution data from several mines in Australia have indicated that factors such as hydraulic radius and stope strike length are often not the most significant factors controlling dilution. Rather it is often local stope specific factors such as undercutting of the hanging wall by ore drives or the presence of geological structures in the stope walls that can have the greatest influence on dilution.

Figures 2, 3 and 4 show the relationship of hanging wall (HW) dilution to various factors such as stope strike length, distance of hanging wall undercut and production blasthole spacing for a gold mine in Australia.

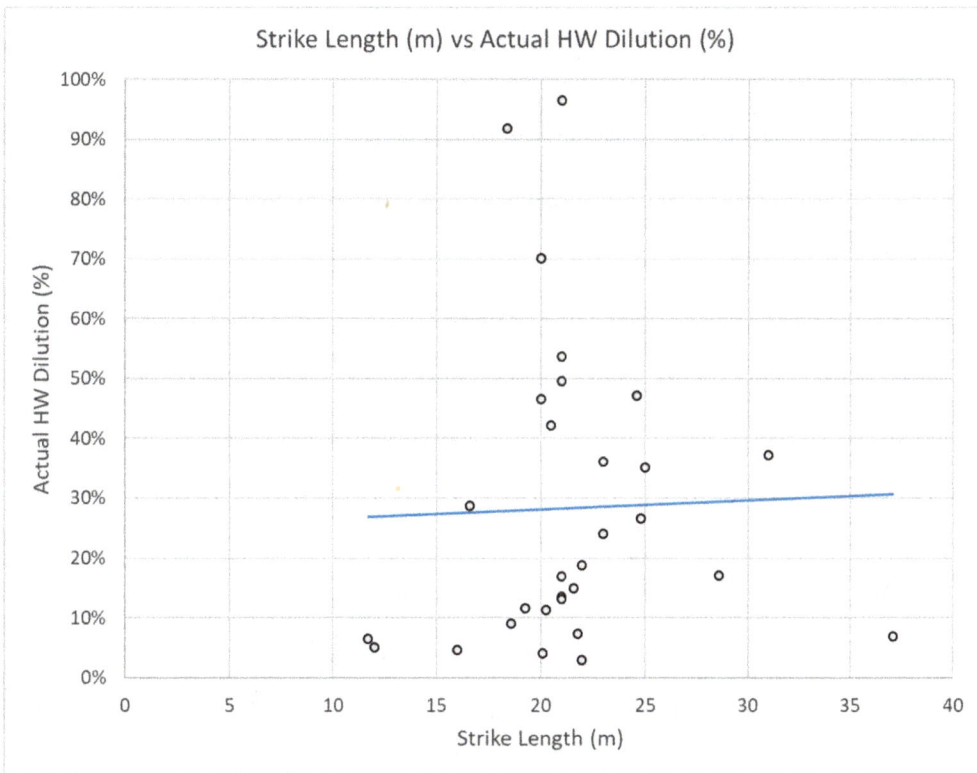

FIG 2 – Comparison of stope strike length and actual hanging wall dilution.

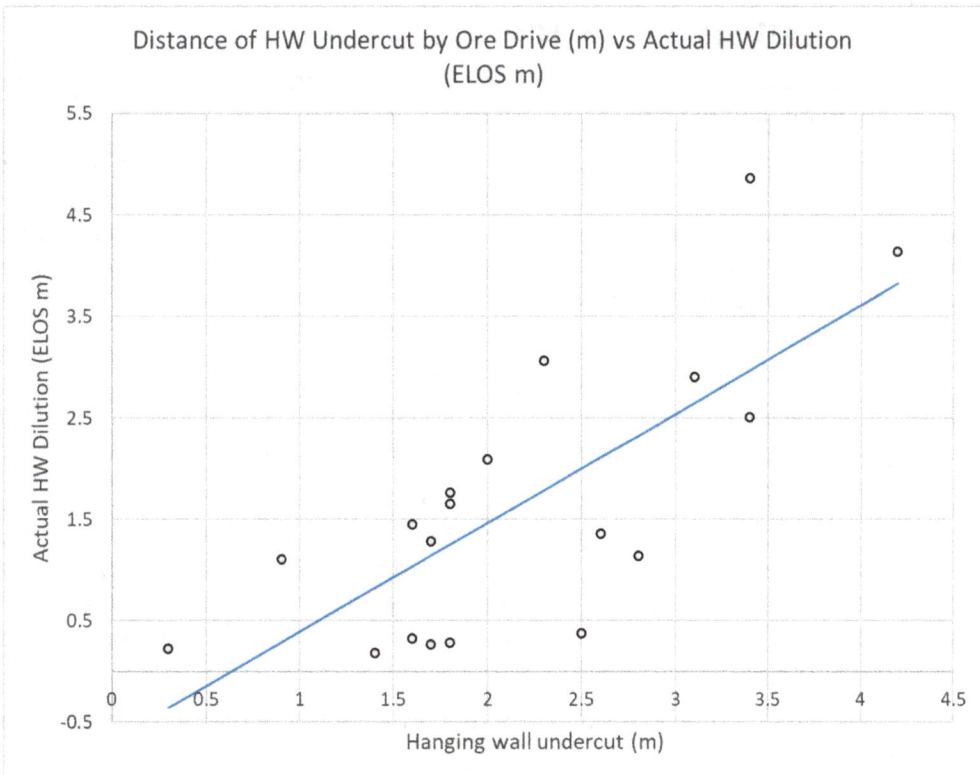

FIG 3 – Comparison of the distance of undercut of the hanging wall and actual hanging wall dilution.

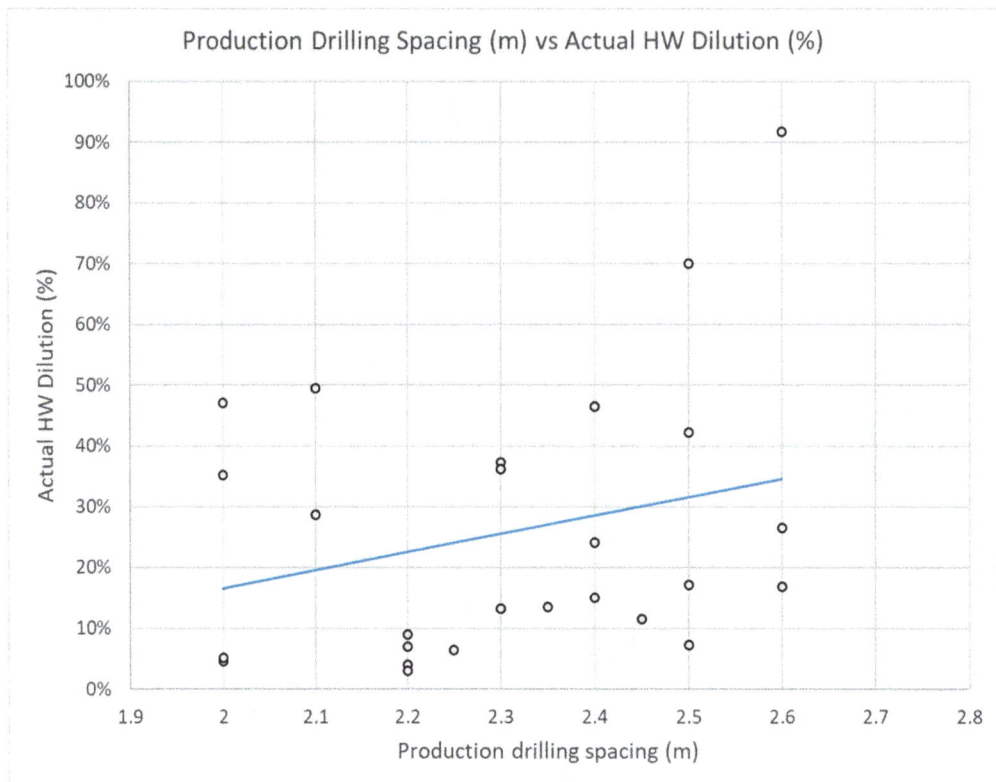

FIG 4 – Comparison of production drilling spacing and actual hanging wall dilution.

ESTIMATION OF DILUTION AND MINING RECOVERY

The estimation and application of appropriate values for dilution and mining recovery is an important aspect of developing an achievable mine plan that allows the value of the resource to be maximised.

If the values for the dilution and mining losses that are incorporated in the mine plan are too low and not achievable in practice, the forecast grade in the mine plan will overestimate the likely actual grade of the ore and loss-making material may be included in the mine plan. Further, additional ore tonnes will have to be mined to achieve the same forecast metal production, resulting in key processing, ore handling and ventilation infrastructure being undersized for the planned metal production rate.

On the other hand, if the dilution and ore loss factors included in the mine plan are overestimating the actual values, potentially valuable material may become sterilised and key infrastructure may be oversized resulting in unnecessary capital expenditure during the project construction stage.

The methods for estimating the appropriate dilution and mining recovery factors vary significantly between an undeveloped underground mine that is being studied for potential development and an operating underground mine that has an operating history to inform the future estimates. It is important to appreciate that the level of risk and uncertainty associated with estimates of dilution and mining recovery will be significantly higher in an undeveloped deposit that does not have operational history to inform the estimates.

Without the operating history, undeveloped mines rely on a variety of methods to estimate the likely dilution and mining recovery including:

- Comparisons with similar mines that have comparable rock mass conditions, mining methods and mining geometry.

- Empirical methods such as the stability graph method as updated by Capes (2009).

- Numerical modelling of the planned excavations to estimate the likely depth of failure.

In an operating mine the recent actual values for dilution and mining recovery are often the most useful tool to estimate the future values for these factors assuming they have been reliably collected and analysed.

INFLUENCE OF STOPE WIDTH ON DILUTION

The external dilution and mining recovery experienced by a stope is often a function of the drill and blast methods used to define the stope wall, the quality of rock mass that makes up the stope boundary walls and the stress environment that the stope is in. These drill and blast methods and geotechnical conditions typically generate an average thickness of overbreak and underbreak on each stope surface that is independent of the width of the stope.

The terms 'equivalent linear overbreak/slough' (ELOS) and 'equivalent linear lost ore' (ELLO) developed by Clark and Pakalnis (1997) represent the average thickness of overbreak and underbreak on a stope surface. ELOS and stope width can be used to estimate dilution as a percentage and this relationship is shown in Figure 5. Figure 5 shows that, for a given ELOS value, dilution when measured as a percentage, increases significantly as the stope width decreases.

Stimpson, Johnson and Hesse (2021) includes the results of a drill hole accuracy survey that shows the total blasthole deviation in a survey of 550+ holes averaged 4.9 per cent. In a 20 m blasthole this equates to 1.0 m of blasthole deviation. If this blasthole deviation results in an ELOS value of 0.5 m averaged over each sidewall of a stope it would result in external dilution values of 20 per cent in a 5 m wide stope and 10 per cent in a 10 m wide stope.

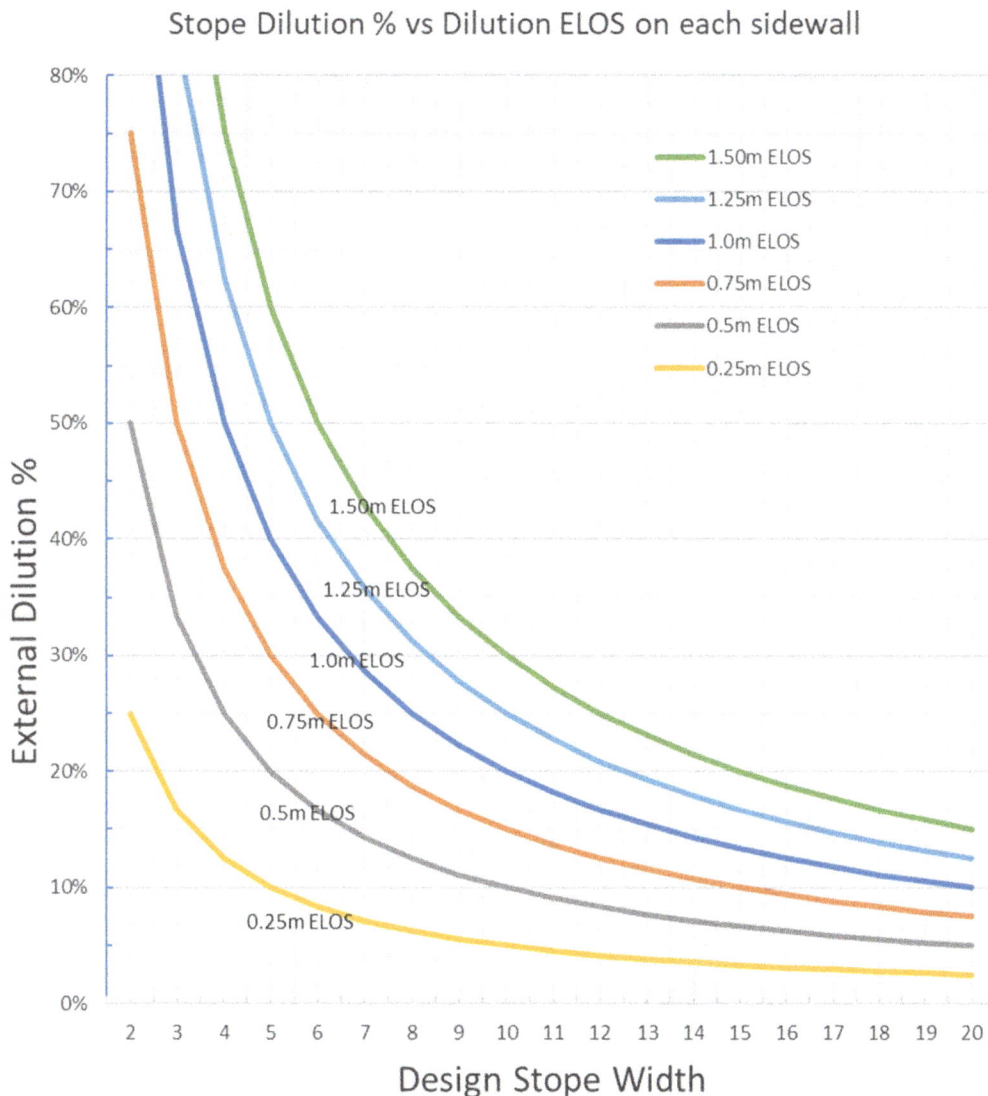

FIG 5 – External dilution percentage and design stope width for various ELOS values.

The actual values for external dilution and mining recovery compared with stope width from 15 open stoping mining operations in Australia is shown in Figures 6 and 7. The actual data illustrates a reduction in dilution and increase in mining recovery with increasing stope width.

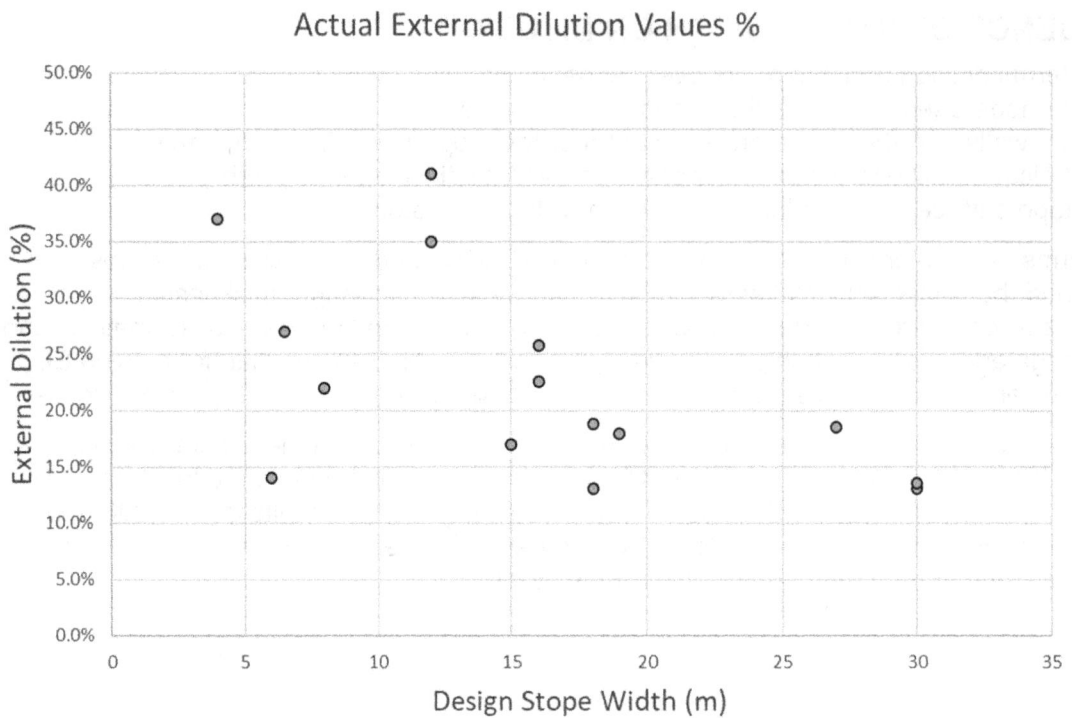

FIG 6 – Actual external dilution data compared with stope width.

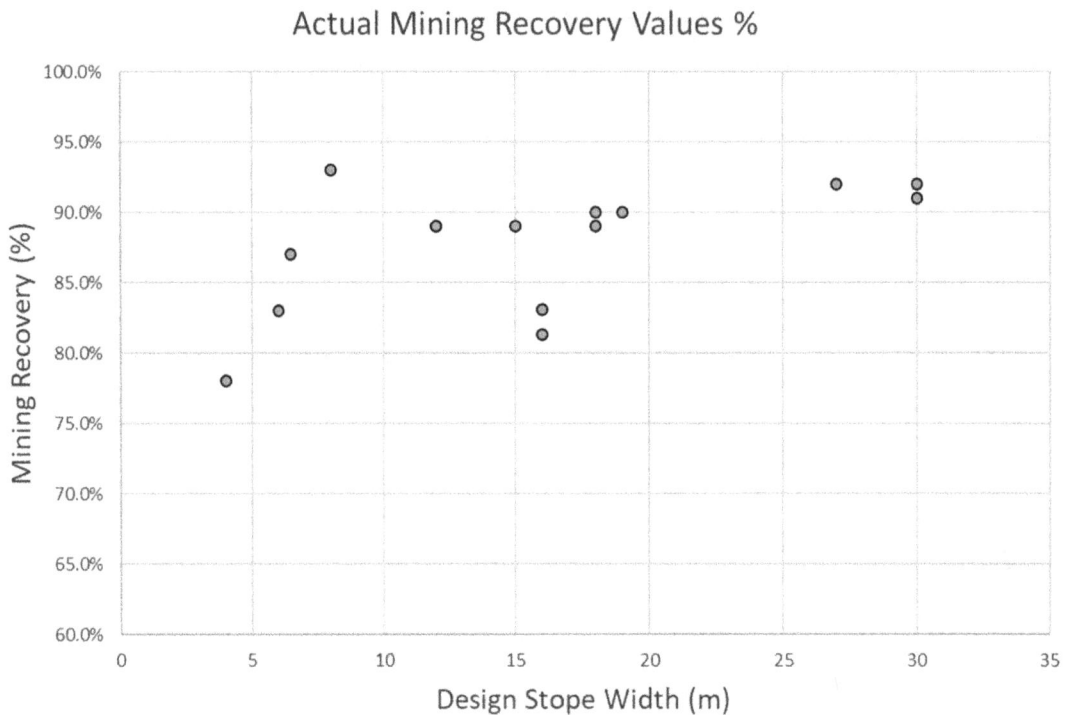

FIG 7 – Actual mining recovery data compared with stope width.

THE VALUE IMPACT OF DILUTION AND MINING ORE LOSSES

As efforts are focused on reducing dilution in a mine there is often a subsequent similar reduction in mining recovery. This has been recently observed at several mines in Western Australia and has unfortunately resulted in an overall loss of value for the operations. This effect is illustrated using example data in Table 1. The third and fourth columns of Table 1 show a 12 per cent reduction in cash flow for an operation which equally reduces dilution and mining recovery by 10 per cent. The subsequent columns in Table 1 illustrate the change in cash flow with various variations in external dilution and mining recovery. Using the example data for each 1 per cent reduction in external dilution the cash flow improves by 1 per cent while each 1 per cent improvement in mining recovery improves the cash flow by 2.3 per cent.

While the results of this evaluation are dependent on the particular grade and cost profile of each mine it does illustrate that improving mining recovery can often be a more effective way of improving cash flow than reducing dilution.

TABLE 1

Variation in cash flow with changes in dilution and mining recovery.

External dilution	%	20%	10%	20%	10%	30%
Mining recovery	%	92%	82%	95%	92%	92%
Design tonnes	t	500 000	500 000	500 000	500 000	500 000
Design grade	g/t	4.0	4.0	4.0	4.0	4.0
Mined tonnes	t	560 000	460 000	575 000	510 000	610 000
Mined grade	g/t	3.3	3.6	3.3	3.6	3.0
Operating costs	$M	75.6	67.9	76.8	71.8	79.5
Metallurgical recovery	%	95.0%	95.0%	95.0%	95.0%	95.0%
Gold sold	oz	56 200	50 091	58 032	56 200	56 200
Revenue at $2000/oz	$M	112.4	100.2	116.1	112.4	112.4
Cash flow	$M	36.8	32.3	39.3	40.6	32.9
Percent change in cash flow	%	0%	-12.3%	6.8%	10.5%	-10.5%

CONCLUSIONS

Dilution and mining recovery are two of the key modifying factors used in developing a mine plan and estimating the future ore production from both planned and operating underground mines. The actual dilution and mining recovery values experienced during mining can have a significant influence on the value generated from a mine and its overall profitability.

The use of a consistent and logical definition for dilution and mining recovery at all stages of the mining cycle is an important foundation for the effective management of these factors.

Analysis of dilution and mining recovery data from several mines in Australia has indicated that it is often local stope specific factors such as undercutting of the hanging wall by ore drives or the presence of geological structures in the stope side wall that can have the greatest influence on dilution.

The particular drill and blast parameters and the stope wall geotechnical conditions often control ELOS values experienced during the mining of a stope. The ELOS values experienced and the stope width can be used to estimate dilution as a percentage value for future stopes.

While dependent on a particular site's grade and cost profile, focusing efforts on improving mining recovery rather than reducing dilution can be a more effective method of improving cash flow.

REFERENCES

Capes, G, 2009. Open Stope Hangingwall Design Based on General and Detailed Data Collection in Rock Masses with Unfavourable Hangingwall Conditions, PhD Thesis, University of Saskatchewan, Saskatoon.

Clark, L M and Pakalnis, R C, 1997. An empirical design approach for estimating unplanned dilution from open stope hanging walls and footwalls, paper presented to 99th CIM Annual General Meeting, New Frontiers for the Next Century, Vancouver.

McCarthy, P L, 2014. Mining Dilution and Losses in Underground Mining, in *Mineral Resource and Ore Reserve Estimation,* second edition, chapter 5, pp 415–418 (The Australasian Institute of Mining and Metallurgy: Melbourne).

Stimpson, R, Johnson, B and Hesse, B, 2021. Increased drilling accuracy leads to significant improvement in ore recovery and dilution, in *Proceedings Underground Operators Conference 2021*, pp 389–400 (The Australasian Institute of Mining and Metallurgy: Melbourne)

Tatman, C R, 2001. Mining Dilution in Moderate to Narrow-Width Deposits, in *Underground Mining Methods* (eds: W A Hustrulid and R L Bullock), chapter 70, pp 615–626 (Society for Mining, Metallurgy and Exploration Inc: Englewood).

A review of the most common underground slot, winze, rise or raise designs

E Wargem[1] and D Tozer[2]

1. Senior Technical Consultant, Dyno Nobel Asia Pacific, Perth WA 6100.
   Email: edward.wargem@ap.dynonobel.com
2. Technical Manager, Dyno Nobel Asia Pacific, Perth WA 6100.
   Email: david.tozer@ap.dynonobel.com

ABSTRACT

The conventional method of underground mining involves a process known as drilling and blasting to excavate valuable ore bearing minerals through the load and haul process. Drilling and blasting processes are completed for both the development of the tunnels used to access the orebody and for mining out the orebody itself. The former being called development drilling and blasting, and the latter production drilling and blasting. In either development or production drilling and blasting, the effective recovery will be largely reliant on the extraction of the initial void created in each blast cycle. In development tunnelling, the initial void is commonly known as a Burn Cut. The initial void created in production drilling and blasting has many names such as a Slot, Winze, Rise or Raise. There are drill patterns and charge designs used to create these initial voids in the production cycle. This paper discusses common slot or winze designs for production drilling and blasting; and various critical design criteria for these. The document also aims to provide a process for completing desktop reviews of current patterns employed on-sites and provides an initial guide in a process for improving the recovery of a slot, winze, rise or raise in underground mines.

SUCCESFUL INITIAL BLASTS, THE KEY TO SUCCESFUL UNDERGROUND DRILLING AND BLASTING

As underground mines operate at increasing depths to extract ore, operating costs are increasing making it pivotal for miners to operate sustainably and efficiently. Production targets must be met in a timely manner to ensure operating and capital costs are minimised and the intended return on investment is delivered. The optimum mining of an orebody in an underground open stoping operation is fundamental to productivity. Effective mining in bulk underground mining is largely reliant on the recovery of the initial void created by drilling and blasting. These initial voids are known by various names such as the longhole rise, slot, or winze of the stope. These initial blasts in stoping or caving operations can be drilled by either top hammer drilling equipment or in-the-hole drilling equipment or in some cases a combination of both. Diligent drilling and blasting of these will facilitate efficient mining of the designed open stope ensuring production of the orebody remains within mining schedules at a minimum cost. When an operation struggles in opening the initial void of a stope or a cave, it causes a negative impact on productivity, mine schedules and mineral recovery and impacts the overall economics of the mining operation. Each mine thus needs to understand the parameters that make the sites initial slot or winze blast successful and to understand what aspects can be changed to suit newer rock conditions and site-specific constraints.

Whatever the bulk underground mining method employed, whether it be open stoping, sub level caving or block caving, the success of the initial blast is dependent on the successful extraction of the slot. Once this initial void is opened, the rest of the drilled-out rings can be charged and fired successfully. To simplify the explanation, an open stoping operation will be used to explain the process of production blasting and the term slot used for the initial void created in the stope. In a sub level open stope that uses downhole drilling and charging, a stope is drilled out using longhole drilling equipment. The stope can be split into two parts, the first being the slot (longhole rise, winze etc). This is the initial part of the stope that is mined out to create a sufficient void for the bulk of the stope to be blasted into. The remaining bulk portion of the stope usually comprises of the main production rings. Figure 1 shows a long section of a simple downhole and uphole stope design differentiating the slot from the main rings.

LONGSECTION VIEW OF A DOWN HOLE STOPE

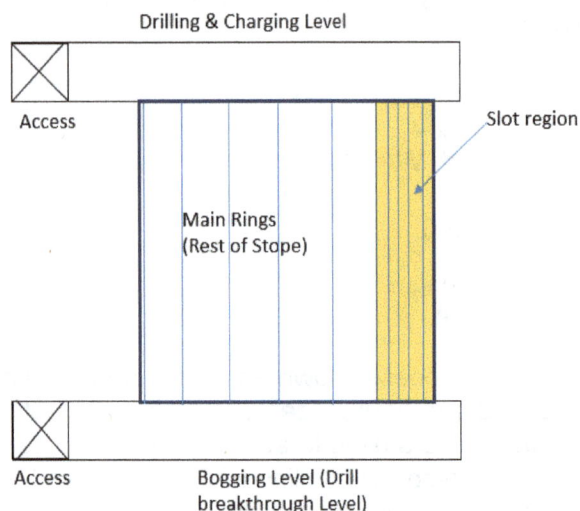

Drilling & Charging Level

Access

Slot region

Main Rings
(Rest of Stope)

Access

Bogging Level (Drill
breakthrough Level)

LONGSECTION VIEW OF AN UP HOLE STOPE

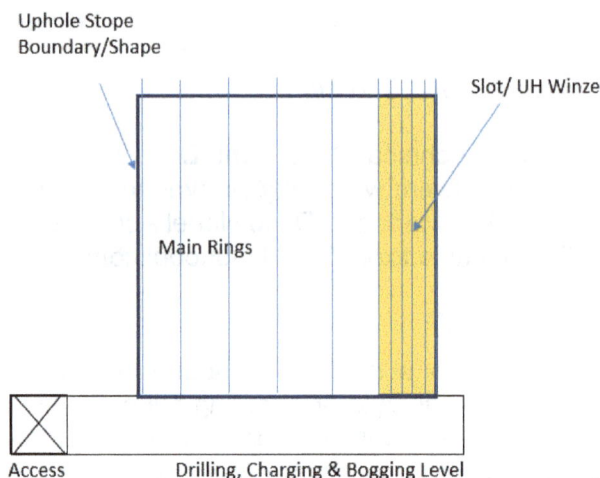

Uphole Stope
Boundary/Shape

Slot/ UH Winze

Main Rings

Access

Drilling, Charging & Bogging Level

FIG 1 – Long section view of a downhole and an uphole stope showing the slot area and the main rings area of a stope.

Depending on stope geometries and drilling platforms, the main rings are usually radially drilled blastholes that are charged and blasted into the void created by the slot blast. If the slot blast is unsuccessful, it is highly likely that the main ring blasts will also be unsuccessful. An unsuccessful slot blast will sometimes be referred to as a frozen or bridged slot as the material is deemed to be frozen in place or creates a bridge shape with the void and the host rock. This frozen or bridged blast will then force the mine to employ recovery drilling and charging options in an attempt to recover this frozen or bridged material. It is therefore critical, especially in uphole stopes that every effort is made to ensure these initial blasts are successful. It is very difficult and, in some cases impractical, to recover a frozen or bridged uphole slot blast. This can result in ore sterilisation that not only affects the mine production and ore recovery, but also delays the mining cycle.

CRITICAL DESIGN PARAMETERS IN SLOT DESIGNS

Slot, longhole rise, winze definition

The mining industry over the years has seen many different design standards for slot, winze, longhole rise or raise designs that have become widely accepted in many mine sites. The slot or longhole rise is characterised by a series of drilled out holes in a small area of the stope. This area is usually in a 3 mW × 3 mL or 4 mW × 4 mL box with varying hole lengths depending on stope shape. Other dimensions are also used in the industry but to a lesser degree. One set of drill holes in the slot box would have a smaller diameter hole that will be charged with explosives. The second set of drill holes will have a larger diameter hole and will remain uncharged. These larger diameter uncharged holes are commonly referred to as reamer holes or relief holes. These uncharged relief holes will act as an internal void into which the earlier firing blastholes in the sequence in the slot will blast when initiated. When the first blasthole in a blast event is initiated the material will fragment and heave into the nearest open void. In a slot blast, the open voids are the larger diameter uncharged relief holes and the drive void below or above the slot box.

In cross-section of the slot design, the blastholes that will be charged are strategically located near the uncharged relief holes. The blastholes will be charged and blasted in a sequence that will allow the charged holes closest to the relief holes to fragment and heave into these larger diameter holes before falling into the drive void below. Downhole slot designs (those drilled and charged as downholes) can be mined out in multiple charge and blast events. These different events are commonly referred to as lifts. These lifts are charged and fired in a similar fashion to a vertical crater retreat blast. However larger lift heights are usually taken due to the nature of the slot design and the volume of the available drive void below and above (in final lift or cap blasts) in the charged area. In an uphole stope, a stope that is drilled and charged upwards from a single development drive, the slot is charged and fired in one blast in large open stoping. Air leg raising uses different methodology and will not be discussed in this paper.

There are several factors that determine the effectiveness of a slot design and these can be used to both assess or review existing mine designs or to help draft a new design for a new mine. These factors, if adequately addressed in the design phase, can assist in improving the success of a slot blast. These design factors include, but are not limited to:

- critical void ratio and overall blast void ratio

- explosive energy distribution

- redundancy in design

- blasthole shielding

- blasthole delay timing

- blasthole sequencing

- uncontrollable factors in drill and blast design, for example available equipment (can be controllable if the mine is just starting up) or geological, geotechnical and hydrological conditions.

These are some of the important concepts to review when analysing a slot design. These factors will aid in ensuring a slot or winze design will be consistently successful over the life of the mine. A successful slot blast is one that creates an open avoid as per the design shape. That is the cavity scan that is taken after mining this slot closely resembles the design shape created by the drill and blast engineer. Careful consideration will need to be made regarding site geology, mining method and equipment limitations when determining a slot design. Certain uncontrollable factors like site geology or equipment limitations will also influence the slot pattern used. Equipment limitations refer to both drilling and charging equipment, for example, can the drilling equipment drill the required blasthole and reamer hole diameter at the required lengths? Can the charging equipment effectively charge these drilled holes? These factors are much like the design considerations when optimising a burn cut design for development or tunnel advance profiles. The following parameters are briefly explained in the following section. The authors have selected the most common and simplest forms of these calculations and methods to use in design reviews for the purpose of this document. Note that there are other calculations that exist as well and a more in-depth discussion is beyond the scope of this paper.

Critical void ratio/void space ratio and overall blast void ratio

Critical Void Ratio (CVR) or Void Space Ratio (VSR) is the first important concept to analyse when reviewing a slot design as this parameter will determine if the blasthole to relief hole distance is sufficient to allow for both material swell and heave in the cut area. The CVR or VSR formula is the ratio of the sum of the area of drill holes (both charged and uncharged) versus the difference between the area of the cut and the sum of the area of the drill holes in the cut.

$$Critical\ Void\ Ratio\ (\%) = \frac{Area_{Empty\ Holes} + Area_{Blast\ Holes}}{Area_{Cut} - (Area_{Empty\ Holes} + Area_{Blast\ Holes})} x\ 100$$

This ratio will help to clarify if the cut area has enough void for the initial blastholes in the sequence to break into.

For successful slot design's, the shot hole must be close enough to 'easily' blast into the relief holes or reamer holes and must be timed slowly enough to allow it to heave and fall into the existing void before the next blastholes in the sequence are initiated. In most areas the slot is charged with electronic detonators which provide accurate timing and assist in preventing cut-offs. Good practice suggests that void ratios for cross-sectional area using the above formula are above 15 per cent, Hagan (1998). Hagan also recommends that 15 percent of the cut should be relief holes although he does not specify how much of the face is included in the cut area.

It is generally accepted in the industry that if the shot hole in a slot is too far away from a nearby void or relief hole, it struggles to move the blasted material into the relief hole area. This over confinement of the shot hole would create excessive ground vibration and likely freeze the blast. It also increases the risk of other failure mechanisms in a blast such as sympathetic detonation, dead pressing of the explosives and charge column dislocation resulting in misfires in nearby blastholes. Due to the nature of the slot design and its high drilling and charging intensity in a small area, these phenomena are common in many slot blasts. Detection of such failures have become more evident as mines introduce near field blast vibration monitoring into their stoping reconciliation process.

The authors of this document consider the area of the cut to be the area in which the blastholes immediately surround the relief holes for longhole rises. For raise bore drilled slots, the area of the cut would either be between the shot hole closest to the large diameter relief hole and the relief hole itself or the initial blastholes forming the inner box shape (or circular shape) and the relief hole. The relief holes in these raises bore designed slots range from 750 mm to 1.1 m in diameter and tend to be sufficiently large to provide a high critical void ratio in the cross-sectional area.

It is important not to confuse the CVR or VSR with the overall void ratio for the blast. The overall void ratio (commonly just referred to as the void ratio) is the ratio of the existing void volume versus the volume of the material to be basted.

$$Overall\ Blast\ Void\ Ratio\ (\%) = \frac{Blast\ Shape\ Volume}{Existing\ Void\ Volume} x\ 100$$

The void ratio is expressed as a percentage and also accounts for the broken ore's swell factor after blasting. The material swell factor is the increase in volume when this in situ material is blasted. An accepted rule of thumb is that if this swell factor is not known, then a conservative figure of 25–30 per cent be used. There are several methods to calculate this void ratio however the authors preferred method is described here.

The overall void ratio will determine if the volume of material to be blasted will fit in the available or existing void. It is recommended to maintain this ratio below 90 per cent to prevent freezing or bridging the blast. If the overall blast void ratio more than 100 per cent, there will not be enough available void for the volume of blasted material to heave into. This will result in an over confined, frozen or bridged blast and will likely cause recovery issues when mining this design.

This calculation is a quick and easy check to assist drill and blast engineers to evaluate blast shapes after completing a drill design.

Shielding and shadowing of initial blastholes in sequence

Another critical parameter in slot or winze designs is shielding and shadowing. Shielding provides protection along a direct line of site between the first blasthole in sequence and the next blasthole in sequence. This is achieved by designing and drilling a relief hole in between these two charge holes. It is recommended to provide initial shielding from the first hole in sequence and the next charged hole as this minimises the risk of sympathetic detonation and or charge column dislocation.

Shielded designs provide protection for the earlier firing blastholes in the sequence, refer Figure 2. Should there be an issue or concern with out of sequencing blasting from these inner cut blastholes, a shielded design could be an option to alleviate or minimise this risk. Structures encountered during drilling could also cause blastholes to interact out of sequence and thus the slot area should be adjusted if the area is highly structured.

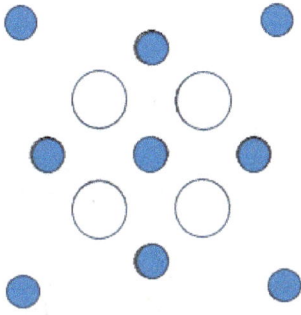

Not Shielded burn cut for a development round (45mm blasthole with 89mm relief hole)

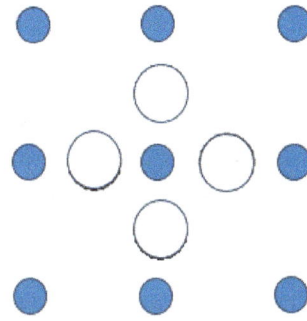

Shielded burn cut for a development round (45mm blasthole with 89mm relief hole)

FIG 2 – Non-shielded and shielded burn cut designs for a development round drilling pattern.

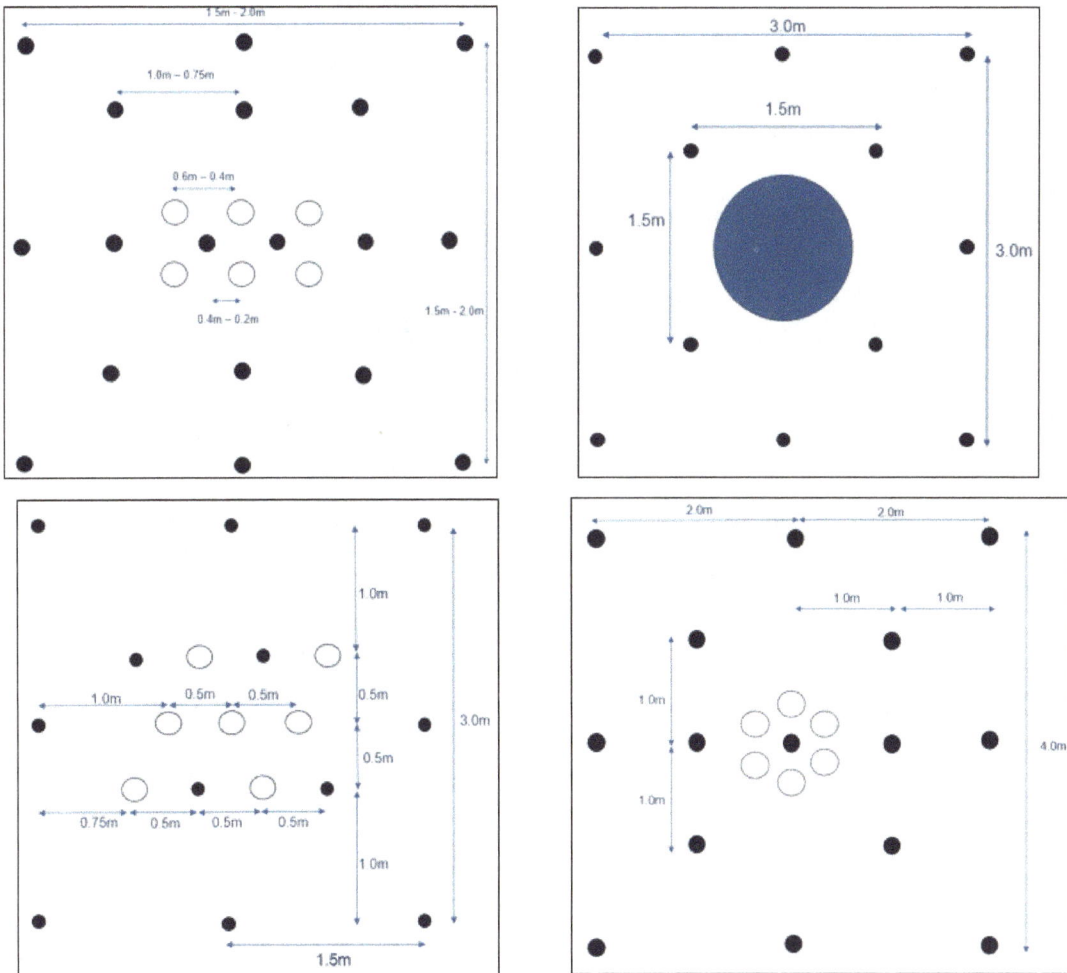

FIG 3 – Various frequently used underground slot designs.

Powder factors and energy distribution models

Powder factors and explosive energy distribution models are a good indicator of the total mass of explosives used per unit volume or mass. Powder Factors are a ratio of the average weight of the explosive used per unit weight of rock (or volume of rock) whereas Explosive Energy Distribution Models are a simple means of showing how the powder factor varies across an underground blast design. This provides an indication of whether a site is overcharging or undercharging these respective blast patterns.

The use of powder factors in the underground drilling and blasting process is a quick first check in the process however it does not factor in constraints in the underground environment, including but not limited to drilling and bogging horizons, different blast types (slot or longhole rises, development blasts, main rings blasts etc). A better approach would be to analyse a design from an explosive energy distribution model or from an individual blasthole damage radius aspect. Simulating 2D or 3D explosive energy distribution models would show designers a visual indication of energy being distributed both within a ring and between rings. This is critical in identifying areas where there is a larger or smaller than ideal energy distribution.

In an ideal scenario, the blastholes in a ring are spaced out such that adjacent blastholes undertake the same amount of work in both fragmenting and heaving blasted material into the existing void. Both must be spaced out far enough to provide sufficient area of influence in a volume of rock and close enough to provide hole-to-hole interaction. In most underground drilling horizons, the drill hole collars are closely spaced together, and, in such instances, it is best to stagger the uncharged collar. Using Energy Distribution Models, design engineers can optimise this uncharged collar and therefore reduce the risk of sympathetic detonation from nearby charged holes.

As shown in Figure 4, deviation of drill holes within the slot can create a uneven and sub optimal explosive energy distribution. Some holes may be overburdened and for others the high explosive energy distribution could increase the risk of sympathetic detonation, column dislocations or explosive product de-sensitisation. Such a result will become apparent when near field blast vibration monitoring data is analysed.

Explosive Energy Distribution Simulation (EED) completed in JK Sim Blast for 3m x 3m slot box with a 760mm Rhino drilled relief hole. 102mm blast holes charged with 0.91g/cc emulsion for a blind up hole. Drill holes surveys and mud maps were used to simulate a more realistic EED model.

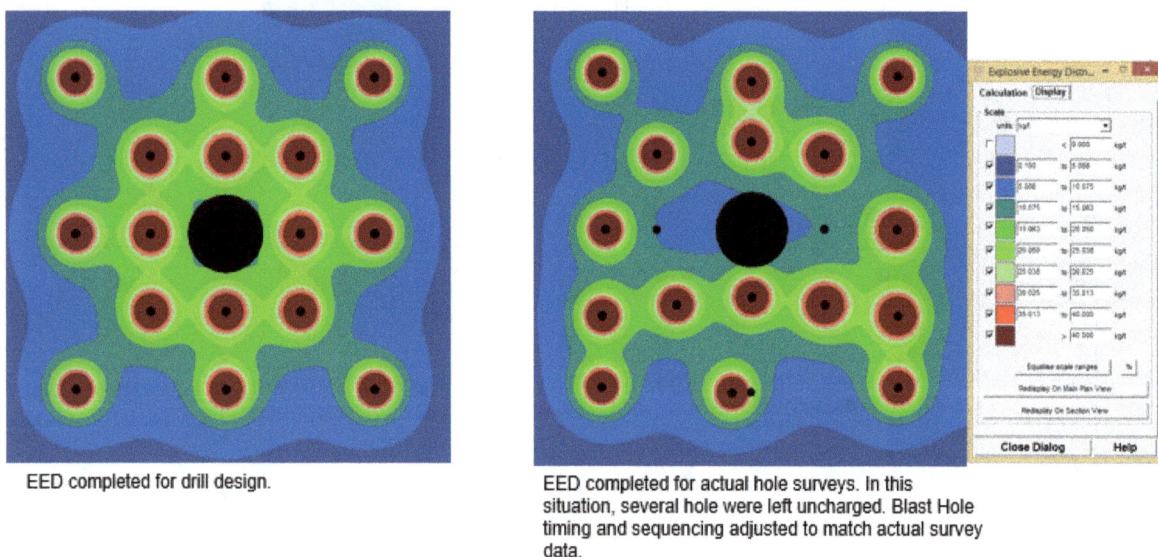

EED completed for drill design.

EED completed for actual hole surveys. In this situation, several hole were left uncharged. Blast Hole timing and sequencing adjusted to match actual survey data.

FIG 4 – Explosive energy distribution for drill design versus actual hole location.

Slot blasthole design timing and sequencing

It is vital for timing within the slot blast to be relatively slow compared to the main production ring blasts. This is to allow for effective clearing of broken material before the next holes in the sequence are initiated to help minimise material freezing or bridging within the shot. This timing will depend on the length of the slot and the inclination of the slot design. A general rule of thumb is to use a timing range from 20–60 milliseconds per metre of charge column in the blasthole. This range can be varied to allow for different inclinations of the rise and site-specific constraints such as relief hole diameters, slot design pattern, geology of the host rock and historical performance of slot designs on-site. The earlier firing blastholes in the slot design should be the slowest in the blast and later firing blastholes can be sped up. This means that once the inner box of the slot pattern has been cleared, the remaining blastholes in the outer box can be sped up to a quicker timing range than the rule used for the inner box. For example, a mine might use 40 milliseconds per metre of charge in the inner box for the first five blastholes. Once these five blastholes have been initiated, the design timing is sped up to 25 milliseconds per metre of charge. This is acceptable as it is still within the industry rule of thumb and it is not uncommon in many underground mines to use this sort of delay timing rule within the slot region of a stope.

Blasthole delay timing of slot patterns varies from site to site and careful Quality Assurance and Quality Control monitoring ought to be undertaken should a site wish to change or alter an existing set of timing parameters. In addition to standard site post blast reconciliation processes, best practice monitoring would also include near field blast vibration monitoring to clearly identify if holes in the pattern are performing to expectation.

In an ideal blasthole firing sequence, whether for a slot blast or a ring blast, the blastholes that have the best breakout angles should be the first to fire, that is, those that are closer to the free face or void. In a slot blast, the shot hole or the first blasthole in the sequence will be the one that is closest to the relief holes. The next blastholes in the sequence should fire into the newly created void and subsequent blastholes will follow this theory opening the void to a point where the initial slot box has been fully opened. This type of sequencing will optimise the actual drilling underground. In areas where in hole surveys are not completed to measure actual drilling, mud maps or stringing of holes can be used to identify which blastholes are closer to the relief holes. Should excessive drill hole deviation occur or be a concern on-site, drilling of another shot hole might be required. Although this does push out stoping schedules, it will give the slot blast the best chance of success and in most instances be less detrimental than a frozen/bridged blast.

Uncontrollable and controllable factors in drill and blast design

Slot designs, like production ring designs, have certain controllable and uncontrollable variables that influence the design. Uncontrollable variables include but are not limited to rock mass characteristics such as rock density, presence of geological structures, plasticity, and rock strength; as well as groundwater, mining method, ore orientation and ore grade; and equipment restrictions. Controllable factors include hole diameter, hole length, pattern dimensions, drilling configuration, explosive type, confinement, stemming material, uncharged collars, initiation sequence, blasthole delay timing and blast size. Certain variables listed in both categories could exist in either category depending on what part of the mining cycle they fall. For example, if the designer is in the feasibility stage of mining, mining method, stope heights and equipment selection (equipment limitations) can be controllable factors however these generally become uncontrollable once in production.

Understanding these levers and at which point they can be pulled will aid in completing optimal site-specific drill and blast patterns. Factoring in these variables in slot designs and using the guidelines stated previously in this document, drill and blast engineers can design patterns that will have high success rates in recovery, providing repetitive and consistently positive results.

DESIGN RECONCILLIATIONS – COMPLETING THE *PLAN, DO AND REVIEW* PROCESS

Design versus actuals

A vital but often overlooked part of the drill and blast design and implementation process is the reconciliation process. The reconciliation process closes the loop of the *plan, do and review* cycle for mining of open stopes. It will identify what was done right, what was done wrong and helps to identify areas of improvement in the mining process. This review is vital to ensure learnings from each stope in a mine is recorded so that the information can used to tailor newer designs or help to mine newer stopes with similar traits. It will also record important performance actuals such as overall recovery and dilutions of open stoping operations.

Most mines will use mine records such as void scans, bogging reports, drilling, and charging returns (paperwork from mine operations) to review the design versus the mined-out area. Information such as dilution, underbreak, overbreak and any adverse issues encountered will be highlighted and discussed. Important drilling and blasting key performance indicators can be obtained from this including:

- design drill and powder factors

- drilling and charging rates

- dilution

- overbreak and underbreak

- cavity scan versus design shape

- actual equivalent linear overbreak slough (ELOS).

These parameters should be captured in the blast reconciliation process to assist drill and blast engineers, shotfirers and other personnel make informed decisions regarding site specific issues in the blasting processes. This data should also form the basis for any future drill and blast optimisation project.

Near field blast vibration monitoring

With conditions being both site specific and often changing within a mine, it is highly recommended to include near field blast vibration monitoring as part of the drill and blast reconciliation process. Near field blast vibration monitoring has come a long way in the underground space with newer technologies to suit different monitoring projects. Equipment manufacturers can supply equipment to measure blast induced ground vibration in either velocity or acceleration. In near field blast vibration monitoring, the monitor is usually installed 20–60 m away from the blast. At this distance, the vibration signals measured have a higher frequency and lower duration than those measured further away. The objective of this type of vibration monitoring is different from most far field environmental monitoring where the structural integrity of a building or infrastructure is needed to be protected. For further field monitoring, ground vibration induced by blasting is monitored near critical infrastructure to ensure that there is no permanent damage caused by the blasting process. In near field vibration monitoring, it is the blast performance that is of greater interest.

Vital information can be obtained regarding explosive product performance, blasthole detonation and blasthole interaction due to sequencing and delay timing. Blast induced ground vibration data retrieved from specialised monitors can provide a detailed review of the blast and give quantitative feedback on individual blasthole performance. Using scale distance laws and site constants, peak particle velocity or acceleration measurements can be used to determine the performance of each blasthole and compare the expected results. This, in conjunction with bogging inspections, fragmentation analysis, cavity scan results and operator return will give the drill and blast engineering team the ability to evaluate the effectiveness of the design and to compile a more in-depth stope reconciliation. It will also strengthen the site's ability to investigate unsuccessful blasts and identify key failure mechanisms and improvement opportunities.

Near field blast vibration monitoring is critical to understanding slot blast performance. Whether reviewing an existing design or attempting to use a new design, the data from vibration monitoring studies will aid in analysing the overall blast outcome and give insights into improvement initiatives. Using historical data from site along with near field blast vibration studies, drill and blast engineers can adjust drill hole spacings and burdens, blasthole delay timing within the slot box, and blasthole sequencing to improve rise design mining.

ACHIEVING AND MAINTING KEY PERFORMANCE INDICATORS IN STOPING

The critical design parameters mentioned earlier in this document are used in many of the common slot patterns found in the underground mining industry in Australia. These patterns are usually adopted by mine engineers and experienced mining professionals from experience on previous operations or from company standards and are introduced into newer mines taking into account the controllable and uncontrollable drill and blast design factors stated earlier. These slot designs are proven and have a generally high success rate if designed and implemented correctly.

It is important to note that the slot, winze, rise or raise design selected as standard for an underground mine must show consecutive successful results. These results must be reviewed and updated to track the success rate of the initial blasts and the stope as a whole. The reconciliation methodology mentioned in the previous section is a vital part of this process.

The main requirement for the slot blast is to open an initial void for the rest of the stope or the cave to fire into. It is good practice to always complete void scans after bogging of the slot or rise blast to ensure there is sufficient void to progress with charging and firing the next rings in a blast. This is especially important in blind uphole slot blasts.

The theory and methods discussed within this document are from multiple sources, accepted industry rules of thumb and best practices and are illustrated here to assist with sharing knowledge in the critical design parameters of a slot, rise, winze or raise design. There is no *one design fits all solution* and close attention must be paid to the drill and blast design fundamentals when selecting or reviewing a slot design for a mine.

ACKNOWLEDGEMENTS

The authors would like to thank the Dyno Nobel Asia Pacific Technology Group for their support and approval to publish this paper.

Near field blast vibration monitoring for slot or winze blast analysis

E Wargem[1] and D Tozer[2]

1. Senior Technical Consultant, Dyno Nobel Asia Pacific, Perth WA 6100.
   Email: Edward.Wargem@ap.dynonobel.com
2. Technical Manager, Dyno Nobel Asia Pacific, Perth WA 6100.
   Email: David.Tozer@ap.dynonobel.com

ABSTRACT

The optimum mining of an orebody in an underground mining operation is fundamental to the mine's ability to operate both sustainably and efficiently. The effective mining of an open stope or a cave is largely reliant on the recovery of the initial void created by conventional drilling and blasting. These initial voids are known by various names such as a Longhole Rise, Slot or Winze and can be mined either as a downhole or as an uphole. Regardless of the underground mining method, the process of the design and implementation of this initial void is largely uniform across many operations with various patterns of slot or winze designs used across these different operations. Diligent design, implementation and review strategies in drilling and blasting of these will facilitate cost-effective mining with minimal disruption to mine schedules and production targets. When an operation struggles in opening the initial void in underground mine blasting, it has a negative impact on productivity, mineral recovery, and the overall economics of mining. Near field blast vibration monitoring is a vital tool to use in analysing slot blasts and will provide additional feedback into the reconciliation part of the drill and blast process. It is one that is often overlooked in the mining industry in many operations due to many factors such as lack of skilled personnel, or a lack of required resources. The data obtained from near field blast vibration monitoring can be used to provide additional feedback into blast reviews and help in any drill and blast optimisation projects. This paper highlights key aspects of underground drill and blast reviews and the use of near field blast vibration monitoring for post blast analysis and shares learnings from various projects that the authors have been involved in within Australia and abroad.

THE IMPORTANCE OF REVIEWING BLASTS

The review process of underground development and production drilling and blasting is a vital step to reconcile and identify areas of improvement in the design and execution of either the production or development blasts in an underground mining operation. The reconciliation process closes the loop of the *plan, do, review and act* cycle for mining of open stopes (Wargem, 2020). This review data is important as it ensures that learning and experiences from each blast event are recorded and reviewed. The recorded data will also provide the basis of overall performance actuals for these blasts (dilution, recovery, overbreak, underbreak etc).

There are many methods, processes and tools that can be used to complete drill and blast reviews such as stope reconciliations or development overbreak and underbreak studies. However, an often-overlooked process is near field blast vibration monitoring and its impacts on all preceding items. The use of vibration monitoring for underground production blast reviews will be discussed here. Vibration monitoring can also be used for environmental compliance monitoring and to assess compliance to regulation limits, damage criteria for structural response, and others. However, these aspects will not be covered in this treatise.

WHAT IS NEAR FIELD BLAST VIBRATION MONITORING

Near field blast vibration monitoring is vibration monitoring of blast induced ground vibration very close to the blast. This ranges from 10–60 m away from the blast. This type of monitoring will utilise vibration monitors and sensors capable of sampling at high frequency, and must be able to withstand the magnitude of the ground movement at the monitoring location. At this distance, the vibration signals measured have a higher frequency and lower duration than those measured further away (greater than 100 m away). This type of monitoring can be used to verify blasthole initiation in a production slot, winzes, longhole rises or raises, or main production rings. The data can also be used in conjunction with existing theory and knowledge of vibration analysis to verify if blast designs are suitable for the site-specific ground conditions. Both types of studies will be explained further in this document.

What is blast induced ground vibration?

Conventional methods of mining involved the drilling and blasting of rock to liberate the ore bearing material. The material is then loaded and hauled out of the mines using underground load haul dump machines. This material is then fed through a resizing process of crushing and grinding to suitable sizes prior to feeding this into the processing mills. During the blasting process, an explosive charge is detonated. The explosive detonation will release large amounts of energy and gas that will break and move the rock (provided suitable conditions exist). As the distance increases from the blasthole being detonated, the energy will cease to break and heave the rock and will radiate out in a form of complex seismic waves and air overpressure. Air overpressure will not be discussed in this paper. The seismic ground waves that emanate from the blast are classified in two main groups. Those that either travel within the body of the earth or on the surface of the earth. The former being known as body waves and the latter known as surface waves.

Introduction to seismic wave and theory of explosives in underground mine blasting

It is important to understand the basics of seismic waves when analysing blast induced ground vibration and the basics of wave form motion physics (Bollinger, 2018). The body waves that travel within the earth after a blasthole has initiated include the Primary wave (P-Wave) and secondary wave (S-wave). The P-wave travels in the direction of the seismic wave and is the fastest. The S-wave travels perpendicularly to the direction of travel. Surface waves are slower than P- and S-waves and travel on the surface of the earth. There are two types of surface waves known as Love Waves and Rayleigh waves. They are larger and more destructive than the body waves, but are generally not observed in the underground mining environment. Figure 1 shows the main types of seismic waves mentioned. These seismic waves can be recorded with appropriate seismometers. It is important to note that seismometers can be either uniaxial or triaxial and care must be taken in selecting a fit for purpose vibration monitor for the project.

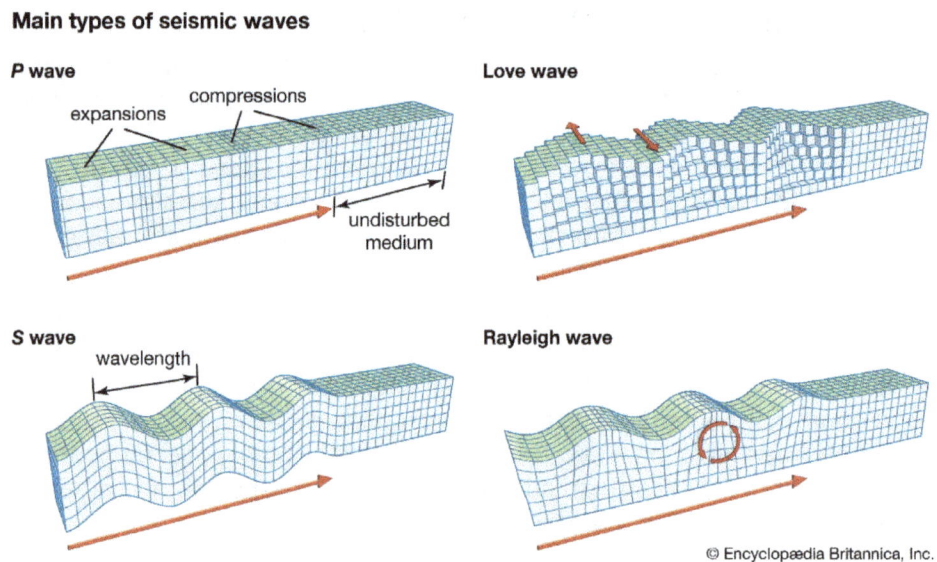

FIG 1 – Main types of seismic waves (Encyclopædia Britannica, 2022).

A triaxial seismometer can record all waves in their specific orientation and hence is best suited for more in-depth monitoring programs.

Drilling and blasting in mining is an important process in the mining cycle where the ore bearing rock is liberated from the host rock into fragmented sizes that would assist in loading and hauling this material out of the mine to the processing plant. The rock must be drilled then charged with explosives in patterns that will ensure that explosive energy is effectively used to fragment this ore into sizes that will allow for optimal loading and hauling and subsequent size reduction at the processing stage. There are many theories regarding the major breakage mechanism when a charged blasthole is detonated, however it widely accepted that the detonation of explosives in a blasthole is an almost instantaneous chemical and mechanical reaction that releases large amounts of pressure at high temperatures. The high-pressure gases formed from this reaction acting on the walls of the drill hole transmit shock waves to the surrounding rock mass. The shock waves crush the material in the immediate vicinity of the blasthole and starts the process of creating new micro fractures. The combination of this gas pressure expansion and fracture propagation will mobilise the material to a free face when the suitable conditions exist. Beyond a certain point, past the critical damage zone where the surrounding rock is crushed and radial cracks in the rock are created, the energy from the blast is observed only as seismic waves. These waves will transmit the energy from the source and radiate outwards from the source until the energy is fully dissipated. With the right equipment and processes and a knowledge of the rock mass, the seismic waves can be measured using specialised equipment and the data used to analyse blast designs.

Monitor type and data integrity

In near-field blast vibration monitoring (typically 10–60 m away from blast), the objective is to detect, record, then analyse blast induced ground motion close to the blast. In the near-field, high frequency sensors are required. Accelerometers or geophones can be used to measure these seismic waves as acceleration or velocity as function of time. The authors of this document prefer to use geophones that measure particle velocity, as it is the preferred means of reporting damage criterion and in line with Australian Standards AS2187.2-2006 (Australian Standard, 2006). The Texcel GTM vibration monitor with a 28 Hz uniaxial geophone is used by the DynoConsult group in Western Australia. This monitor records ground vibration as a function of time.

Monitors such as the Texcel GTM and the uniaxial geophone sensors can be configured via the software suite to adjust the maximum vibration range, maximum time of each blast vibration recording and frequency of sampling per second. Triaxial seismometers can also provide a full spectrum of seismic/explosive waveforms however for practicality in underground applications for near field blast vibration monitoring, the uniaxial seismometers are sufficient. It is then important to have some prior knowledge of the blast design maximum timing, expected vibration range and frequency of measurements as a baseline for sensor configuration. These settings can be adjusted to match the blast design and factor in the distance that the monitor is placed away from the source of the vibration (ie the blastholes being detonated). Depending on the requirements of the vibration monitoring program, whether compliance monitoring for vibration limits or for drill and blast design reviews, the appropriate selection of equipment, sensor and its configuration is vital. Equipment manufacturers, or technical experts in this field will be able to assist with various generic settings and fit for purpose sensors.

USING NEAR FIELD BLAST VIBRATION MONITORING TO ANALYSE SLOT OR LONGHOLE RISE BLASTS

Analysing a near field seismograph result for mine blasting

The main benefits of near field blast vibration monitoring for underground mines include, but are not limited to, verifying individual blasthole initiation, reviewing performance of drill and blast design standards, and analysing and identifying the cause of unsatisfactory blast results. For many underground mines in the world, the main data used to verify drill and blast designs (whether for production or development) is to reconcile the design shapes against the void survey scans. A complete stope reconciliation process typically includes review data from geologists, geotechnical engineers and operational reports from drilling crews and underground charge crews. All this information is compiled and presented as a full review of the selected blocks. This data can be used to review existing standards and identify areas of possible improvement. Some mines have the capability to complete downhole surveys prior to finalising charge designs. In these instances, this data could be used in the review and optimisation process as well. This standard reconciliation process is in a lot of cases sufficient, however, there is no actual quantitative data to prove that each blasthole designed, drilled, charged and blasted was effectively used in the blasting cycle. Many mines that have permanent seismic monitoring systems installed to monitor seismicity also record vibration events due to blasting however the far field monitoring technique only records the event and in most cases the data is not suitable to view individual blastholes being detonated in the blast event. Near field blast vibration monitoring is the process that can provide actual blast data and is a key input used by the DynoConsult group in their work in the underground mines throughout Australia.

The induced ground vibration from a blast can be recorded and analysed to complete various review processes as already mentioned. Figure 2 shows a vibration trace of a blind uphole slot blast in an underground sub level open stoping mine. The x axis of the vibration trace is the time (in seconds) and the amplitude in the y axis is the particle velocity (in mm/s). Each separate group of peaks and troughs in the waveform is measurement of the ground vibration at the monitoring location due to a blasthole (or holes) being detonated. The highest peak or the lowest trough in each group of wave forms is the peak particle velocity (PPV) measured for that blasthole. The PPV is measured in millimetres per second and is the peak vibration level generated by the blasthole being detonated. This measurement is important in environmental impact monitoring as there are regulatory limits on vibration limits and these are often quoted in PPV levels.

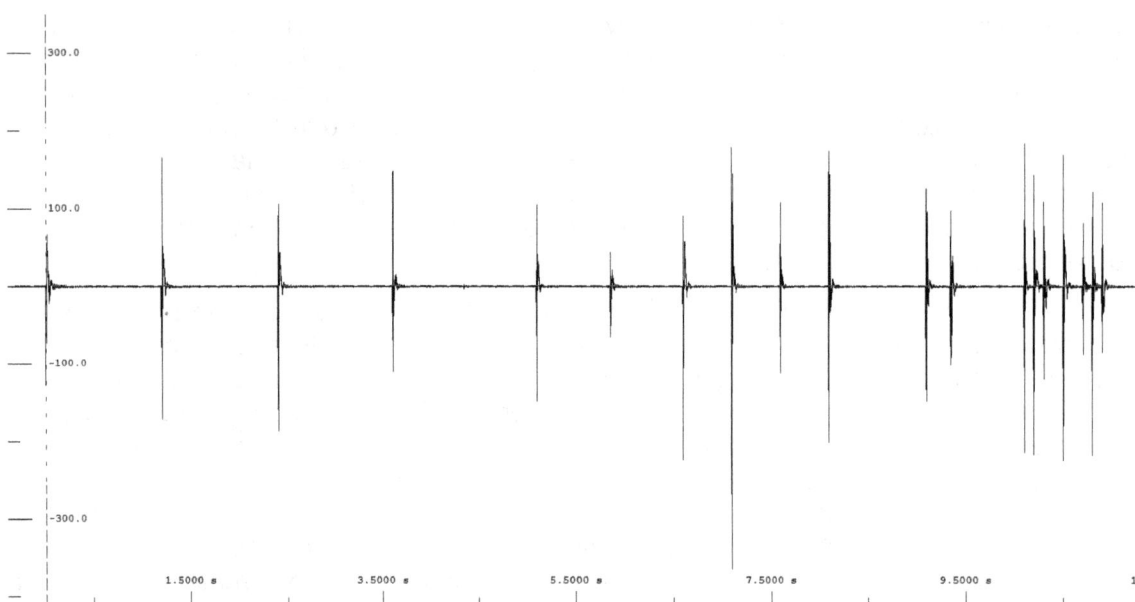

FIG 2 – Vibration trace of an uphole slot or longhole rise blast from the Texcel GTM monitors.

The PPV recorded at certain distances from the blast will be affected by the maximum instantaneous charge per delay, vibration frequency, rock characteristics, blasthole conditions, coupling ratio, explosive type and the distance from the monitoring location to the blast site.

In the absence of sufficient vibration data to construct site specific ground vibration estimation calculators, charge weight scaling laws can be used. There are many PPV calculations, however, the equation stated below is used as it aligns with the estimation of ground vibration as stated in the Australian Standards AS 2187.2-2006 (2006).

$$V = K \left(\frac{R}{Q^{1/2}} \right)^{-B}$$

Where:

V	is the PPV
R	is the distance between blast site and measurement location
Q	is the maximum instantaneous charge
K and B	are constants related to the site and rock properties for estimation purposes. The site decay and confinement constants K and B can be determined after sufficient data is collected.

Once the near field blast vibration monitoring program has been employed in a mine, whether to determine individual blasthole detonation (for testing of newer explosive product etc) or for production blasting reviews; a range of expected PPV readings can be determined from set distances away from the blast site and in different rock types in the mine. The readings from the vibration trace, like that shown in Figure 2, combined with data from the charge design (or actual charge return sheets) can be used to show whether a blasthole has initiated or not.

Once a trigger threshold has been exceeded on a vibration monitor, the monitor will start to record the whole blast event. The start of the trace will be considered as the first blasthole in the sequence. This first blasthole in the sequence is known by different names such as the shot hole or initiating point (older mining terminology from the use of older explosives). In a slot or longhole rise blast, the shot hole will have the same charge mass as the rest of the blastholes in the 3 m × 3 m or 4 m × 4 m box, however, they will be blasted in a specific sequence with slower delay timing in the inner box before speeding up the blast timing to open up the outer box. Looking at the formula for PPV calculation stated above, the expected PPV for each blasthole detonated in sequence in the slot area should measure the same and should be displayed on the trace with respective blasthole delay as per the design.

To expand on this concept further, an example of a 28 m blind uphole longhole rise design like that shown in Figure 3 will be used.

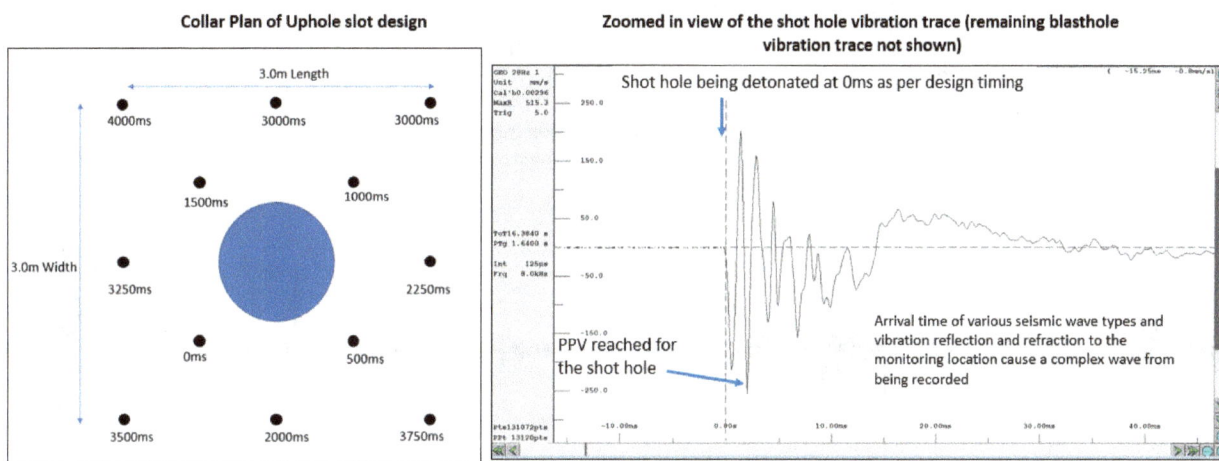

FIG 3 – Example collar plan of slot design on left and a zoomed in view of a vibration trace shot hole of a slot blast.

The blind uphole slot has 12 blastholes drilled with 102 mm bit and these surround a single 760 mm relief hole. Each of these blastholes has approximately 174 kg of explosive (considering emulsion density and uncharged collars). A vibration monitor with a uniaxial geophone is placed 20 m away from the blast (in a safe location) to record the blast induced ground vibration. The shot hole records a PPV of 255 mm/s. The remaining 11 blastholes fired in sequence should record around 255 mm/s for their PPV (amplitude in the trace file) if they complete the same amount of work to break and heave the rock into the existing void. There are however other factors to consider such as the instantaneous void being created and material swell factors as well as the break out angles of each blasthole in sequence. The blast will be determined successful if a majority of the blastholes are recorded, their expected PPVs are visible in the vibration trace at their respective designed delay timing and a void scan proves the blast achieves the design objectives (design versus actual). It is not uncommon to see missing traces of blastholes charged, or lower than expected amplitudes of PPV from blast vibration recordings. There are many causes of this including but not limited to sympathetic detonation, charge column dislocation, product desensitisation, explosive product shock damage and even combinations of these blasthole explosive failure mechanisms. Near field blast vibration monitoring will also provide feedback to mining personnel of possible misfires and to alert respective personnel to complete necessary post blast inspections prior to or during bogging out of this blasted material.

This data can be used to better analyse blasthole spacing, burden, delay timing and sequencing. If blastholes are fired too fast in sequence, provided the maximum instantaneous charge (MIC) is kept constant and drilling is as per design, the vibration trace will show an average increase in PPV measured. The MIC is the maximum quantity of explosive charge detonated in one delay timing within a blast. If the blastholes are fired too quickly and a stope blast freezes or bridges (unsatisfactory blast when material is left frozen in a stope after blasting), the vibration trace will show this increase in PPV. However, other factors such as suboptimal drilling or charging design can be another issue and hence must be investigated as well. This quantitative data from near field blast vibration monitoring provides the user with more in-depth analysis.

Figure 4 shows an example trace of an unsuccessful uphole slot blast. The uphole slot blast did not achieve full height and the near field blast vibration data was used with design and implementation data from the day of charging to provide a better understanding of what caused the unsuccessful slot firing. From this blast analysis, operator feedback and vibration trace analysis, the conclusion was that the shot holes within the inner cut area of the slot were wet and the ANFO had desensitised at the top of the blastholes. The blastholes fired earlier in the blast sequence did not pull to height resulting in insufficient void for the later firing blastholes in the sequence to blast into. The vibration trace at the end of the seismograph shows the PPV for each blasthole measuring higher than expected values indicating low void firing and hence bridging of the top portion of the slot.

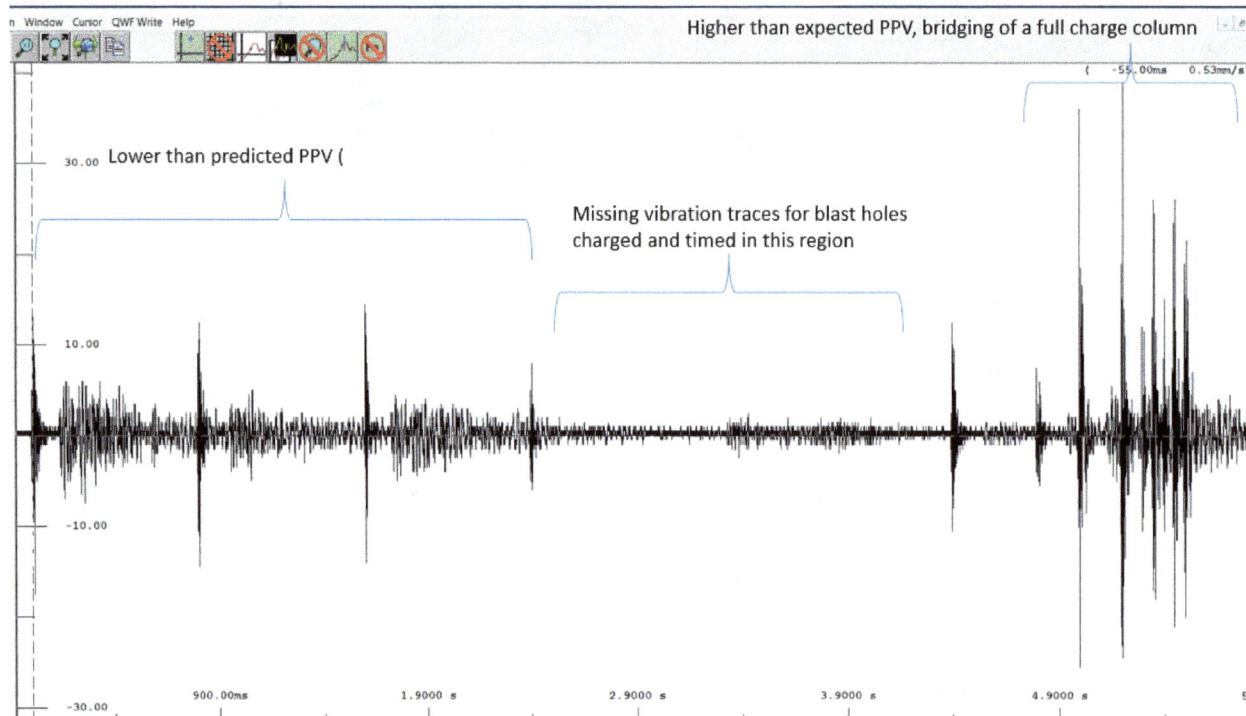

FIG 4 – Example of an unsuccessful uphole slot blast seismograph. The blast froze and the vibration trace assisted site personnel to better understand what went wrong during the blast event.

Using data for continuous improvement initiatives

The authors recommend completing several near field blast vibration measurements of current slot or longhole rise blasts using the site design and implementation standards to provide a suitable site baseline. Once this baseline study has been completed, any improvement initiatives identified from the stope reconciliation process can be implemented. Continuous vibration monitoring should then be employed to measure the results of any changes to the blast design under the baselined site-specific conditions and properties of the rock mass.

For example, if a certain slot design continues to provide unsatisfactory blasts (like that shown in Figure 4), vibration monitoring can be used to identify if there is an issue with delay timing within the design. If continuous monitoring shows clear signs of bridging due to fast blast delay timing, the timing can be slowed down to provide better relief of material prior to detonating the next blastholes in sequence. This can be used in conjunction with other drill and blast review processes to determine the success of the current mine design and to also identify areas of improvement in the design and implementation phase. The more data site has, the more informed decisions can be made in the drill and blast process.

ACKNOWLEDGEMENTS

The authors would like to thank the Dyno Nobel Asia Pacific Technology Group and the Texcel Ltd Australia for their support and approval to publish this paper.

REFERENCES

Australian Standard, 2006. AS 2187.2-2006, *Explosives – Storage and use, Part 2: Use of of explosives*. Sydney: Standards Australia.

Bollinger, G, 2018. *Blast Vibration Analysis,* New York: Dover Publications Inc.

Encyclopædia Britannica, 2022, August. *Seismic wave: main types*. Retrieved from Encyclopædia Britannica: https://www.britannica.com/science/earthquake-geology/Shallow-intermediate-and-deep-foci#/media/1/176199/210259

Wargem, E, 2020, September 16. Stope and Raise Design and Timing_Final_DynoConsult_UGD&B_Workshop2020, Mt Isa, Queensland, Australia.

Black Rock Cave (BRC) – overcoming challenging ground conditions and mitigating the impacts to production

H Wright[1], B Hales[2], A Young[3], J Harris[4] and L Williams[5]

1. Senior Mining Engineer, Black Rock Cave, Mount Isa Mines, Mount Isa Qld 4825. Email: huw.wright@glencore.com.au
2. Senior Geotechnical Engineer, Mount Isa Mines, Mount Isa Qld 4825. Email: brett.hales@glencore.com.au
3. Mining Engineer, Black Rock Cave, Mount Isa Mines, Mount Isa Qld 4825. Email: albert.young@glencore.com.au
4. Senior Mining Engineer, Minserve, Brisbane Qld 4000. Email: jordanharris@minserve.com.au
5. Project Manager – Black Rock Project, PYBAR Mining Services, Mount Isa Qld 4825. Email: luke.williams@pybar.com.au

ABSTRACT

Black Rock Cave (BRC) is a sublevel cave which was developed in 2018–2020 and has been in production since November 2020. BRC is part of the Mount Isa Mines complex and extracts a leached orebody comprising of high-grade chalcocite ore, situated below the historic Black Rock Open Cut workings. The mine is owned by Mount Isa Mines and operated under contract with PYBAR Mining Services.

In 2021 a number of rockfalls occurred throughout the mine and were increasing in frequency. This culminated in a major rockfall on the decline in June 2021, after which the decision was made to withdraw all persons from the mine, initiate rehabilitation and conduct further investigations to deduce a safe method to provide long-term support.

The rockfall investigations have shown that the corrosive nature of the highly leached rock mass can consume a split set in less than 18 months. A review of potential corrosion resistant ground support methods concluded that steel split sets have a safe lifespan of six months at BRC, and that all workings must have corrosion resistant support installed prior to exceeding the safe life of the friction bolts. The review recommended the use of 4 m cable bolts as the most suitable method of corrosion resistant support.

This new support system, and the requirement to rehabilitate all mine workings within the leached orebody, had an impact to the schedule and an effect on both equipment and manning levels at BRC. Accelerated jumbo rehab allowed access to the production levels within two months of the rockfall. Cable bolting then followed the jumbo bolting in all leached areas. The direct impact has been ten months of reduced production, due to night shift closures of the decline to enable cable bolting.

The impact to the mine has been minimised by the small flexible team that enabled the mine plan to change rapidly and often to accommodate the challenging conditions. In less than two months from the rockfall event, production had resumed. Within ten months of concluding that the mine required; full rehabilitation, a new support system and a full re-schedule, normal mining operations had restarted, with pre-event productions rates being achieved.

DEVELOPMENT PHASE

BRC was designed to extract a leached copper orebody from underneath the historic Black Rock Open Cut (BROC). Due to the leached nature of the orebody and very poor rock mass, the decision was made to utilise the sublevel caving method rather than sublevel open stoping which is traditional at Mount Isa Coper Operations (MICO).

The drawpoint (DPT) and footwall drive locations were constrained by historic mining. With the mine development being locked-in, and the productivity that could be achieved by completing all development upfront, the tender for the development stage of BRC included all levels of the mine. PYBAR Mining Services were awarded the development contract and a portal was mined from the Black Star pit to the north of BROC, with the first cut being taken in May 2018.

The initial 200 m of development of the I66 Decline (DEC) is in un-weathered rock and so resin bolts and mesh could be used to provide long-term support. The remainder of the mine is within the leached zone. It was found that resin bolts cannot be installed in the leached ground. A geotechnical review (Shiels, 2019) stated:

> Posimix bolts have been observed not to work in the leached zone due to breakout of the borehole annulus. The high clay (kaolin) and thick rubble infill on the joints and bedding planes, when water is introduced, is rapidly washed away resulting in instability.

> The stiff bolt (pre-grouted friction bolt) has been found to work well in the leached zones, coupled with air-mist, retract drill bits and using a resin bolt size bit (38 mm) to drive the friction bolt in. In certain instances, only collaring the hole is completed, the bolt is then installed through the shotcrete with percussion to achieve adequate friction.

No cable bolter was included in the development contract due to the limited number of cable bolts anticipated when tendering. Intersections cables were drilled with the jumbo and hand pushed/grouted. Due to the rate of cable installation able to be achieved, it was not considered practical to utilise cable bolts as part of the standard ground support system. The grout in the friction bolts was to provide some level of corrosion resistance and further investigations would determine the most suitable support system within the leached rock mass.

Production was initially scheduled to commence at sublevels 5C and 4C on the upper panel (Figure 1). With the low draw from 4C this level would progress rapidly and due to the small pillar to surface this would instigate a sizeable surface subsidence imprint. The subsidence zone of BRC was predicted to impact:

- Enterprise Mine change-house
- mining operations, fixed-plant and mining technical services offices
- Enterprise Mine surface substation
- mines rescue station.

The sizeable logistical operation to relocate this surface infrastructure was not able to match the mine schedule. To allow mining to start on schedule, it was determined that production could start on the lower panel, with the subsidence zone constrained within the BROC pit. Starting on the lower panel had the added benefit of allowing high-grade tonnage to be brought online early. In the revised mine schedule, the lower panel would retreat to a safe pillar with the upper panel and be paused whilst the upper panel went into production (Figure 1).

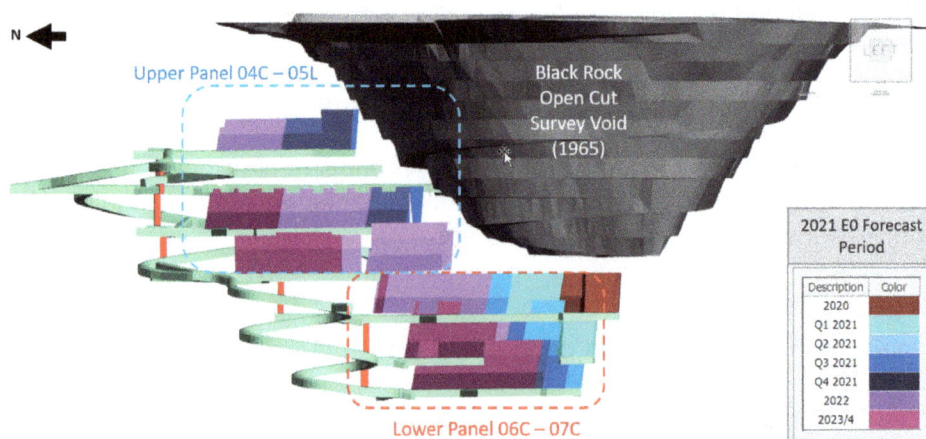

FIG 1 – BRC cross-section with 2021 schedule overlay.

PRODUCTION PHASE – EVENTS PRECEEDING THE ROCKFALL

Development concluded in November 2020 and after the tender process, the production contract was awarded to PYBAR Mining Services to extract the orebody. Caving commenced in the lower panel on 6C and breakthrough to the base of the BROC pit occurred in December 2020.

With production activities occurring in the lower panel, the upper panel had not been accessed regularly since development of the drawpoints finished in late 2019. With production scheduled to start in 4C and 5C in Q3 2021, damage mapping was conducted in the first half of 2021. This identified several rockfalls in 5C (Figure 2.7) which were successfully rehabilitated.

The damage mapping identified several small areas of 4C DPT's that required pre-emptive rehabilitation prior to being subjected to production related vibration. The rehabilitation had commenced and had progressed to the slot drive when at the start of shift on the 20/06/2021 it was found there was a rockfall in 4C DPT3 (Figure 2.2). Over the next two hours a crack then opened at the access of the same drive (Figure 2.3). Both locations had been inspected by an experienced Geotechnical and Mining Engineer and were believed to be stable with no visible indicators of ground or support deterioration.

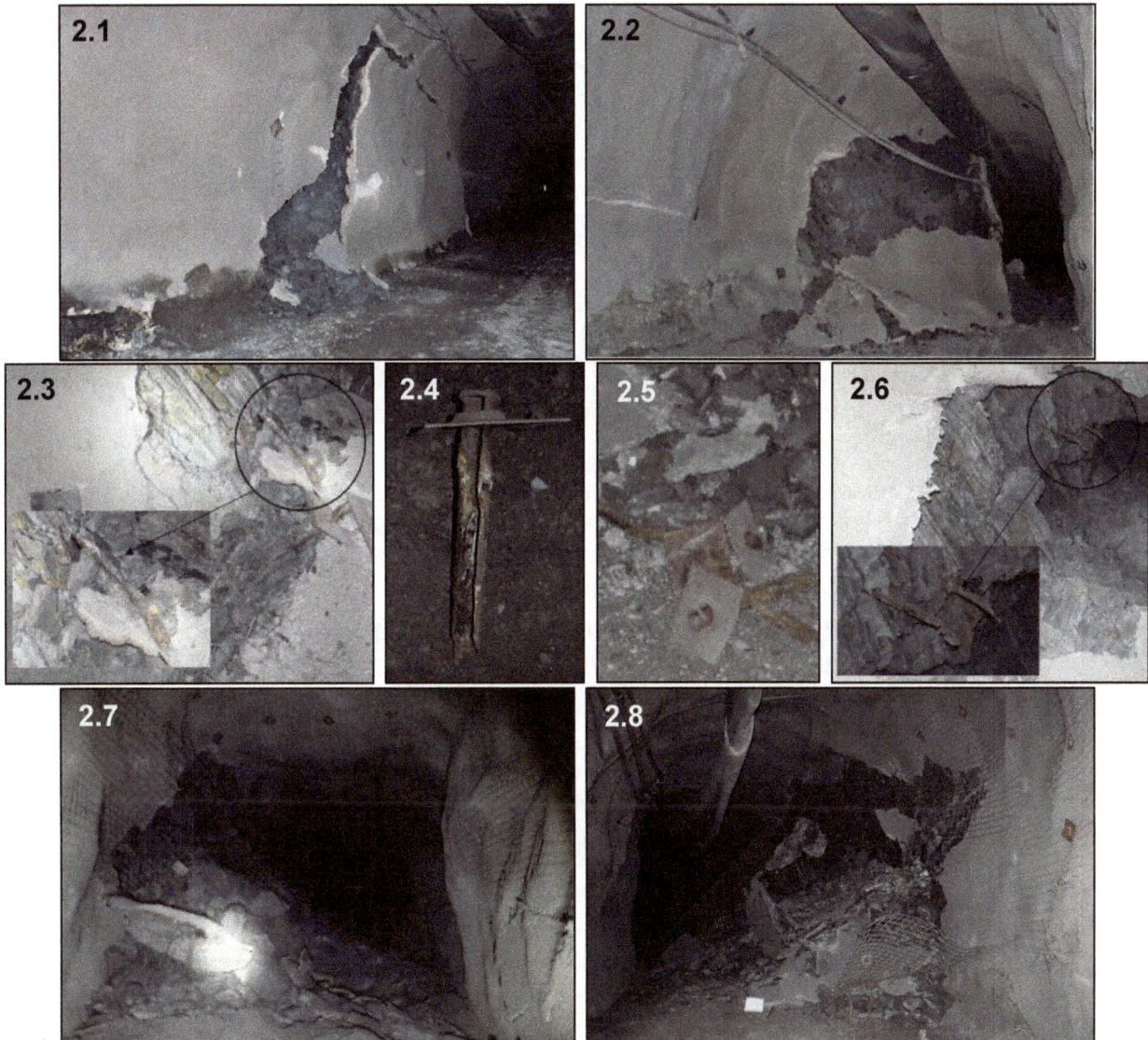

FIG 2 – Rockfall and corrosion photographs from 2021 – (2.1) 4C DPT1 rockfall 1: June 2021; (2.2) 4C DPT1 rockfall 2: June 2021; (2.3) 4C DPT1 rockfall 2: corroded friction bolt; (2.4) 7C R60 DEC rockfall: corroded friction bolt; (2.5) 7C R60 DEC rockfall: corroded friction bolts (2); (2.6) 4C DPT1 rockfall 2: corroded friction bolt (2); (2.7) 5C DPT3 rockfall: May 2021; (2.8) 7C R60 DEC rockfall: June 2021.

The incident investigation from the 4C rockfalls found that the corrosion of friction bolts (Figure 2.3/2.6) was the primary cause of failure. With no external signs of corrosion or failure the confidence in our inspections ability to identify areas requiring rehab was reduced. A series of actions were raised to try and aid in identifying areas where support could be compromised and rectify the problem. These actions included:

- Produce a corrosion resistant support recommendation for BRC in the leached rock mass (utilise a consultant if required).

- Identify areas of BRC where there is low confidence in the installed ground support and issue PDD's (Primary Development and rehabilitation Design) to rectify.

Work had begun on these actions when a rockfall occurred in the R60 decline during night shift on the 30/06/2021 (Figure 2.8). Further evidence of corrosion was seen in the bolts (Figure 2.4/2.5) yet no external signs of damage were present on the day shift prior, when the area was inspected. This prompted management to make the decision to withdraw from the mine and deem all ground within the leached rock mass 'unsupported' until an effective re-entry plan and corrosion resistant ground support system was available.

GROUND SUPPORT REVIEW

With no practical means to identify corrosion at depth or imminent sudden failure, all support within the leached rock mass were deemed to be compromised. Rehab was initiated to begin replacing this support and allow access to production levels. Concurrently with rehab; a review of the rockfalls, current ground support and corrosion resistant support requirements was commissioned from Mining and Civil Integrity Testing.

From review of the rockfalls, Hassel and Villaescusa (2021) concluded that:

> The rock falls showed that extreme corrosion is occurring on the non-galvanised friction and stiff bolts after a relatively short installation times of 18 months. This indicates that very high rates of corrosion are occurring. The corrosion damage is reducing the thickness of the steel causing the bolt to fail by breaking, opposed to pulling out from loss of frictional resistance. Corrosion damage is occurring from just inside the bolt collar.

> Bolt plates generally show only light corrosion. This means a surface corrosion criterion by visual inspection cannot be used to determine the internal corrosion damage. Borehole camera inspections and pull testing appear to be a suitable method to estimate bolt serviceability.

> Mesh encapsulated in shotcrete and mesh that is exposed on the excavation surface does not appear to be corroded except in a few areas where groundwater flow reaches the drive surface.

> The rockfalls have generally occurred in areas that appear dry. But precipitates indicate the existence of previous groundwater flow. Several rockfalls are associated with known faults and all rock falls occurred in the walls of drives. Due to the nature of the sublevel cave, higher vertical stresses would be expected in the pillars and these would be creating displacement in the walls. The demand of some failures exceeded the design capacity of the ground support with some failure at depths greater than the bolt length. Cable bolts are required to provide embedment beyond the failure depths as well as high retention capacity. The reinforcement should be extended down the walls to the floor.

Recommendations for ground support requirements (Hassel and Villaescusa, 2021) in the leached rock mass were:

- 50 mm fibre reinforced shotcrete.

- Frictions bolts should extend to 1 m from the floor (extra bolt on each sidewall).

- Galvanised friction bolts and mesh to provide initial support.

- Frication bolts have a safe lifespan of six months after which time corrosion resistant support is required.

- Jacked 4 m plain strand cables should be installed for corrosion resistant support on a 1.6 m spacing with a 1.5 m ring spacing.

- Spraying of 50 mm fibre reinforced shotcrete over the mesh from floor to floor would enable load to be transferred more effectively between bolts.

Based on this review, a new support standard was created for within the leached rock mass with a focus on requirement to install full rings of cable 4 m plain strand bolts to act as the corrosion resistant support within the six month safe lifespan of friction bolts.

Further avenues of investigation have included, a corrosion trial with multiple different bolt types in different locations around the mine and a groundwater monitoring program. Several trail bolts have been used at locations within BRC to determine if there are more effective methods of primary support.

RECOVERY

Based on the results of the ground support review and an assessment of installed support within the mine, the areas of the mine within the leached rock mass were split up into three categories based on the required rehab (Figure 3):

1. <u>Friction bolts required</u>: the drive will not be required to be open for more than six months, so fits within the safe lifespan of a friction bolt at BRC.

2. <u>Cable bolts required:</u> the drive is supported by friction bolts currently within their safe lifespan. Install cable bolts prior to the support 'expiring'.

3. <u>Friction and cable bolts required</u>: support is older than six months and so requires full rehabilitation.

In total this comprised of 3.7 km of rehabilitation. Of this total, 3.2 km of the drives required cable bolting which equates to 89 km of installed cable bolts.

To help run scenarios to enable the most efficient production re-start, the scheduler for Black Rock was released of all other duties to enable a focus on recovery. The scenarios concluded that whilst high productivity rings were available in the lower panel of the mine, the rehab required to access these rings was more time consuming than rehabbing to 5C and starting the upper panel. The mine schedule was then re-cut for top-down mining with just-in-time rehab for each production level.

To accommodate the additional rehab, the decision was made to increase mining personnel to allow the twin-boom jumbo to run both day shift and night shift and to have a nipper on each shift. These increased personnel numbers would be maintained after the rehab, to recover the lost metres from the development plan.

The rehab linear advance for the months of July and August was 982 m and 898 m respectively which allowed the mine to restart production less than two months after the rockfall event in the decline. The upper panel started production on 5C on August 10 and subsequently 4C on September 10.

With jumbo rehabilitation progressing at an elevated rate, a backlog of cable bolting started to build. As BRC had no cable bolter, it was deduced that the quickest way to source a cable bolter was to complete a re-build on a MICO cable bolter that was due for replacement. During the sourcing and re-build, several production drillers and loader operators were deployed to MICO's other operations to train on cable bolters. When the cable bolter came out of the rebuild, operators with previous cable bolter experience were able to be supplemented by those that had been in training. Cable bolting started in earnest in October 2021 and a second cable bolter was sourced from anther Glencore mine in late October. To enable production activities to be maintained, the cable bolters were scheduled to work above or on the production levels during night shift and below the production levels on day shift so as not to interact with trucking. With extended stockpiles, this enabled production rates of 70 per cent of the pre-event rates to be achieved.

Using this schedule, cable bolting was able to be completed within the six month lifespan of the split sets. The decline and level cable bolting was completed to below 5B by April 2022, ten months after the rockfall. This enabled full mine production to be resumed with pre-event production rates achieved in April 2022 and a mine record production in May 2022.

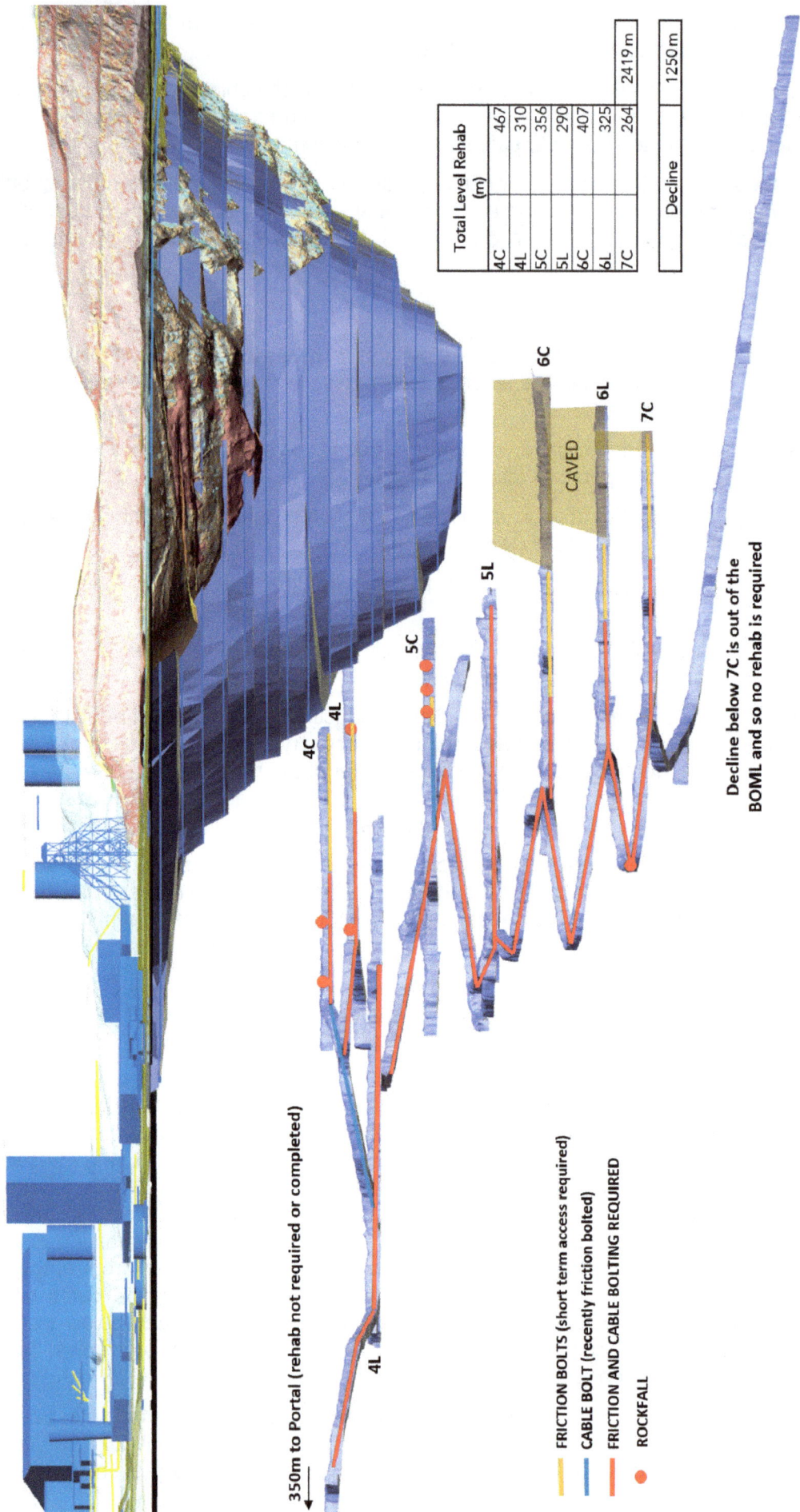

Total Level Rehab (m)

4C	467
4L	310
5C	356
5L	290
6C	407
6L	325
7C	264
	2419 m

Decline	1250 m

Decline below 7C is out of the BOML and so no rehab is required

350m to Portal (rehab not required or completed)

FRICTION BOLTS (short term access required)
CABLE BOLT (recently friction bolted)
FRICTION AND CABLE BOLTING REQUIRED
ROCKFALL

FIG 3 – BRC cross-section with required rehab (July 2021).

CONCLUSIONS AND LEARNINGS

The rockfalls at BRC were a result of extreme corrosion of friction bolts within a leached rock mass. When the resin bolts were found to be ineffective in the poor rock mass, the rate of corrosion was not understood. This meant the investigation into a new corrosion resistant support system was not assigned the priority it warranted, leading to rockfalls in June 2021.

Management made the correct decision to withdraw all persons from the mine and deem areas within the leached rock mass 'unsupported' when confidence in the effectiveness of the ground support system came into question. This ensured that no injuries have occurred due to ground support failure at BRC.

After receiving the new corrosion resistant ground support system from the geotechnical engineers it became apparent that 3.9 km of drives needed rehab, with 3.2 km of the rehab also requiring cable bolting. Management was quick to identify that current resourcing and equipment levels were not sufficient. With the assistance of PYBAR, two jumbo operators, four nippers and four cable bolt operators were quickly mobilised to site. To supplement the cable bolt operators, loader operators were sent to other operations to complete cable bolter training. Two cable bolters were acquired from other Glencore sites and were commissioned and operational within four months of the event.

After multiple scenarios were analysed, the mine schedule was able to be recut to prioritise access to the top panel and subsequent top-down production with just-in-time rehabilitation for the production levels. This enabled production to resume within two months of the rockfall event. The cable bolting schedule was fine-tuned to enable day shift production and night shift closure of the decline. This enabled a production rate of 70 per cent of pre-event levels to be achieved whist completing the rehab. With rehab having been completed to below the production levels within ten months of the event, full mine production was able to resume with record production achieved in May 2022.

The successful recovery was achieved, as the small and flexible team at BRC were given the authority to make decisions as required. With a solid grasp of the schedule, the engineering team were able to make quick and effective decisions to facilitate effective allocation of resources.

Significant learnings from this event have been:

- When entering a new orebody, it is recommended to complete groundwater testing to give an indication of corrosion rate.
- Once within a new orebody, corrosion testing should be implemented to provide early warning of any ground support issues.
- To complete this testing, suitable dedicated geotechnical engineering resourcing is required.
- Empowering your teams on the ground to make major decisions allows for a flexible and rapid response to any issues.

ACKNOWLEDGEMENTS

Werner Murach (Superintendent – MICO Drill and Blast) for his support and guidance. Sam Roberts (Mine Manager – MICO) for his review works.

REFERENCES

Hassel, R and Villaescusa, E, 2021. Black Rock Cave ground support corrosion review, MCIT report to Mount Isa Mines.

Shiels, A, 2019. Black Rock sub-level cave geotechnical review, Mount Isa Mines internal report.

Collaboration between suppliers, operations, and alternative industries

Performance of fully mechanised rock reinforcement using self-drilling anchors with pumpable resin technology

P Jere[1], R Battison[2] and M Arnold[3]

1. Principal Geotechnical Engineer, Master Builders Solutions, Perth Airport, WA 6105. Email: precious.jere@mbcc-group.com
2. Senior Geotechnical Engineer, BHP, Perth, WA 6000. Email: richard.battison@bhp.com
3. Underground Operations Manager, St Barbara, Leonora, WA 6438. Email: matthew.arnold@stbarbara.com.au

ABSTRACT

Mechanically coupled rock bolts encapsulated in cementitious or polyester cartridge resin grouts have been widely adopted for various underground rock reinforcement applications. Resin grouts are considered a superior product, however, both systems have their advantages and limitations, with the relative performance, quality installation cost and cycle time as the most notable parameters.

Both applications have optimal rock bolt combinations and preferred ground conditions.

Cementitious grouts are favourable for use with secondary deep embedment anchors. These are installed using various equipment with a series of autonomous tasks required to complete the installation cycle. The cartridge resin system can be utilised by a bolting or face boring rock drill to complete the full support installation cycle. This reduction in equipment changeouts mid-cycle generally deems it the more practical application and popular choice for primary support.

Self-drilling anchors (SDAs) and hollow bolts have been a viable alternative to traditional rock reinforcement in variable and challenging ground conditions. Using a pumpable post installation grouting system, SDAs have proven to aid the quality of installation, but due to the additional equipment and cycle steps associated with application, the cycle time is increased.

In recent years, fully mechanised bolting equipment designed to utilise a combination of self-drilling anchors and a thixotropic pumpable resin installed in a single pass has entered the market. This innovation is anticipated to deliver the advantages described above, with fewer limitations, and significantly improved development and rehabilitation rates.

St Barbara Mining and Byrnecut Australia, supported by Master Builders Solutions, were first to conduct extensive trials and implement this rock bolting approach in an Australian hard rock underground mine and record the actual performance.

INTRODUCTION

Mechanically coupled rock bolts encapsulated in cementitious and polyester cartridge resin grouts have been widely adopted for various underground rock reinforcement applications. Resin grouts are considered a superior product, however, both systems have their advantages and limitations, with the relative performance, installation cost, quality, and installation rates as the most notable parameters.

Pumpable resin for anchoring has readily become an attractive option, conceptually offering more advantages on the quality and durability of installed ground reinforcement. Aftermarket retrofitted applications systems have been available for some time, however they have not gained much popularity as they generally require additional maintenance and add complexity to the process.

Recently fully mechanised bolting equipment designed to utilise a combination of self-drilling anchors and a thixotropic pumpable resin installed in a single pass has entered the market. This innovation is anticipated to deliver the advantages described above, with fewer limitations.

Byrnecut Australia, supported by Master Builders Solutions, were first to trial the Epiroc Boltec M with pumpable resin technology. This bolting rig was commissioned in mid-2020 and utilised Master Builders Solutions MasterRoc Rock Bolt Anchor (RBA) 380.

To best comprehend the performance of this system, a benchmark/comparative approach with the traditional system was undertaken. This report will highlight the disparities of the system and subsequent performance.

To best capture any changes in development rate, the performance of the systems was compared over several months.

EQUIPMENT AND CONSUMABLES

Machinery

Prior to trialling the purpose-built bolting rig, the equipment utilised for ground support installation on-site consisted solely of twin boom development rigs. To best comprehend the disparities, the equipment features comparison in Table 1 was derived (Epiroc, 2022a, 2022b).

TABLE 1

Comparison of equipment parameters.

Function and ground support handling features	Twin boom Jumbo	Epiroc Boltec M
Source	Epiroc (2022a)	Epiroc (2022b)
Primary function	Face drilling	Rock Bolting
Auxiliary function	Scaling	Resin injection
Bolt installation	Semi-mechanised installation	Fully mechanised
Magazine capacity	NA	10
Bolt length 1	0.4–3 m	1.5–3.5 m
Bolt length 2	NA	>70% of Bolt length 1
Size of face plates, rectangular	NA	max 150 × 150 mm
Size of face plates, round	NA	max Ø 200 mm
Mesh handling	Boom	Welded screen mesh handling arm
Boring tool	Drill steel	Self-Drilling Anchor (SDA)
Optimised borehole diameter	76 mm Max	30–40 mm
Auto boring	Capable	Capable
Max working height	7.5 m	9.5 m
Grouting function	Manually feeding of resin cartridge	Pumpable resin system
Resin storage	N/A	320 L
Extension drilling	Extension speed rod	SDA Couplings

Rock bolts

The R32 and R28 self-drilling anchors were used in the trial as a substitute for the previously used static and dynamic primary ground support. In addition, coupled R32 SDAs were used to substitute secondary support cable bolts in intersections and other developments. The selection of the SDAs ensured an equivalent or superior performance, to adhere with the Ground support management plan. Table 2 is a summary of the ground support properties.

TABLE 2

Rock bolt parameters.

Parameter	Bolt 1: Solid Rebar	Bolt 2: Solid Bolt	Bolt 3: R32 SDA	Bolt 4: R28 SDA
Source	DSI Underground (2022)	Normet (2020)	DYWIDAG (2022)	DYWIDAG (2022)
Description	Static steel rebar with spiral mixer at toe	Yielding steel bar with pedals along length	Rope threaded hollow bolt	Rope threaded hollow bolt with smooth 'de-bonded' section
Profile	Threaded rebar	Smooth bar with Anchors	Threaded hollow bolt	Threaded hollow bolt with smooth section
Length	2400 mm	2400 mm	2400 mm	2400 mm
External diameters	22.2 mm	22 mm	32 mm	28 mm
Internal diameter	NA	NA	18 mm	9 mm
Mass per metre	2.47 kg	3 kg	3.4 kg	3.1 kg
De-bonded length	NA	Yes	NA	Yes
Sacrificial drill bit	NA	NA	R32/Ø51 mm	R28/Ø38 mm
Plate dimension	Dome plate 200 × 200	Dome plate 200 × 200	Dome plate 150 × 150	Dome plate 150 × 150
Yield load	185 kN	175/190 kN	280 kN	160 kN
Ultimate load	215 kN	230/250 kN	360 kN	220 kN
Elongation (%)	12%	15%	12%	18%
Maximum drilling depth	NA	NA	20 m	16 m

Grouts

The grouts being utilised on-site were polyester resin capsules and cementitious grouts, which were substituted with the MasterRoc RBA380 pumpable resin grout. Table 3 is a summary of the grout properties (Master Builders Solutions, 2021).

TABLE 3

Grout parameters.

Parameter	Units	Polyester Resin Capsule	MasterRoc RBA380
Source		Rocbolt Technologies (2020)	Master Builders Solutions (2021)
Components		2	2
Description		Sausage shaped cartridges hosted in a transparent film, with one resin containing mastic and the other a catalyst hardener. Available in various diameters, lengths and set speeds.	MasterRoc RBA 380 is a fast reacting, two-component, instantly-thickening, solvent-free polyurea silicate anchoring resin specifically designed for rock bolting.
Chemistry PTA	-	Polyester mastic	Sodium silicate, modified
Chemistry PTB	-	Catalyst harder	Modified isocyanate (pMDI)
Mixing ratio	[vol.]	Pre-defined	1:1
Mixing type	-	Bolt rotation	14 cross mixing elements
Mixing time	[sec]	8–20	N/A
Mixing restrictions	-	Do not over spin	Optimise mixing potential to site conditions
Pumpability	-	N/A	≤30 m
Thixotropic time	-	Immediate	Immediate
Penetration capacity	-	Undefined	≥0.14 mm
Setting time	[sec]	[20°C] 15–240	[23°C] 30–360
Setting time control	-	Pre-defined	Mixing potential and working temperatures
Compressive strength	[Mpa]	>60	>35
Working temperature	[°C]	10°–30°	≥5°
Shelf life	[months]	6–12	24
Storage temperature	[°C]	0°–20°	5°–35°
Storage conditions	-	Cool dry, ventilated	Cool dry, ventilated
Stability	-	Unstable above 38°	Unstable above 60°

INSTALLATION CYCLE

A series of tasks and automated functions are required to complete an install on either system. As the bolter is fully mechanised, hosting a bolt magazine and onboard resin tanks enables the operators to complete ten installs before the manual reloading of bolts in the carousel. In addition, up to 145 installs can be completed before resin tank refill is necessary. Below is a summary of the steps required to complete a full installation.

Conventional system | jumbo | multi-pass

- Generally utilising both booms: One with drill steel and the other for bolt/resin capsule loaded manually.
- Pre-drill borehole to recommended depth and diameter.
- Clean borehole. Flush with water during steel retreat.
- Manual load resin via inserter tube.
- Insert resin capsule in correct orientation, ie fast set at toe.
- Manually load bolt.
- Insert bolt into borehole, rotate to break resin film and mix components.
- Hold to recommended durations.
- Tension bolt.

Note 1: It is critical that hole depth and diameter are within specifications.

Note 2: Over or under mixing are detrimental to quality of the product.

Mechanised system | bolter | single pass

- Load magazine and drifter with ten × SDAs.
- Drill borehole at recommended percussion and rotation, flush with water to clear sacrificial drill bit.
- Drifter detaches from SDA and retreats to allow injection nozzle to advance and connect to collar.
- Inject resin through hollow bolt, till returns are observed at the borehole collar.
- Hold for recommended duration for curing.
- Tension bolt.

Note 3: Resin volumes can be pre-set or manually controlled.

TIME IN MOTION STUDY

For this study, only active time to complete the first ten installs of five separate bolting tasks without an off-sider present were recorded. This was to capture the bolters full magazine install against the conventional practice both with mesh handling and tensioning included. The data was compiled from a combination of live observation and an automated installation log record. Figure 1 shows a time comparison for bolt installation of the two machines.

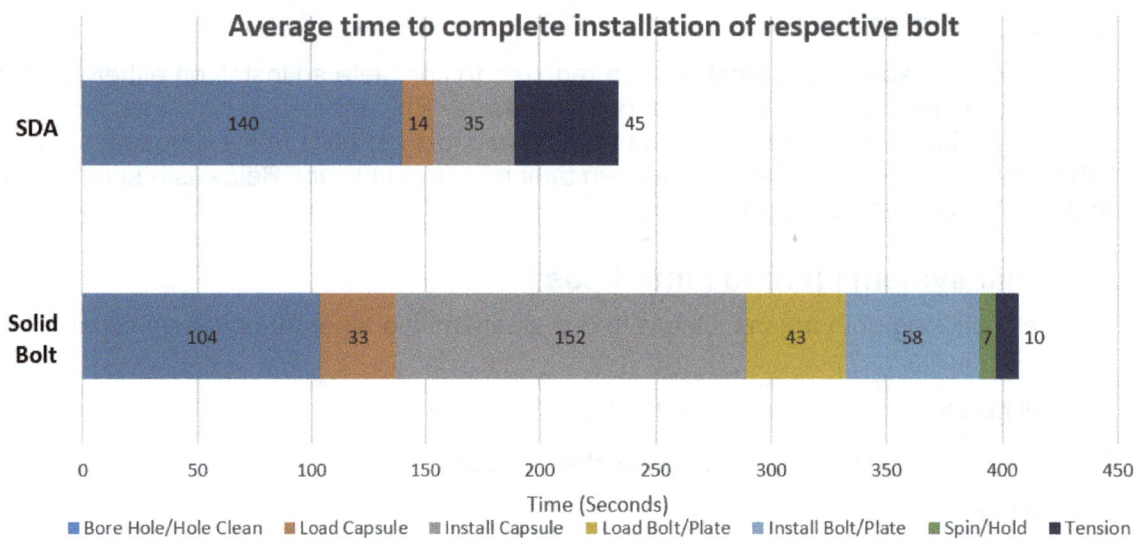

FIG 1 – Time comparison for bolt installation.

COST COMPARISON

One factor that must be considered when looking at new technology for bolting headings is the economics in terms of cost, not only of consumables, but also cycle times. If the process is cheaper but takes a lot longer to complete, then there will be a net loss from the change. The opposite is also the case, the process might cost more but if development turnover rate can be increased then the net benefit can be positive.

To assess the economics of the SDA and resin pumping system versus conventional resin cartridge bolting a direct comparison of the consumables cost has been conducted as well as a review of the time required to install each bolt.

The ground support for a standard decline cut of 4.3 m in length at Gwalia was chosen to carry out the cost comparison as this is the primary bolting scenario the Boltec was planned to be utilised for. Figure 2 shows the two different ground support standards that were prescribed for the standard decline cut and which have been used for the cost comparison.

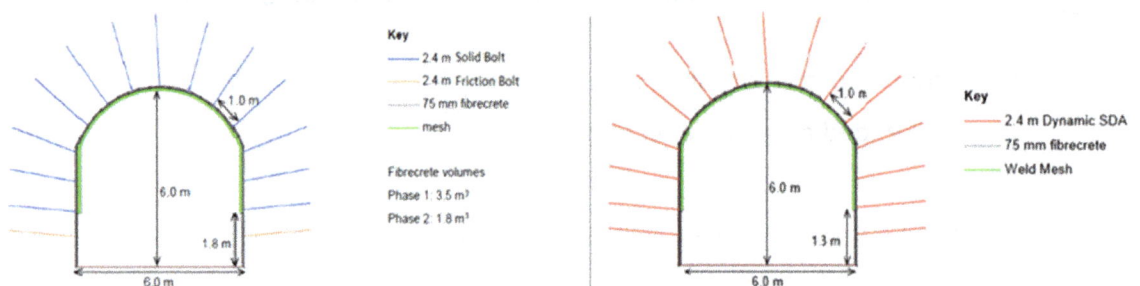

FIG 2 – Ground Support Cross-section for a 6 m × 6 m Arched Development Drive for use with:
a) conventional jumbo, b) Boltec.

The cost of each ground support element used in the following tables is based on the pricing at the time the study was carried out in 2020. In addition, the cost of ground support has been reduced by a common factor to allow for comparison, while not disclosing confidential information. Table 4 shows the cost of supporting a 6 m × 6 m × 4.3 m arched development cut with solid bolts and self-drilling anchors.

TABLE 4

Ground support consumables required and cost to support a 4.3 m long, arched, 6 m × 6 m development cut utilising a conventional jumbo and the Boltec M.

Ground support element	Amount required	Total per metre	Total per 4.3 m cut	Cost [*factored]
2.4 m Solid bolt + Plates per ring	13	13	52	$59.36
Resin Cartridge	13	13	52	$20.88
2.4 m Modified Split Sets + Plates per ring	2	2	8	$6.52
Fibrecrete (m³)	75 mm thickness	1.2	5.1	$54.58
Mesh required (m²)	12.1	12.1	52	$10.47
Total cost per 4.3 cut using conversional jumbo			**Total**	**$151.81**
Ground support element	Amount required	Total per metre	Total per 4.3 m cut	Cost [*factored]
2.4 m SDA + Plate and Drill Bit	13	13	52	$84.34
Pumped Resin (l)	28.6	28.6	114.4	$26.31
Fibrecrete (m³)	75 mm thickness	1.2	5.1	$54.58
Mesh required (m²)	12.1	12.1	52	$10.47
Total cost per 4.3 cut using Boltec M			**Total**	**$175.70**

PERFORMANCE COMPARISON

Several parameters were observed post installation to enable comparison of the two systems. Visual observations and destructive testing were completed, and results are outlined below.

Pull testing

As part of the site-specific QA/QC, a percentage of the installed bolts were subject to monthly pull testing. Below is a summary of the performance of the two systems under similar static load test conditions.

Bolt 2: Solid bolt

11/06/2020   125 tested 85 per cent pass [level 1700H]

1/03/2021   85 tested 93 per cent pass [levels 1700R, 1720R,1740R]

Bolt 3: SDA R32

16/06/2020   41 tested 100 per cent pass [level 335H]

Bolt 4: SDA R28

16/03/2021   35 (R28) tested 100 per cent pass [levels 1580, 1740R and 1760H]

Note 1: Refer to Table 2 for bolt descriptions.

Note 2: Installation of Bolt 1 has been discontinued and has been fully replaced with Bolt 3.

Full encapsulation

Full encapsulation is a critical factor for bolt performance and durability. Insufficient encapsulation is predominately observed at the collar. Visual observations and measurements was planned on random installs to enable comparison of the two systems as part of the study. To achieve these observations, the nuts, and plates of completed installs require removing and the collar profile examined and photographed. Unfortunately, this data couldn't be compiled due to travel restrictions and conflicting schedules. However, it is very evident that the degree of encapsulation using cartridge resin is very susceptible to the ground condition and integrity of the hole. Thus the pumpable resin system achieves are more consistent and superior encapsulated install.

AUXILIARY APPLICATIONS

During the extensive trial, alternative applications of the Boltec M were explored. The aim was to fully comprehend the equipment's capabilities to complete complex tasks that would otherwise be time consuming, manually intensive and/or require equipment change out.

Extension drilling

Extension drilling is a function where rated couplings are used to connect the SDA to allow for longer installs. This function was utilised to substitute secondary support cable bolts in the development. During the trial a combination of up to 4 × 2.4 m [9.6 m] bolts were installed, and the cycles, consumables and elapsed time were recorded in Table 5. The extension drilling function offered significant reduction in active cycle time with four intersections, requiring 9 × 4.8 m installs each completed in just over an hour, a fraction of current common practice.

TABLE 5

Extension drilling.

Sample size	Active development	4 INT × 9 ext bolts = 36 installs	2 INT × 12 ext bolts = 24 installs	Rehab deep anchor = 9 installs
Sample dates:	May 2020 to May 2021	Nov 2020 to Jan 2021	Nov 2020 to Jan 2021	Jan 2022 [13/01/2022]
Extension length (m)	**2.4 m**	**4.8 m**	**7.2 m**	**9.6 m**
Bolt 1	2 min 22 sec	2 min 22 sec	2 min 22 sec	2 min 30 sec
Bit change:		N/A	2 min 14 sec	N/A
Coupling:		35 sec	35 sec	35 sec
Bolt 2		3 min 05 sec	3 min 10 sec	3 min 29 sec
Bit change:			2 min 34 sec	2 min 32 sec
Coupling:	N/A	N/A	35 sec	35 sec
Bolt 3			3 min 50 sec	3 min 27 sec
Bit change:				3 min 7 sec
Coupling:			N/A	42 sec
Bolt 4				4 min 2 sec
Injection:	9 sec	15 sec	35 sec	48 sec
Cure/hold:	35 sec	35 sec	35 sec	35 sec
Average install time:	**3 mins 6 sec**	**6 mins 52 sec**	**16 mins 30 sec**	**22 mins 22 sec**

Pillar support

Coupled SDAs were successfully utilised as cable bolt substitutes to wrap fretting, preconditioned and damaged pillars with OSRO straps in production ore drives and permanent infrastructure. This utilised the similar concept described in extension drilling section above for horizontal installs.

Rehabilitation in fractured ground

One crucial advantage of the Boltec M one-pass installation is evident when installing ground support in fractured ground. Typically, installation rates are significantly reduced in these conditions when using the conventional multi-pass system. Advantages of the one-pass installation include reduced impact of hole collapse, improved encapsulation and subsequently reduces exposure of operators to compromised ground conditions. It was observed that using the SDA no significant decrease in installation rates occurred and integrity of bolt installation was maintained.

Ground consolidation injection and cavity filling

The common practice of ground consolidation involves pre-drilling holes, installing a lance and a packer, then manually injecting the resin to predefined cut-off pressures. Using the Boltec M, an SDA and a SDA collar plug, this process can be mechanised, allowing resin to be injected into a borehole at relatively higher working pressures compared to bolting. This induced the conditions to permit the resin to penetrate fractures and cavities in country rock, rather than freely flowing out of the collar. Similar predefined cut-off pressures are observed during this process.

ADVANTAGES AND LIMITATIONS

The following advantages and limitations were observed during the trial.

Full encapsulation

Due to the nature of the installation SDAs and pumpable resin, full encapsulation of the bolt install can be achieved on a significantly higher proportion of installs. This is due to difficulties placing the cartridges and variable borehole volumes not being accounted for when using solid bolts and resin capsules. Substandard resin installs can generally be visually observed in development drives.

Improved corrosion protection

Improved full encapsulation rates enable superior corrosion protection through reducing the exposed surface area of the bolts.

Hole data logging

The Boltec M has the capability to log installation data. The parameters captured include heading ID, date and time, operator, penetration rates, bolt length, void interception, and resin volume. This data was downloaded, reviewed, and reported as an added quality assurance measure.

Reduced resin waste

Compared to the conventional jumbo ground support installation using resin capsules, the pumpable resin system on the Boltec M has significantly reduced potential for waste generation. The resin waste is limited to overflow of hole when full encapsulation is achieved, which can be managed through presetting volumes or through manual control by the operator. The resin capsules often create higher volumes of waste, as they are difficult to manipulate during borehole placement and are often damaged and discarded. In addition, the shelf life and storage requirements of the pumpable resin are more favourable for the underground mining environment, as previously noted in Table 3.

Scaling capabilities

Unlike the conventional jumbo, the level of scaling that can be achieved with the Boltec M is very limited due to the equipment set-up. While bolting, light scaling can be completed as required. Ideally, the rig is designed to bolt headings that have been pre-scaled, thus requiring equipment change-out.

Rock bolt type

The Boltec M is primarily designed to use self-drilling anchors; however, modifications can be made to allow for installation of different ground support such as split sets and solid bolts using a multi-pass system. This change-out requires significant downtime and would not be practical where alternative functions are required frequently.

Rock bolt tensioning

It is evident that using the independent tensioning arm (hiab) of the Boltec M can be detrimental to cycle time. It was observed that operators had difficulties aligning to couple the nut post completion of injection. During the trial it was concluded that the amount of pressure exerted on the nut and plate during injection against the excavation surface which is maintained until resin cures achieved an acceptable level of tension. Tests were conducted that confirmed that upwards of five tons of tension had been induced thus the tensioning arm would only be required if site tension requirements exceeded this limit.

CONCLUSIONS

Pumpable resin technology is beginning to be utilised more widely in the mining industry as a means of speeding up ground support installation cycles and to strive for better quality bolt installations. The study comparing the two different ground support installation practices carried out at the St Barbara Gwalia Mine has highlighted the pros and cons of this new technology.

The Boltec M has been shown to decrease ground support bolt installation times by around 42 per cent. This reduction in installation times has far reaching economic benefits for mines which are constrained by development as it can lead to an increase in development heading turnover. In addition, when the Boltec M is used to install extension bolts instead of cable bolts then the time saving benefits of this rig can be increased. Most mines wait 12 hours for cable bolt grout to set before taking turnouts in an intersection. With SDA's drilled with extension drilling capabilities the turnover rate can be brought down to under three hours as any grout setting delays are eliminated. If managed properly the potential time savings can well outweigh any additional cost for ground support consumables.

The Boltec M has been designed to provide improved operator safety and better ergonomics by reducing the amount of manual handling and mounting/dismounting of rigs required by the inclusion of a bolt carousel. This will also have benefits in managing operator fatigue in high temperature environments as longer periods can be spend in ambient conditions.

When utilising resin cartridges there can be considerable wastage of product due to installation issues which can be linked to operator skill and/or ground conditions. This wastage is reduced using the Boltec M pumpable resin system as there is no cartridge to tear open and resin is delivered to the back of the hole where it is required.

The Boltec M has been designed specifically for installing ground support. This is a drawback compared to the conventional twin boom jumbo as it limits the capabilities of the equipment. The conventional twin boom jumbo can carry out all required tasks for developing whereas the Boltec M cannot be readily used for scaling and boring headings. This means additional equipment is required to support the Boltec M which is not necessarily required for a conventional jumbo.

QA/QC is also an important part of managing ground support installation practices. The updated on-board data recording systems on the Boltec M allow for review and analysis of bolting operations after they have been carried out. This type of analysis can assist in operator training and ongoing improvement in how the equipment is utilised.

Variable ground conditions can slow down ground support installation with conventional jumbos sometimes requiring a change of bolt type to systems that do not use resin. Generally, friction bolts are the favourable bolts in such conditions, which have a lower performance and durability. The Boltec M removes this issue as it can continue to install SDA bolts irrespective of ground conditions with the amount of resin being pumped into the ground tailorable to the conditions encountered.

From the comparison study the Boltec M would be of most benefit to an operation which is constrained by its development and has the capacity for a dedicated bolting rig, as there are the capabilities to improve development cut turnover. The benefits of this outweigh the consumable costs. The findings from this paper can be used by an operation to consider whether such a system would be of benefit for them. The applicability of the Boltec M for operations should be reviewed on a case by case basis with the current requirements of the operation weighed up against the advantages and limitations of this equipment.

ACKNOWLEDGEMENTS

The authors would like to acknowledge the support of Byrnecut Australia, St Barbara Mining and Master Builders Solutions.

REFERENCES

DSI Underground, 2022. Rock Bolts – Solid Bar. Available from: https://www.dsiunderground.com.au/products/mining/rock-bolts/solid-bar/posimix-debonded-resin-bolt-system [Accessed: 1 April 2022].

DYWIDAG, 2022. DYWIDAG Geotechnical Product Range. Available from: https://assets.ctfassets.net/wz1xpzqb46pe/6s259beLSy5Afjc1VPkE8H/3e67082861cb091d19a8a985943b77e3/dywidag-geotechnical-product-range-en.pdf [Accessed: 12 April 2022].

Epiroc, 2022a. Boomer M. Available from: https://www.epiroc.com/en-ba/products/drill-rigs/face-drill-rigs/boomer-m [Accessed: 12 April 2022].

Epiroc, 2022b. Boltec M. Available from: https://www.epiroc.com/en-ba/products/rock-reinforcement/rock-bolting-rigs/boltec-m [Accessed: 12 April 2022].

Master Builders Solutions, 2021. MasterRoc RBA 380. Available from: https://assets.master-builders-solutions.com/en-au/masterroc-rba380-tds.pdf [Accessed: 14 April 2022].

Normet, 2020. D-bolt dynamic rock bolt, Technical data sheet. Available from: https://www.normet.com/wp-content/uploads/2016/09/d-bolt-tds-global-20200511.pdf [Accessed: 28 April 2022].

Rocbolt Technologies, 2020. Fasloc Resin Cartridges. Available from: http://www.rocbolt.com/_files/ugd/beb733_06e7cccf4e694cb4b71c561f630575fa.pdf [Accessed: 8 April 2022].

Reverse stoping using wireless initiation at CSA Mine

B Small[1], A Bermingham[2], R Goodwin[3], S Watson[4] and M Ireland[5]

1. Technical Services Superintendent, Glencore CSA, Cobar NSW 2835.
   Email: ben.small@glencore.com.au
2. Drill and Blast Engineer, Glencore CSA, Cobar NSW 2835.
   Email: alan.bermingham@glencore.com.au
3. Technical Services Engineer, Orica Ltd, Kurri Kurri NSW 2327. Email: rory.goodwin@orica.com
4. Senior Technical Services Advisor, Orica Ltd, Kurri Kurri NSW 2327.
   Email: sean.watson@orica.com
5. Mining Manager, Glencore CSA, Cobar NSW 2835. Email: matthew.ireland@glencore.com.au

ABSTRACT

CSA Mine, named after the nationalities of the mines first owners being Cornish, Scottish, Australian, is one of Australia's deepest and highest-grade copper mines. A combination of the modified Avoca method and shrinking was a standard practice where production rings were fired against and bogged to waste fill. This method resulted in dilution and involved rehandling of waste material to refill the bogged stopes in the Western region and the QTS Central.

The introduction of pre-charging blastholes using a wireless initiating system (WebGen™) has enabled CSA to design and fire stopes in reverse. This involves the use of a pre-charged Temporary Rib Pillar (TRP) against the waste fill material in a previously mined stope. The temporary pillar reduces dilution and improves conventional bogging as material is blasted towards the drawpoint rather than towards the waste fill material. Additional drill, prep and blast re-work is reduced or completely removed as rings can be charged prior to, in conjunction with, or after, charging of the slot. This is a safer practice as the charge crew spend no time working adjacent to a brow or open void.

A TRP trial was conducted in the western orebody at CSA. The risk of exposure to a brow or open void while charging underground, and the bogger rehandling of waste material was successfully eliminated.

Once reverse stoping was established in modified avoca stopes at CSA, two TRPs were implemented in challenging ground conditions in the QTS Central region. The TRPs were utilised in a retreating four-way intersection to extract a central diminishing pillar to close the level out. The ore recovered in this stope provided 5 per cent of CSA's copper for the year. Without pre-charging with a wireless initiating system, two 5 m pillars totalling 6936 t of ore at 7 per cent either side of the stope would not have been recovered.

CSA has implemented the practice of reverse stoping in modified avoca stopes using TRP for ten stopes since the successful trial. The development of additional blasting techniques enabled by pre-charging with wireless initiating systems is currently underway in the QTS North orebody at CSA Mine.

INTRODUCTION

This paper details the results of TRP trials completed through collaboration between Glencore and Orica in the QTS Central and Western orebodies at CSA Mine. QTS is named after Queensland, Tasmania, South Australia.

CSA Mine

CSA Mine is one of Australia's highest grade copper mining operations, and part of Glencore, one of the world's largest diversified natural resource companies (Glencore, 2022). At 1.9 km deep, it is one of Australia's deepest underground mines. The CSA deposit is located 9 km north of the township of Cobar in Central Western NSW and has been mined in two major periods since it was discovered over 125 years ago (McDermott, Smith and Jeffrey, 1996).

Figure 1 shows the QTS Central and the Western orebodies which consist of smaller lenses compared to the QTS North. The QTS Central and Western provide 20 per cent of the mine's yearly ore tonnes, while the QTS North provides the remaining 80 per cent.

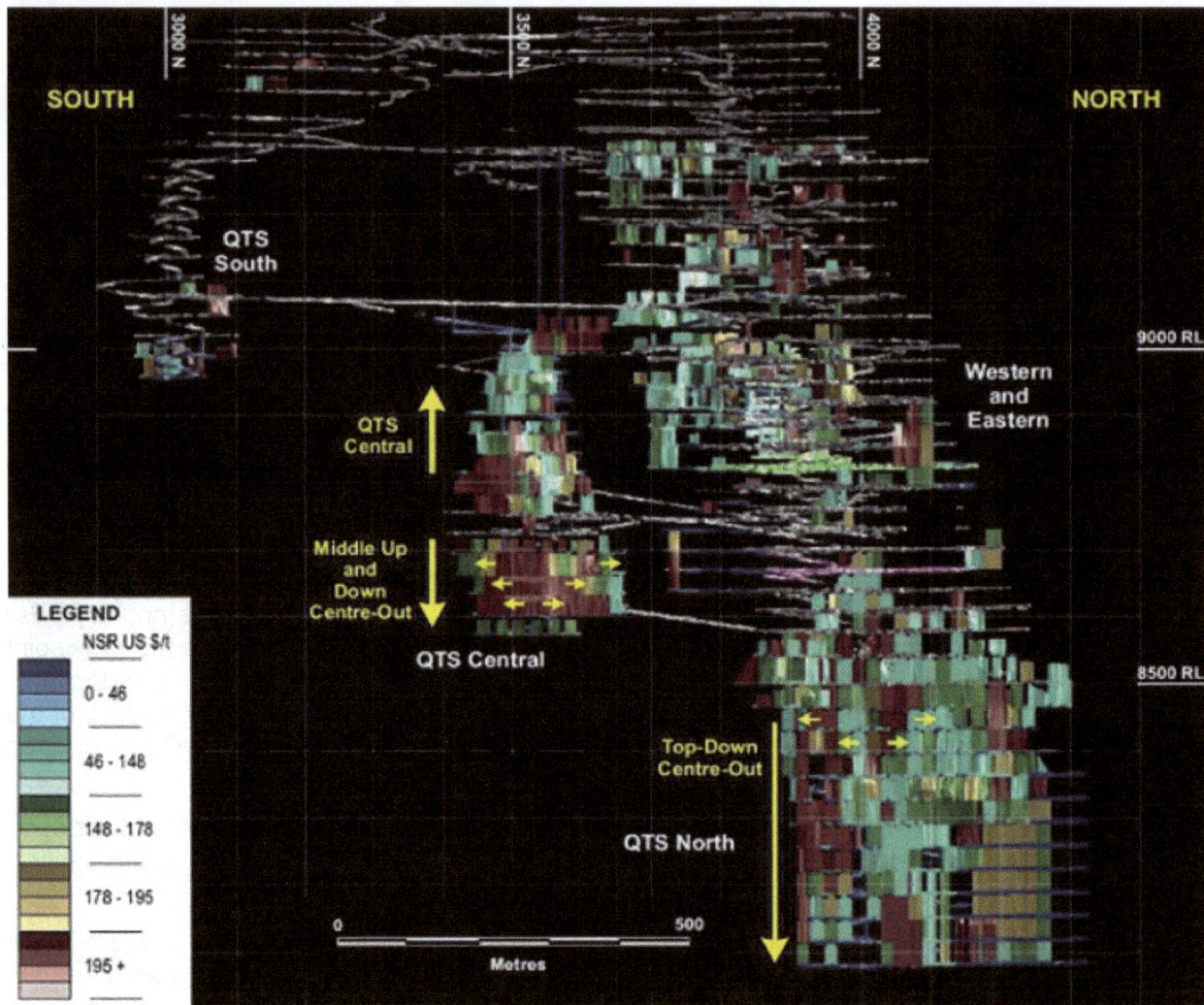

FIG 1 – CSA Mine overview of QTS Central, QTS North and Western regions (Behre Dolbear Australia (BDA) Pty Ltd, 2022).

QTS Central and Western Mining method

The level interval in both the QTS Central and Western systems is 30 m and orebody widths are between 6–10 m. The modified avoca method typically commences mining in the central area of the ore zone and progresses towards both ends of the ore drives. A slot is established to create the initial void for subsequent stope firing and rings are then fired in slices to a stope length as allowed by the stability assessment and then bogged clean. Figure 2 shows the empty stope is then filled with waste rock from the upper sublevel and then mucked out to a natural angle of repose and subsequent rings are blasted to a free face for the first blast in the next stope (BDA, 2022).

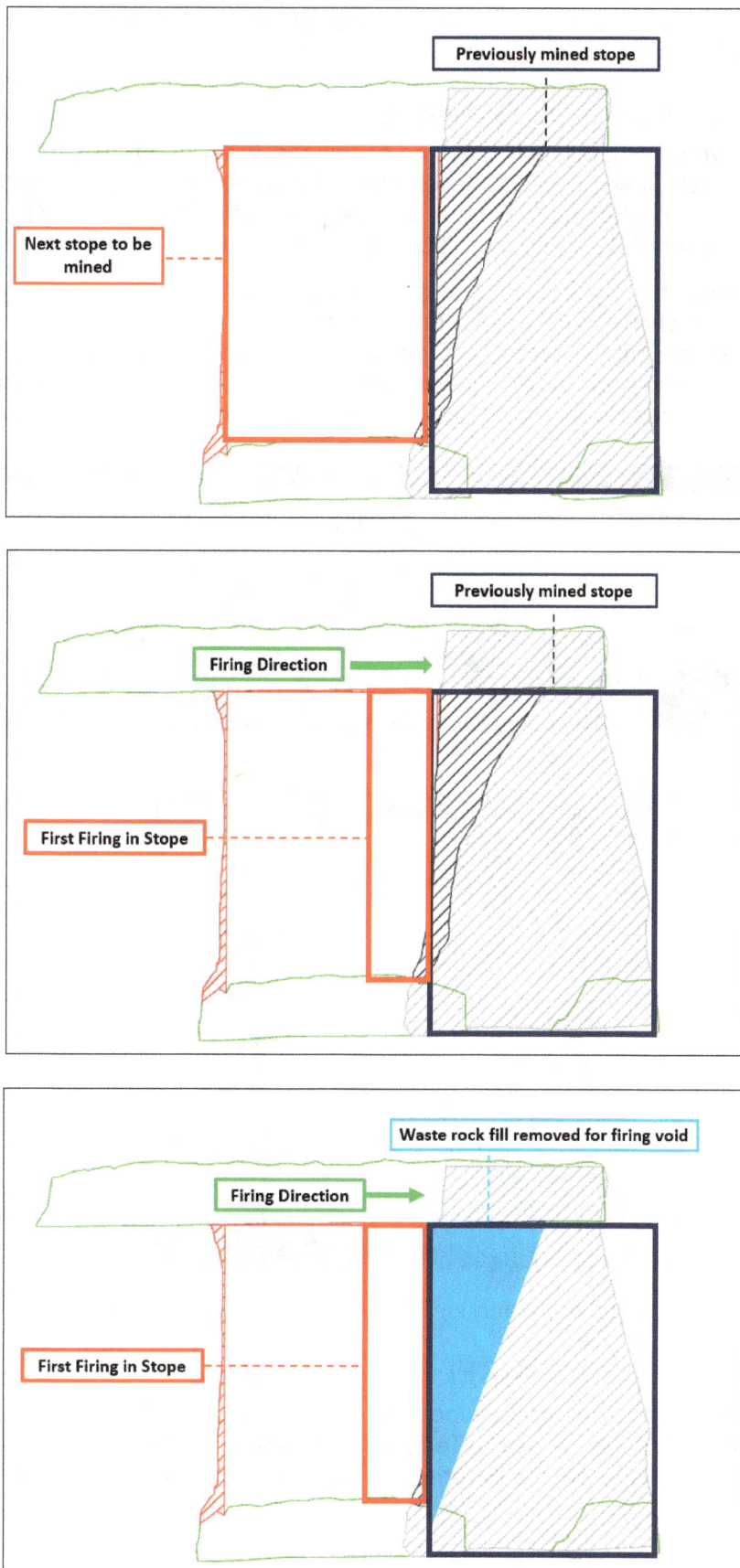

FIG 2 – Conventional modified avoca blast sequence.

This method is used in narrow orebodies such as QTS Central and relies on backfilling with waste rock to provide support as the ore is progressively blasted and removed from the stope (BDA, 2022). While an effective mining method at CSA, benefiting from always having a degree of wall support from the waste rock (longhole open stopes must be completely extracted before paste filling can commence), the avoca stopes can suffer from higher ore losses, increased waste dilution due to firing into waste rock, and backfilled waste rehandling (BDA, 2022).

Central diminishing pillar to close level out

QTS Central is known for its challenging ground conditions, with overbreak experienced regularly in the Eastern and Western walls. The 8721 level in the QTS Central has been mined from the northern and southern drives to an intersection (Figure 3). The intersection stope 8721_QC1_352, contained approximately 5 per cent of CSA's copper to be produced in 2022.

With waste fill stopes on either side and poor ground conditions surrounding the stope, the conventional approach shown in Figure 3 required two 5 m pillars. These pillars are required to limit waste dilution and enable safe access to charge and fire 'S2' after enough void was created by firing 'S1'. 'S1' and 'S2' are large shots which increases risk of poor performance but are required to extract maximum ore before access is restricted due to the open void created.

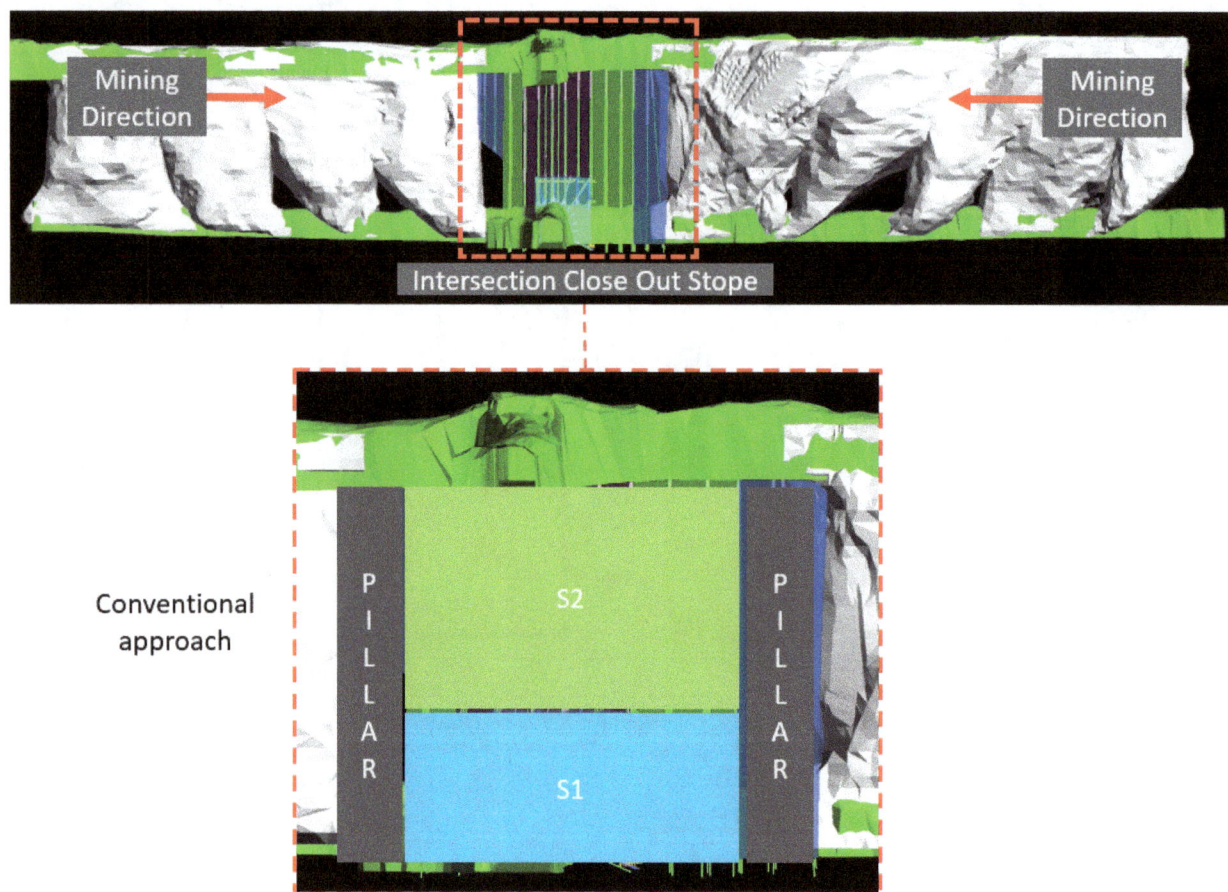

FIG 3 – Conventional approach for 8721_QC1_352 intersection close-out stope.

REVERSE STOPING USING WIRELESS INITIATION

An opportunity to improve the modified avoca mining method of the QTS Central and the Western orebody was identified through the use of Temporary Rip Pillars (TRP). TRP are pre-charged with a wireless initiating system and enable an operation to fire stopes safely in reverse without leaving behind valuable pillars of ore.

Wireless initiation with WebGen™

The WebGen™ system includes wireless in-hole primers which are initiated by a firing command that communicates through rock, water and air. Unlike traditional 'wired' systems where a firing command travels from the blast box through harness wire and into the detonator, the WebGen™ system communicates with the in-hole primer via ultra-low frequency signals called magnetic induction (see example in Figure 4). The blasting sequence is 'stored' in the Primer during encoding which is performed when charging the blast (Orica, 2022).

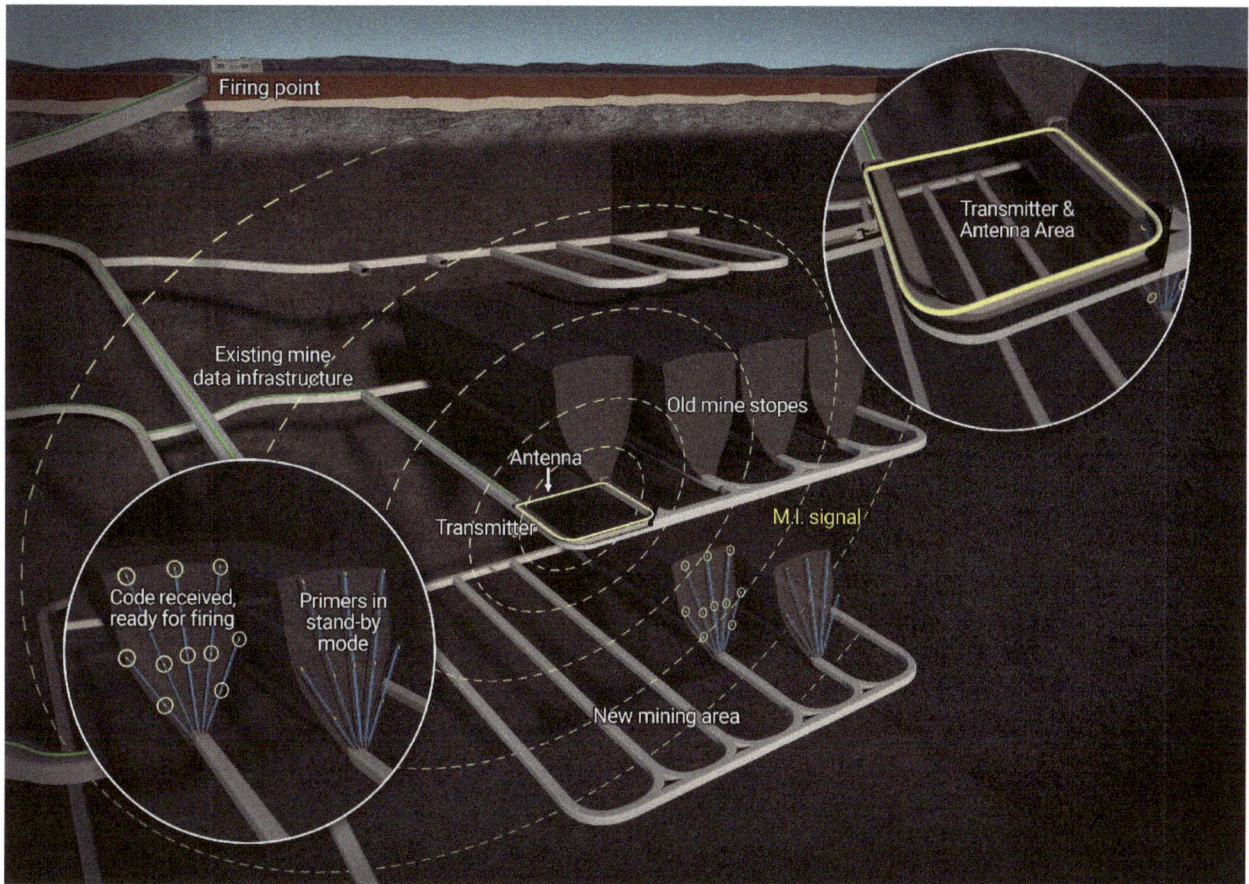

FIG 4 – Communicating with detonators wirelessly through magnetic induction (Orica, 2022).

Pre-charged Temporary Rib Pillars (TRP)

Without wires to initiate detonators, re-entry is not required by the blast crew into the stope once a void is created and subsequent blasts within a stope can be pre-charged. Pre-charging with the WebGen™ wireless initiating system is a technique introduced by the drill and blast team at CSA which has enabled the reversal of firing direction from conventional stoping. A key component of the reverse stoping technique is the pre-charged TRP. The TRP holds back the waste rock in the adjacent backfilled stope, minimising dilution until most of the ore has been bogged.

Figure 5 shows the blasting sequence of a reverse stope including pre-charged rings and a TRP shown as 'S5'. This included a slot fired and bogged in two lifts at the front of the stope (adjacent to the drawpoint) followed by three pre-charged production shots at the back of the stope. Figure 5 considers the same stope shape shown with conventional sequencing in Figure 2 for comparison of the two methods.

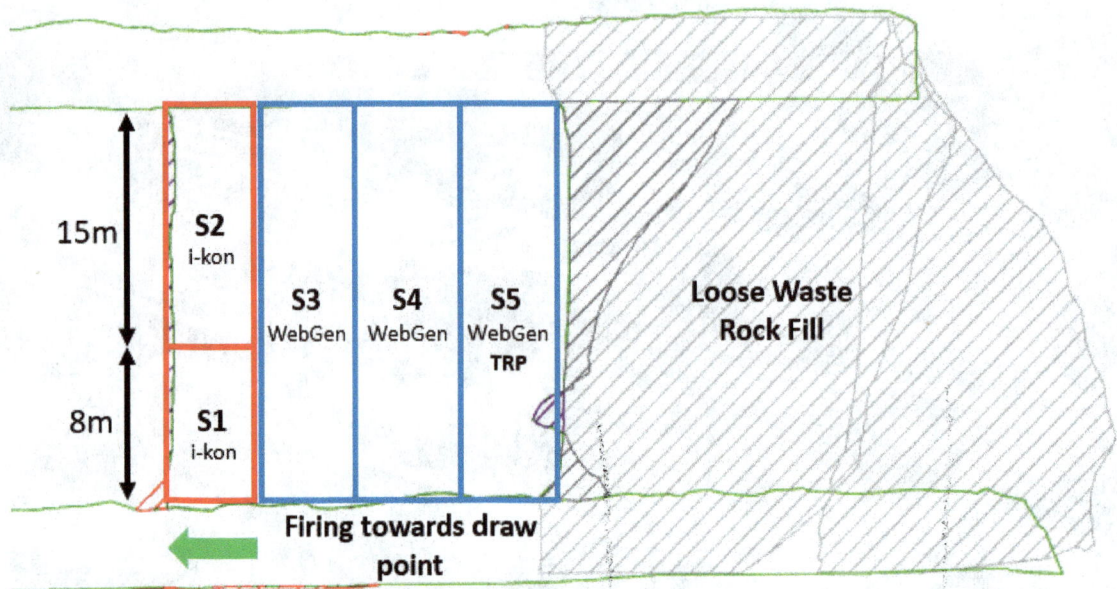

FIG 5 – Modified avoca stope fired in reverse using rings pre-charged with WebGen™.

Pre-charging TRPs to fire stopes in reverse had the potential to:

- Eliminate personnel exposure to open voids or damaged brows by pre-charging S3, S4, and S5 with WebGen™ while charging and firing S2 with i-kon™.

- Improve efficiency of drilling and blasting by reducing visits required to blast each stope.

- Eliminate rehandling and shrinking 30 per cent volume of the previous stopes loose waste rock fill.

- Increase availability of operational equipment (boggers and trucks) by eliminating shrinking.

- Throw blasted material towards the drawpoint, increasing manual bogging which is more productive than remote bogging.

- Reduce the amount of waste rock fill dilution.

- Enable the safe extraction of ore pillars.

In order to implement TRPs additional drilling would be required as well as an increase to blasting consumables cost. Firing stopes in reverse also reduces the opportunity to recover any underbreak if S1 and S2 don't perform.

TEMPORARY RIB PILLAR TRIALS

The TRP trial in the Western region, was conducted between 25 November and 3 December 2021. The pre-charged rings included 52 WebGen™ units and was charged along with the second i-kon™ initiated lift shown as 'S2' in Figure 5. The WebGen™ rings were encoded as three separate firings with two full production rings for each as shown by the collar plan in Figure 6. The timing sequence was designed with the option of firing three discrete shots or a combination of two or three shots as a mass blast. Firing each shot independently allowed ore to be recovered without waste rock fill material diluting the muck pile. Firing two shots or all three shots together was an option in case of any geotechnical failure during bogging.

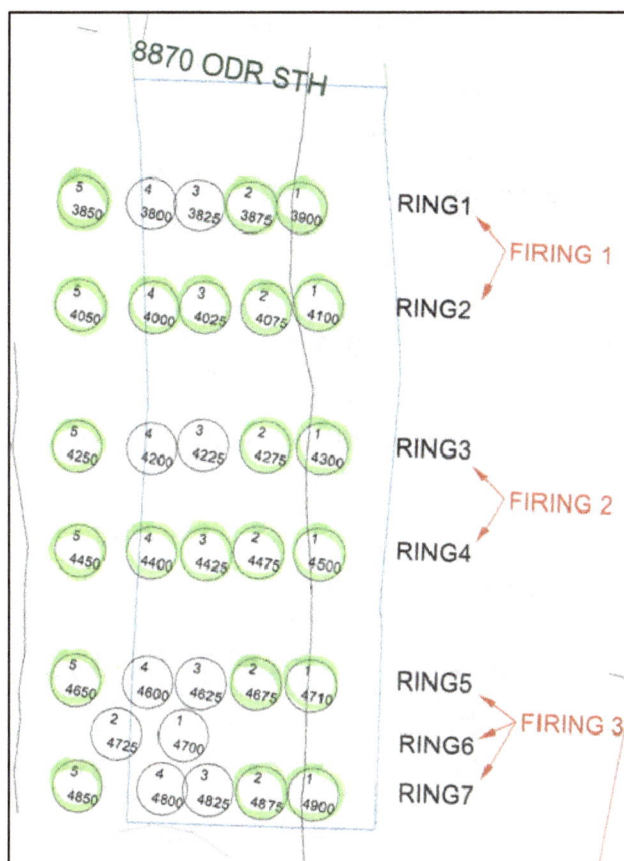

FIG 6 – Collar plan of the WebGen™ shots.

Blast crew exposure to void or brow

Table 1 shows the number of blasting activities required of the trial reverse stope 8840-W4-384 and a nearby conventionally fired stope 8840-W4-382. The two stopes were comparable in size and location.

TABLE 1

Personnel exposure to brow or open void comparison.

	8840-W4-382 (i-kon™)	8840-W4-384 (WebGen™)
Blasts	2	5
Blast preparation activities	2	2
Charging activities	2	2
Total blasting activities	6	2
Total activities near void or brow	4	0

Pre-charging with a wireless initiating system required four fewer visits to the stope and eliminated all six exposure events of blast crews to brows or open voids making charging significantly safer.

Waste rehandle from shrinking

One of the major disadvantages of the modified avoca method is the double handling of waste material. The 8840-W4-382 conventional stope required re-handling of 3214.8 t with 147 loader buckets and four trucks. The reverse stope method did not require shrinking so these resources were able to be utilised elsewhere at CSA. This provides significant productivity benefits, yet to be fully quantified, on rehandling and moving operational bottlenecks.

When stope widths increase to more than 6 m, adjacent waste rock fill rills and confines the toe. Blasting this has often resulted in toe material being left behind. To unconfine the toe and blast the material out successfully, full removal of the previous waste rock filled stope would be required. This would increase the unsupported span of the future stope further impacting wall stability and dilution. Since the introduction of reverse stoping in the wider modified avoca stopes the toe material is no longer left behind improving the recovery.

Manual bogging

Reverse firing allows each pre-charged blast to be thrown towards the drawpoint rather than into waste rock fill. This improves productivity of boggers as more ore can be loaded safely without Tele-remotes which is significantly slower. The trial stope results were compared to the average results of the three previous stopes fired conventionally on the same level. Table 2 shows a decrease of 5.82 per cent in tele-remote bogging portion and in equipment exposure.

TABLE 2
Comparison of ratio of bogging techniques.

| | Average of 8840-W4–378, -380 and -381 (i-kon™) | | 8840-W4-384 (WebGen™) | |
	Buckets	Ratio	Buckets	Ratio
Manned	263	23.76%	223	29.58%
Tele-remote	844	76.24%	531	70.42%

Waste fill dilution

Previous muck pile shapes in stopes 380 and 381 did not prevent waste rock fill diluting the muck pile. Reverse stoping with a TRP allowed the waste to be held back and minimise waste rock fill in the muck pile. The average waste rock fill dilution of the 380 and 381 stopes was 13.1 per cent. Table 3 shows the trial stope had 8.8 per cent less waste rock fill dilution than the average of comparable stopes on the same level.

TABLE 3
Comparison of waste rock fill dilution.

	8840-W4-380 (i-kon™)	8840-W4-381 (i-kon™)	8840-W4-384 (WebGen™)
Waste Rock Fill Dilution (% of total volume)	14.9%	11.3%	4.3%

Stope performance

Eastern wall conditions in the western region are problematic for stope performance. Table 4 shows 8840-W4-384 resulted in significantly more overbreak than comparable conventional stopes. This was due to a failure in the eastern wall caused by an unidentified sheer zone running parallel. This reduced the overall grade of the stope from a designed 2.9 per cent Cu grade to 2.29 per cent Cu grade and it is understood that this would have occurred regardless of blasting technique.

TABLE 4
Comparison of design totals and underbreak.

	8840-W4-380 (i-kon™)	8840-W4-381 (i-kon™)	8840-W4-384 (WebGen™)
Recovery of Designed (% of total volume)	88.2	87.3	82.1
Rock Overbreak (% designed volume)	16.6	21.4	47.8

Increased overbreak resulted in targeted bogging volumes being achieved earlier. This resulted in a portion of the temporary pillar not being recovered. To improve recovery, a modification to the temporary pillar design is recommended. Decking the last two rings and sequencing the bottom decks to fire completely into the muck pile before the top decks. This could provide better throw of the bottom decks enabling the ore to be more accessible in the muck pile and decrease the risk of leaving ore behind.

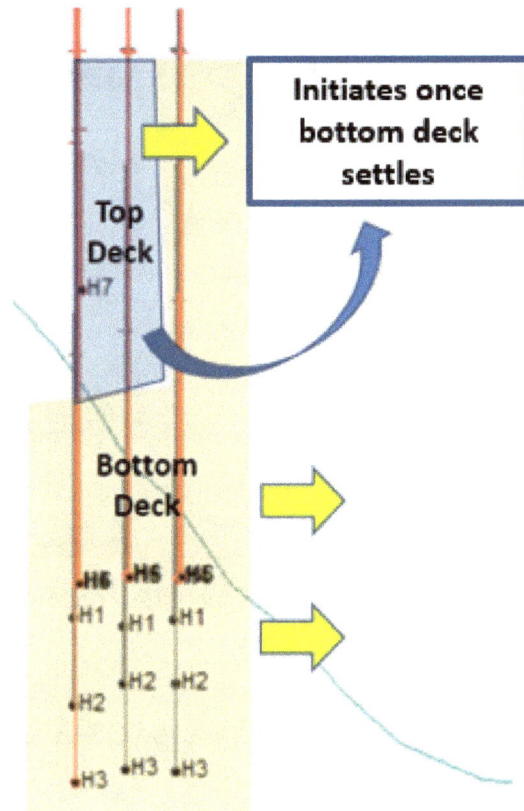

FIG 7 – Modified charging and timing techniques for TRP.

Drill and blast cost

Conventional modified avoca stopes at CSA use shrinking to create sufficient void to fire and only requires a slot for the first stope on the drive. Reverse stoping requires a slot adjacent to the drawpoint for each stope to provide adequate void for the pre-charged rings resulting in an increase in drilling. Table 5 shows the 8840-W4-384 trial stope compared with the average of three comparable stopes mined conventionally. The trial stope required 23 per cent more drill metres and the explosives consumables cost $1.53/t of ore more than the conventional average.

TABLE 5
Comparison of production drill metres required.

	Average of 8840-W4-380, -381 and -382 (i-kon™)	8840-W4-384 (WebGen™)
Production Drilling (meters)	1148.6	1396.5
Cleanout Drilling (meters)	77.6	27.1
Primers (i-kon™)	100	66
Primers (WebGen™)	0	52
Subtek™ Emulsion (kg)	7190	7789
Explosives Consumables Cost	$1.63/t	$3.17/t

Central diminishing pillar recovery

The 8721_QC1_352 intersection close out stope had significant access limitations and poor ground conditions. Pre-charging with two TRPs (Figure 8) enabled smaller shots to be fired limiting geotechnical failure risk and maximising recovery of the pillars. The stope was divided into two i-kon™ and four pre-charged WebGen™ blasts fired between 15 May and 25 May 2022.

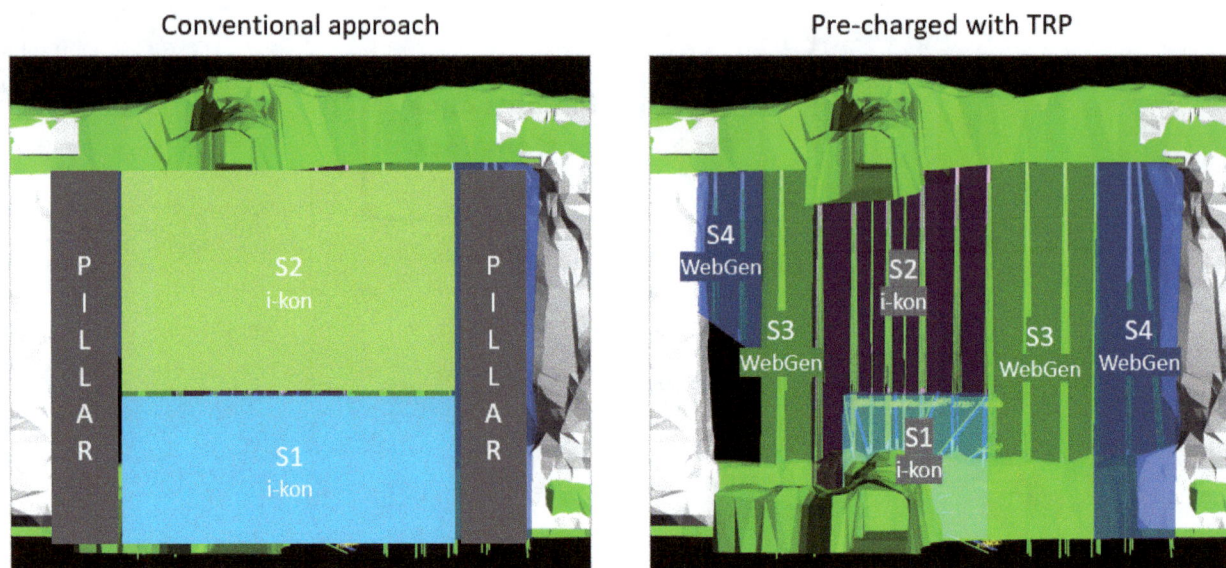

FIG 8 – Section view of blasts in the 8721_QC1_352 stope.

Stope 8721_QC1_352 carried an increased risk of ground failure once a void was created due to a combination of poor ground conditions around a four-way intersection. Eliminating exposure for charge crews after a void was created significantly improved CSA's ability to safely mine this stope.

After all firings, there were no major ground failures or mass ore loss. Table 6 compares actual reconciliation data from firing with TRPs compared with ideal performance of the conventional approach as shown in Figure 8.

TABLE 6

Comparison of 8721_QC1_352 stope reconciliation with theoretical conventional approach.

	Conventional (theoretical)	With TRP (actual)
Volume	7591	9901
Tonnes	23 381	30 553
Cu Grade (%)	8.07	6.99
Cu (t)	1887	2136
Cu ($AU)	23 530 047	26 632 730

Overbreak reduced the grade of the 8721_QC1_352 stope by 1.08 per cent. Despite the decrease in grade, the TRPs enabled the recovery of an additional 249 t of copper with a market value of A$3 102 683.

CONCLUSION

Conventional blasting techniques in modified avoca stopes required removal of 30 per cent of tight firing volume from the adjacent waste rock filled stope to provide adequate void. This process required re-handling 3214.8 t with 147 loader buckets and four trucks in a conventional stope that was used as a baseline. Pre-charging with the WebGen™ wireless initiating system has enabled

CSA mine to use techniques that reduce exposure of operational personnel to risk and improve scheduling and productivity.

A trial modified avoca stope was fired in reverse using pre-charged TRPs. Reverse stoping eliminated the need for operations to rehandle waste, enabling better utilisation of mining equipment. Due to the reversal of the conventional direction of firing, manual bogging was possible for 5.82 per cent more of the total bogged tonnes of the trial stope. This also reduced waste rock fill dilution by 8.8 per cent. Pre-charging eliminated charge crew exposure to open stopes or damaged brows. By reducing visits required to blast each stope, efficiency of drilling and blasting activities improved.

The trial stope required an additional 23 per cent drilling and a blast consumable cost increase of $1.53/t of ore. A failure in the eastern wall of the trial stope caused by an unidentified sheer zone running parallel did not enable an improvement in reconciliation results.

After the trial, pre-charged TRPs were implemented in a central diminishing pillar to close out a four-way intersection on a modified avoca level. The stope was in the QTS Central, one of CSA's most geologically difficult areas. Overbreak reduced the grade of the 8721_QC1_352 stope by 1.08 per cent. Despite the decrease in grade, the TRPs enabled the recovery of an additional 249 t of copper with a market value of A$3 102 683.

Modifications to the TRP design have been recommended including decking and sequencing the last two rings to optimise the muck pile shape and limit dilution. To date, ten stopes have been blasted in reverse using TRPs at CSA. Another project to reduce 'hammer head' development metres has commenced through pre-charging with a wireless initiating system in QTS North.

ACKNOWLEDGEMENTS

Permission to publish this paper and ongoing support by the team at CSA Mine, Glencore plc and Orica Ltd is respectfully acknowledged.

REFERENCES

Behre Dolbear Australia (BDA) Pty Ltd, 2022. SEC S-K Independent Technical Report Summary – CSA Copper Mine, Australia [online]. Available from: <https://www.sec.gov/Archives/edgar/data/1853021/000110465922067125/tm2217160d1_ex96–1.htm>

Glencore, 2022. CSA Mine – Who we are. Available from: <https://www.glencore.com.au/operations-and-projects/csa-mine/who-we-are> [Accessed 20 July 2022].

McDermott, J, Smith, C and Jeffrey, S, 1996. Geology of the CSA Deposit (The Australasian Institute of Mining and Metallurgy: Melbourne).

Orica, 2022. WebGen™ Worlds First Truly Wireless Initiating System [online]. Available from: <https://www.orica.com/wireless> [Accessed 20 July 2022].

Feasibility studies and
mine design

E26 Lift 1 North block cave development and construction

R Cunningham[1], S Melloni[2] and M Greenaway[3]

1. Manager of Mining, CMOC Northparkes Mines, Parkes NSW 2870.
   Email: rob.cunningham@au.cmoc.com
2. Development Superintendent, CMOC Northparkes Mines, Parkes NSW 2870.
   Email: sergio.melloni@au.cmoc.com
3. Construction Manager, CMOC Northparkes Mines, Parkes NSW 2870.
   Email: matt.greenaway@au.cmoc.com

ABSTRACT

The E26 Lift 1 North Project took 37 months from the approval on 1st January 2019 to the commencement of production on 1st March 2022. This was five months ahead of the feasibility plan. When compared to the previous E48 Block Cave construction 12 years prior, which was of a similar size and design with ten extraction drives, crusher and conveyors, the E26 Lift 1 North project has been delivered at 29 per cent less capital cost.

Key metrics achieved by the project include 12.7 km of development, 240 km of production drilling for drawbells and undercut blasting, 107 drawbells fired ready for production as well as only three Lost Time Injuries compared to 16 during E48 construction.

The Lift 1 North Project Team has established a significant point of difference with the delivery being completed by the 'Owners Team Model' not Lump Sum or Turn-key. E26 Lift 1 North is now supplying the Northparkes processing plant for the next eight years.

This paper discusses major milestones achieved throughout the process, modifications to processes from previous block cave construction at Northparkes as well as novel approaches to work to achieve the successful delivery of the project.

INTRODUCTION

As part of achieving Northparkes Mines' vision of 'a century of mining', a new block cave located to the North-East of the existing E26 Lift 1 was envisaged. This project, designated E26 Lift 1 North (E26L1N) was given initial approval for construction on 1st January 2019. From this initial approval it was 37 months to the commencement of production on 1st March 2022 – five months ahead of plan determined during the feasibility study. Throughout this process Northparkes 'challenged the impossible' and delivered an outstanding result.

Northparkes has experience in delivering block cave projects, with E26 Lift 1, E26 Lift 2 and E48 Block Caves also being developed since the 1990s on-site. The E48 Block Cave, which was of a similar size and design with ten extraction drives, underground crusher and conveyors was delivered 12 years prior. The E26 Lift 1 North project has been delivered at 29 per cent less capital cost compared to the E47 Block Cave.

Some of the key metrics for the delivery of the E26 lift 1 North mine include:

- 12.7 km of development.
- 240 km of production drilling for drawbells and undercut blasting.
- 107 drawbells fired ready for production (with monthly completion rates approaching industry best practice).
- 2615 t of emulsion explosives were consumed for development and cave activation.
- 1.81 Mt of ore moved for development and cave activation.
- 110 889 t of shotcrete and concrete sprayed or placed for the project.
- 590 t of steel were placed for the Materials Handling System (MHS):
  - MHS 350 t and Crusher 240 t.

- 30.1 km of electrical cable installed.

- 4400 m of roadways construction, with 90 per cent being panels which were poured and cured on-site (these are large pavers that can be removed and replaced in the future, having dimensions of 2.5 m × 3.5 m).

- The HSE performance has been exceptional with only three Lost Time Injuries:

  o 2006 – E26 Lift 2 had 23 LTI's.

  o 2010 – E48 had 16 LTI's (previous Block Cave constructions at Northparkes).

- There were no reportable environmental incidents.

- The project fostered a culture of reporting incidents and improvements.

The Lift 1 North Project Team has established a significant point of difference with the delivery being completed by the 'Owners Team Model' not Lump Sum or Turn-key.

E26 Lift 1 North is now supplying the Northparkes Processing plant for the next eight years.

HSET PERFORMANCE

The HSET site framework under the umbrella of the Site Safety Management Plan was implemented, maintained, and monitored by the HSET team aligned to the project risks and needs.

The total number of personnel that worked on L1N was 496 individuals. The breakdown of personnel is captured in Table 1.

TABLE 1

Breakdown of personnel across the project.

Mining Development and Production	94
Roadways	16
Mobile Equipment Maintenance	21
Structural Mechanical Piping (SMP) / Civils / Electrical	167
Northparkes Technical Staff	49
Northparkes Contractors (Others)	149
Total	496

The Northparkes L1N project has performed well in most aspects of the key identifiable areas associated with Health, Safety, Environmental and Training (HSET). In comparisons with statistics associated with the previous block cave development and construction at E26 Lift 2 and E48, it was identified that improvements have been achieved.

The HSE performance has been exceptional with only three Lost Time Injuries (LTI) during the 37 months. Previous block cave construction at Northparkes had the following safety statistics:

- 2006 – E26 Lift 2 had 23 LTI's 2010 – E48 had 16 LTI's. There were no reportable environmental incidents. The culture of reporting incidents and improvements was improved on previous projects.

After the removal of the major Structural, Mechanical, Piping (SMP) contractor the L1N HSET team redesigned the induction and implemented a Safety Reset Induction process. The project was reset with the addition of extra training to be rolled out to all supervisors and workers aligned with L1N. The below presentations were specifically designed to reset the project and refocus on zero harm operations:

- Zero Harm ABC Risk Pulse (all participated).

- E26L1N Construction General Induction reset (all participated).

- E26L1N Leaders Supervisors Reset (supervisors and leaders only).

After this process the project realigned and started the new journey of self-perform to construct a new block cave and material handling system for Northparkes. The final handover of the crusher and material handling system to Northparkes operational teams occurred in March of 2022. No serious injuries or incidents were reported in the 11 month period and the refocusing process was a considered to be a success.

MINING

Development mining of the E26L1N project commenced in January 2019, shortly after being approved in December 2018. The owners team model was chosen, utilising a core group of skilled development mining personnel that Northparkes had been growing for over a decade, minimising the need for contractors, and placing full responsibility on the development team, supported by Northparkes L1N Projects, Mining Technical, Mining Production, and Mobile Equipment Maintenance teams. The owners team model enabled improved cost control, a reduction of contractors and external parties, faster decision-making (without contract variations) and the agility to change processes or controls as required.

Due to the E26 sub level cave still being in operation when E26L1N commenced, a major ramp up of team numbers and skills was required to meet planned development metres per month. Recruitment of personnel and procurement of equipment proved to be a slow and difficult process in a competitive mining industry that currently exists in Australia. This led to the team not meeting planned development metres in 2019. In 2020 as the team increased in skilled personnel and a reset on key priority areas that enabled the team to surge ahead and ultimately complete development mining three months ahead of schedule on 25th April 2021 with 12 514 m delivered at a cost of $6636/m, improving on established industry benchmark costs for block cave development.

The development of E26 Lift 1 North has been a positive experience with enormous learnings for the future development of block caves and sub level caves at Northparkes. The final design has seen significant changes and improvements from the original feasibility design without detrimental effect to schedule and cost.

Longhole production drilling

Longhole production drilling commenced in May 2020 and was completed in December 2021. Total production metres required for the undercut and drawbells increased due to design changes made during the development mining phase, including changing from herringbone to El Teniente style requiring more undercut rings, drilling design changes around CV7 and the undercut start-up methodology, unplanned probe holes, geotechnical monitoring and equipment holes, electrical holes and drain holes. A total of 240 km of drilling was completed with two production drills for a large portion of the time.

Box-hole drilling of the drawbells from the extraction level is critical for creating the initial void space when firing the drawbell. The vertical box-hole is 700 mm in diameter and 17 m long, although the vertical height varied depending on the undercut location as there were shorter holes and longer holes because the extraction level height varied. Box-hole drilling was completed four weeks early and did not impact any sequenced works on the extraction level.

E26L1N cave activation

Undercut and drawbell firings commenced in April 2021 and were completed in February 2022. The last few months saw 18 drawbells per month fired to bring the project in ahead of schedule. These rates are approaching industry best practice and were achieved through reduced swell bogging and no restrictions from roadways installation or drawbell preparation. The drawbell was fired as a single blast using Orica ICON detonators.

The firing plan for the post-undercut section was first conceived and planned in the feasibility study, but did not consider in detail the firing sequences where it would be required to fire uphole and downhole sections continuously. During the detailed planning of the firings, success was achieved by using Orica's wireless detonators to pre-charge the areas and fire in sequence with bogging activities. Most of the undercut was fire using Orica ICON detonators.

The 21 steps from extraction level development to production bogging of the swell tonnes from the drawbell firing shown in Figure 1 outline the process followed by the Northparkes team to achieve first production from each drawpoint.

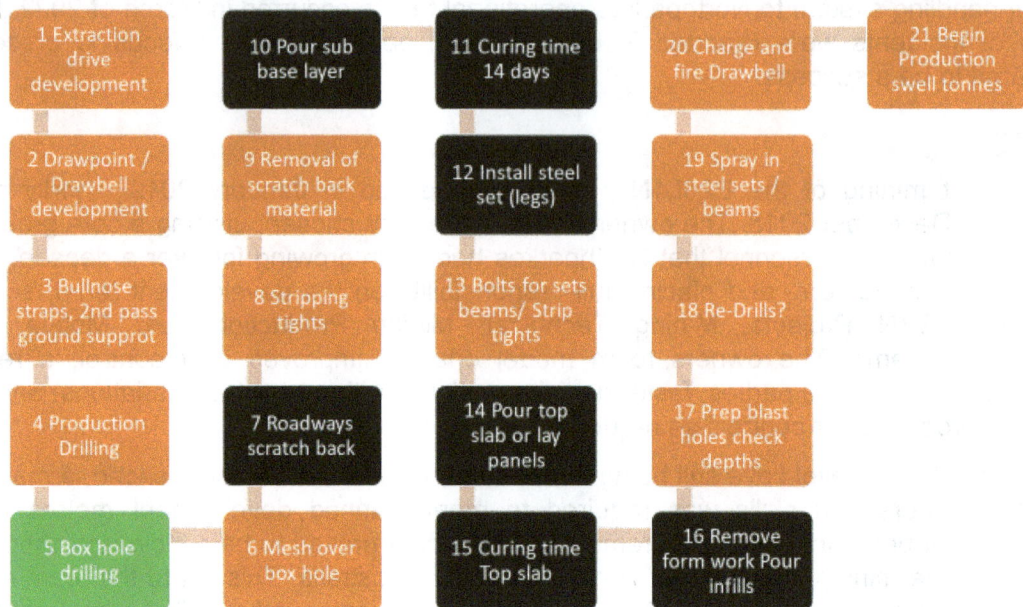

FIG 1 – Extraction level interaction diagram.

Hydrofracturing

Hydrofracturing was implemented to pre-condition the rock mass above the undercut level to promote cave propagation. This practice has been used successfully at Northparkes on two previous occasions within the E48 Block cave. Hydrofracturing pre-conditioning is achieved by pumping water into the rock mass under high pressure to infiltrate into the fine rock structures and cracks. This high-pressure injection of water weakens the local area around the holes and encourages caving as the cave propagates upwards. A program of eight drill holes were designed and drilled to conduct the hydrofracturing from surface.

Roadways construction

Another first for the Northparkes owners team model was the execution of the E26L1N extraction level roadways scope of work. In all previous block cave construction projects at Northparkes, this work has been contracted out. Again, the decision was made to use the owners team model to reduce complexity and cost.

The Northparkes roadways team commenced work in December 2020 and had the same start-up issues as the development mining team with a lack of skills and equipment leading to below planned results in the first three months. However, once the mobilisation of personnel especially experienced concreters and equipment was complete, compliance and exceedance to the advance rate was achieved bringing the project in three weeks early and with a $900 K saving compared to plan.

The owners team model was extremely advantageous and was a key reason the project was delivered within the time frame and costs outlined above. The owners team model allowed wholistic control of the extraction level area by Northparkes. This allowed decisions to be made based on overall priorities without the need to manage contractors and risk negative contract variations because of delays to areas from the mining team's plan. When changes were required to be made regarding work fronts or priority tasks, they could easily be evaluated and implemented, resulting in excellent compliance to all L1N development KPI's including production drilling, steel set installation and drawbell firing.

The design of L1N roadways shown in Figure 2 were concrete panels 2.5 m × 3.5 m, constructed and cured on surface then transported down to the extraction level and installed on the 32 MPa mass concrete base like a large paver. This was different to previous Northparkes block caves, which were *in situ* pours of 32 MPa mass concrete base with a 250 mm to 300 mm *in situ* 80 MPa concrete pavement of up to 20 m in length. These panels were used across most of the extraction level but *in situ* 80 MPa pours were completed in low traffic areas.

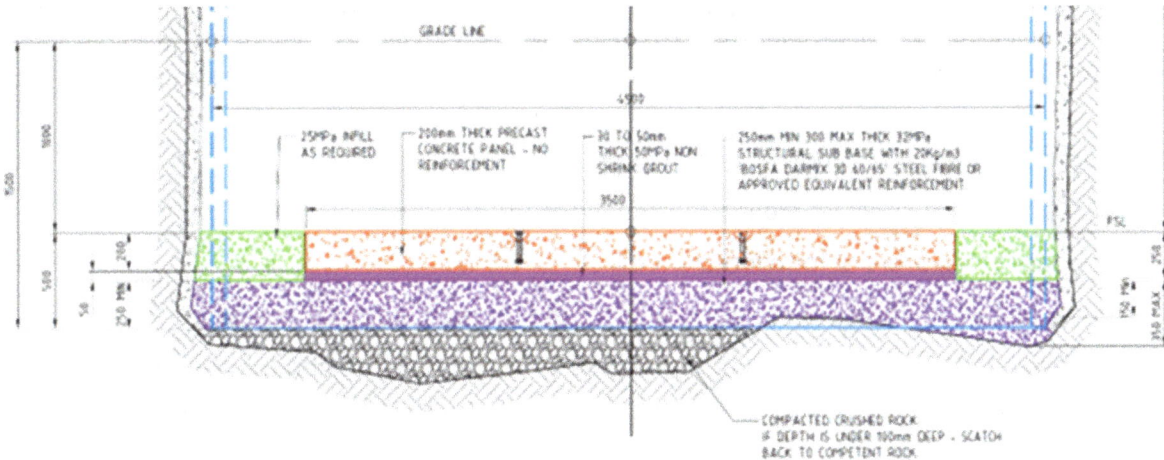

E26 L1N ELA - EXTRACTION DRIVE PAVEMENT DESIGN - PRECAST PANELS

FIG 2 – E26L1N extraction level road pavement design.

Ventilation

A significant ventilation upgrade project for site was delayed by 2.5 years which complicated the ventilation strategy for E26L1N. Some key changes were made to achieve a safe working environment:

- Installation of the Lower Decline (LD4) vent doors allowed for development of Extraction Level Access (ELA) West without disruption to the amount of equipment working in the active sublevel cave in E26.

- Existing booster fans located in E26 Lift 2 were reversed to re-distribute the air from E48 into E26 which allowed for an increase in trucking numbers.

- Two development fans were installed in the L1N Orepass, acting as booster fans for all L1N development, allowing for significantly more development headings to be opened as well as access travel ways.

- An additional fan was added to the top of the Brazen drive to increase the airflow into the CV16 single heading development.

Throughout this process, targeted Diesel Particulate Matter (DPM) monitoring also allowed additional trucking capacity.

All these changes were managed through management of change processes and regular monitoring confirmed the safety of all workers. Once the vent upgrade was completed the increased air to the mine was redistributed across E26 and E48 opening more ventilated work areas.

Air blast control plugs

The feasibility study acknowledged the risk of air blast in L1N. There were six air blast plugs required, totalling approximately 600 m³ of shotcrete. This was completed as part of the minimum requirement and would need increased measures should a greater airgap occur. The design is a 5 m plug of concrete keyed into the ground with rock bolts protruding 1 m and mesh tied to the bolts and shotcrete sprayed into the plug.

MATERIAL HANDLING SYSTEM CONSTRUCTION

Description of works

The E26L1N Material Handling System (MHS) comprises of civil, structural, mechanical, piping and electrical material, plant and equipment required to extract 900 t/hr of crushed ore from the E26L1N orebody to the existing material handling system at Top Of Bins (TOB), or better known as the shuttle conveyor (CV008).

The underground materials handling system includes but is not limited to:

- Direct tip primary crushing station (shown in Figure 3).
- Permanent rock breaker for oversize ore.
- Conveyor CV016 conveying system, feeder, and tramp metal removal station.
- Conveyor CV017 conveying system and CV016/017 transfer station (shown in Figure 4).
- Conveyor CV017/CV008 shuttle conveyor transfer station (shown in Figure 5).
- Shuttle conveyor CV008 upgrade modifications.
- Associated services infrastructure including fire detection system.

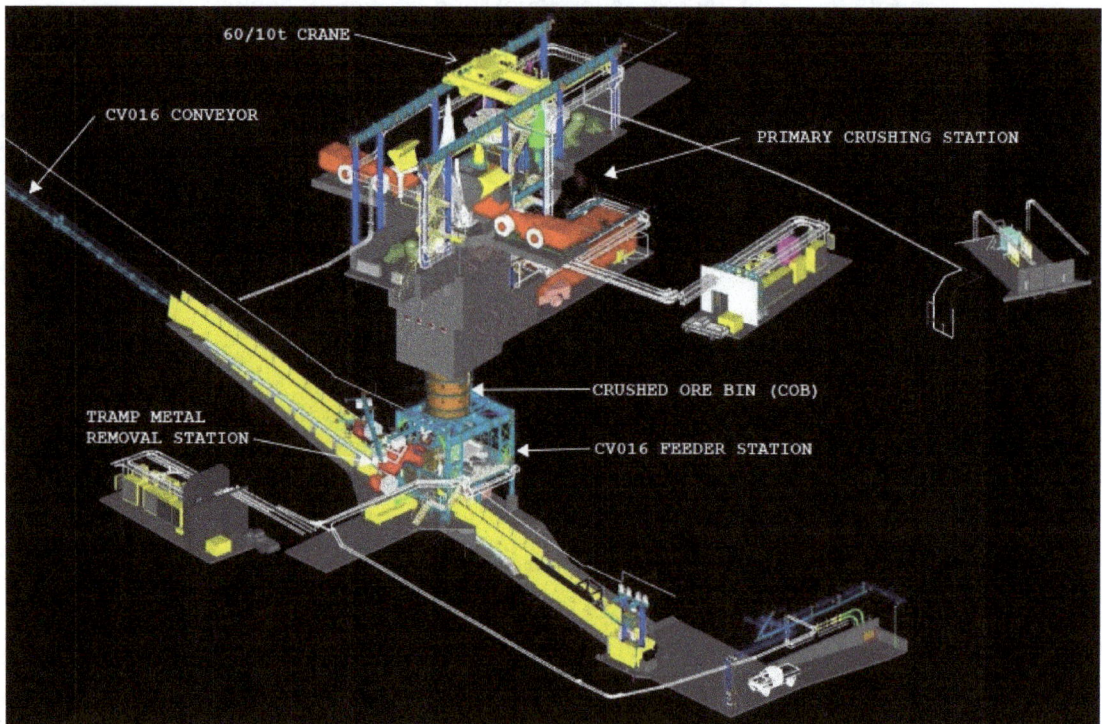

FIG 3 – Crushing station, tramp removal and CV16 feeder station.

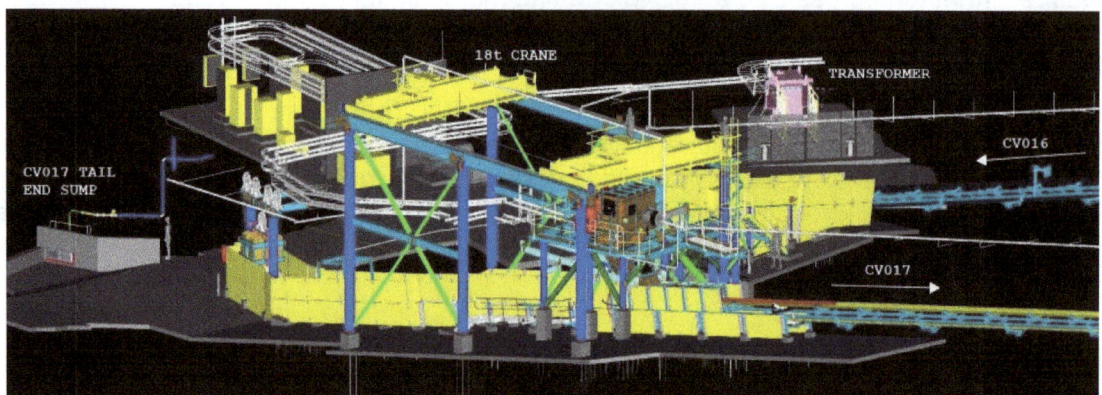

FIG 4 – CV016/017 transfer station.

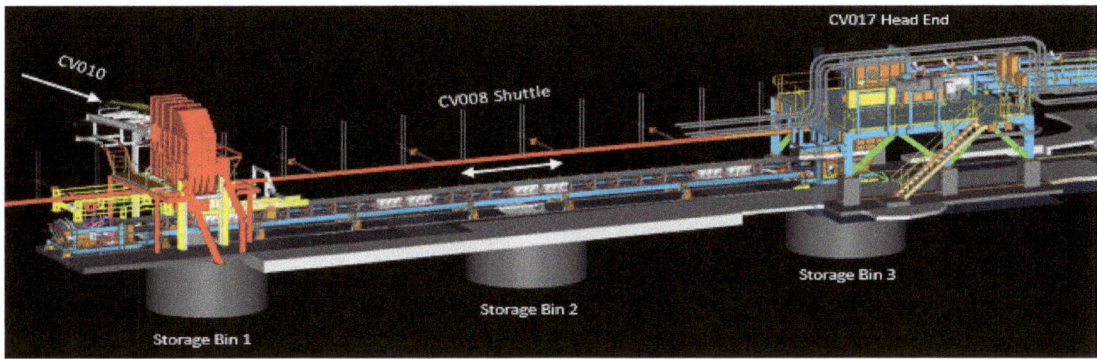

FIG 5 – CV008 shuttle conveyor and CV008/017 transfer station at top of bins (TOB).

The SMP, civil and electrical scope of works were put out for tender in mid-to-late 2020 and were awarded based on a lump sum unit rate with electrical as 'time and materials'.

Civil and SMP works were packaged up under the one external contractor under a lump sum model due to benefits highlighted in the tender and recommendation to award process and works commenced in July 2020 and progressed through to January 2021 when works were suspended by the project leadership team due to several safety and progress concerns. The civil and SMP works remained on hold while Northparkes and the contractor undertook legal discussions to ensure a clear path forward into execution to ensure a safe delivery of the remaining works.

Northparkes Mines made the decision to no longer progress the on-site works relating to the civil and SMP lump sum contracts. At this time, Northparkes made the decision to complete the works based on a cost-plus basis under the direction of Northparkes itself.

The electrical construction activities and installation of the materials handling system was executed by an internal execution team following a strict time and materials engagement contract with an external labour provider. The approach worked well and was delivered on budget and in line with SMP installation.

The project team secured additional resources to increase its in-house project delivery capabilities and sourced trades, plant and equipment from local contractors and labour providers. The schedule was re-baselined, and an ambitious operations handover date was selected for February 2022.

Northparkes were accountable for all contractor management, schedule management, technical instruction, supervision and all HSET requirements and responsibilities aligned with a self-perform model. The transition from lump sum to a self-perform model with a newly engaged contractor group comprising AWCON (Civils), MCA (Crane and Conveyor Fabrication), specialised crusher services (L1N Crusher Install), JT Electrical, Usher and Co (Surveying) and an assortment of smaller contractors started in March of 2021 and all works were completed and commissioned in February 2022.

The high-level schedule in Figure 6 outlines the overall materials handling construction, including the tender periods, recommendation to award period, procurement and installation and commissioning stages of the project.

Activity	2020												2021												2022	
	Jan	Feb	Mar	Apr	May	Jun	Jul	Aug	Sep	Oct	Nov	Dec	Jan	Feb	Mar	Apr	May	Jun	Jul	Aug	Sep	Oct	Nov	Dec	Jan	Feb
Tender																										
Contract award civil early works																										
Civil early works mobilisation																										
Civil early works																										
Contract award SMP early works																										
SMP Early works procurement																										
SMP early works construction																										
Contract award civil																										
Civil works																										
Contract award SMP																										
SMP detailing																										
SMP fabrication																										
SMP installation																										
Crusher Assembly Surface																										
Contractor standown																										
NPM Crusher assembly surface																										
NPM Crusher underground ass																										
NPM Civil works																										
NPM SMP Detailing																										
NPM SMP Fabrication																										
NPM SMP Installation																										
Contract Award Electrical																										
NPM Electrical Mobilisation																										
NPM Electrical Construction																										
NPM Commissioning																										

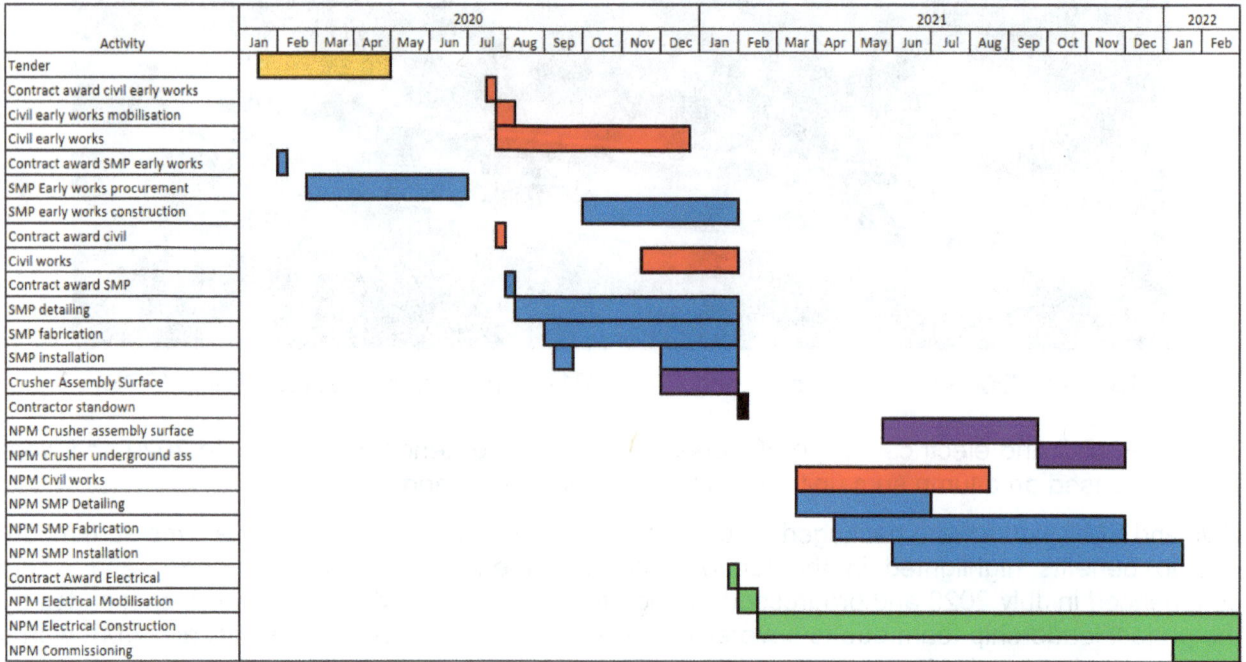

FIG 6 – Civil, SMP and electrical schedule through to project completion.

Civil construction

The civil works for the L1N project was awarded to a local construction and services contractor under three contracts. These being: Early Works, Crusher, and Conveyors. Initial works around the Crusher are shown in Figure 7 and an example of the conveyor work is shown in Figure 8. The contractor was directed to complete the early works contract first, as it related to the overhead crane commissioning in the crushing chamber, which was on critical path for project delivery. At an early stage in the project, it became evident that the contractor did not have the experience and project management resources that were required to complete the civil works for and underground crushing and conveying system.

FIG 7 – ROM crusher level.

Underground Operators Conference 2023 | Brisbane, Qld | 27–29 March 2023

FIG 8 – CV017 tail end concrete reinforcing.

Civil under owner model

By January 2021 it was clear that the contractor would not be capable of completing the more complex civil works within the crusher contract. This scope included pouring the major concrete elements that would support the crusher itself, including the suspended ROM and motor slabs. This type of concrete works requires experienced planning, execution, and oversight to complete successfully, and a mistake made during a pour can pose a significant safety and financial risk. It was determined that the risk of allowing the contractor to complete the remaining civils scope was too high, and works were halted.

In the absence of a definitive path forward, the project team began planning to self-perform the remaining package of works. The contractor's Project Engineer was identified as a key resource with invaluable knowledge of the civil works. The working relationship that had been established with this individual led to their acceptance of a position on the project team to assist in completing the remaining civil scope on the project.

To complete civil works related to the crusher, the project team decided to engage a contractor who had originally tendered for the project but had been determined too costly. This contractor may have cost more, but they had extensive experience constructing underground crushers of the same design and a good reputation in the business. This contractor was signed under an owner managed agreement to complete the remaining concrete and rebar scope under the Early Works and Crusher contracts. The new contractor and project team immediately started working together to re-baseline and plan the remaining civil works, drawing on their experience and constructability advice to determine the fastest and best way to deliver the project. After a few short weeks of planning and mobilisation, the new contractor had commenced works on the crusher civil works.

It quickly became clear that the new contractor was very familiar with civil construction works in an underground environment, and their management team was more than capable of delivering the project. The project team was able to form an effective relationship with the Contractor that allowed visibility of the project progress, and problems to be proactively identified and resolved. The result was the successful delivery of the remaining crusher civil works within the re-baselined time frame and without any major quality or safety incidents.

The project team also worked with the contractor to allow for an early start on the SMP and electrical scope and help recover some of the schedule slippage.

For the completion of the conveyor civil works, the project team decided to engage the mine's preferred civil works provider in combination with the previous Sydney based steel provider. While these contractors did not have the same level of experience as the contractor completing the crusher works, they both had a good reputation within the project team, and it was determined they were well suited for the remaining scope under the conveyors civil contract.

Under the new owner's team, the remaining conveyor civil works was re-baselined and planned to allow installation of the SMP and Electrical works to commence ahead of the previous schedule. Conveyors works were closely managed by the project team, and despite several challenges related to the mining profile and IFC design, the project was delivered successfully.

Structural, mechanical, and piping (SMP)

Crusher installation and associated equipment

The critical path construction sequence always identified the installation of the 60 t crusher crane as a priority. The 60t crane, concrete, steel, and equipment was to be installed and operated to give the construction team an operating crane to help with construction activities and the installation of the Crushed Ore Bin (COB). This crane is shown in Figure 9.

FIG 9 – Crane equipment installation.

The crusher surface works included fitting of the welded and bonded liners, main shaft assembly and crusher test assembly. The crusher was then disassembled into the spider, top, mid, and bottom sections, and main shaft for transport underground. Each section was mounted on a skid and was towed by an underground haul truck and supported at the rear by a low loader as shown in Figures 10 and 11.

FIG 10 – Bottom shell ready for transport UG.

FIG 11 – Spider section entering portal.

Electrical construction

The original scope of works was generated by Northparkes as a lump sum, all-inclusive model to allow external bidders to accurately include all required effort-based tasks and provide a 'best case cost' if the scope was adhered to correctly. The lump sum offer excluded any delays or additional scoped items, so it was clear that this was to be considered 'best case' due to the existing project challenges such as with the SMP and Civil works, under and overbreak challenges in the crusher chamber and work front congestion faced by the nature of the works in an underground environment.

With the known underground construction area challenges and risks associated with working in tight and congested areas, it was clear that access for electrical construction would be challenging. To reduce the risk of working in and around other disciplines electrical work fronts that were not influenced by others were prioritised and planning a daily basis for other congested fronts with an afternoon meeting conducted to plan for the next day to reduce start of shift delays.

The following aspects of the electrical construction worked well and in favour of the Company and should be included in any major future projects:

- One team approach (contractor and Northparkes) on schedule of rates contract with strict cost control.
- Dedicated HSE/Training, admin and materials coordinator roles by contractor on-site as well as dedicated resources for the project locked in for commissioning assistance.
- Dedicated Northparkes electrical engineer role for quality reviews and construction inspections, supported by strong leadership of experienced Northparkes Construction supervisors on 7:7 roster.
- Dedicated EWPs and light vehicle for electrical works.
- Two week lookahead and schedule tracking tool.
- Electronic approval process for redline mark ups alongside an electronic ITR Register for tracking all test sheets and progress.
- Free issuing tooling to the construction contractor and retaining ownership at the end of the works for future projects.

PROJECT CONTROLS

Project controls team

A Superintendent equivalent role of Senior Project Controller was added to the reporting line to the Project Manager in the project organisational structure following the Feasibility Study. A Cost Controller resource was also added to the project controls team for increased cost analysis resourcing. The project controls team was optimally resourced in the organisational structure shown in Figure 12.

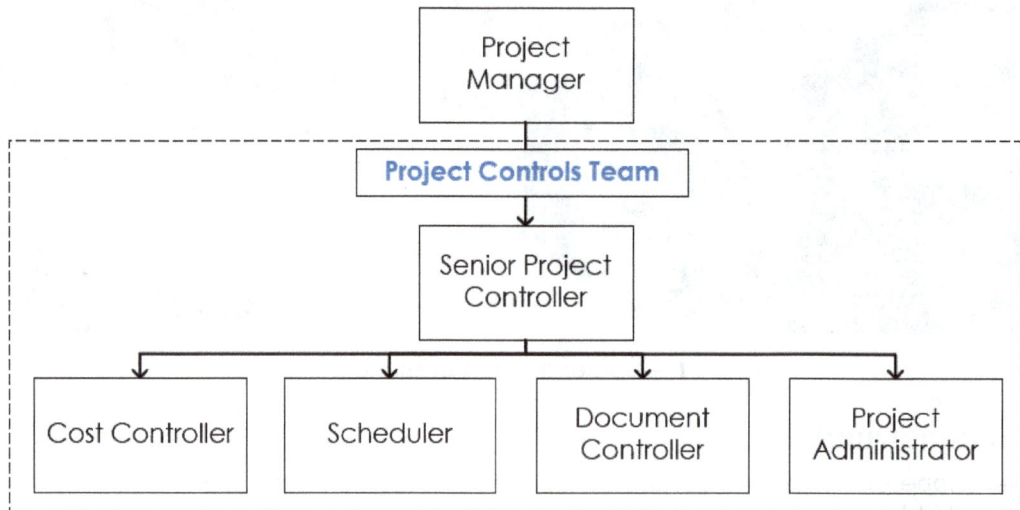

FIG 12 – Project controls organisational structure.

The resourcing of the team diminished throughout the project with the loss of the Cost Controller role in October 2020. Only the Senior Project Controller and Project Administrator remained until the project transitioned to OPEX on 28 February 2022 (when the last drawbell was mined).

Key learnings

Cost estimation

The project transitioned from the Feasibility Stage into the Implementation Stage with an ambiguous WBS structure and significant cost estimating spreadsheets which resulted in allocations of some budgets to areas without clear traceability to the Feasibility Study cost estimation. Project cost estimation should be constructed from the full WBS structure intended to be used in the project and WBS owners assigned according to budget accountability in the budget structures. Commonality between the structures should also be implemented and the project master schedule should also be cost loaded.

Budget

Project budget was allocated at the commencement of the project and was available at an overall level. The process of forecasting monthly budget and weekly cost review meetings and obtaining approval became cumbersome without a dedicated cost controller and sight was lost on forecasting the overall project cost to complete. Weekly reviews were too frequent for a meaningful review for actuals and forecast to complete due to the dynamic nature of raising commitments and commitments actualisation. As a result, the following learnings were clear:

- A cost review with forecast accuracy tracking monthly of each department would be a more reasonable frequency, supported by a project controls procedure to document project controls governance across the site projects portfolio.

- A dedicated cost controller role should be maintained for added cost support and detailed cost analysis.

Schedule management

The project schedule was an integrated master schedule that captured engineering, procurement, contracts and MHS construction and commissioning activities and high-level mining development activities in P6 Primavera software, supported by a scheduler resource assigned to the Project Controls Team. Lookahead schedules were produced weekly with a full schedule update also produced weekly. A separate scheduler resource was also assigned to the Mining Team for long-term mining development scheduling in Deswik software. This process could have been more effective with the following improvements:

- Tracking of the owners MHS construction schedule was limited due to no owner's team QA/QC resourcing. A dedicated resource for QA/QC evaluations for validation are necessary for future defence of claims if required.

- Discrepancies in a schedule provided by a contractor should always be highlighted in correspondence to the contractor, particularly to highlight errors of actual start and finish dates and causes of delay.

- The level of activity detail included in the Feasibility Study Master Schedule could have been improved. As an example, the crushing station and MHS construction schedule only included one activity each for the civil works and commissioning of the crushing station and conveyors.

Documentation and record keeping

The E26L1N project highlighted a need for improvement on documented processes and forms for future projects. A few of these learnings are presented as follows:

- A process for Project Change Notifications (PCN) should be documented and developed as part of the project controls procedure.

- A project document should be drafted at the start of the project detailing operational, engineering and quality processes to ensure clarity of construction processes from commencement of the project.

- Templates should be established prior commencement of the contractor works to allow all project personnel to record and save details:

  o Templates should be designed for Non-Conformance Reports (NCR's), Requests For Information (RFI's), deviation requests, notices of dispute, notices of delay, notices of claim, notices of variation and minutes of meetings.

  o Meeting minutes (including shift start meetings) should be signed by the contractor representative and the company representative and include a data field for delays.

  o Templates should not be editable by the contractor and include time bar details.

  o A numbering convention should be established for each document.

  o Training and enforcement of the documents should be enforced by a document controller.

- A project folder structure that includes all files and documentation for the entire project should be set-up from commencement.

Other learnings

Various other learnings from the Project Controls team are listed as follows:

- KPI's/benchmarks need to be established in the Feasibility Study for evaluation in project execution, with consideration for each yearly capital plan for major projects that an earned value KPI might be more appropriate.

- Systems such as Aconex Connected Cost, Power BI and MASTT reporting need to be explored further as they provide clean and simple reporting to complex financial analytical work and reduce the requirement for spreadsheets and therefore spreadsheeting errors.

CONCLUSIONS

CMOC Northparkes Mines successfully delivered the E26L1N project over the course of 37 months at significantly reduced capital cost to previous, similar sized block cave projects at site. This project was executed successfully to deliver a production environment that will supply ore for the a number years in the future.

Development of the mine was achieved with industry leading costs per metre and with an exceptional safety performance with only three Lost Time Injuries over the course of the project (also a substantial improvement over similar previous projects on-site).

Though the project was successful, a number of improvements and learnings were made from the commencement through to completion and for future projects. These related to the management of key contract partners, particularly in the Civil, SMP and Electrical areas where difficulties throughout the project had potential to throw schedules and costs out of line of plan. The successful adoption of an Owners Team Model was a key transition in the approach and has laid the groundwork for future projects. Likewise, a number of other improvements to the project process have been identified and should be adopted going forward.

ACKNOWLEDGEMENTS

The authors wish to acknowledge the work and contributions of the following personnel from CMOC Northparkes Mines over the course of the project and the creation of this paper:

- Richard Plowes Project Manager
- Jeremy Neill Construction Manager
- Andrew Wright Electrical Construction Manager
- Maz Rees Controls Superintendent
- Lee Bailey Project Engineering Superintendent
- Natalie Simpson Senior Assistant Project Support.

Design of raise caving operations in LKAB mines

T Ladinig[1,2], M Andersson[3], M Wimmer[4] and P Gams[5]

1. Senior Researcher, Montanuniversitaet Leoben, Leoben 8700, Austria.
   Email: tobias.ladinig@unileoben.ac.at
2. Senior Geotechnical Engineer, LKAB, Kiruna 98186, Sweden. Email: tobias.ladinig@lkab.com
3. Consultant Systems Engineering, Combitech AB (within the Saab corporate group), Trollhättan
   46153, Sweden. Email: marianne.andersson@combitech.com
4. Manager Mining Technology, LKAB, Kiruna 98186, Sweden.
   Email: matthias.wimmer@lkab.com
5. Junior Researcher, Montanuniversitaet Leoben, Leoben 8700, Austria.
   Email: patrick.gams@unileoben.ac.at

ABSTRACT

Raise caving is a novel cave mining method, which has been proposed for mass mining at increasing mining depths in LKAB operations. The method makes use of the raise mining technology and an active stress management approach. For the implementation the method is split into two phases, namely a de-stressing phase and a production phase. In the de-stressing phase, a slot-pillar system is established. The purpose of slots is to provide stress shadows and the purpose of the large pillars is to control stress and seismicity. The production phase follows after the de-stressing phase. Required infrastructure in the production phase is developed delayed in stress shadows and bulk extraction takes place in de-stressed zones. Eventually the pillars crush and they are extracted in the production phase. A comprehensive research and development program has been launched to develop the raise caving method towards implementation. As part of this program a full-scale test site of raise caving will be conducted as well as raise caving operations are planned for depth extensions in LKAB mines.

This paper discusses the design of raise caving operations in practice at the LKAB mines. In particular, the purpose of the paper is to focus on requirements for the development and design of a raise caving operation. The development of raise caving is discussed from a systems engineering perspective as well as from a management perspective. The functional requirements in the de-stressing and production phase are outlined and discussed in the paper. The discussion is on basis of the main elements of the raise caving system, such as slots, stopes, pillars and infrastructure. Critical rock engineering design requirements in the de-stressing and production phase are highlighted and dependencies between individual main elements are identified. The analyses show that pillars are especially important for the success of the mining method. Overall, the analyses show that raise caving is a complex system and that its development demands a multidisciplinary approach.

INTRODUCTION

Raise caving is a novel mining method, which has been specifically developed to address encountered rock mechanics and operational issues, such as rock bursts, excessive damage to infrastructure and long development times, in (deep) mass mining. By doing so, raise caving enables to reduce the risk considerably and provides an attractive alternative for sublevel caving and block caving.

The raise caving method was developed to address rock pressure problems at depth encountered in LKAB mines, which have caused operational issues in the past. LKAB operates two of the largest underground mines worldwide, and has always played a pioneering role to develop, test and implement new mining technology. To remain profitable from ever increasing depths in an extremely competitive iron ore market, a highly productive mining process is vital. Numerous examples of both incremental and radical process changes contribute to today's high-performing sublevel caving (SLC) operation (Wimmer and Nordqvist, 2018). The development and potential implementation of raise caving is such a radical process change.

A detailed description of the raise caving method, its background, its principles and its advantages, is given by Ladinig *et al* (2021, 2022) and Ladinig, Wimmer and Wagner (2022). A brief outline of most important aspects of the mining method which are relevant for this paper, is provided.

There are two different variants of raise caving, namely a de-stressing variant and a block caving variant. The block caving variant is also referred to as integrated raise caving. The de-stressing variant targets deep mining conditions and its emphasis is on controlling the rock pressure systematically and strategically, whereas the integrated raise caving variant focuses mainly on operational aspects, namely an efficient and fast ramp-up of a large-scale caving operation. This paper deals only about the de-stressing variant of raise caving. The reason for this is that the de-stressing variant of raise caving is considered and investigated as a possible mining method for mining at great depths in LKAB mines. Hence, the term raise caving refers in this paper to the de-stressing variant of raise caving.

Outline of raise caving method

The raise caving method relies on two well-established technologies, which are:

- raise mining technology
- active stress control approach.

Raise mining technology

In modern raise mining, raises are used to excavate vertical or steeply dipping excavations in an efficient manner in a bottom-up sequence. After a raise has been drilled by means of raise boring and supported, a small hoist system installed on top of the raise. This hoist system is used to move a platform with machinery inside the raise. The platform and machinery are used to conduct the drilling and blasting work. Figure 1 shows the machinery for drilling and charging, which is currently under development. A high degree of remote-control and automation is possible in this set-up, because the bored raise is an excavation with a relatively simple (circular) cross-section and because the hoist system allows to position the platform and machinery well along the length of the raise. After blasting the material falls into the excavation and is drawn from the excavation at drawpoints, which are mostly situated at the bottom of the excavation. The drill fan layout allows to adapt the cross-section of the excavation and the shape of the excavation roof. A description of the modern raise mining concept is given by Gipps *et al* (2008), Gipps and Cunningham (2011) and Ladinig (2022). The raise mining method has been successfully applied.

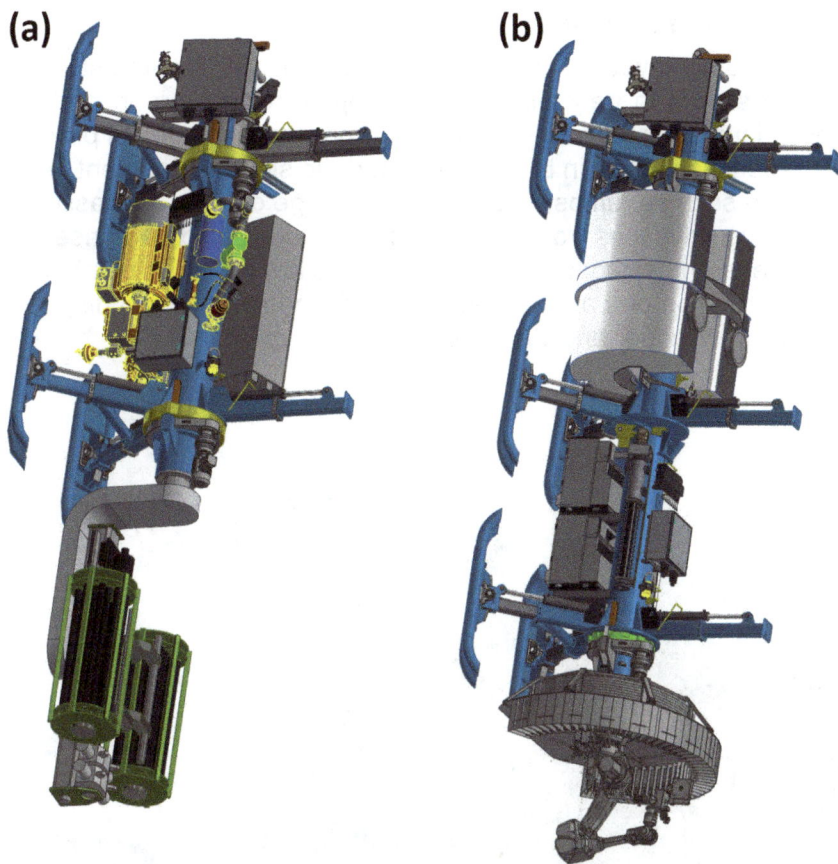

FIG 1 – Raise mining machinery: (a) drilling module; (b) charging module.

Active stress control approach

The active stress control approach emphasizes on controlling the stress magnitudes and mining-induced seismicity in active mining areas to prevent the occurrence problems related to mining-induced stress. The mine layout and mining sequence are decisive for the active stress control approach (Ladinig, 2022). The layout and sequence must ensure that stress magnitudes and mining-induced seismicity and corresponding dynamic loading are limited such that active infrastructure and active stopes can be kept stable and safe without undue amount of support. This implies that high stress magnitudes and seismically active areas must be constrained to locations in the mine, which are distant from active infrastructure and active stopes so that high stress magnitudes and seismic energy release do not cause damage to active infrastructure and stopes. An active stress control approach has been successfully applied in deep South African gold mines for many decades (Jager and Ryder, 1999; Ryder and Jager, 2002). Advanced layouts and sequences have been developed to extract gold reefs up to a depth of nearly 4000 m. Central for these layouts and sequences is to control stress magnitudes at active stope faces and near active infrastructure, to provide stress shadows for active infrastructure and to control the release of seismic energy. However, the applied layouts and sequences are specifically designed for the prevailing mining environment, in particular the predominantly occurring, gently dipping, tabular gold reefs. In recent years a similar approach has been applied in a massive deposit, which is formed of several closely spaced gold reefs (see Watson *et al*, 2014; Andrews, Butcher and Ekkerd, 2019). Horizontal de-stressing cuts are extracted first, which generate stress shadows for subsequent mechanised extraction of wide stopes.

Raise caving utilises a similar active stress control approach to the approach adopted in the South African gold mines. Stress shadows are created first by extracting so-called de-stressing slots. Afterwards, large-scale mineral extraction commences in provided de-stressed ground. The active stress control approach of raise caving is suited for steeply-dipping or massive deposits, in which horizontal stress magnitudes need to be reduced.

Raise caving – layout and sequence

Figure 2 shows a conceptual layout of raise caving in a thick, tabular and steeply-dipping deposit. De-stressing slots are shown in dark blue colour, stopes in dark green and light green colour, drawbells in orange colour, slot raises in yellow colour, production raises in pink colour, raise levels in light blue colour, production levels in black colour and the slot development level in brown colour. Long-term infrastructure such as ramps, shafts, main haulage drifts or orepasses are not shown for overview reasons. Mining advances from left to right in Figure 2 and comprises two phases, a so-called de-stressing phase and a production phase. The following bullet points give an overview of both phases. More information on both phases can be found in corresponding references (Ladinig *et al*, 2021, 2022; Ladinig, Wagner and Wimmer, 2022).

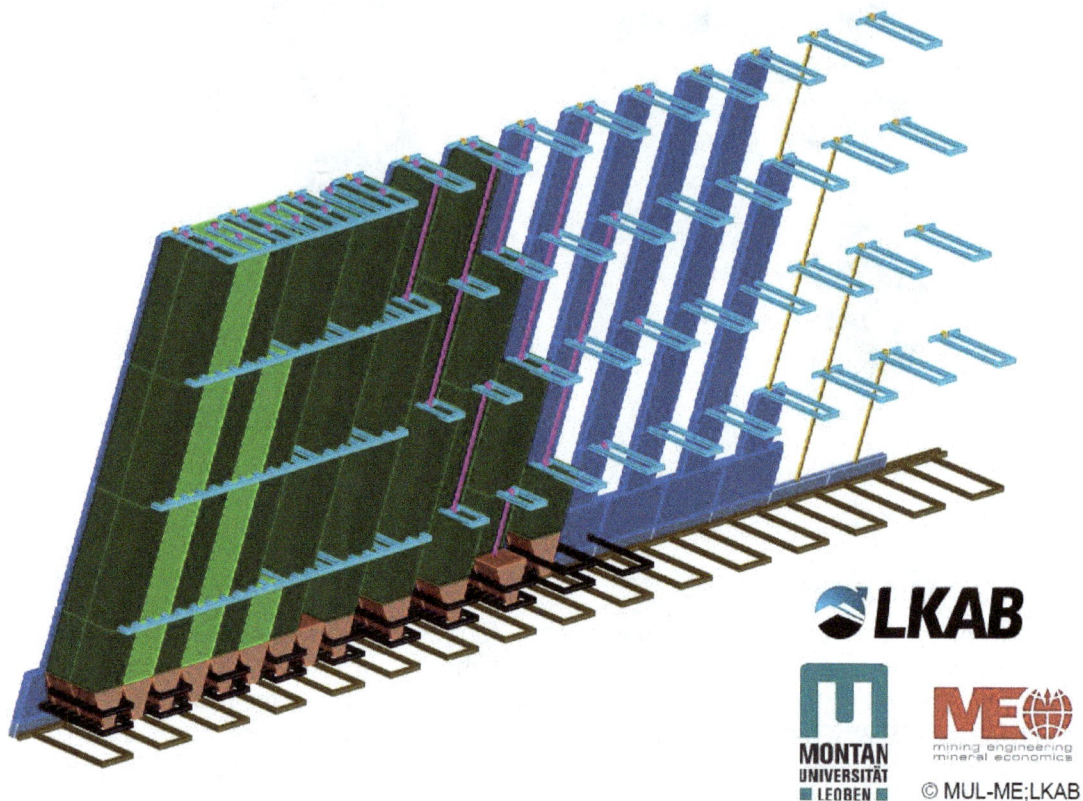

FIG 2 – Conceptual layout of raise caving in a tabular, thick, steeply-dipping deposit. View is from footwall towards hanging wall.

Other key aspects for the design of a raise caving operation are discussed later in this paper.

- <u>De-stressing phase:</u> The de-stressing phase is the first phase. Its objective is the implementation of the active stress control approach. The de-stressing slots are developed at the hanging wall contact from the so-called slot raises by means of raise mining technology. The de-stressing slots are always filled with blasted rock mass, as only the swell is drawn after each blast. The de-stressing slots provide stress shadows for the infrastructure, which is developed in the subsequent production phase. Large pillars with a width of 30 m to 50 m are left between slots. The purpose of these pillars is to control stress magnitudes and mining-induced seismicity in the de-stressing phase, because the de-stressing phase is high stress mining.

- <u>Production phase:</u> The production phase is the second phase and follows after the de-stressing phase. The objective of the production phase is the large-scale mineral extraction as well as the continuation of the active stress control approach. Production raises are developed in de-stressed ground provided by de-stressing slots and large stopes are extracted from these raises. The blasted ore is drawn through large drawbells at production levels and through drawpoints, which are established on former raise levels. As long as drilling and blasting of a stope continues, only the swell is drawn so that stopes are throughout filled with blasted ore. After the stope is completely blasted, the stope is drawn empty. While stopes are drawn empty, hanging wall caving is initiated and caved material fills up progressively. If a former cave exists above the raise caving operation, which is the case in LKAB mines, an alternative may be to connect the raise caving stopes with the former cave so that primarily formerly caved material fills up stopes progressively, while they are drawn empty, and so that consequently surface subsidence impact can be reduced. Blasted stopes provide further stress shadows for the extraction of next stopes. Furthermore, the extraction of stopes behind de-stressing slots triggers crushing of pillars. As a result, regional stress and energy changes occur. Crushed pillars are de-stressed too and can be extracted. Abutment stresses develop at the boundaries of mined-out stopes. For this reason, the extraction of de-stressing slots must take place some distance ahead of the extraction of stopes so that the abutment stresses near mined-out stopes do not affect the extraction of de-stressing slots adversely.

Design of a raise caving operation

In summary, the raise caving method is a complex system, where individual elements, such as de-stressing slots, pillars, stopes, as well as rock mechanics and operational aspects, such as stress control, production target and required time for development, are strongly interrelated and dependent on each other. For the design of a raise caving operation, it is important to understand the functional requirements of each element and their dependencies. Out of these functional requirements and dependencies result constraints, which need to be considered in the design.

The objective of this paper is to discuss these functional requirements and dependencies and resulting constraints. Central functional requirements related to rock mechanics are outlined and discussed for the de-stressing and production phase. An emphasis is put thereby on selected aspects of the active stress management approach and the corresponding requirements on the layout and sequence. Due to space limitations in this publication, the level of discussion is on a general basis and details cannot be shown further. Finally, system engineering and requirements of the development of raise caving from a management perspective are made.

SPECIFIC REQUIREMENTS ON THE MINE LAYOUT AND MINING SEQUENCE

One of the key critical success factors for complex system development is to capture the relevant requirements at the earliest stages of the development project. Requirements originate from different stakeholders, so the early identification and involvement of critical stakeholders is essential. The main requirements to fulfil is to develop a mining method that provide safe environment for personnel, stable production, environmental footprint accepted by the society, and financial sustainability. Hence, the requirements are both functional and non-functional.

On a large scale, the functional requirements can be divided into rock mechanics related and production related. The rock mechanics related ones are relevant for the implementation and continuation of the active stress control approach in the de-stressing and production phase, and make provisions for safe and stable preparation and production. The production related ones are relevant for meeting the production target in terms of quantity and quality, which make provisions for financial sustainability. In the de-stressing phase, in which the active stress control approach is implemented, rock mechanics aspects are most important, because only very small amounts of ore are mined in this phase. In the production phase, both the rock mechanics and the production aspects become central. The active stress control approach must be continued, and the production target must be met.

Requirements in the de-stressing phase

In the de-stressing phase, a slot-pillar system is created. The main elements of this slot-pillar system are slots and pillars as well as the infrastructure required to extract slots. The functional requirements of the individual elements are:

- Slots:
  - providing stress shadows for subsequent large-scale mineral extraction.
- Pillars:
  - controlling stress magnitudes and energy changes

- o stabilising hanging wall to prevent premature cave initiation in the hanging wall.
- Infrastructure:
  - o providing access for slot extraction.

The fulfillment of the functional requirements impacts the design of individual elements. Relevant design aspects are discussed further.

Slots

Slots shall have a tabular shape, because tabular shaped excavations are best suited for providing stress shadows. The terms used for describing slot dimensions are slot width, slot thickness and slot length. The terms slot width and slot thickness are shown in Figure 3 in a plan view of vertically oriented slots. The slot length is the extension of the slot in the out-of-plane direction in Figure 3.

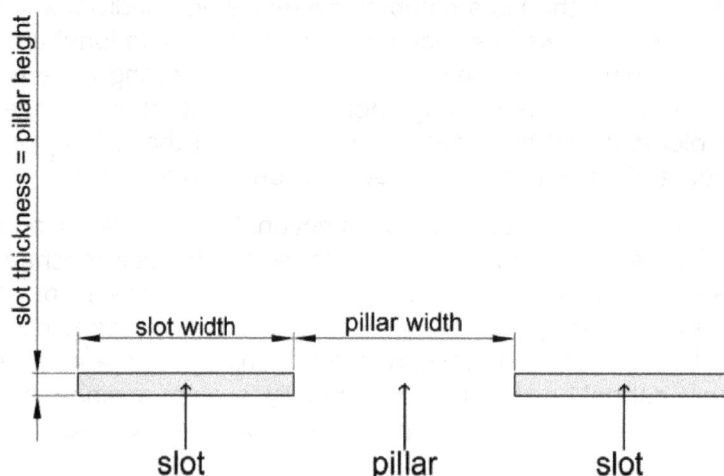

FIG 3 – Plan view showing the terms used for describing slot and pillar dimensions. Slots are oriented vertical.

Slots reduce the stresses in direction of slot thickness. Hence, a de-stressed zone develops in front and behind slots as shown in Figure 4, which shows the horizontal stresses in direction of slot thickness in a horizontal cross-section. The slot dimensions are 50 m in width and 10 m in thickness, the pillar dimensions are 50 m in width and 10 m in height. The stress distribution is derived with FLAC (Itasca, 2016) with a linear elastic rock mass behaviour and for a hydrostatic primary stress state of 80 MPa. Investigations showed that the spatial extent of the stress shadow depends strongly on the slot width. The wider the slot is the larger the spatial extent of the stress shadow becomes. Hence, wider slots are more favourable. The slot length does not have such a pronounced influence on the stress shadow due to the comparatively large length of slots.

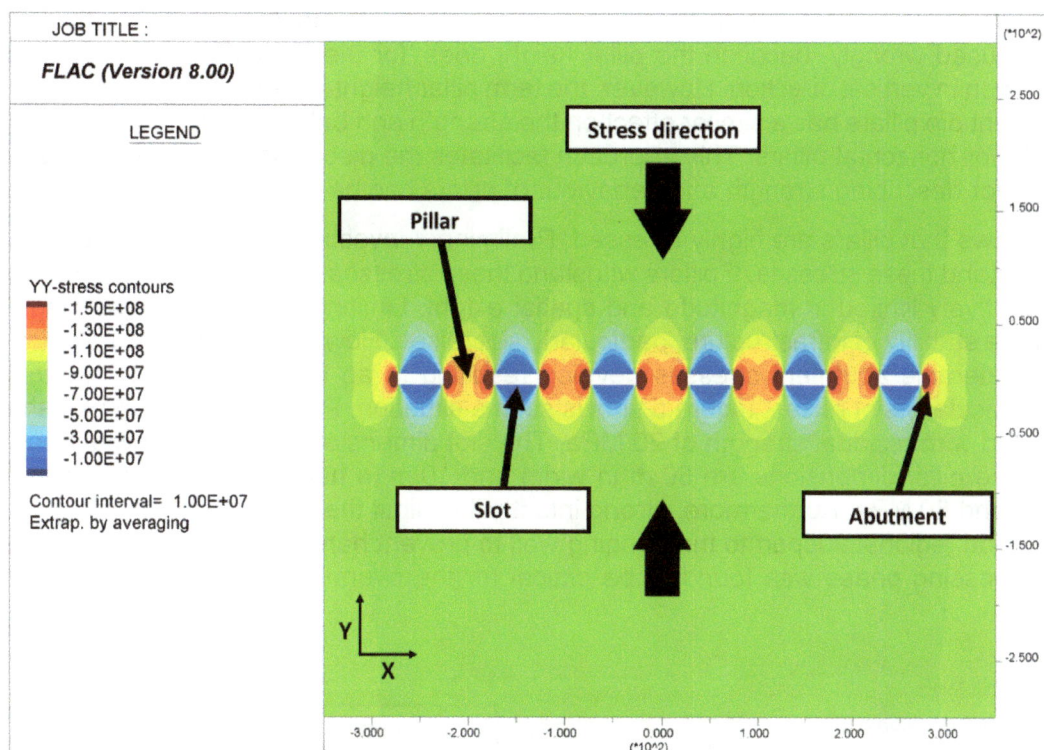

FIG 4 – Horizontal stress distribution in a slot-pillar system.

Besides the spatial extent of the stress shadow, the direction of the major principal stress is important. Generally, slots shall be oriented such that stresses in the critical direction (ie the major principal stress orientation) are decreased. This means that the slot thickness and critical stress direction are oriented in the same direction. The critical stress direction is defined by the primary stress field and the mining-induced stress changes on a regional scale. The foregoing point is of particular relevance for tabular deposits. Due to extensive mineral extraction along strike and dip in these deposits and the corresponding regional stress changes, the critical stress direction is perpendicular to strike and dip, namely in direction of the deposit thickness, and differs from the *in situ* stress field.

There are design constraints for slots. The slot walls and the slot roof must be stable. Excessive overbreak or caving of the slot may result in major operational issues, such as cutting of blastholes in the slot roof, boulders causing hang-ups in the slots or caved material preventing the formation of sufficient swell volume below the roof for the subsequent blast. The slot width has a governing effect on slot stability. The wider the slot becomes, the lower is the stability of the slot. The slot length does not have such a pronounced effect, because the length is already significantly larger than the width. Operational considerations, such as the maximum length of drill holes or the maximum spacing at the end of drill holes, can limit the slot width too. However, this limitation could be solved by developing a wide slot using more than one raise. Another design constraint for slots is the requirement that blasted rock mass must be able to flow in the slots towards drawpoints. Ore flow is required to create swell volume below the slot for the subsequent blast. Hang-ups in the slot can prevent the creation of swell volume below the roof and resolving such hang-ups may be difficult or impossible due to the limited access. Hence, the formation of hang-ups in slots must be avoided as far as possible by design. The achieved fragmentation by blasting and the slot thickness is particularly important. The slot thickness shall not be below a critical threshold, below which the formation of hang-ups becomes likely. This critical threshold is subject of ongoing investigations. Hence, the slot thickness is strongly determined by ore flow mechanics.

Pillars

Pillars are positioned between slots and their long axis is oriented along dip of the deposit. Hence, pillars are so-called dip pillars. Figure 3 shows the terminology used for describing pillar dimensions. The pillar width is the horizontal distance between two adjacent slots, the pillar height is equal to the slot thickness and the pillar length is the dimension of the pillar along dip. The term pillar height may seem to be used wrongly, because the pillar height does, for the present pillars, not describe the pillar extension in vertical direction. However, the term pillar height is used, because the 'pillar height' for the present dip pillars has a similar effect on the strength and behaviour of dip pillars as the (real) pillar height for horizontal pillars. This approach facilitates the discussion on pillars, because widely used terms for describing strength and behaviour of pillars can be used.

Figure 4 shows that pillars are highly-stressed. Preliminary investigations show that pillars must be able to withstand these stresses. If pillars withstand these stresses, the abutment stresses near slots can be effectively limited in magnitude and spatial extent. Limiting abutment stress magnitudes is critical for the stability of infrastructure, particularly slot raises. However, if pillars are overloaded and crush, considerable abutment stresses develop resulting in an adverse impact on infrastructure stability, particularly slot raises. Figure 5 shows such a situation. Pillars crushed and crushed pillars are simulated with residual strength of 20 MPa. The slot dimensions are 50 m in width and 10 m in thickness, the pillar dimensions are 50 m in width and 10 m in height. The primary stress state is hydrostatic and 80 MPa. Furthermore, strong intact pillars limit the seismic energy release and they provide a good regional support to the hanging wall to prevent hanging wall caving. Pillar behaviour in the de-stressing phase was found to be critical for the overall applicability of the raise caving method.

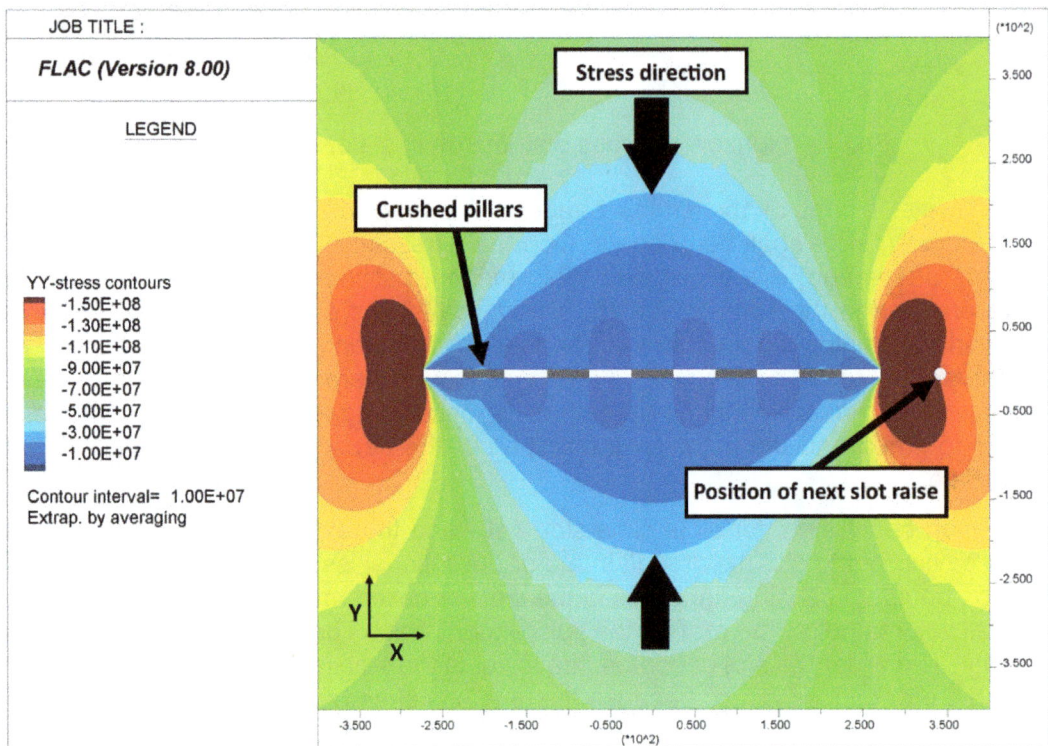

FIG 5 – Horizontal stress distribution in a slot-pillar system in case of pillar crushing.

For achieving the objectives of pillars in the de-stressing, the design shall ensure that pillars remain largely intact in the de-stressing phase and that potential overloading of pillars is limited to the pillar boundaries or short sections of the pillar. Therefore, central parameters are the pillar dimensions, in particular the pillar width-to-height ratio, and the extraction ratio in the slot-pillar system. The pillar width-to-height ratio governs the pillar strength and behaviour. The strength of the pillar increases with increasing pillar width-to-height ratio and the post-peak behaviour transforms from a crushing to a yielding to a strain-hardening behaviour with increasing pillar width-to-height ratio. The extraction ratio in the slot-pillar system has a dominant effect on the stresses in pillars and can thus be used to control the pillar stress magnitude in the design process. The extraction ratio (ER) is determined by the slot width (W_S) and pillar width (WP); see Equation 1. The higher the extraction ratio the larger the pillar stresses become.

$$ER = W_S / (W_S + WP) \qquad (1)$$

As the pillar height is given by the slot thickness, which itself is governed by ore flow considerations, the pillar width is the remaining design parameter for pillars. Gams (2022) analysed the impact of pillar dimensions, slot dimensions and pillar behaviour. His investigations show that pillars with a width-to-height ratio above five are preferable in the de-stressing phase. However, such pillars may be relatively wide, which may cause difficulties for pillar extraction in the production phase.

However, at this point it should be noted that the pillar strength and behaviour of pillars with the width-to-height ratio and the dimensions utilised in raise caving is not well known and thus rather uncertain (Ladinig, 2022). For this reason and the importance of pillars in raise caving, comprehensive investigations on the strength and behaviour of pillars are ongoing in the raise caving project.

Infrastructure

Infrastructure for the development of slots are slot raises and tunnels and it is required on several levels. The stability of the infrastructure must be ensured for the development of slots. The extraction of slots is high stress mining, thus infrastructure cannot benefit from stress shadows and it is exposed to primary stress magnitudes and abutment stresses near slots. Some infrastructure may also be overstoped during slot extraction, resulting in significant stress relaxation near this infrastructure. For infrastructure stability, it is central to limit abutment stress magnitudes and moreover dynamic loads resulting from mining-induced seismicity. The layout of the slot-pillar system, the position of the infrastructure and the extraction sequence are critical. Despite adapting the slot-pillar system and the extraction sequence there may still be significant loads on infrastructure and a heavy support system may be necessary.

Requirements in the production phase

The production phase follows after the slot-pillar system was created. Large-scale mineral extraction takes place in the production phase. The main elements in the production phase are stopes, drawbells, pillars, abutments and the infrastructure required for stope extraction. The objectives of the individual elements are:

- Stopes including drawbells:
  o extracting ore
  o providing stress shadows for large-scale mineral extraction
  o controlling stress magnitudes and energy changes.
- Pillars:
  o controlling stress magnitudes and energy changes
  o providing support to hanging wall rock mass.
- Abutments of stoping areas:
  o transferring significant stress magnitudes.
- Infrastructure:
  o providing access for large-scale mineral extraction.

The fulfillment of the functional requirements impacts the design of individual elements. It is beyond the scope of this paper to discuss all relevant design aspects. Instead, the role and importance of pillars in the production phase is outlined briefly. Overall, the pillars were identified as the most critical aspect for the application of the raise caving method.

Regional impact of stope extraction

In contrast to the extraction of slots, where the stress and energy changes are more localised to areas near slots, the extraction of stopes has a regional impact on the stress situation as well as on energy changes. A critical point is that besides the regional stress and energy changes the active stress management can be maintained.

The stress shadows in the production phase are generated by the extracted stopes. Extraction of stopes changes the position of stress shadows. The position of stress shadows and the spatial extent of stress shadows are thereby governed by the stope dimensions and the overall stope layout. In general, if stopes are extracted in the stress shadow of slots, the position of these stress shadows can then be found near the stopes.

Besides the impact on the position of stress shadows near stopes, extraction of stopes has a further regional impact on the stress situation and an impact on the pillars between slots and stopes. The pillar strength is impacted and pillars will eventually crush and hence also de-stress; compare next section. Finally, the de-stressed pillars are extracted. As a consequence, abutment stresses as well as stress shadows of large regional extent form as shown in Figure 6, which shows a situation, after stopes and pillars between stopes have been extracted. The slot dimensions are 50 m in width and 10 m in thickness, the pillar dimensions between slots are 50 m in width and 10 m in height and the stopes have a cross-section of 50 m × 50 m. The primary stress state is hydrostatic and 80 MPa. The regional abutment areas have a significant impact on the pillars between adjacent slots and the mining activities in the de-stressing phase. Namely, they increase the stresses acting in pillars between slots as well as on infrastructure utilised in the de-stressing phase. For this reason, it is important that the extraction of slots takes place such a distance ahead of stope extraction that the created abutment stresses do not affect infrastructure of the de-stressing phase, in particular slot raises, adversely. Moreover, regional energy changes take place in the production phase, which have an impact on the occurring mining-induced seismicity. For all of these stress and energy changes as well as the corresponding mining-induced seismicity, the pillars and the subsequent pillar extraction have again a central role. The impact of extracting stopes and pillars on the regional stress state and the released energy are of interest in the development of raise caving.

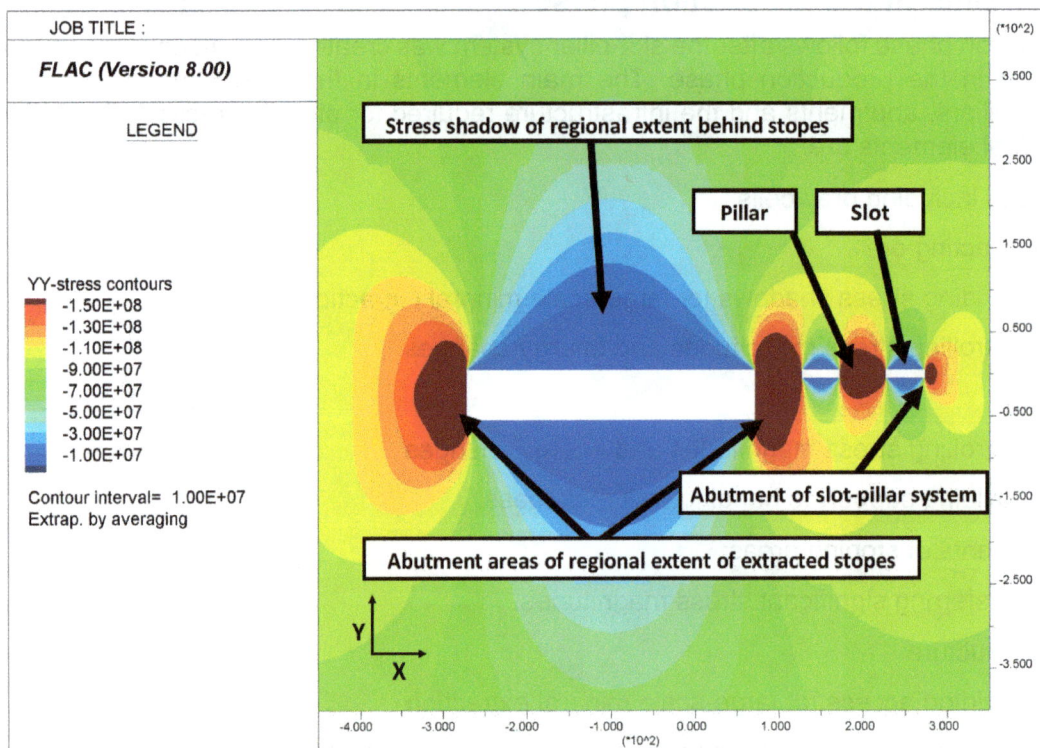

FIG 6 – Horizontal stress distribution in the production phase.

Pillars in the production phase

After the de-stressing phase pillars are highly stressed. Accordingly, infrastructure and stopes cannot be developed or extracted, respectively, in the highly stressed zones in and near the pillar; compare also Figure 4. A significant amount of the deposit is located inside high stress zones of pillars. Hence, at least some of the pillars must be extracted to achieve an acceptably high extraction ratio. For pillar extraction two principal options are available, namely:

1. Extracting the pillar actively from production raises, which are positioned behind neighbouring de-stressing slots in their stress shadows, by means of drilling and blasting.

2. Causing pillar crushing. The process of pillar crushing must be stable. The relation between the post-peak strain softening curve of the pillar and the stiffness of the loading system are therefore decisive as demonstrated by Salamon (1970). The stress magnitudes in the pillar are decreased considerably because of pillar crushing. The residual strength of pillars is important as well, because it determines the stress magnitudes inside crushed pillars. Afterwards the pillar can be extracted with a production raise, which is positioned inside the now de-stressed pillars.

The different possibilities to implement these options are shown in Figures 7 to 10 schematically in a horizontal cross-section. The slot and pillar width are 50 m each and the slot thickness and pillar height are 10 m each. The first option, namely extracting the pillar actively by means of drilling and blasting, may be difficult to realise in practice due to the length of the required drill holes; compare Figure 7. Furthermore, drill hole stability in the highly stressed pillars can be problematic. Hence, causing stable pillar crushing seems to be the left-over option. Pillar crushing can be achieved in following ways:

- Stopes are only extracted behind de-stressing slots and the extraction of stopes results in decreasing the pillar width-to-height ratio by increasing the pillar height; compare Figure 8. Decreasing the pillar width-to-height ratio decreases the pillar strength and it changes the pillar post-peak behaviour. Of particular importance is that the strain softening rate is increased.

- Stopes are extracted behind the de-stressing slots and with these stopes also a part of the pillar is extracted actively; compare Figure 9. The effect on the pillar is the same as in the previous way. The actual stope layout and dimensions can be adapted to create pillars of specific width-to-height ratios and to decrease the pillar width-to-height ratio in a planned and specifically engineered way.

- Pre-conditioning techniques are applied inside the pillar to decrease the rock mass strength. Therefore, the drill holes are drilled from the production raises into the highly stressed pillar to deploy techniques such as hydraulic fracturing or confined blasting; compare Figure 10. The anticipated effect is that the pillar strength is decreased because of a decrease of the rock mass strength. The actual amount of decrease is though not known and hence difficult to estimate. Furthermore, the impact of the pre-conditioning on the post-peak behaviour of the pillar is not well known.

- The regional abutment stresses from the nearby stoping activities cause pillar crushing of the pillar closest to the stoping area; compare Figure 6. In this case, the pillar strength and pillar stress strain behaviour remain largely unchanged, because the width-to-height ratio of the pillar is not modified to cause pillar crushing.

Which of the outlined possibilities to extract pillars between slots is best suited in the raise caving method and in the mining environment is currently being investigated as part of the project.

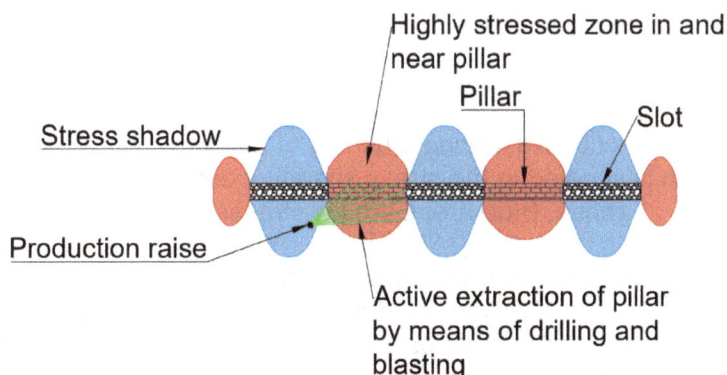

FIG 7 – Horizontal cross-section illustrating extracting a pillar actively by means of drilling and blasting from a production raise, which is situated in a stress shadow behind slots. Red areas are high stress zones. Blue areas are de-stressed zones.

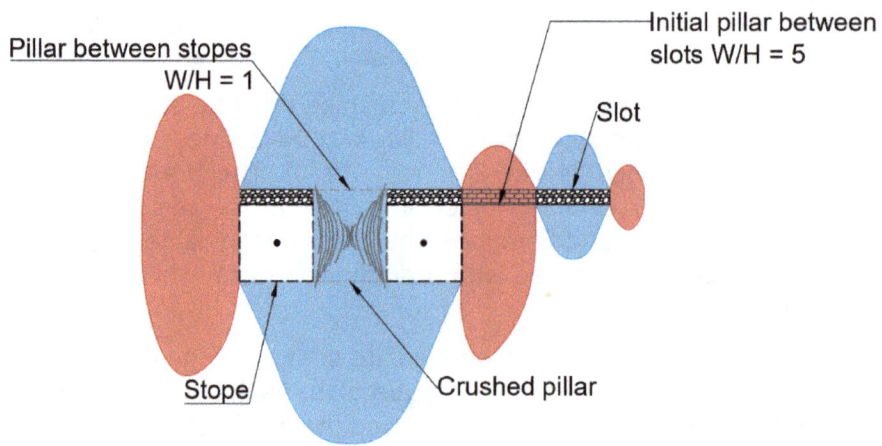

FIG 8 – Horizontal cross-section illustrating crushing of a pillar by stope extraction behind slots, which cause a reduction of the pillar width-to-height ratio (WP/HP). Red areas are high stress zones. Blue areas are de-stressed zones.

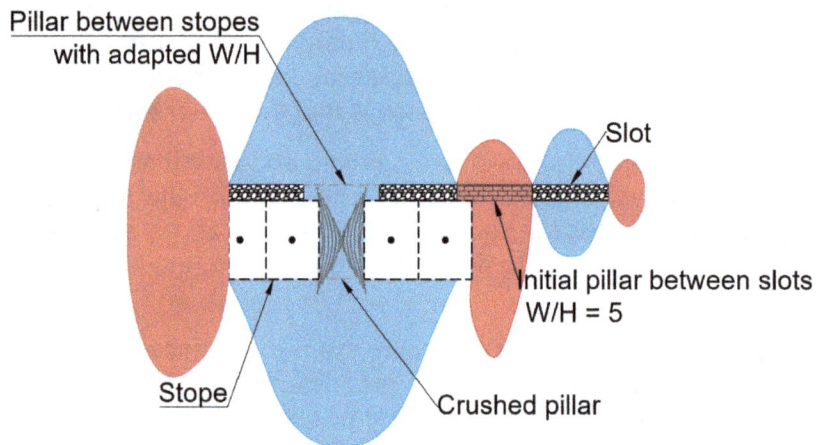

FIG 9 – Horizontal cross-section illustrating crushing of a pillar by stope extraction behind slots, which cause a reduction of the pillar width-to-height ratio (WP/HP). The stope shape is utilised to adapt the dimension and shape of the crushing pillar. Red areas are high stress zones. Blue areas are de-stressed zones.

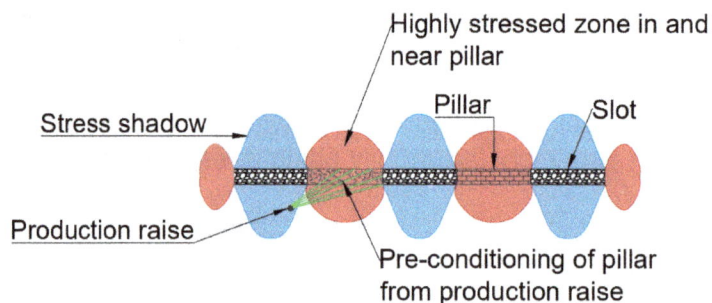

FIG 10 – Horizontal cross-section illustrating crushing of a pillar by pre-conditioning the pillar from a production raise, which is situated in the stress shadow behind slots. Red areas are high stress zones. Blue areas are de-stressed zones.

The considerations on pillars show that pillars are vitally important for a successful raise caving operation. However, the design of pillars is difficult. Reasons for this are that the functional requirement on pillars change considerably between the de-stressing phase, the production phase at an early stage and the production phase at an advanced stage. Moreover, that the knowledge regarding strength and behaviour of pillars utilised in raise caving is very limited (see Ladinig, 2022). Thus, pillars are central topic in the raise caving development project.

Importance of the extraction sequence

The extraction sequence has been mostly neglected so far. However, in practice the extraction sequence has a central and important role. The extraction of slots and stopes dominates the large-scale stress and energy changes in terms of their spatial and temporal occurrence, whereas the time of infrastructure development has a strong impact on whether infrastructure is exposed to high stresses or large dynamic loads. Consequently, the extraction sequence adds a time component to the rock pressure situation and thus it makes the stress management to a 4D issue. The impact and role of the extraction sequence is part of ongoing analyses in the raise caving project.

DEVELOPMENT OF RAISE CAVING IN PRACTICE FROM A MANAGEMENT PERSPECTIVE

Throughout underground mining history, LKAB has almost exclusively used SLC, which has been regarded a productive and cost-efficient method. Provided a large-scale, mainly transverse application in fairly steep orebodies, it is also a very forgiving method in terms of recovery (Wimmer and Nordqvist, 2018). However, over the past 15 years, rock mechanical conditions have become increasingly adverse and managing energy release has become a controlling factor for mine production (Ladinig, Wimmer and Wagner, 2022). Before 2008, the Kiruna Mine could be described as a rock factory and the geotechnical conditions did not have a noticeable or adverse effect on the production. A new situation came into place after a seismic event in 2008 causing a fatality (Dahnér, Malmgren and Boskovic, 2012). Stress related problems, in particular mining-induced seismicity and rock bursting, have since been constantly increasing. In May 2020, a major seismic event with a local magnitude of 4.2 caused severe damage to parts of the mine (Dineva *et al*, 2022; Boskovic, 2022). Only very few mines in the world have ever experienced such a large seismic event and in the event. Only the event timing, which occurred shortly after production blasting during the ventilation period, prevented multiple fatalities.

While still planning to resume the production from the affected damaged area, production targets are becoming increasingly difficult to achieve. Currently, mine planning is brought on a stable course with sustainable and achievable production targets based upon geotechnical considerations. In view of the planning of production at much greater depths and below current main levels, the legitimate question arises though whether a significant production can ever be achieved with SLC again. In addition, other technical challenges, such as ventilation, haulage and hoisting or logistics, related to mass production at great depth and future sustainability requirements must be mastered. The development of new mining methods is a possible solution. However, new mining methods are sparsely developed. Costs and risks involved are immense. Success may ultimately imply large merits due to complete process changes. In many cases only modifications to existing methods are made, which may be not sufficient to address the current challenges in deep mass mining within LKAB.

The raise caving mining method itself can be an abstract term for non-mining people, but in fact the right selection of mining method could be all decisive if a specific project is profitable or not. This is particularly important for LKAB when it comes to the development of entirely new underground mines such as in the Svappavaara ore fields (Sormunen and Saiang, 2022) or the Per Geijer deposit. The motivation for radical process changes requires from the very beginning continuous persuasion work with different decision-makers – technical and non-technical people – at different levels. Vital is a proper communication strategy and material, which correctly conveys the technical advantages and emphasizes the potential for opportunities on a holistic picture. For this reason a video animation of raise caving including the main method steps was produced as well as a large 3D model was printed in a scale of 1:1500. Furthermore, the way of communication itself plays a decisive role too, because individual persons or departments may understand changes or correction as their personal mistake made in the past. SLC might have very well been the ideal mining method given the past conditions. Fundamental for mining is though that there is no perfect set-up that works throughout the life of a mine, as mining conditions change with depth and with progressive mining. Hence, it is all decisive to make necessary changes and corrections timely in the long-term extraction of large orebodies as found in LKAB mines.

Undeniably, raise caving has many potential advantages and could become a real game changer (Ladinig *et al*, 2022). Particularly noteworthy is the potential establishment of a favourable stress environment, which should lead to a stable production. By contrast, the suggested SLC adaption for greater depths (Quinteiro and Nordqvist, 2022) does not explicitly tackle the root causes of seismicity. However, the extensive experiences from using SLC cannot be compensated for, and it is therefore of outmost importance to create a high level of trust in the new method. Depending on the actual mine site, decisions on the expansion of mines below current main levels need to be taken as early as Q4 2025 and likely the choice of mining method(s) will become the pivotal question. Diligent desktop work and convincing results from a test site in due time are therefore indispensable (Karlsson, Ladinig and Grynienko, 2022). Thereby the choice of the test site deserves special attention. The risks of disrupting the SLC must be weighed up with the required representativeness of test conditions. This requires an active dialogue between many different parties to find an optimum, but also an environment of ownership across departments to enable prestige less development work. Finally, learnings on destress and yielding pillar concepts may very well also become applicable for future SLC designs.

A new mining method requires many different aspects to be investigated. The development of the mining method itself is deliberately decoupled from the machinery. However, in both cases transparent requirements must be set *a priori*, and these must also be prioritised. Provided a diligent test planning and documentation, the results achieved can be compared to alternative approaches, which should be evaluated in the very same manner. Long-term development work always requires perseverance, and usually people and organisations would change along its way. This poses a further challenge to properly comply with the pre-set requirements.

CONCLUSIONS

The design of raise caving is discussed from a system engineering and management perspective. It is found that the development of the method is complex and that it requires to involve both, technical and non-technical personnel. Besides the overall view from system engineering and management, technical aspects, mainly on the rock engineering design, are highlighted. It is again shown that raise caving is also a complex system on a technical level, where many different functional requirements and dependencies must be combined to a working system. From a rock engineering point of view, the mine layout and mining sequence are critical for success. Therein, pillars have a prominent role. Reasons for this are that the functional requirement of pillars change considerably from the de-stressing to the production phase and that the current knowledge on pillar strength and behaviour is very limited. In order to achieve a functioning raise caving system, fundamental research and development are necessary. A comprehensive research and development program has been launched by LKAB for this reason. The outline and discussion on requirements in this paper assist to allocate resources in the development of the method appropriately to ensure an efficient, targeted development.

ACKNOWLEDGEMENTS

This paper originates from the raise caving development project. The authors would like to thank LKAB for funding this work, for allowing the publication as well as for the close and joint collaboration in the raise caving development project.

REFERENCES

Andrews, P G, Butcher, R J and Ekkerd, J, 2019. The geotechnical evolution of deep level mechanized distress mining at South Deep, in *Proceedings of the Ninth International Conference on Deep and High Stress Mining*, (ed: W Joughin), pp 15–28 (The Southern African Institute of Mining and Metallurgy: Johannesburg).

Boskovic, M, 2022. Challenges of resuming the production after a major seismic event at LKAB's Kiirunavaara mine, paper presented at the 10th International Symposium on Rockburst and Seismicity in Mines (RaSiM), Tucson.

Dahnér, C, Malmgren, L and Boskovic, M, 2012. Transition from Non-seismic Mine to a Seismically Active Mine: Kiirunavaara Mine, paper presented at the ISRM International Symposium – EUROCK 2012, Stockholm.

Dineva, S, Dahnér, C, Malovichko, D, Lund, B, Gospodinov, D, Agostinetti, N P and Rudzinski, L, 2022. Analysis of the magnitude 4.2 seismic event on 18 May 2020 in the Kiirunavaara mine, Sweden, paper presented at the 10th International Symposium on Rockburst and Seismicity in Mines (RaSiM), Tucson.

Gams, P 2022. The role of pillars in raise caving, MSc thesis, Montanuniversitaet Leoben, Leoben.

Gipps, I and Cunningham, J, 2011. ROES® – Automated Rock Extraction, in *Proceedings Second International Future Mining Conference*, pp 35–39 (The Australasian Institute of Mining and Metallurgy: Melbourne).

Gipps, I, Cunningham, J, Cavanough, G, Kochanek, M and Castleden, A, 2008. ROES® – A Low-Cost, Remotely Operated Mining Method, in *Proceedings Tenth Underground Operator's Conference*, pp 147–156 (The Australasian Institute of Mining and Metallurgy: Melbourne).

Itasca, 2016. FLAC, version 8.0 (by Itasca Consulting Group Inc., Minneapolis, Minnesota)

Jager, A J and Ryder, J A, 1999. *A handbook on rock engineering practice for tabular hard rock mines* (The Safety in Mines Research Advisory Committee (SIMRAC): Johannesburg).

Karlsson, M, Ladinig, T and Grynienko, M, 2022. Test mining with raise caving mining method: one-time chance to prove the concept?, in *Proceedings of Caving 2022: Fifth International Conference on Block and Sublevel Caving* (ed: Y Potvin), pp 667–682 (Australian Centre for Geomechanics: Perth).

Ladinig, T, 2022. A contribution towards the practical implementation of stress management concepts in underground mining, PhD thesis, Montanuniversitaet Leoben, Leoben.

Ladinig, T, Wagner, H, Bergström, J, Koivisto, M and Wimmer, M, 2021. Raise caving – a new cave mining method for mining at great depths, in *Proceedings Fifth International Future Mining Conference*, pp 368–384 (The Australasian Institute of Mining and Metallurgy: Melbourne).

Ladinig, T, Wagner, H, Karlsson, M, Wimmer, M and Grynienko, M, 2022. Raise Caving – A Hybrid Mining Method Addressing Current Deep Cave Mining Challenges, *Berg – und Hüttenmännische Monatshefte*, 167(4):177–186.

Ladinig, T, Wimmer, M and Wagner, H, 2022. Raise Caving – A novel mining method for (deep) mass mining, in *Proceedings of Caving 2022: Fifth International Conference on Block and Sublevel Caving* (ed: Y Potvin), pp 651–666 (Australian Centre for Geomechanics: Perth).

Quinteiro, C and Nordqvist, A, 2022. Preliminary results from tests using sublevel caving with 40 m sublevel height at LKAB, in *Proceedings of Caving 2022: Fifth International Conference on Block and Sublevel Caving* (ed: Y Potvin), pp 699–712 (Australian Centre for Geomechanics: Perth).

Ryder, J A and Jager, A J, 2002. *A textbook on rock mechanics for tabular hard rock mines* (The Safety in Mines Research Advisory Committee (SIMRAC): Braamfontein).

Salamon, M, 1970. Stability, instability and design of pillar workings, *International Journal of Rock Mechanics and Mining Sciences and Geomechanics Abstracts*, 7(6):613–631.

Sormunen, M and Saiang, D, 2022. Future mining at LKAB Svappavaara: potential to combine caving and stoping methods, in *Proceedings of Caving 2022: Fifth International Conference on Block and Sublevel Caving* (ed: Y Potvin) pp 417–432 (Australian Centre for Geomechanics: Perth).

Watson, B P, Pretorius, W, Mpunzi, P, du Plooy, M, Matthysen, K and Kuijpers, J S, 2014. Design and positive financial impact of crush pillars on mechanised deep-level mining at South Deep Gold Mine, *Journal of the Southern African Institute of Mining and Metallurgy*, 114(10):863–873.

Wimmer, M and Nordqvist, A, 2018. Present-day sublevel caving functionality uncovered – what's next?, in *Proceedings of the 12th International Symposium on Rock Fragmentation by Blasting*, (eds: H Schunnesson and D Johansson), pp 469–480 (Luleå University of Technology).

The design and implementation of the Kibali gold mine off-shaft development

T Peters[1], I Thin[2], I Traore[3] and O Mulume[4]

1. FAusIMM, Principal Mining Engineer, Piran Mining Pty Ltd, Perth, WA, 6076.
   Email: tim.peters@piranmining.com
2. MAusIMM, Principal Geotechnical Engineer, KSCA Geomechanics Pty Ltd, Perth, WA, 6076.
   Email: iain.thin@kscageomechnics.com.au
3. FAusIMM(CP), Group Underground Mine Planning Manager, Barrick, Africa and Middle East.
   Email: ismail.traore@barrick.com
4. Fixed Plant Superintendent, Barrick, Kibali Gold Mine, DRC.
   Email: olivier.mulume@barrick.com

ABSTRACT

The Kibali Gold Mine is currently comprised of several open pits, a 3.5 Mtpa underground mine and a 7.5 Mtpa processing plant. Development of the underground operation commenced in December 2012, with initial access established through twin surface declines, followed by the commencement of a blind sink 8 m diameter rock hoist shaft in March 2013. Stope production started in December 2014 with ore and waste trucked to surface. The 755 m deep blind shaft sink was completed in August 2015 together with the cutting and equipping of three main station levels. Two of these levels provide access for the establishment of the underground materials handling system as well as the main dewatering system for the mine.

This paper covers the process of determining the optimal location of the Kibali hoisting shaft relative to project-specific critical factors, and then focuses on the design and implementation journey associated with the off-shaft development that essentially encompasses the Material Handling System (MHS) for the mine. The journey starts from the findings of the previous owners Feasibility Study (FS), a subsequent Optimised FS, followed by detailed engineering through to implementation. It was identified during implementation that a significant change in the designed off-shaft layout was required following detailed underground geotechnical and hydrogeological investigations. In addition, the paper provides details on the MHS performance since commissioning to steady state, and on key learnings relating to the overall project process.

INTRODUCTION

The Kibali Gold Mine, (a joint venture company between Barrick formerly Randgold Resources Limited, who manage and operate the mine, AngloGold Ashanti and Société Miniére de Kilo-Moto), is now one of the largest gold mines in Africa, located in the far north-east of the Democratic Republic of Congo (3° 13' N, 29° 58' E). By Australian standards, the location of the mine is considered remote, being 1800 km west of the closest port of Mombasa in Kenya.

The topography of the project area consists of undulating terrain, with altitudes ranging between 700 m and 1500 m above sea level. The typical terrain is illustrated in Figure 1.

FIG 1 – Typical view of the terrain in the immediate vicinity of the Kibali Project (2011).

PREVIOUS FEASIBILITY WORK – SHAFT AND OFF-SHAFT

A conceptual shaft haulage layout was first proposed in 2009. This layout was established on broad rules of thumb as part of a trade-off assessment. This work was based on the design physicals established as part of a Feasibility Study conducted by previous owners.

As shown in Figure 2 the off-shaft design detail was limited at this stage.

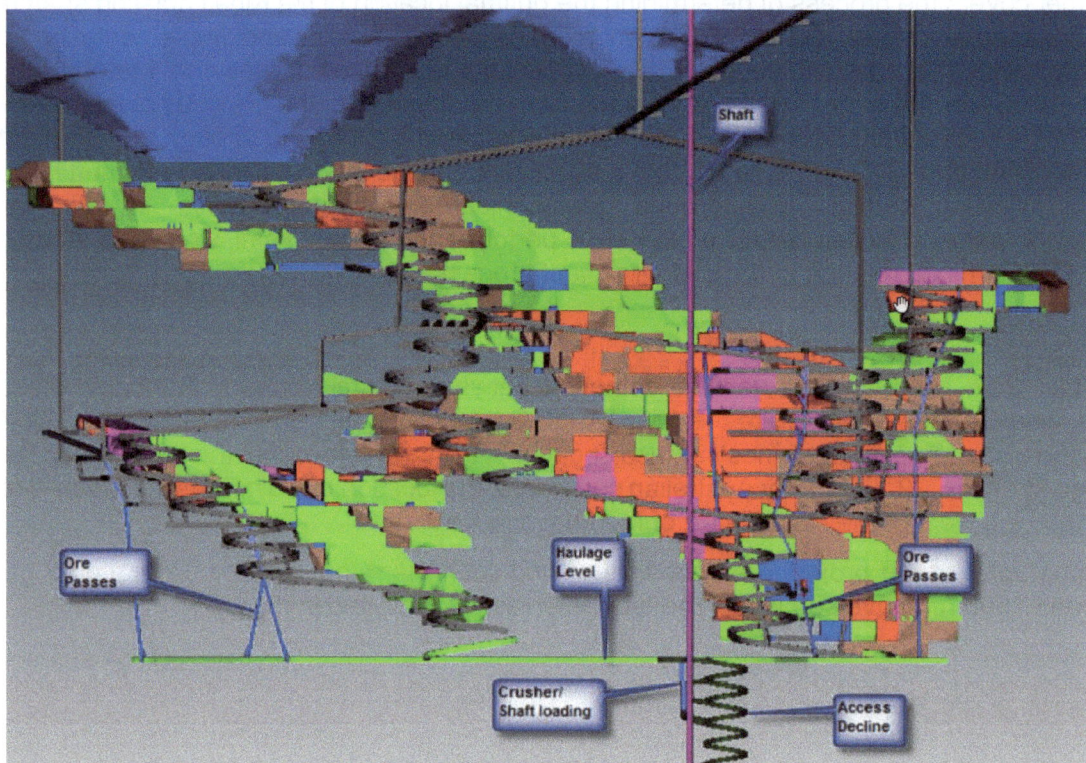

FIG 2 – 2009 trade-off assessment design.

When Barrick (Randgold) took control of the project in December 2009 the previous study work was reviewed in detail. The outcome of this review was to initiate an Optimised Feasibility Study (OFS). The intent being to then implement the study without further delay.

The OFS scope in relation to materials handling (MH), was to proceed with a hoist shaft layout. This initiated a significant program of data gathering to support a robust MHS design considering the limited study detail previously completed. To provide a catch-up phase, several trade-off studies had to be undertaken and completed in parallel with the OFS timeline.

SHAFT LOCATION FACTORS

A fundamental component of the proposed MHS for Kibali was to select the optimal shaft position. During the OFS the key factors determined as project critical for the shaft location were identified as follows:

- mineral resource limits
- geotechnical conditions
- hydrogeological environment
- hydrology
- topographical setting
- process plant location
- Blast Exclusion Zone (BEZ)
- future expansion of the underground operation
- provision of mine services.

A ranking process was subsequently conducted by the study team to assist in determining the optimal location. This considered two main characteristics of each factor: these being criticality and flexibility. For example, the BEZ is a critical input in that there can be no justification in placing the shaft within this zone. Whereas when considering future mine expansion, its criticality is low, although there may be cost implications.

Since it is possible that several locations may produce a similar final scoring a cost-benefit analysis was then conducted to identify the final position from the short-listed locations.

Figure 3 shows the proposed general layout of the underground mine development with the approximate limits of the KCD open pit which in part covers a low-lying area that was previously a man-made lake, originally used to provide water for historical mining activities in the area.

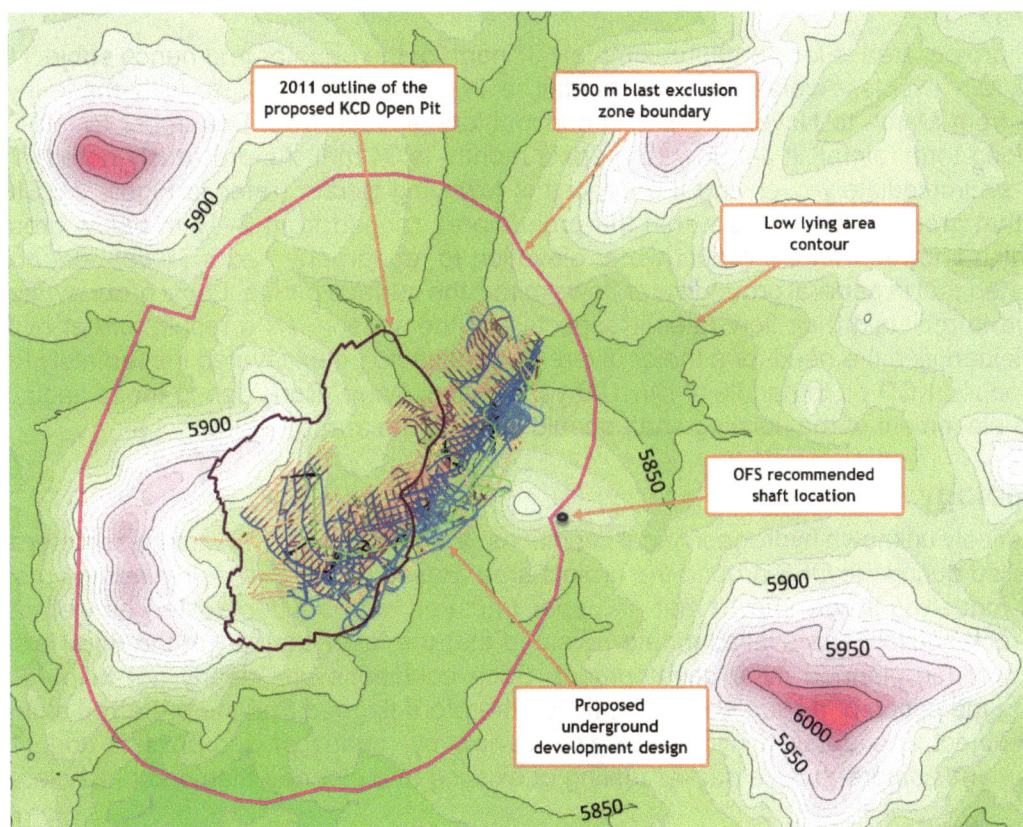

FIG 3 – Topographical plan of the mine area overlaying the OFS underground design and KCD pit outline.

A brief discussion on each of the key location criteria is provided.

Mineral resource limits

At an OFS stage of study, the level of confidence in the mineral resource limits is expected to be high. With a good understanding of the geological footwall position, it is possible to confidently layout the offset required from the footwall to the proposed shaft barrel location. This process considers the various MHS design features required between the tip point and the shaft, such as crushers, ore bins, loadout points, etc.

Establishing the shaft as close as possible to the known centre of gravity of the mineral resource will also ensure optimal efficiency of the underground systems used to transport ore and waste to the shaft. At this stage of study, geological knowledge will also provide an indication of the likely direction of resource growth. Depending on corporate strategy this may be considered when selecting the hoist shaft position.

Geotechnical conditions

Based on the geotechnical drilling conducted during the OFS, the overall rock mass conditions within the footwall and hanging wall were assessed. The footwall conditions were seen to be more favourable in terms of competency where the quality was classified as Fair to Good. Hence locating the shaft in the footwall was selected, fortunately, this aligned with the other criteria.

Longer term geotechnical items to also consider is the relaxation of the rock mass adjoining the KCD pit and the influence this may have on the shaft barrel.

Blast Exclusion Zone (BEZ)

The KCD pit mines the up-plunge component of the deposit. The BEZ offset applied was a 500 m buffer from the proposed crest of the KCD pit outline. This zone is indicated in Figure 3. The BEZ is considered a high-risk area for placement of the headframe or winder house. Hence the shaft must be located outside of this zone.

Hydrology

The Kibali project area is located approximately 3° north of the equator and hence subject to severe equatorial rain events. Average annual rainfall is in the order of 2.5 m, with the rainy season extending from March to November with a nominal 'dry' season from December to mid-February, although long-term rainfall in January (the driest month) is 34 mm. As previously noted, part of the surface area immediately overlying the deposit is low-lying terrain (refer to Figure 3). Study work revealed that groundwater levels within the project area ranged from 2–17 m below ground level. Analysis also showed that the groundwater elevation to topography had a 76 per cent correlation. This indicated that in general groundwater flows mimic the surface profile. Local groundwater springs were found to maintain their flow rates irrespective of the season. As a result, further hydrological study work identified the need for a major diversion channel to be excavated immediately to the east of the proposed KCD pit to handle a 1-in-100 year rainfall event. Considering these findings, it was concluded placement of the hoisting shaft should avoid these areas.

Hydrogeology

Due to relatively unknown hydrogeological conditions, a series of vibrating wire piezometers (VWP's) were installed during the OFS around the project area. Based on the data retrieved and resultant 3D numerical modelling it was shown that an upper fractured rock aquifer existed as well as a lower fractured rock aquifer. The available data revealed that the latter aquifer (at the time) extended to the bottom of the planned mine development. Such conditions are clearly not favourable when considering the option of a blind shaft sink. It was therefore flagged that the preferred shaft location must be subject to a detailed hydrogeological review by means of geo-physics (identifying key structures), and comprehensive packer testing of existing and planned diamond drill holes.

Topography

Ideally, a hoisting shaft should be located where minimal surface earthworks are required. It is not so much the operation of the shaft but the construction of a shaft that requires the most real estate. This can be a conflicting factor when compared to near-surface geotechnical conditions where a location on high ground may be seen as preferred in minimising the pre-sink requirements due to a lesser depth of weathering. It is also important to consider the accessibility of the shaft area from the perspective of connecting to services, as well as personnel and equipment. Ideally, the aim is to minimise distances from existing or other planned mine infrastructure.

A significant related factor was the planned KCD pit. The project schedule showed that the shaft would be in construction while the KCD pit was in operation. Although the immediate shaft area may be outside of the BEZ, related access roads, service connections etc, also needed to be considered and placement determined by topographical features.

Process plant location

Positioning the shaft close to the proposed processing plant provides obvious benefits in terms of minimising overland transport capital expense (CAPEX) and operating expense (OPEX) commitments. In addition, aligning the shaft loadout toward the plant run-of-mine area is considered optimal provided the gradient between the two areas is not excessive.

Provision of mine services

With the decision to install a hoisting shaft, it makes good economic and environmental practice to also use this significant asset to provide key services to the underground operation. Key services can be primary ventilation (normally intake), high voltage power (ring main connection), main dewatering column, communications (typically primary fibre optic cable), compressed air (optional), slick line (essential for timely underground construction works) and by no means least, a second means of egress. Clearly, if the shaft is located too distant from the centroid of the main mining area the benefit of placing these services via the shaft will diminish both in efficiency and cost-benefit.

Final shaft position

Based on the analysis conducted, the OFS recommended placing the shaft on the eastern slope of the 'shaft terrace' (as shown in Figure 3). This position satisfied all of the criteria detailed, although follow-up hydrogeological and geotechnical investigations were flagged to be completed prior to implementation.

OPTIMISED FEASIBILITY OFF-SHAFT DESIGN

Based on the shaft being located as shown in Figure 3 the OFS then determined the necessary off-shaft development to connect to the rest of the mine infrastructure. The proposed off-shaft development consisted of three main levels. These being:

- Mid Shaft Level
- Haulage/Production Level
- Crusher Level.

The relative position of these levels to the mine design is shown in Figure 4. It is noted the shaft barrel is approximately 350 m offset from the footwall haulage drive.

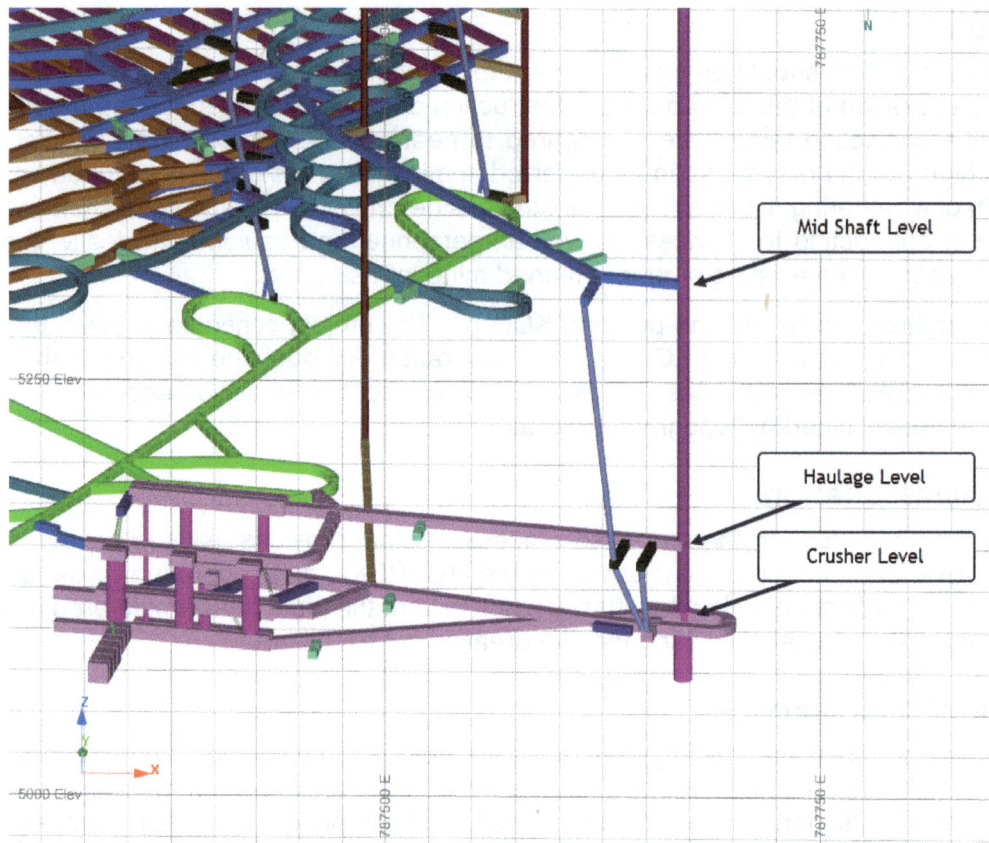

FIG 4 – Orthogonal view looking north of the OFS hoist shaft and off-shaft development design.

Post OFS actions

Immediately following acceptance and approval of the OFS by the joint venture several key actions were initiated as part of the detailed engineering design phase. In reference to the mining team these actions included:

- Completion of further surface geotechnical drilling to enable assessment of the proposed winder house and shaft headframe foundations.

- Geotechnical drilling for assessment of the various surface vent shafts.

- Detailed trade-off analysis of the preferred method of materials handling on the main haulage.

- Trade-off assessment to increase the shaft diameter from 7 m to 8 m.

- Trade-off to assess the benefit of installing a Mary-Ann cage within the hoist shaft.

- Review the option of accessing the shaft bottom via a decline versus an internal ladder system.

- Reassess mine dewatering capacities based on new hydrogeological data and modelling. This directly influences the off-shaft development relating to the main pump station, clearwater dams, clarifiers, mud handling etc.

- Detailed risk assessments by the owners' team specifically relating to the blind shaft sink.

As a result of these agreed work programs in relation to the shaft and off-shaft area the following changes were subsequently incorporated prior to detailed design and implementation:

- The shaft barrel and winder house were realigned marginally to better suit the ground conditions within the shaft terrace area with specific regard to the winder foundation specifications. This resulted in relocating the shaft barrel approximately 10 m to the north of the position identified in the OFS.

- The proposed shaft diameter was increased to 8 m (finished).

- The shaft barrel design was updated to include the installation of a MaryAnn double-deck cage.

- The off-shaft design was amended to include a decline to access the shaft bottom.

- The main haulage was redesigned from a truck-based layout to an automated rail-based system. This subsequently led to significant changes in the main tip areas and the separation of the twin crushers into separate crusher chambers.

- The inclusion of an additional exhaust vent-raise from surface located approximately 95 m from the main shaft.

These changes versus the OFS design for the Mid Shaft, Haulage Level and Crusher Levels are illustrated in Figures 5, 6 and 7 respectively.

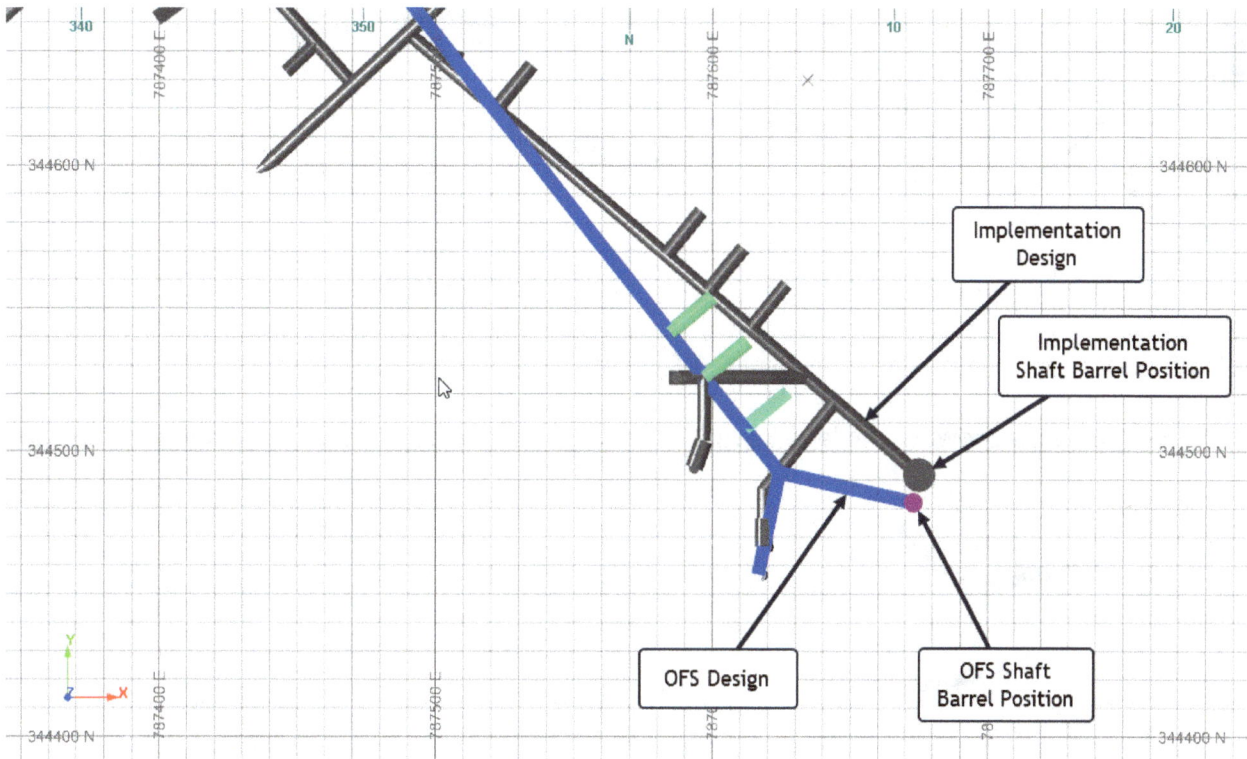

FIG 5 – Plan view showing mid shaft OFS design (coloured) and the implementation design (grey).

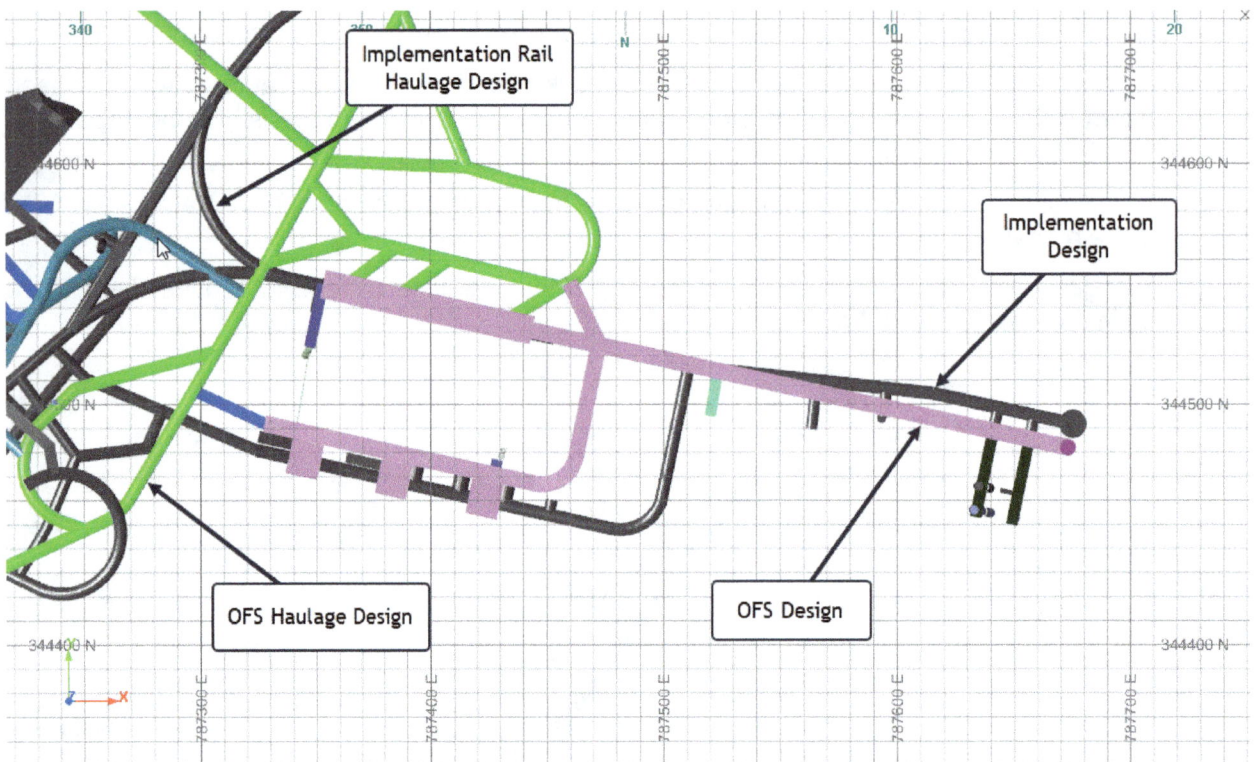

FIG 6 – Plan view showing Haulage Level OFS design (coloured) and the implementation design (grey).

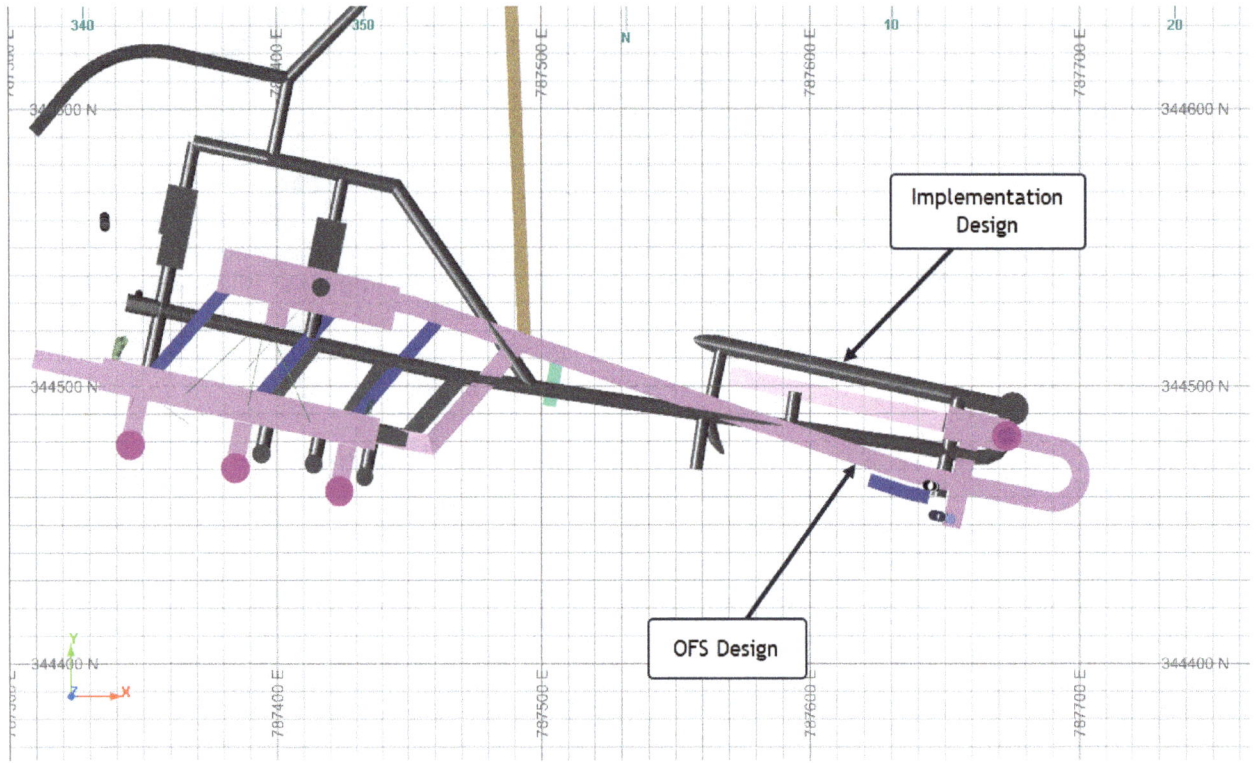

FIG 7 – Plan view showing Crusher Level OFS design (coloured) and the implementation design (grey).

GEOTECHNICAL REVIEW

The MHS geotechnical drilling initially started (in 2012) with a dedicated geotechnical drill hole that followed the entire designed trace of the shaft. This hole was drilled from surface, with the recovered core geotechnically logged. In addition to this source of data, an acoustic televiewer survey was also completed which provided structural data. Combined, these two data sets were used to complete a shaft stability assessment. In parallel a 3D Mining Rock Mass Model (MRMM) was developed which encompassed the entire underground planned development (including the MHS) and stope production areas. The development of the 2012 MRMM (which represented the first model for the project) utilised all available drill hole data at that time (covering geotechnical and geological drill holes from both surface and underground).

While the stability assessment confirmed the planned location of the shaft was suitable in that no instability issues were identified, the initial MRMM highlighted the presence of three geotechnically significant structures. One of these major structures was identified in the north-eastern part of the mine (away from the proposed MH area). This structure ranged from 0–100 m in width suggesting the zone continued beyond the limit of the available data.

At this time no further surface drilling from the shaft terrace was possible due to the increasing level of shaft construction activities. In addition, it was concluded that attempting to target the MHS area with further drilling from the shaft terrace would unlikely provide meaningful data due to drill hole inclinations limiting the ability to provide adequate angles of intersection. A decision was therefore made to continue with the geotechnical drilling from underground via the off-shaft development once suitable drill sites became available.

By 2014, off-shaft development was progressing on the Mid Shaft and Haulage levels, and with the proposed creation of two crusher chambers rather than the originally proposed single one (refer to Figure 7), it was decided to undertake targeted geotechnical drilling in the areas that would house the two chambers (Figure 8).

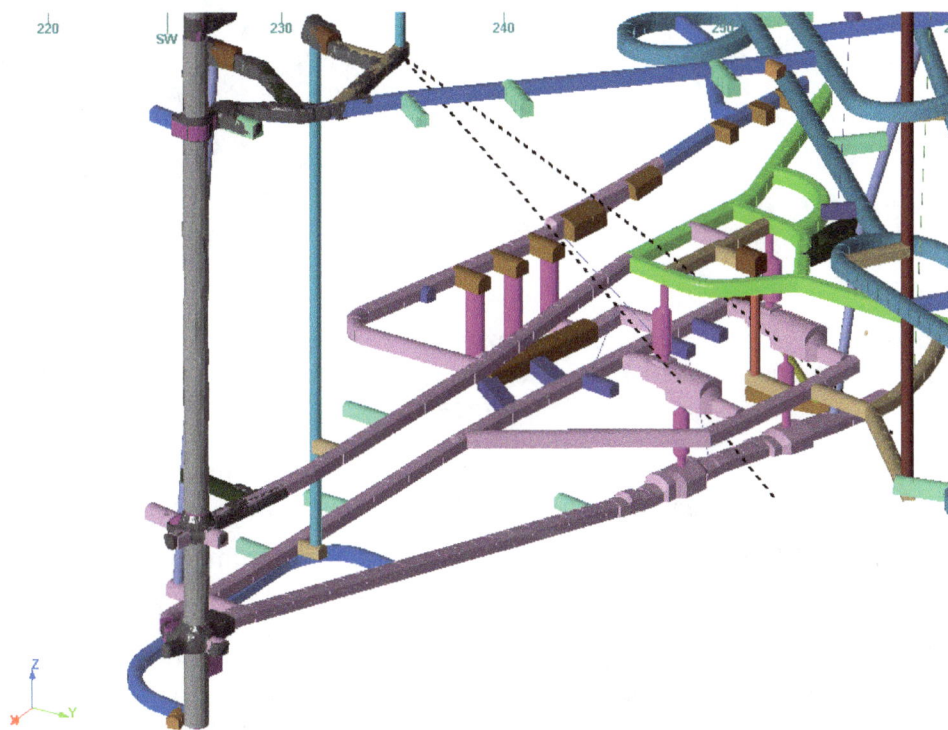

FIG 8 – Isometric view looking south-west showing the completed shaft barrel and off-shaft development (grey solids), the planned development and proposed targeted geotechnical drilling from the Mid Shaft Level for the two crusher chambers.

The subsequently recovered core from these two holes were geotechnically logged and the results assessed to understand the rock mass conditions for the proposed crusher chamber locations. This revealed the presence of previously undiscovered geotechnically significant structures intersecting the planned crusher chamber locations.

In reviewing the interpretation of these newly identified geotechnical structures it was apparent that the presence of such structures intersecting oversized excavations would result in long-term instability issues. This instigated the need for further targeted drilling and geotechnical analysis to gain more information and understanding about the rock mass conditions and geotechnically significant structures in relation to the key MHS infrastructure.

In light of this new geotechnical information, the location of the crusher chambers had to be urgently reassessed. This approach is in line with the hierarchy of controls, ie it is better to re-engineer the design to avoid features that pose an increased risk to a project (proactive), rather than try and fix up the consequences later (reactive). Consequently, a further geotechnical drill program was designed (Figure 9) and implemented in late 2014, with the drilling undertaken from the off-shaft Haulage and Crusher Levels.

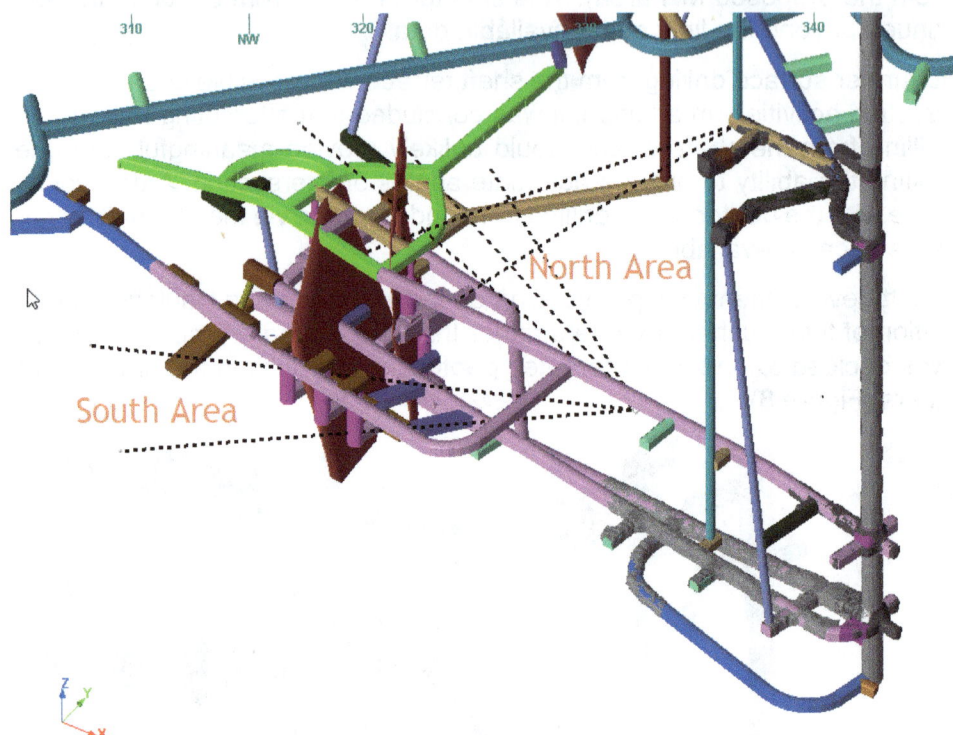

FIG 9 – Designed drill holes to investigate the extent of structures within available real estate to the north and south of the base case MHS layout. New geotechnical structures shown by the red solid surfaces.

The aim of this drilling was to assess the optionality of relocating the main MHS development either to the north or south of the implementation design layout (as shown in Figure 7).

The targeted drilling of the northern area was undertaken and completed first, with the results highlighting challenging conditions; not only for the drilling itself (repeated intersections of groundwater with moderate flow rates, but under high pressure) but also a large exposure of fractured and water bearing Banded Iron Stone (BIF) lithology and further geotechnically significant structures (example shown in Figure 10). Although surface based geotechnical drilling had previously identified the BIF lithology, it was considered competent without any indication of pressurised groundwater.

While groundwater was likely to be present in these early holes the steep inclination and depth from the surface precluded the exposure of groundwater under pressure. Whereas the shallow dipping underground holes not only provided additional drill hole inclination (subhorizontal) but also revealed groundwater pressures and flow rates, improving the project knowledge for this critical area of the mine.

The subsequent drilling towards the south revealed no geotechnically significant structures and far less exposure to the BIF.

FIG 10 – Structure intercept KUG008 314.80 to 316.40 m.

MHS design changes

Considering the newly identified geotechnical structures (identified in January 2015), and the localised characteristics of the BIF, an urgent review of potential options for the off-shaft development was undertaken by the mine design team in Western Australia, in conjunction with the engineering design team in South Africa, the implementation team on-site and the shaft sinking contractor.

While the development off-shaft had already commenced by this time, it did not mean that the mine was committed to the current base case position, although the CV02 decline would have to remain on its current azimuth (refer Figure 11). A joint detailed analysis between the mine design and engineering design teams, concluded that the transfer point from the CV01 to CV02 conveyors would provide a suitable point of rotation for any modification in design of the MHS layout (refer Figure 11).

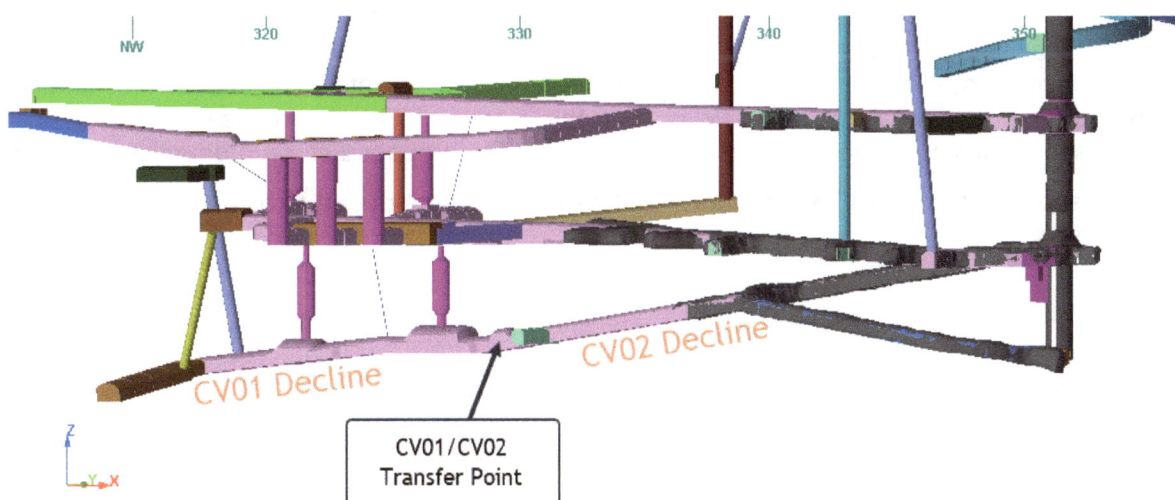

FIG 11 – Isometric view looking north-west showing CV01/CV02 point of rotation.

DESIGN ASSESSMENT

While the additional geotechnical drilling program continued, various design options were assessed in parallel. This enabled the teams to be engaged throughout and ensured appropriate stakeholder engagement.

Following review of numerous design options together with the updating of the MRMM and structural models (in April 2015), the decision was made to rotate the MHS layout by 60° to the south of the implementation design.

This rotation would move as much of the MHS excavations as practically possible away from the BIF lithology and geotechnically significant structures, particularly the numerous oversized chambers involved in the design.

With the agreement to rotate the MHS development to the south, a rigorous change management process was implemented to ensure this fundamental variation did not result in unforeseen consequences for the long-term operability of the proposed MHS. At the time of the decision to rotate the design, it was identified that various engineering modifications would also be required.

During the period of geotechnical drilling conducted from the Haulage Level, the off-shaft development was halted on the Haulage Level and the Crusher Level, apart from the ramp to the shaft bottom. It was therefore time critical for the project to rapidly complete the variation in the detailed design required for implementation. By early May 2015, it was determined that off-shaft development could recommence on the Haulage Level with the exception of the development of the CV02 Decline, since engineering changes were still required on the CV01/CV02 transfer point. Fortunately, this area of development was not on the critical path for the early commissioning of the MHS.

Similarly, development could also recommence on the Crusher Level. This was a key heading since it leads to the main pump station excavation for the mine, as well as the water clarifiers and clear water dams.

Modifications identified as part of the change management process included:

- CV01/CV02 Transfer excavation design.
- Updating of the clarifier/trommel excavations.
- Finalisation of the clear water dam dimensions.

Geotechnical verification of design changes

Once the design modifications were completed, it was decided to confirm its appropriateness from a geotechnical perspective through a further round of 3D mine-wide numerical modelling. The modelling incorporated the most up-to-date geotechnical input data available at that time. The outcome of the numerical modelling was that the rotated MHS design layout would not be compromised by mining-induced effects associated with the proposed stope production sequence.

Figure 12 shows the key MHS excavations that sit outside the induced stress field of the adjacent stoping blocks.

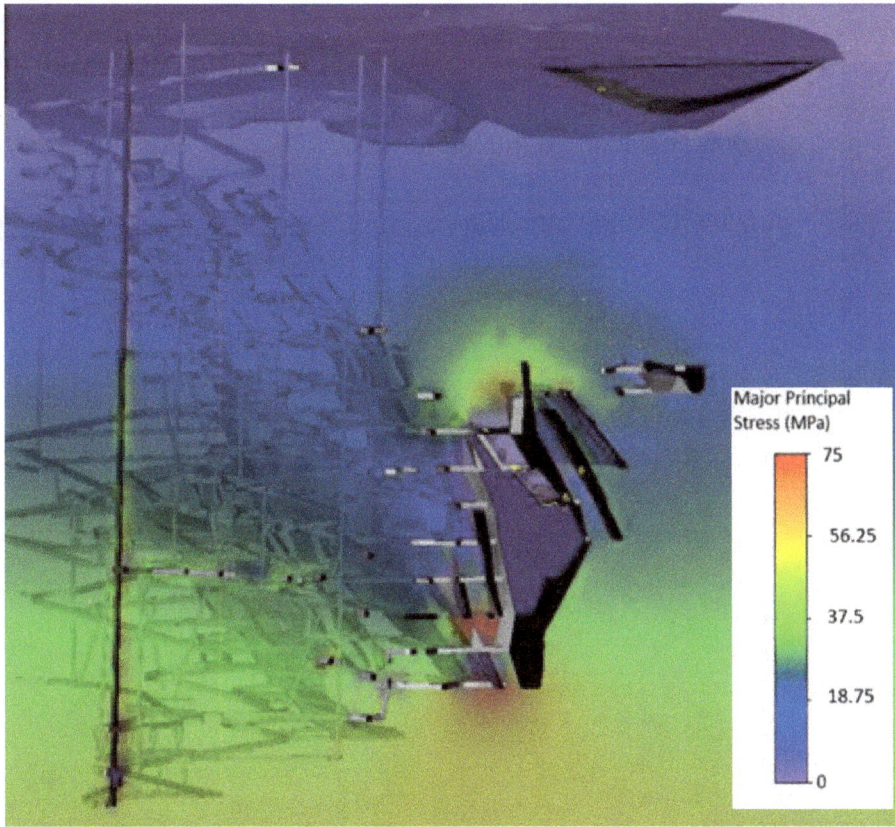

FIG 12 – Cross-section showing the numerically modelled major principal stress in relation to the main haulage shaft at the end of mine life.

FURTHER DESIGN MODIFICATIONS

Service drive

Early in Q2 2016 an update of the off-shaft development plan was undertaken. This update included both the mine development and subsequent equipping of the Load-out Level (CV01), which was identified as on the critical path for ultimate commissioning of the underground MHS.

Due to slower than scheduled development advance in the Conveyor Decline the updated schedule reported a completion date of the Load-out Level in January 2017, whereas the budget (Figure 13) had a completion date of mid-November 2016.

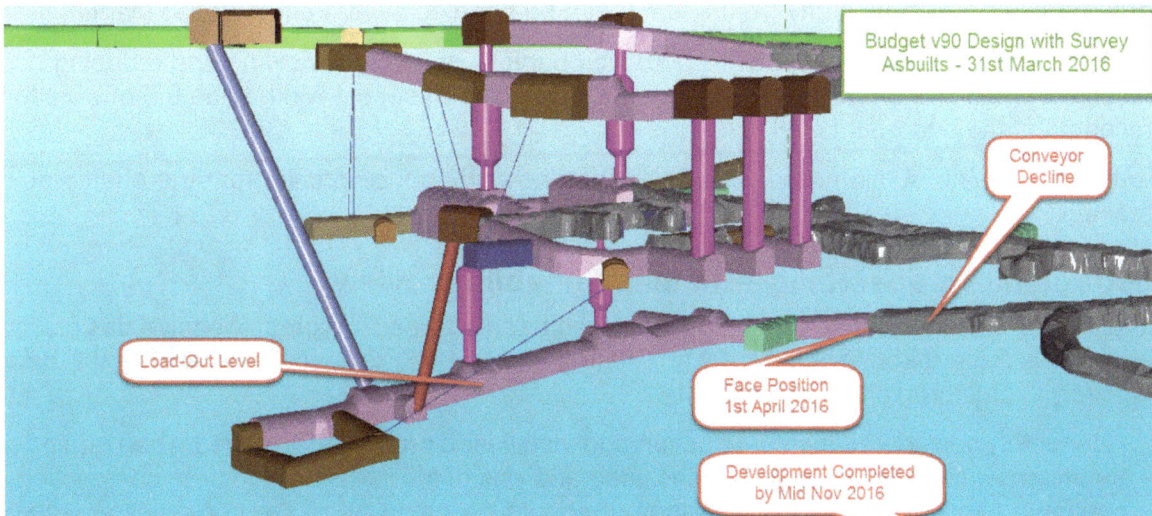

FIG 13 – Updated budget MHS design.

When adding in the equipping schedule activities (civil, structural, mechanical, electrical and instrumentation), the final completion date for the MHS was forecast to be mid-September 2017.

A review was subsequently undertaken by the owner's design team and shaft contractor to assess options to permit earlier commencement of the equipping works, and hence reduce the lead time to shaft commissioning.

This review process identified an option of mining an additional drive that would run parallel to the Load-out Level (Figure 14). This drive was named the Service Drive, since it was considered it would also benefit long-term servicing activities required on the load-out systems by providing direct access to the three load out areas.

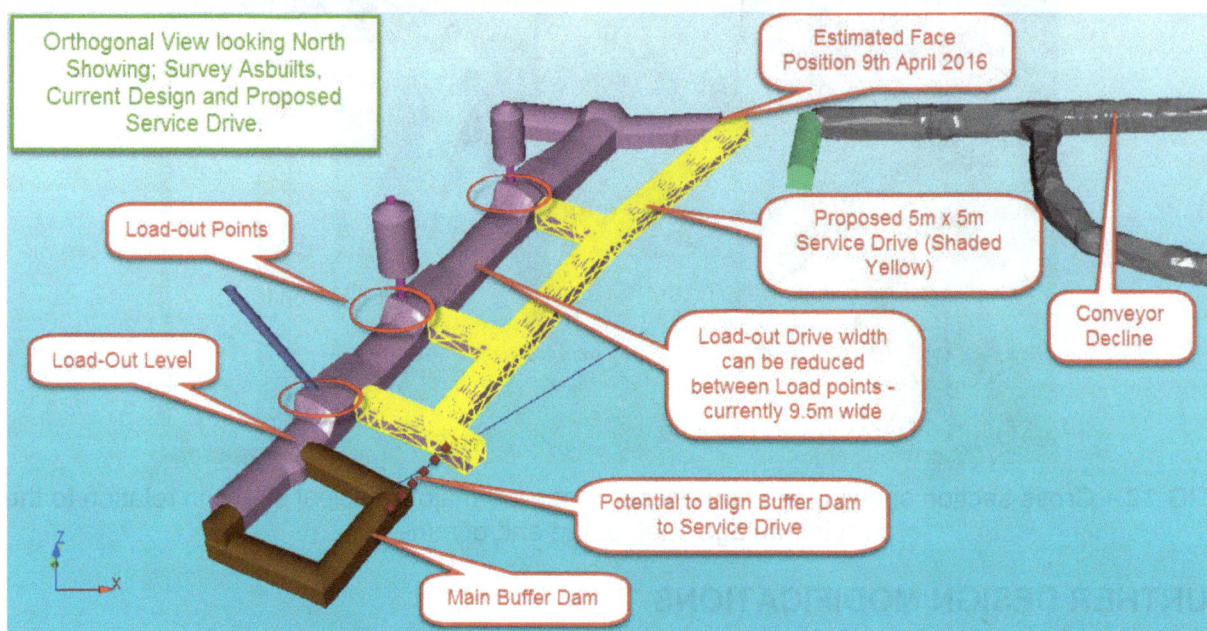

FIG 14 – Proposed design update to the MHS post rotation.

As shown in Figure 14 it was identified that construction work (civils), could then commence as early as mid-August 2016 on the CV01/CV02 Transfer, since the Service Drive provides a bypass function to the active development heading. This process could then be repeated at each load-out point.

It was also identified that if the Service Drive was connected to the Buffer Dam area, loaders can then clean and maintain the dam without having to tram past the CV02 conveyor at any point. Likewise, the relevant gradients can be designed to ensure the Load-out Level remains free draining into the Service Drive, hence any fissure water can be handled away from the fixed plant.

The cost of the Service Drive (~US$750 000) was partially offset by reviewing and reducing (where practical) the width of the CV01 drive, since the original design set the width to be 9.5 m wide for the entire length of the Load-out Level.

Subsequent analysis of this proposal showed that the early ounce profile from the shaft would be maintained.

Constructability, operability and maintainability review

Once all of the modifications to the MHS layout had been finalised, a comprehensive design review workshop was undertaken in South Africa. The key focus of this review was to ensure the detailed designs (civil, structural, and mechanical):

1. Adequately considered the construction and installation processes involved when equipping in an underground environment. This included analysis of the installation logic in conjunction with confirming asbuilt excavation drawings. It should be noted all of the construction work was completed via the shaft. This work therefore required details relating to ensuring slinging of every component down the shaft was scheduled and assessed.

2. Provided for suitable functionality for efficient and safe operation of the MHS taking into account the various operating constraints and general working conditions.

3. Included the practical issues faced when undertaking significant maintenance procedures, such as crusher rebuild, chute replacement, feeder removal, etc.

This design review process took the combined team more than a week to complete and resulted in numerous items being marked up for revision upgrade. It is considered the outcomes from this workshop avoided numerous additional hours in construction and ensured a better operating system was ultimately installed.

INSTALLED MHS OVERVIEW

Final commissioning of the integrated MHS commenced in Q3 2017 which initially utilised waste material. The handover of the commissioned MHS occurred in November 2017.

Figure 15 shows the annualised hoist output from the point of fully commissioned status until the time of writing this paper (August 2022). In comparing the actual tonnage values to the OFS scheduled hoist tonnes it is seen that the MHS has outperformed by a material margin.

This increase in performance is credited to the fact that the MHS design was subject to a program of continuous review throughout the post-OFS period which allowed for identified risks to be managed and mitigated in an efficient manner. Likewise potential system improvements were also addressed. The benefits of this approach were only made possible due to the availability of rapid decision-making and hence implementation of the changes identified. All of this occurred without changing any of the key component specifications established in the OFS.

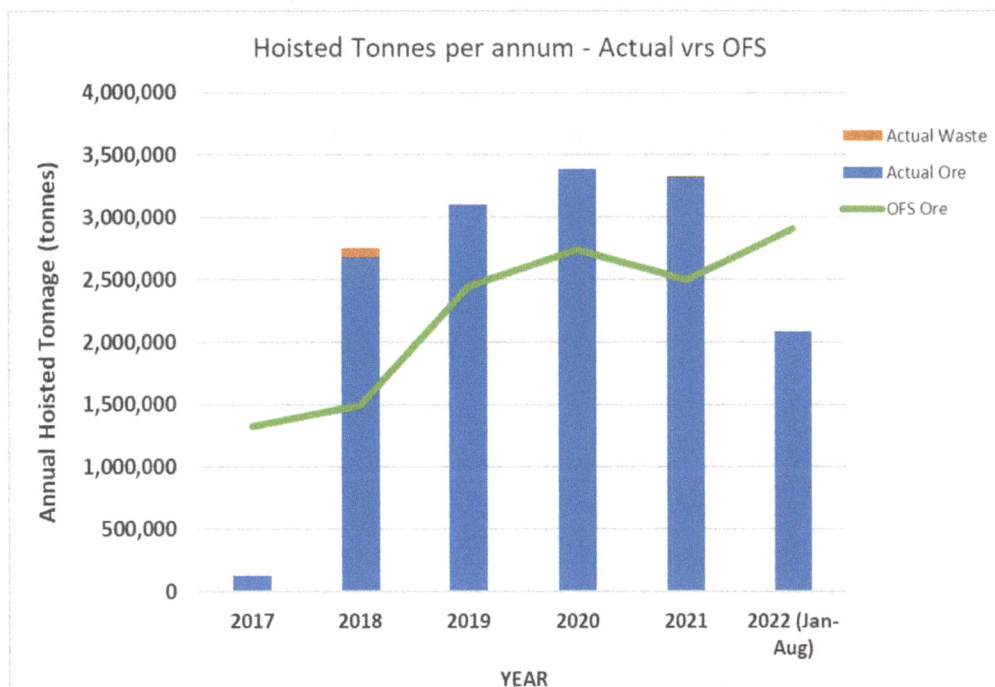

FIG 15 – Graph of total hoisted tonnes per annum from commissioning to date versus the 2011 OFS estimate.

After commissioning of the MHS, various initiatives were undertaken to further optimise the overall MHS performance.

LEARNINGS

In reviewing the journey taken and experienced at Kibali, from feasibility study, through detailed design and final implementation the following key learnings are presented for the benefit of future projects with regard to underground MHS.

- Drilling from the surface has limitations when considering discrete subvertical structures, in particular with reference to targeting oversized critical MHS infrastructure excavations. The project schedule should allow for underground drilling within close proximity to the target at shallower angles compared to surface hole inclinations.

- Limited surface real estate constrains the drilling window. When construction activities commence this constrain will grow.

- Limited dewatering capacity in a blind sink shaft will restrict off-shaft drilling in the event that material flow rates of fissure water are intersected.

- An expectation of change must be accepted during project implementation, this may involve additional design work being undertaken in parallel. The viability of a project may be threatened if the implementation is delayed due to an excessive data-gathering period (hydrogeological and geotechnical).

- There is a balance between having sufficient data to proceed with an acceptable level of risk versus the time required in attempting to define all the unknowns (particularly if attempted to be conducted solely from the surface).

- It is considered that the success of the design changes was achieved due to the close collaboration between the design team (off-site), implementation team (on-site), shaft contractor, and the senior management and executive.

A professional trust established early in the project between the executive and design teams enabled highly effective decision-making throughout the project. This ensured minimal delay between highlighting and solving design challenges, consequential minimising additional costs associated with contractor downtime.

ACKNOWLEDGEMENTS

The authors would like to thank Barrick for permitting the writing and publication of this paper.

Health and safety, and navigating social licence

Integrated multi-disciplinary auditing and benchmarking of catastrophic hazards for underground mining operations

A R Penney[1] and J J L du Plessis[2,3]

1. Technical Manager/Director, AMC Mining Consultants (Singapore) Pte Ltd, Singapore, 018937. Email: apenney@amcconsultants.com
2. Manager Group HSEC Audit, Glencore Holdings, South Africa, 2076. Email: jan.duplessis@glencore.co.za
3. Extra Ordinary Professor, University of Pretoria, South Africa, 0002. Email: jan.duplessis@up.ac.za

ABSTRACT

Safety management, both personal and that associated with catastrophic hazard, helps to define the success of mining companies. Personal safety management, often categorised by slips, trips and falls, has been widely addressed, with numerous approaches successfully adopted. Catastrophic hazard can lead to a single accident (eg ground collapse, underground fires etc) causing multiple fatalities. Catastrophic hazard management in mining is an emerging discipline, and as such there are various pitfalls to demonstrate its effectiveness and success; it is much more than simply meeting regulatory requirements.

Various standards, guidelines, and leading practices are available for reporting key performance indicators. While measuring and reporting safety related key performance indicators (KPIs) such as Total Recordable Injury Frequency Rate (TRIFR), Lost Time Injury Frequency Rate (LTIFR), are important, they often lack detail, and give little insight into the effectiveness of catastrophic hazard management strategies and controls. Catastrophic hazards are low likelihood, high consequence events, and as such demand a different management approach. Companies cannot use the suite of personal safety metrics to help demonstrate management of catastrophic hazards.

In-field verifications have revealed that senior management commitment to safety is often clear, well organised, and resourced. However, there are often some inconsistencies between what management promotes and plan, to what the operational teams understand and implement. Communication and clear understanding of expectations were the common themes identified to help empower supervisors to reinforce requirements of catastrophic hazard management. A lack of specific, measurable, and time-sensitive goals and objectives for the short- or medium-term, making it challenging to track progress was another key observation.

Development of an integrated, multi-disciplinary auditing process executed by an experienced team of technical subject matter experts, provided an in-depth understanding of the requirements of catastrophic hazard management. Areas such as underground fire prevention, explosive storage, electrical safety, mobile vehicle interactions, ground control, inrush, and shafts and winders management have substantial overlap and intertwined processes, and similar strategies for emergency responses. Auditing of catastrophic hazards and safety management systems compliance is only the first step in safety assurance. Completing operational performance audits in technical areas and benchmarking the site verification of safety management documents and verifications forms a critical part to clearly demonstrate industry leading safety practices. No longer is it the external inspection or regular hazard management plan review requirements that dominate, but rather the culture of continual hazard identification, management, controls verification, underpinned by improvement and follow-up verifications of the corrective actions.

INTRODUCTION

Catastrophic hazard assessments are not a new topic to the mining industry, but the application of clear principles and guidelines are continuing to develop. Extensive and varied published works, guidelines, and regulatory requirements are available globally, making a consistent approach often difficult to adopt (NSW Department of Primary Industries – MDG010, 2007; Safework Australia, 2012; US Department of Labor 2015; Standards Australia, 2018). The ability to demonstrate effective management of these catastrophic hazards are often overlooked in the traditional reporting of standard safety KPIs, such as TRIFR, LTIFR, etc. These measures are by no means redundant; in fact, they are often the key statistics to show in well-structured mining operations where no catastrophic events have occurred in a substantial time period. However, they do not truly demonstrate if the catastrophic hazards are effectively identified and managed, or if an element of good luck has prevailed (Standards Australia, 2010). Challenges exist in measuring and tracking the catastrophic hazard management data when focused on standard safety KPIs. Commonly used safety KPIs do not provide the information to measure the effective management of catastrophic hazards. Typically, personnel with these safety reporting responsibilities are experienced in health and safety matters, but may not have the technical experience in the identified catastrophic hazard subject matter areas to truly understand the industry practices and requirements of the appropriate controls.

In 2015, the International Council on Mining and Metals (ICMM) published their Risk Management Principles and Guidelines (ICMM, 2015) resulting in substantial rejuvenation of how many leading mining companies approach catastrophic hazard management. These guidelines included methods to demonstrate and verify the effectiveness of the controls to prevent and mitigate a material unwanted event from occurring. While certain aspects of this guideline were not materially different to processes already in place, the introduction of verification processes to check various stated systems (critical controls) was a substantial step towards effective demonstration of catastrophic hazard management. These ICMM methods build on the risk management requirements contained in ISO 31000:2018 (Standards Australia, 2018).

All mining companies, and personnel working at mining operations understand the need to have thoroughly considered and executed catastrophic hazard management system. The development and maintenance of corporate policies and Principal Hazard Management Plans (PHMP) are a key component to demonstrate catastrophic hazard management. PHMPs are commonly prepared to meet regulatory requirements, and social license expectations, and outline various practices to follow. Undertaking periodic internal or external reviews of PHMPs is an important part of ensuring these documents remain effective for the operation, and adopt latest industry practices and regulatory expectations. Until recently, most PHMPs did not clearly outline the underlying risk assessments, the main material unwanted event(s), and the controls established to prevent and/or mitigate the unwanted event from occurring. Rather, they summarised current knowledge and various practices for safe work.

In essence, a corporate policy and supporting PHMPs are a commitment to the regulator and key stakeholders that you know and manage the catastrophic hazards. However, demonstrating the effectiveness of these PHMPs can sometimes be difficult. Demonstrating that the stated methods for hazard management contained in the PHMP are in place, and meet the expectations of industry peers is a step more challenging. With implementation of corporate policies and guidelines for expectations subject matter area management plans (ground control, explosives, shaft winder and haulage, water/inrush, emergency response etc), improved methods to ensure good practices can be shared, and evaluation of effectiveness of the documents considered.

Conducting reviews (internal or external) in individual subject matter areas are effective, however, they can be quite time and resource demanding when executed as isolated programs. Many subject matter areas have substantial crossover and similarities between the disciplines, often requiring the same knowledgeable people (or risk owners) to be involved over numerous campaigns. Conducting these reviews as an integrated program has two main benefits; ensuring effective alignment and information sharing between disciplines, and reduces the time demands on operations and team members. Getting the maximum benefit from an external audit, the use of subject matter experts in the chosen fields with international experience to understand industry trends and standard practices is vital.

The whole auditing program is based on completely identifying all relevant failure modes, and verifying that all reasonable causes and hazards associated with these failure modes are considered. Ensuring these hazards are correctly identified and assessed so that the right critical controls can be implemented is a fundamental step in being able to demonstrate appropriate catastrophic hazard management. The final aspects are ensuring that management systems are in place to test, track and report the effectiveness of these critical controls.

In this paper, we present the experiences of using subject matter experts (SME) with risk management leaders to demonstrate the effectiveness of the integrated auditing approach of the catastrophic hazard management programs, the improvements being observed across the business, and examples of benchmarking results across the industry.

THE PROCESS

All companies periodically review and updates corporate policies, guidelines, and procedures covering catastrophic risks. Supporting the policies, Glencore established of the Fatal Hazard Protocols; a list of identified hazards which can lead to catastrophic, multiple fatality consequences. The Fatal Hazard Protocols set out the company's expectations and requirements to demonstrate minimum criteria are established and functional.

An assurance group was established to periodically review and verify the operations are meeting the corporate guidance requirements on catastrophic hazard management. The assurance team developed a process to undertake assurance audits, and report to senior management that the hazards are effectively managed. A team of subject matter experts with international operations experience were assembled to assist with this assurance program.

Collectively, the assurance group developed guidelines and workbooks for the audits for each discipline. These guidelines and workbooks were shared with the operations to ensure they understood the focus and intent of the audits. Adopting appropriate risk management processes, and identifying appropriate controls that are specific, measurable, and auditable is a key component for effective risk management.

There are two approaches commonly adopted when undertaking catastrophic hazard audits. The first is to engage a team who can review processes and documents against appropriate standards and regulations. This style is often a desktop or office facilitated review and aimed at ensuring a strong documentation trail is in place. These teams are generally not completely familiar with the technical discipline, and often lack operational experience. The second is to have an experienced auditing leader, and engage subject matter experts in selected fields who have substantial international operations experience. This second approach is a higher-level commitment, reviewing key documents and supporting risk assessments, completing in-field discussions and reviews, bringing their knowledge and experience in the technical area and operations to the program. These groups can often get bogged down in the technical detail, hence the requirement for an experienced and disciplined auditing leader to maintain the team focus.

Using subject matter experts who are independent from the operation with relevant experience in practical mining operations or a broad nature, complimented with experience on mining studies is critical. Independence from the site allows better insight into systems observed at other operations, and will provide an unbiased view of the performance.

To maximise cross-discipline discussions, and reduce the burden to mining operations, the integration of the following subject matter areas was adopted:

- Emergency Response
- Fires and Explosions
- Strata Control
- Inrush and Outburst
- Shafts and Winders
- Bulk People Transportation.

These disciplines all share common hazards, and often have similar mitigating strategies in the unlikely event that an incident was to occur. With the substantial similarities, group discussions can be held with appropriate team members at the same time to ensure consistent information is shared, and similar conclusions on effectiveness of controls can be agreed.

The auditing process involves a dedicated lead, with specialist knowledge in compliance auditing and broad familiarity on all subject matter areas. The team is then typically comprised of three to five other subject matter professionals.

Ensuring the correct material unwanted event identification is the first item to agree on. The effectiveness of the supporting processes is where external subject matter experts add the greatest value to the program once the unwanted event is known and agreed.

Three main categories of audit outcomes are possible against the assessment criteria and workbooks:

- Finding: A substantial deviation from the criteria, or an identified concern which would cause a catastrophic event to materialise at some point in the future. If a finding is deemed of a critical or imminent nature, this item can be escalated for immediate treatment (known internally as a matter of urgency).

- Observation: A minor deviation from the criteria, or opportunity for improvement where a catastrophic event is unlikely to materialise.

- Positive Confirmation: All items expected to be present, and outlined in site documentation are in place and effective.

Using a comprehensive and consistent auditing process, based on the fatal hazard protocols, supported by consistent guidelines and workbooks, enables audit results to be compared across different operations and jurisdictions more consistently.

THE PROGRESS

The plan for the company-wide underground integrated assessments was as follows:

1. Undertake a pre-reading on requested data and documents, filling out sections of the auditing workbooks. Timing of this stage varies depending on the complexity and maturity of the operation(s), language translations, etc.

2. Undertake the in-field audit. The audit is nominally over a 1–2 week period.

3. Report to site of the audit results, including any recommendations for remedial work to correct any findings.

4. Agree to a corrective action list and execution time frame.

5. Plan the verification audit for nominally 18-months' time.

This process is shown in Figure 1.

The catastrophic hazard management process for underground mining assessments was piloted in 2016. The bulk of initial assessments were conducted during 2018 and 2019, taking on board several key learnings during the 2016/2017 audits. Follow-up verification audits commenced during 2017 to actively track the progress against identified issues. The program continues to run with ongoing verification audits being completed on an 18-month period to ensure ongoing compliance to the corporate policies even if all findings and gaps are appropriately actioned. The implementation across all underground operations took approximately four years. The verification programs have maintained an average 18-month period (short delays experienced in 2020 due to covid interruptions) (Figure 2).

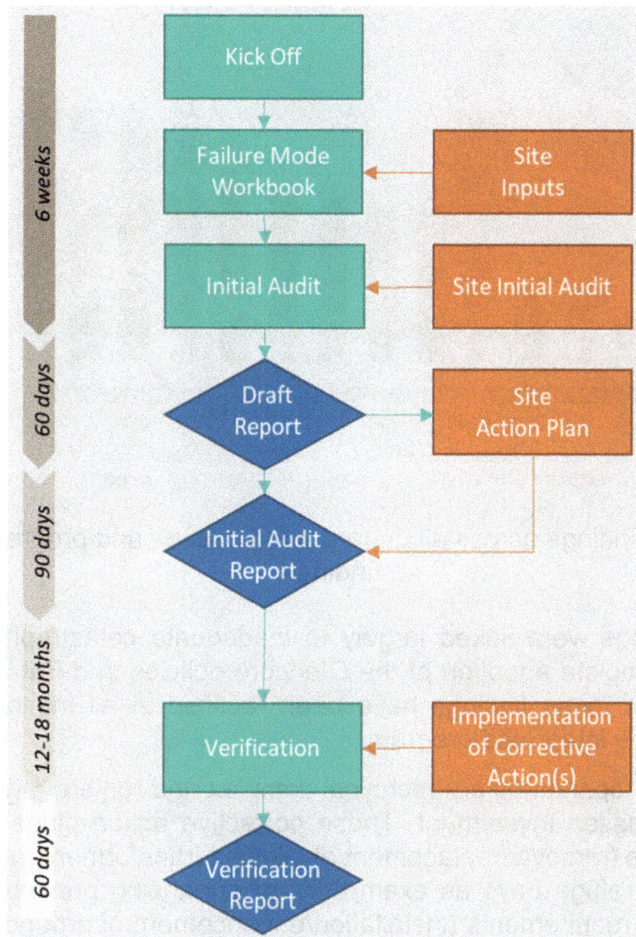

FIG 1 – Summary of the integrated auditing program time frames and deliverables.

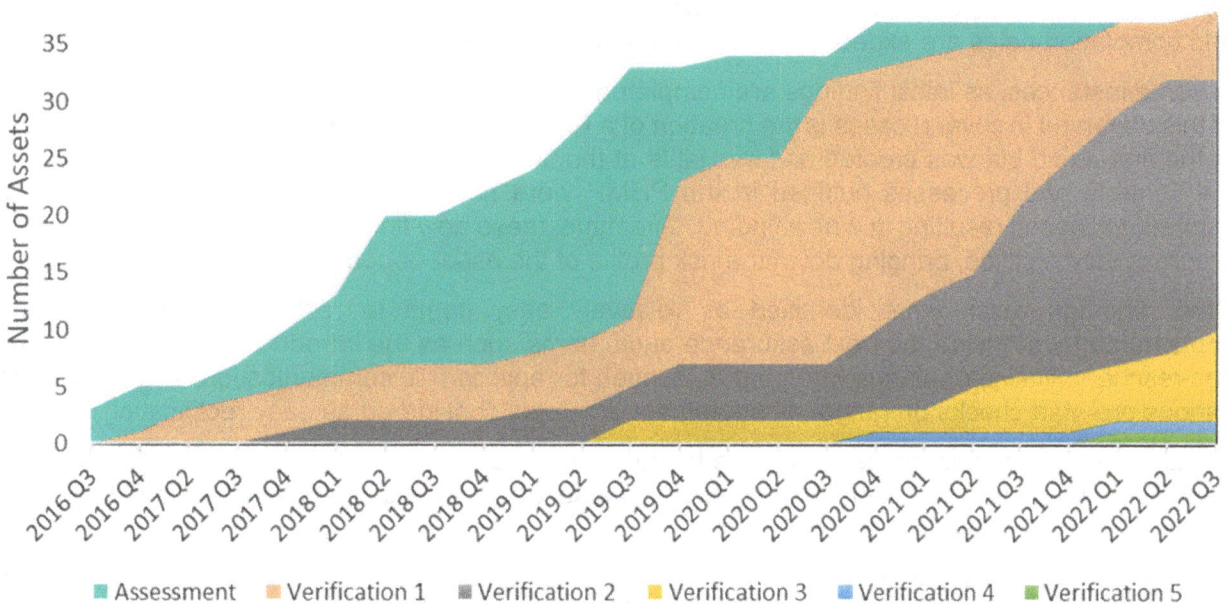

FIG 2 – Cumulative assessment and verification auditing across all underground assets.

The auditing program has produced over 530 findings across the six catastrophic hazard disciplines outlined earlier, for the 37 underground mining complexes (Figure 3). *Note: there are more underground mines in the portfolio, operations which fall under a common operation management team or complex have been combined in some cases.*

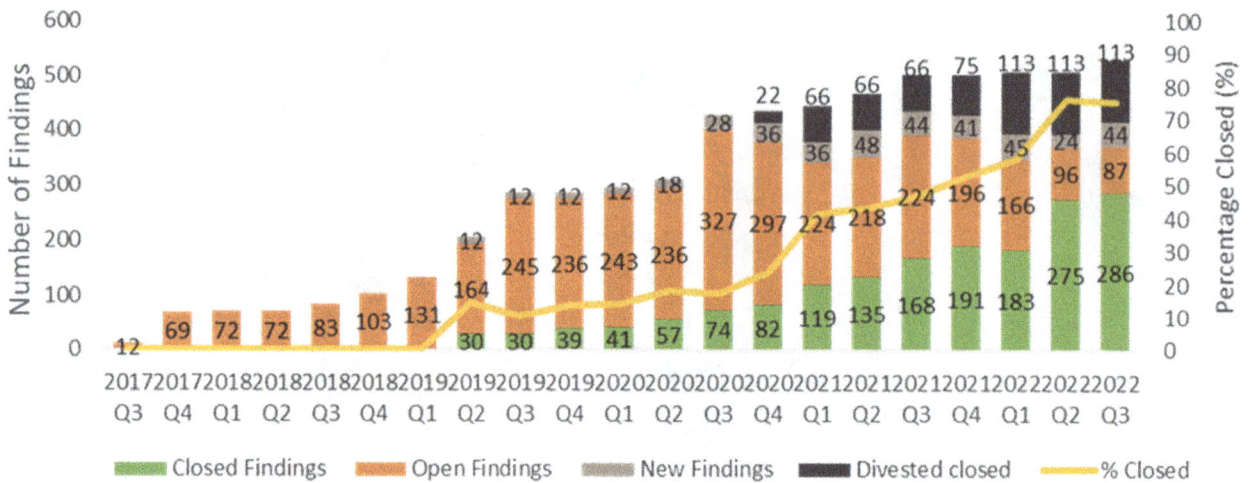

FIG 3 – Cumulative findings across all underground assets, and progress to remediation of findings.

Many of the initial findings were linked largely to inadequate catastrophic hazard management processes, and the incomplete adoption of the Glencore policies and Fatal Hazard Protocols. The overwhelming majority of these findings have been rectified at all mining complexes across all subject matter areas in the integrated program.

Several findings at some operations are technical complex and require multi-year corrective action plans and substantial phased investment. These corrective action plans can require substantial upgrades to infrastructure (removal/replacement of oil-filled transformers, replacement of shaft steel work, creation of safety refuge bays as examples), or enhancing previous accepted practices to being in line with current requirements (installation/enhancement of ground support in unsupported main accessways, understanding and evaluation of groundwater, certifying modifications to mechanical hoisting components etc). A large percentage of these long-term items have remained as an open finding across multiple verification audits (Figure 3). However, progress is being tracked and completion dates are expected by the next round of verifications.

In some instances, as initial findings are completed, a new finding has been identified. An example of this observed in several cases is the creation of a PHMP. In several cases, no PHMP was in place at the first audit, but was created and available at the first verification audit. However, a number of the controls and processes outlined in the PHMP were not developed or effective, often were aspirational goals, resulting in a new finding. Over time, these new findings are being addresses and progressively rectified, bringing down the risk profile of the asset or complex.

Other findings which were identified as relatively easy items to rectify were actioned and implemented well inside the next assurance audit. Items such as the introduction of self-contained self-rescuers, linking of alarms to a central location for appropriate communication, and updates to various pre-start checks on ranges of equipment are items considered as easy rectification.

Steady progress against these findings is being demonstrated. In Q2 2022, a total of 131 findings remain open (of the 530 identified). Many of these open findings are linked to the long-term rectification programs summarised earlier.

As with most large mining businesses, acquisition or opening of new assets, divestment of assets, and closure of mines will change the tracking of these types of metrics. Over the past two years, several assets have been divested or closed for various commercial reasons. For tracking purposes, if the asset has been closed or divested, all open, closed, or new findings have been moved into 'divested-closed' category shown in Figure 3. The purpose of this is to keep an up-to-date record of the findings that remain critical to the ongoing safety of the business.

To provide more detailed information per catastrophic hazard area, compliance assessments against the individual Fatal Hazard Protocol requirements are also conducted. Compliance scores against individual criteria in the catastrophic hazard areas are maintained by the assurance team and the subject matter experts. Each site is evaluated against Fatal Hazard Protocol requirements against two key measures; the system compliance, and the implementation performance. These measures allow a more detailed picture to be presented on the specific areas requiring the most attention (Figure 4). These specific audits help to identify smaller gaps which may not lead to a catastrophic event occurring, but allow ongoing improvements to be made in overall catastrophic hazard management.

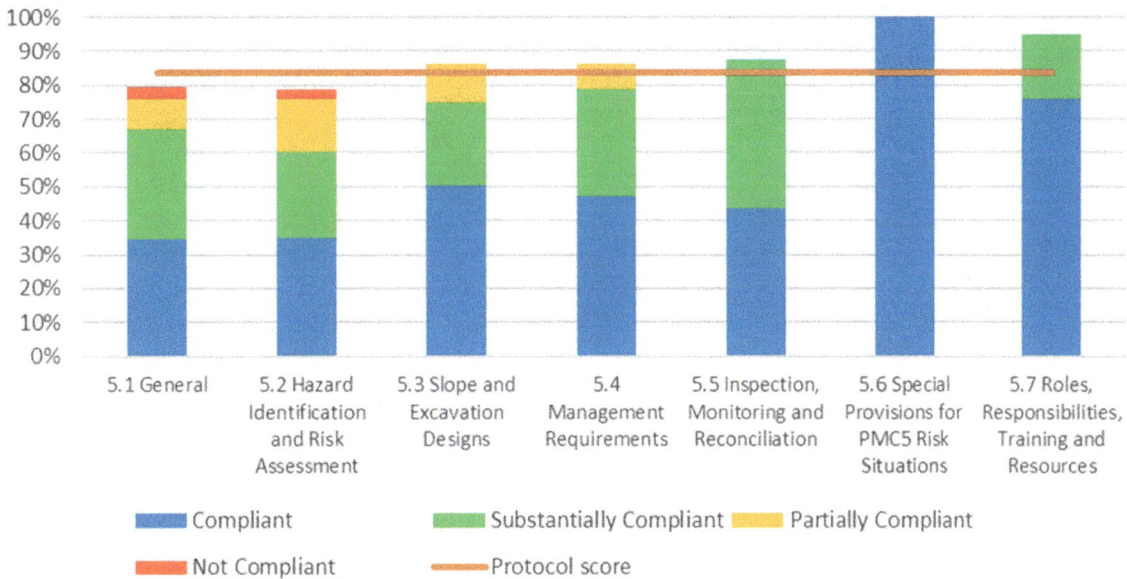

FIG 4 – Compliance results for criteria set out in individual fatal hazard protocols.

Because the sites are audited by the same subject matter team members against the same criteria, this allows a ranking, or benchmarking system to be used internally. While comparing performance internally is a valuable tool for larger mining companies, there is high value is demonstrating performance against a broader data set of similar operations. These larger benchmarked data sets can be further divided to show performance based on geographic region, underground mining style, and other broad filters. In several cases, the SMEs used in this program undertake similar assurance programs for numerous other mining companies globally, and maintain a combined benchmark data set to show the compliance of the site against peers in the industry. While the questions are slightly different in these SME benchmarked programs when compared to the Glencore assurance program, there is substantial crossover that allows easy population of these data sets for more broad benchmarking and integration. An example of the benchmarked data is shown in Figure 5 for 15 operations against similar criteria in this assurance program.

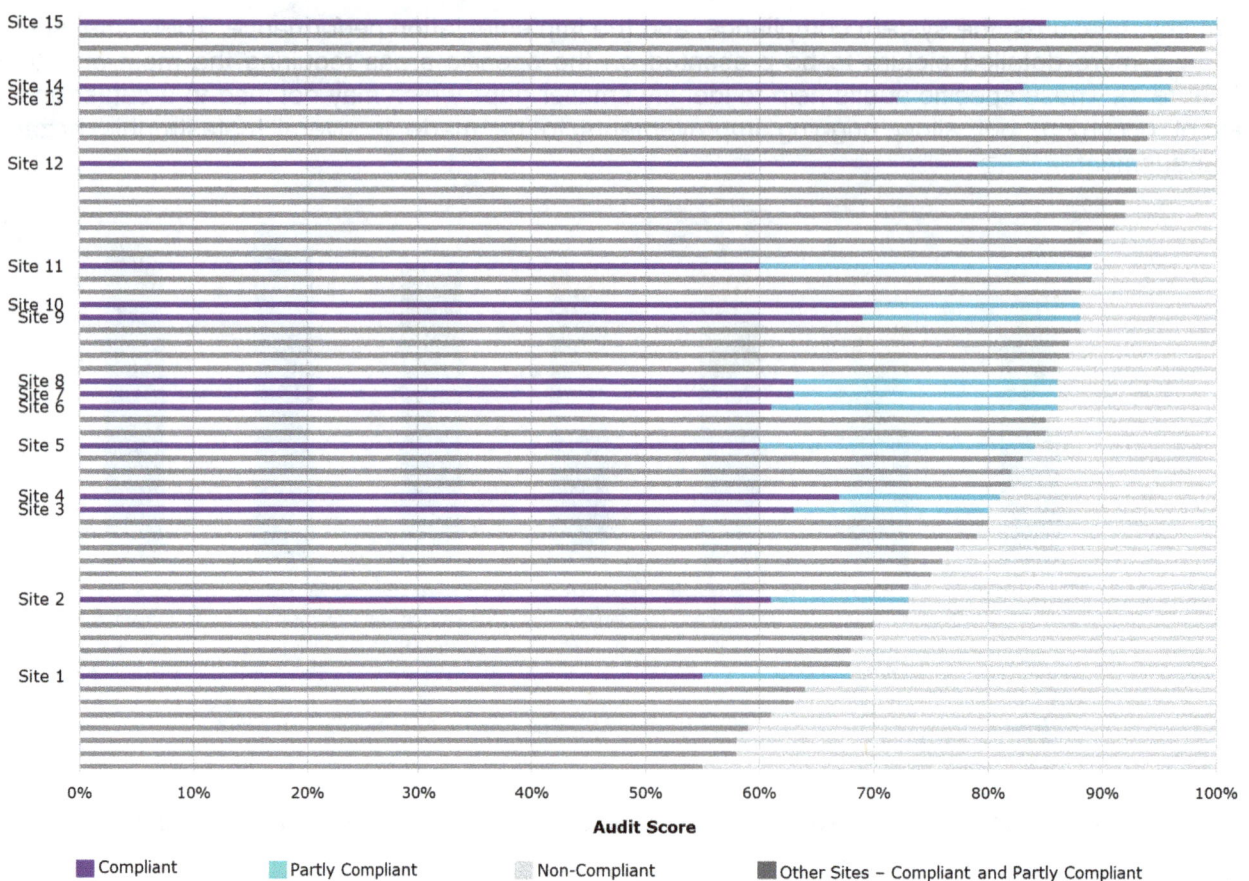

FIG 5 – Benchmark of results from multiple sites to allow comparison of performance against peer operations globally.

This benchmarked data allows the individual sites compliant value, or compliant and partly compliant values to be easily visualised against other sites audited in the industry. Names of all sites are held confidentially. When a site is re-audited, only the most recent value is maintained in the data set. This can allow the site to demonstrate ongoing improvements and performance against industry peers.

What can be seen in Figure 5 is a large spread of overall compliant and partly compliant cumulative results, ranging from 55 per cent to 100 per cent. In all cases, there is a considerable percentage of partly compliant items, often being opportunities for improvements to bring certain aspects into line with industry peers. In all benchmarking reports, the comments against each individual line item are provided, supported with recommendations to bring the item to compliance, and suggested timelines. These benchmark data sets have proven extremely valuable in clearly demonstrating and communicating the individual site performance, and allowing a more open discussion on expectations for improvement.

CONCLUSIONS

There is no doubt that catastrophic hazard management is at the forefront of every mining operation globally. Companies invest heavily in programs to ensure worker safety, many with extremely low TRIFR or LTIFR statistics. Catastrophic hazard management in mining continues to evolve, and as such there are various pitfalls to demonstrate success; it is much more than simply meeting regulatory requirements.

Detailed integrated audits coupled with in-field verifications have revealed that senior management commitment to safety is often clear, well organised, and substantially resourced. However, there was often a lack of specific, measurable, and time-sensitive goals to demonstrate effective catastrophic hazard management; a trend that is substantially changed at Glencore.

Development of a thorough integrated, multi-disciplinary auditing process executed by an experienced team of technical area SMEs with substantial operational knowledge, provides a detailed and complete picture of catastrophic hazard management. The integrated auditing approach ensures a consistent picture is presented to all stakeholders, all while reducing the burden of multiple disruptions of numerous potential audits throughout a year.

Understanding and auditing of catastrophic hazards and safety management systems compliance is only the first step in safety assurance. Completing operational execution audits in technical areas and benchmarking the site verification of safety management documents and verifications forms a critical part to clearly demonstrate industry leading safety practices. No longer is it the external documentation reviews that dominate, but rather the culture of continual hazard identification, management, and controls verification.

ACKNOWLEDGEMENTS

The authors would like to thank Glencore and AMC Consultants Pty Ltd for approving publication of this work. A special thanks to Riaan du Plooy for valuable contributions and review of this paper.

REFERENCES

ICMM (International Council on Mining and Metals), 2015. A guide to good practice: health and safety critical control management, ICMM, London.

NSW Department of Primary Industries (DPI), 2007. MDG 1010, Risk: minerals industry safety and health risk management guideline, Mine Safety Operations Division, January.

Safe Work Australia, 2012. Managing risks to health and safety at the workplace, fact sheet.

Standards Australia, 2010. HB 327:2010 Communicating and consulting about risk, Standards Australia, Sydney.

Standards Australia, 2018. ISO 31000 Risk management–principles and guidelines, Standards Australia, Sydney.

US Department of Labor, 2015. Mine safety and health at a glance, Mine Safety and Health Administration.

Technology and innovation

Dashboards, data sources and a modern data platform

D Bruce[1], J Rossiter[2] and L Best[3]

1. Superintendent – Mine Projects, OZ Minerals, Adelaide SA 5950.
   Email: daniel.bruce@ozminerals.com
2. Superintendent – Site Operations Centre, OZ Minerals, Adelaide SA 5950.
   Email: jonathon.rossiter@ozminerals.com
3. Principal Data Engineer, OZ Minerals, Adelaide SA 5950. Email: luke.best@ozminerals.com

ABSTRACT

OZ Minerals' Carrapateena mine is a Sublevel cave mine located in South Australia, approximately 160 km north of the regional centre of Port Augusta. Carrapateena commenced underground development in 2016, with production commencing in 2019 and ramping up to 4.25 Mtpa by December 2020.

In pursuit of continual improvement, the Carrapateena mining team sought to identify how it could better interrogate its current productivity levels and its drivers (root causes). During this process the team were confronted by a disconcerting number of different opinions (which often conflicted) on our actual productivity performance which led to a need for review.

Although reliable data and a well-designed time utilisation model (TUM) existed, it was still unclear what actions needed to be put in place to improve productivity.

This was for the following reasons:

- Data sources were disparate unless significant effort was applied to compile them.

- Data sources were not available in a format that could be easily viewed or analysed.

- General understanding of the data which was collected and how it could be used was poor.

- Process of regularly analysing performance and forming tangible actions to improve was lacking rigour.

A project was then started with the goal to fix the areas stated above. An initial requirements gathering exercise was undertaken to identify key data sources for site which returned a staggering 42 known systems which were then shortlisted to the key systems that contained operational data and could generate tangible improvements when analysed. This paper discusses the approach to selecting the OZ Minerals Insights Hub as the chosen platform, and the effort and collaboration between cross-functional departments that was involved in transforming the data so it could be displayed in a useful way.

The data sources identified included a fleet management system, two planning databases, a control system historian, and a mobile fleet machine data system. New processes were also introduced to digitally capture daily commentary; this was achieved utilising SharePoint Forms. Data pipelines were then developed in the Insights Hub platform to automate the ingestion and transformation of the raw mining data into curated data models for end user consumption.

Tableau dashboard packages have been created to support facilitation of short interval control, daily review meetings, and weekly review meetings as well as data analysis of key equipment and locations. These dashboards visualise the mining data models in a form that provides the user a clear picture of what has occurred, why it has occurred, and where to focus improvement efforts. This has resulted in a team which is more accountable, more equipped to assess the root cause of performance, and more regularly able to make data driven decisions.

The next step? To build this out from Mine to Mill and create a data indexing tool which will drive self-service reporting, a data agnostic platform and predictive tools that help our people continue to be one step in front.

INTRODUCTION

OZ Minerals is a copper-focused, global, modern mining company based in South Australia. Its growth strategy is focused on creating value for all stakeholders. OZ Minerals owns and operates the Prominent Hill and Carrapateena mines in South Australia, and the Pedra Branca mine in the Carajás Province located in the Para State of Brazil.

The Carrapateena copper-gold mine is approximately 460 km north of Adelaide, and 160 km north of Port Augusta in South Australia's highly prospective Gawler Craton (see Figure 1). The project is located on Pernatty Station and its supporting infrastructure is located within Oakden Hills Station. The Kokatha people are the traditional owners of the land.

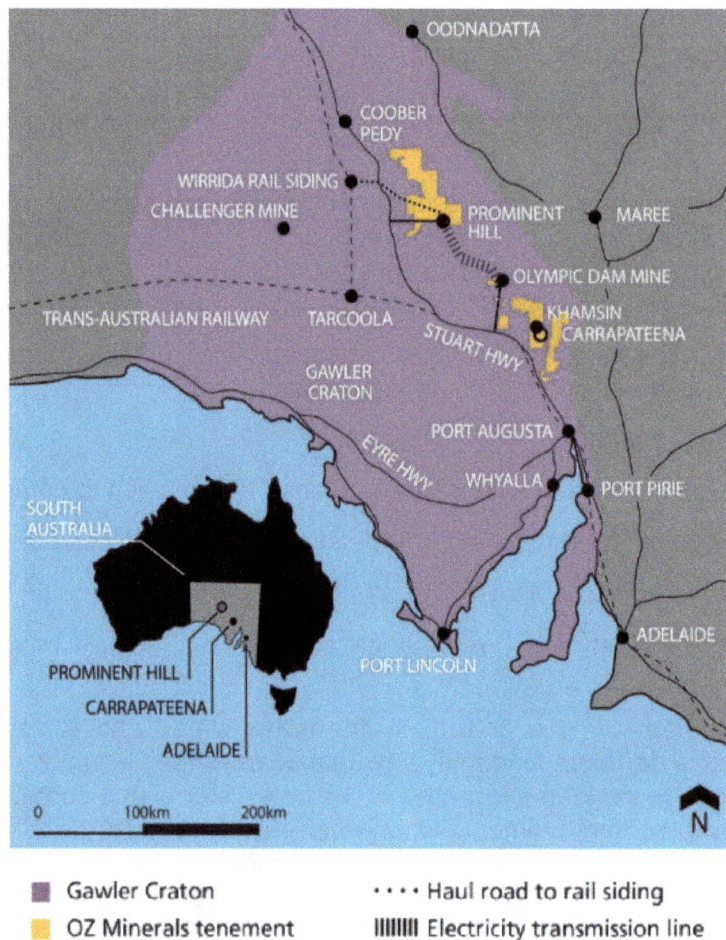

FIG 1 – Carrapateena Location.

First saleable concentrate production for the Carrapateena copper-gold mine in South Australia was achieved in December 2019, and it has ramped up to 4.25 Mtpa over the following 12 months. Ore is processed on-site to produce a copper concentrate containing copper and gold minerals.

PROJECT INITIATION

This project was initiated by several of OZ Minerals workforce across different departments. Everyone witnessed the same problems in their teams regarding data.

Data management at Carrapateena is reasonably advanced with the following in place:

- A Site Operations Centre (SOC), which supports data capture, administration and reporting.

- A Time Utilisation Model (TUM), which defines the way time-based equipment and location data is captured.

- Sophisticated source systems across mobile fleet, fixed plant, and planning systems.

However, at the time of project initiation, the data was significantly underutilised. Carrapateena was reliant on a handful of Excel spreadsheets with regular cadence (daily, weekly, monthly) reports produced by specialised professionals, such as mining engineers or site operations supervisors. These few spreadsheets were difficult to administer, and usually there was only one subject matter expert (SME) for whom was called upon to fix the spreadsheet if problems arose. These SME's would be regularly called upon to modify the already bulky spreadsheets to include new data or visualise the data in a different way.

A deep dive into the reasons for underutilisation of data revealed:

- Data was difficult to access for the following reasons:
  - Users did not have knowledge of which source system data they required.
  - Users needed to be set-up in the source system reporting tool prior to having access.
  - Users needed training in the source system reporting tool to be able to extract data.
  - Users needed to use a difficult process to set-up links to Excel.
  - Source system non-digital.
- Data was difficult to use to make better decisions for the following reasons:
  - Source system reporting tools produce raw data only which often needed to be imported into Excel and refined.
  - Multiple data sources required for data to be useful (eg comparing actuals versus plan).

With these problems at the front of mind, a project team was formed with the following goals:

- Identify the key data sources, which, if freely available in a useful format, would have the biggest impact to performance at Carrapateena.
- Identify how data users wanted to see data presented.
- Ingest the key data sources into a central database.
- Create a suite of freely available and useful data visualisation dashboards.

A simple graphic was developed to convey the goals of the project to a wide audience (Figure 2).

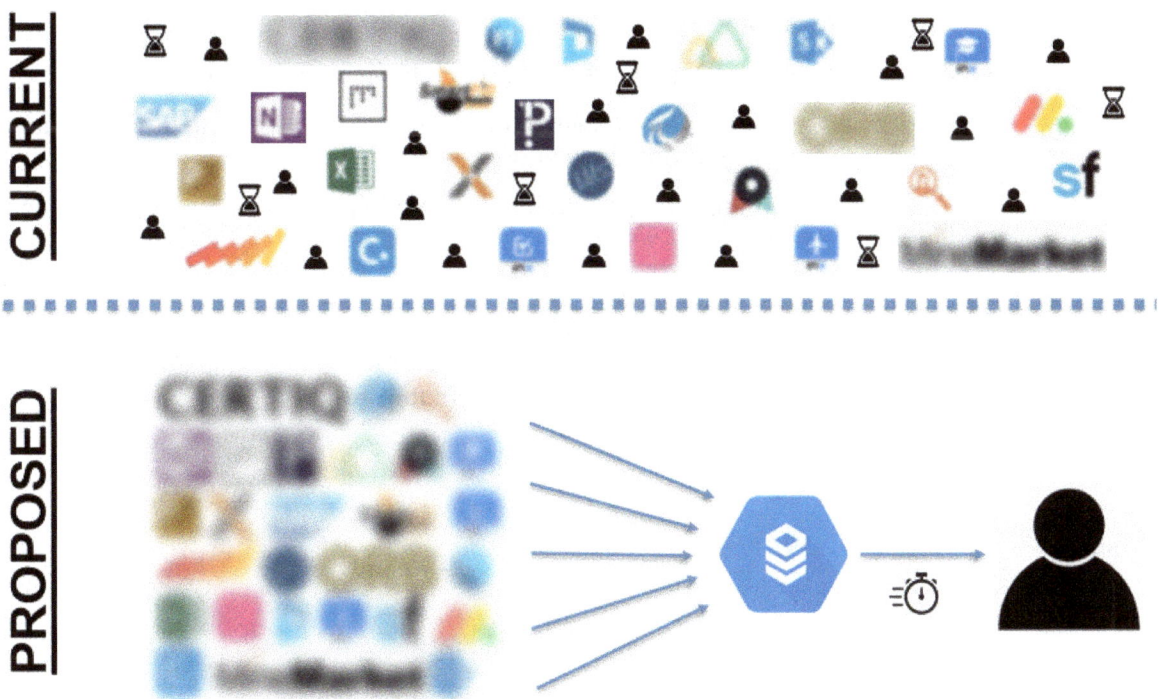

FIG 2 – Data project graphic representation.

With the project goals defined, the cross-departmental project team presented their idea to upper management at Carrapateena and received endorsement to pursue the project. A project charter was created, and a project sponsor appointed. An agile sprint planning work methodology was employed, where the team met weekly to update progress against project milestones and critical tasks.

A risk assessment was completed to identify all threats/opportunities, controls required to be in place, and what the risk rankings were. An excerpt from the risk assessment is shown below to demonstrate the process which was undertaken, the table only shows the threats/opportunities that were identified, however in the adjacent columns were the controls and risk rankings as per a standard risk assessment template (Table 1).

TABLE 1

Excerpt from Risk Assessment.

ID	Threat/Opportunity	
001	Threat	Developed reporting tool produces inconsistent results with source system.
002	Threat	Poor reliability and availability of reporting tool adversely impacts site operations.
003	Threat	Loss of data sovereignty. Security breach of the reporting platform exposes OZL Data to unauthorised parties.
004	Threat	Inability to manage and support ongoing changes with Schema's (ie change in data collection methods).
005	Threat	Project scope is too large to manage.
006	Threat	Project exceeds scheduled delivery.
007	Threat	Project funding not secured.
008	Threat	Inefficient use of OZL Resources.
009	Threat	Department Stakeholders unavailable to work with vendors.
010	Threat	Limited Uptake of the Application.
011	Threat	Poor vendor application support/ongoing support not available.
012	Opportunity	Utilising inhouse resources (Insights Hub) for project delivery.
013	Opportunity	Providing interactive/innovative technology to visualise the dashboards and improve employee data utilisation.
014	Opportunity	Leverage smartphone technology to provide persons with improved access to dashboards.

The project team completed a template which was then distributed to department managers to document their team's data requirements. These were all returned and compiled and formed the basis for the project scope of work. An excerpt is shown in Figure 3.

Please use the form below to list your requirements for your department's dashboard and reporting services

Item	Response		
Department:	e.g. Safety		
Department champion's name:	e.g. Joe Bloggs		
To utilize a self-service report tool, the following is important to me: (list as many things as possible)	e.g. Ability to filter date range e.g. Drag and drop metrics e.g. Ability to access on my phone		
Systems I use: (list all)	e.g. INX InControl, CGR etc.		
Data refresh frequency: (please circle) How often do you need to see updates to your dashboard?	Every Second Every Shift Other:	Every Minute Every Day	Every Hour Every Week

FIG 3 – Excerpt from department requirements template.

SCOPE OF WORK

Forming the scope

A total of 12 department requirements forms were returned which comprised a wide range of business units spanning; Environment, Community, Finance, HR, Integrated Planning, Health and Safety, Mining and Processing.

These were compiled to give two lists, a list of data sources that existed (an excerpt of priority one data sources is in Table 2) and a list of requested data visualisation dashboard packages.

TABLE 2

Data Sources (P1 Only).

#	Data source system	Department requested	Refresh time
1	Fleet Optimisation Tool	Mining	5 mins
2	Fleet Management Data Capture	Mining, SOC, Environment, Geology	5 mins
3	Planning Databases	Mining	5 mins
4	Mobile Machine Data	Mining	5 mins
5	Fixed Plant Control System	SOC, Environment, Processing	5 mins
6	13W Planning Tool	Mining Engineering	Monthly
7	Weekly Planning Tool	Mining Engineering	Weekly
8	Safety System	Mining, SOC, Community, Geotech	Daily

A total of 43 source systems were identified and prioritised according to:

- How much the data source served the site constraint (mining) to make better decisions.
- How much the data source served other business functions to make better decisions.

This resulted in eight priority one data sources.

A total of 64 dashboard packages were requested by the departments and prioritised according to the same criteria as the data sources, resulting in 15 priority one dashboard packages.

Data sources

The key data sources that formed the scope comprised:

- Actuals data collected through data capture software:
  - Through radio call-ups to SOC.
  - Through tablet data entry in machines.
- Actuals data collected autonomously:
  - Materials Handling System data.
  - Mobile fleet machine data.
- Plan data collected through an export to database process:
  - Schedule tool exports ingested to a SharePoint database.
  - Weekly Plan ingested to SharePoint.
  - 13 week Plan ingested to SharePoint.
- Safety reporting data collected through API:
  - Direct from the source system.

To solve a common problem, there was also one additional data source that needed to be created. The problem was, that even with the best data visualisations, some manual commentary is always required from the subject matter experts to more effectively interpret the data. Thus, we added to scope a commentary solution whereby the area owners (Mine Foreman and Supervisors) could add commentary each day on the performance of their area.

Dashboards

Fifteen priority one dashboards were selected, these broadly fit into three categories (Table 3).

TABLE 3
Dashboard Packages (P1 Only).

Dashboard categories	Purpose of dashboard
7 × Equipment Performance dashboards	To visualise historical performance metrics for specific KPI's and make informed decisions on how to improve in the future. Equipment included Jumbo's, Fibrecrete, Produ Drills, Loaders, Trucks, Charge-up and the UG Material Handling System.
3 × Review dashboards	To visualise previous 24 hr/Weekly performance and identify opportunities to improve and/or priority actions that need to be completed along with understanding compliance in development areas and reasons for variance, particularly in headings that impact on future mine excavation plans.
5 × Short Interval Control dashboards	Includes development, production and loader servicing short interval control dashboards. Designed to help trigger important decisions that need to be made in-shift and the validation of data to confirm accurate metrics are being captured.

Scope learnings

The following was learnt during the project:

- When defining the project goals, the time, cost and resources required to ingest and visualise according to all Carrapateena's needs were significantly underestimated. Once this was learnt the scope was reduced to priority one data and dashboard packages only.

- The detail is critical for both dashboards and data sources. Completing a scoping study prior to tendering for the work would have minimised unexpected delays and cost. Basic items to understand include:

  o Data source access provisions (Server, API, Firewalls etc).

  o Data source refresh frequency.

  o Data source structure.

  o Dashboard package size (how many pages) and complexity (what type of visualisations).

- Including in scope the creation of a base data model to define the way that data will be transformed. This ensures that the archiving of all data sources is data source agnostic.

- Including in scope a resource for user acceptance testing and post-deployment bug fixes.

ASSESSING OPTIONS

Data platform and technology

The industry is now inundated with technology offerings in the data and analytics space, it was a challenge to perform an options analysis and product review against these offerings with limited expertise and experience in the implementation and operationalisation of them at site. To understand the full range of technology and services already supported in OZ Minerals, the Carrapateena mining team conducted several workshops with the Insights Hub team to understand how data insights were being delivered for other assets and functions within the business.

Three options were identified for consideration:

1. Utilise the OZ Minerals strategic data platform – Insights Hub.

2. Build an internal, dedicated Carrapateena Data Platform technology capability and team.

3. Outsource an off-the-shelf SaaS solution and support arrangement.

After review, Option 1 was decided upon as the data platform on which to deliver the reporting requirements discussed in the Scope of Work.

Insights Hub

The Insights Hub is a program of work to deliver the foundational capability for OZ Minerals to shift to being a data-driven organisation, with a clear vision on data enablement and strategy, providing OZ Minerals with a data platform for the future.

Technology

Insights Hub has adopted a cloud-native modern data technology stack, utilising best in breed solutions to meet each functional objective.

The data three main functional areas:

1. **The data platform** – On the left in Figure 4 is the data stack comprising of:

   o Cloud native storage – AWS S3.

   o Snowflake Data Platform – The data platform is at the heart of the data stack and is responsible for providing the data processing power, storage, sharing and security.

   o Data processing framework – facilitates the ingestion of data from source systems.

- Data modelling framework – Responsible for providing Data and Analytics engineers the platform to write develop data models, produce the data catalogue and to document and test the data pipelines.

- Business Intelligence – Visualisation tool for providing data insights and analytics to end users.

2. **Data pipeline orchestration** – On the right in Figure 4 is the orchestration platform. This is responsible for management and automation of data pipelines and operational tasks, responsibilities include:

- Providing a UI and analytics interface for DevOps to support the operational environment.

- Scaling on demand as workloads require it.

- Automating alert and monitoring capabilities critical for supporting the platform.

3. **CI/CD** – At the bottom in Figure 4, underpinning the environment and operations the CI/CD function is responsible for automation of build, test and deployment of data pipelines to production, responsibilities include:

- Source control: Capturing versioned documentation and code, enabling technical reviews and formalising software releases.

- Automation: automating the deployment and actions required to deploy a release to an environment.

- Orchestration: Performing *ad hoc* deployment tasks not covered by the Automation.

- Infrastructure-as-code (IAC) manage the automation of infrastructure, environment and security management build and deployment.

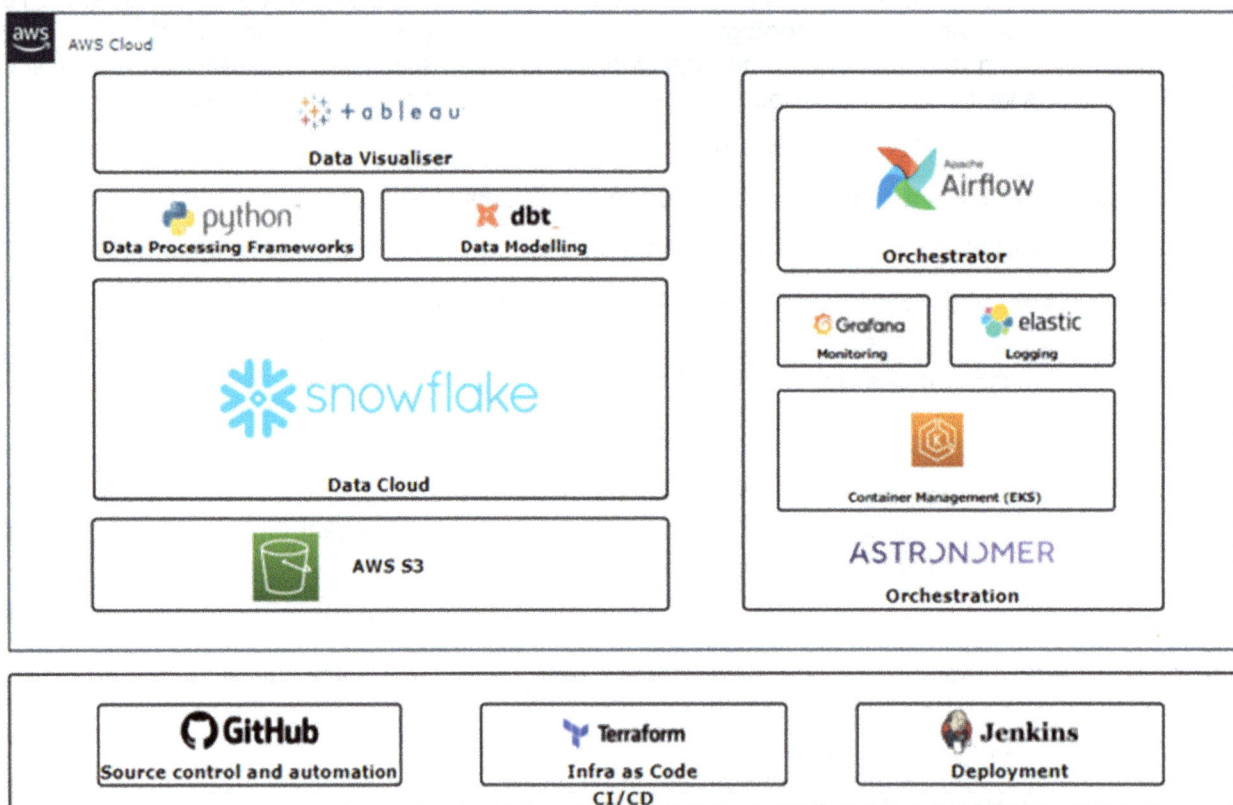

FIG 4 – Insights Hub Technology.

How does it work?

The data flow from data source to data insights follows a consistent pattern (Figure 5):

- On the far-left data producers, data is either entered manually by a person or automated and generated by a particular system.

- The following workflows are orchestrated by the orchestration tool:

  o Firstly, data is extracted from the source utilising the data ingestion framework.

  o Data is then staged in cloud native storage for processing.

  o Raw staged data is automatically loaded into an ODS (Operational Data Store) layer in the data platform.

  o The data modelling layer transforms the raw data into curated data models, automatically tests each pipeline run for data quality.

  o Once the data is transformed, data lineage, documentation and dictionary are automatically generated and surfaced as interactive websites.

- The data consumers (far right), such as Tableau (BI), Self-service users and/or systems connecting to internal or external data shares securely access the data models for reporting and analytics.

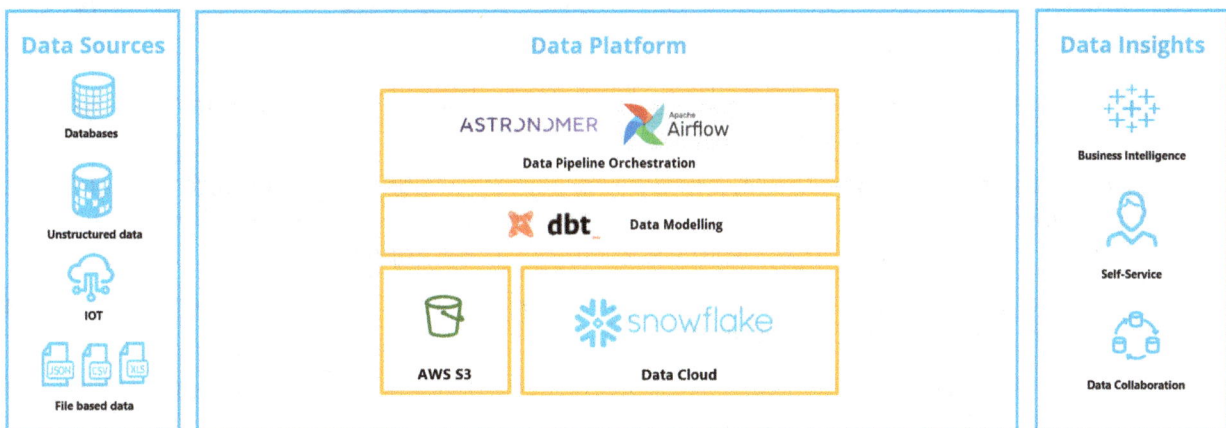

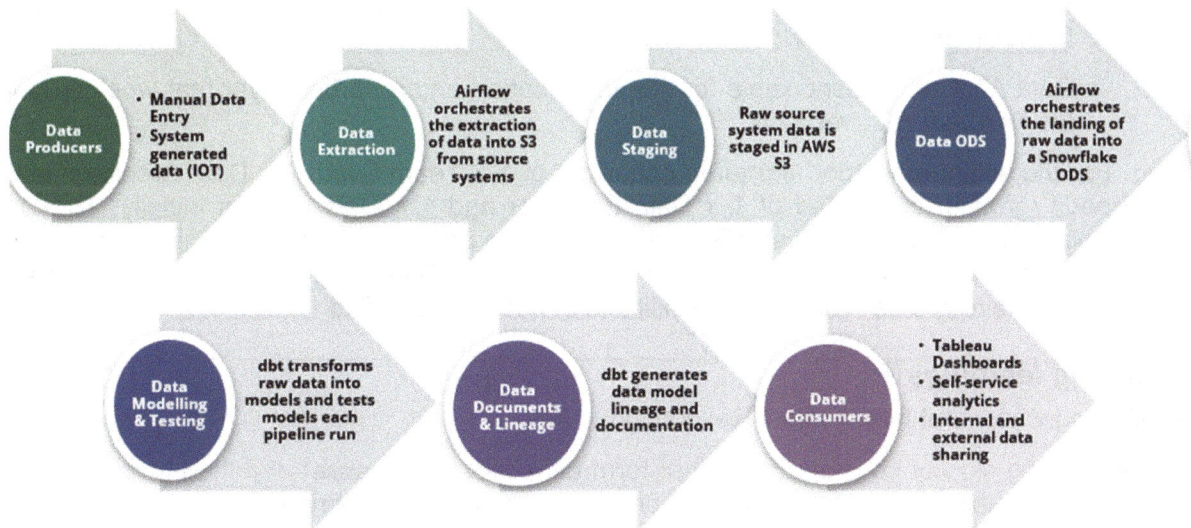

FIG 5 – Insights Hub data flow.

Project resourcing

With a scope of works determined, and preferred business platform in Insights Hub, the next key consideration for the project team was to determine the type, and level of resourcing that would be required to execute the project:

- Internal or external:

- Engagement sessions were held with another OZ Minerals asset that had commenced a similar journey to understand the approach that was taken, and the lessons learned.

- Time estimates were placed against each dashboard and data source to understand the total time required to execute the project.

- As a result, external resourcing was the preferred option for three key reasons:

  - The project required dedicated focus by experienced resources.

  - The site did not have a mature data team, or a full appreciation for the roles required to execute a data related project.

  - Reduced lead time in terms of advertisement, recruitment and onboarding of dedicated internal resources.

- A tender was put out to market with the required scope of works to determine available vendors who could support the project. The responses were compiled into a matrix to help in the shortlisting process and five key vendors were selected to provide further detail on similar projects they had undertaken. An assessment was made for each vendor based on timelines to achieve the scope, costs involved, experience levels and team structure. Once assessed, the project team made the decision to proceed with the preferred vendor.

- Team size and structure:

  - An internal project manager was appointed to support the management of external stakeholders and project scope.

  - During the tender process, vendors were asked to propose the team size and structure that would be required to execute the scope of works within the required time frame. Majority of vendors came back with a similar structure comprising of:

    - Team Lead – Responsible for the team's delivery of tasks against the project scope and maintaining team momentum.

    - Data Architect – Responsible for structuring the data feeds into a landing environment which supported data ingestion.

    - Data Engineer – Responsible for transformation of the data into the required tables and format in preparation for the reporting layer.

    - Data Analyst – Responsible for the visualisation and reporting layer that would support the end user requirements.

  - For the selected vendor, the team consisted of 1 × Team Lead, 1 × Data Architect, and 2 pods (groups) comprised of 1 × Data Engineer and 1 × Data Analyst initially. The Data Engineer and Data Analyst effectively worked in pairs given the close relationship held between the two roles. This was later reduced to 1 × Data Engineer and 1 × Data Analyst while retaining a key role that had skillsets of both – a rare find.

Resourcing learnings

The following was learnt from Project Resourcing:

- Resourcing did not consider the longer-term capability of the asset and the internal role that would fulfill this was brought into the project too late. Early recruitment of an internal role would have complemented external resources and maintained quality control, user acceptance testing, operational handover, and the challenging of technical issues.

- The cost of external resourcing versus building internal resource should be heavily considered. Based on learnings to date, the ideal structure for an internal data team is shown in Figure 6. A data scientist is not crucial for businesses starting out, however should be considered, and worked into future workforce plans:

  - 1 × Data Lead

  - 1 × Data Engineer

- o 1 × Data Analyst

- o 1 × Data Specialist.

- Technical review meetings need to be set-up early with project resources to ensure quality is managed in the same light as task completion. This would have allowed early identification of code that did not meet governance standards.

- Expected time frames for completion varied significantly from original estimates. A scoping study could have been undertaken by the vendor to understand the detail required for each data source. Consideration should also be given to the benefits of a fixed price contract versus a schedule of rates contract that is at risk of project overrun, or initial underestimation of work involved.

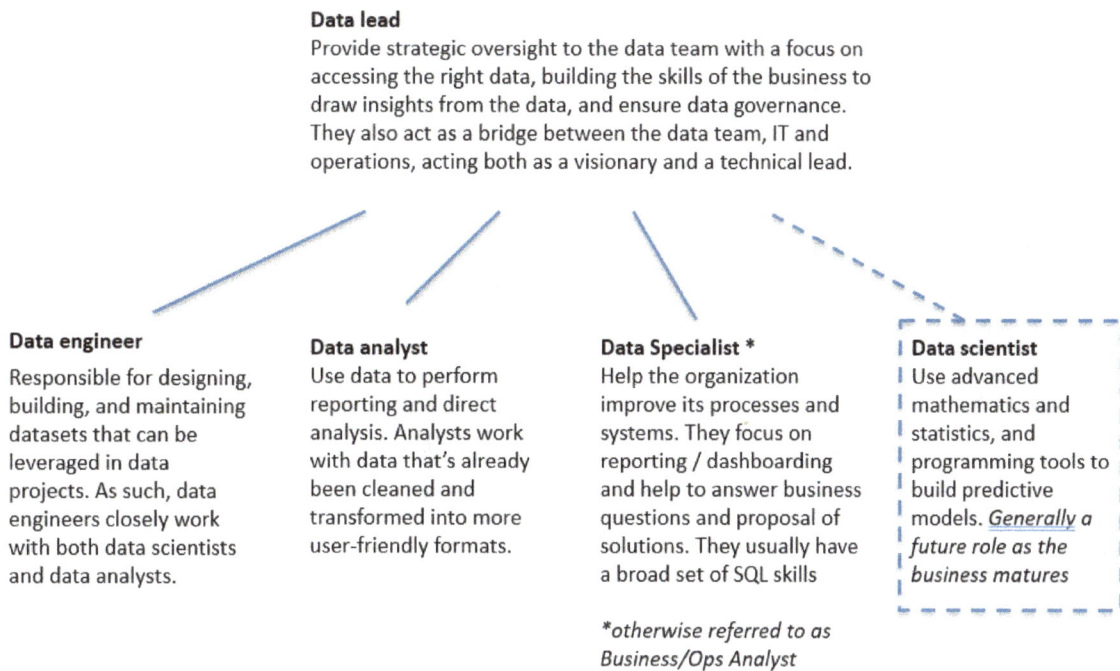

Data lead
Provide strategic oversight to the data team with a focus on accessing the right data, building the skills of the business to draw insights from the data, and ensure data governance. They also act as a bridge between the data team, IT and operations, acting both as a visionary and a technical lead.

Data engineer	Data analyst	Data Specialist *	Data scientist
Responsible for designing, building, and maintaining datasets that can be leveraged in data projects. As such, data engineers closely work with both data scientists and data analysts.	Use data to perform reporting and direct analysis. Analysts work with data that's already been cleaned and transformed into more user-friendly formats.	Help the organization improve its processes and systems. They focus on reporting / dashboarding and help to answer business questions and proposal of solutions. They usually have a broad set of SQL skills	Use advanced mathematics and statistics, and programming tools to build predictive models. *Generally a future role as the business matures*

otherwise referred to as Business/Ops Analyst

FIG 6 – Recommended structure for internal data team.

IMPLEMENTATION

Process

Overview

To deliver an Insights Hub package of work a well governed Software Development Life Cycle (SDLC) process (Figure 7) is followed that guides the delivery team on approach throughout the project life cycle, from design and planning through to production go-live and into operations.

The delivery team comprised of the following internal and external resources:

- Tableau BI specialist (external)

- 2 × Data engineers (external)

- Scrum Master (external)

- Project manager (internal)

- Principal Data Engineer (internal)

- Business Partner and BI lead (internal).

The project delivery followed Agile Scrum methodology working in two-week sprint cycle. Each sprint comprises of:

- Sprint planning where the sprint goal and tasks are defined.

- A sprint review, which the whole project team attends, giving the implementation team an opportunity to present their work to product owner and receive feedback.

- A sprint retro where the team reflects on how the sprint has gone, what went well and what can be improved.

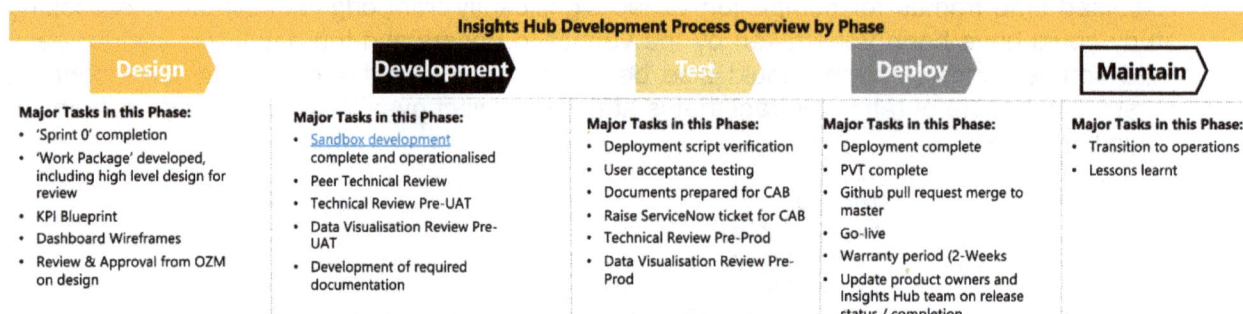

Insights Hub Development Process Overview by Phase				
Design	**Development**	**Test**	**Deploy**	**Maintain**
Major Tasks in this Phase: • 'Sprint 0' completion • 'Work Package' developed, including high level design for review • KPI Blueprint • Dashboard Wireframes • Review & Approval from OZM on design	Major Tasks in this Phase: • Sandbox development complete and operationalised • Peer Technical Review • Technical Review Pre-UAT • Data Visualisation Review Pre-UAT • Development of required documentation	Major Tasks in this Phase: • Deployment script verification • User acceptance testing • Documents prepared for CAB • Raise ServiceNow ticket for CAB • Technical Review Pre-Prod • Data Visualisation Review Pre-Prod	Major Tasks in this Phase: • Deployment complete • PVT complete • Github pull request merge to master • Go-live • Warranty period (2-Weeks • Update product owners and Insights Hub team on release status / completion	Major Tasks in this Phase: • Transition to operations • Lessons learnt

FIG 7 – SDLC process.

Design and planning – Sprint 0

The first stage was the project discovery and design, called Sprint 0.

The purpose of this stage is to:

- Identify and assign a Product Owner from OZ Minerals.

- Identify key stakeholders such as Data and Mining SMEs.

- Form the scope of work and capture this in a work package document, this comprises of:
  - vision
  - problems (current state)
  - benefits (future state)
  - user stories (success criteria)
  - risks (barriers to success)
  - urgency (by when).

- Capture any key KPIs, metrics or calculations that need to be reported and the owners of these in our KPI blueprint work pack.

- Identify data sources, review technical documentation such as API specifications.

- Review the source system data model and identify required data for ingestion.

- Data Engineers produce data model entity relationship diagrams (ERDs) for the proposed data model and data transformation workflows in Draw.io design tool.

- BI Specialists produce wireframes in Balsamiq design tool.

Once the above artifacts are complete this is reviewed by both the data engineering and BI teams before moving into build phase.

Sandbox

Data engineers use sandboxes to develop data pipelines and perform data discovery in their local environment.

The Insights Hub sandbox environment comprises of an Integrated Development Environment (IDE) and the orchestration and data modelling platform running in Docker containers images on user's local machines, the container environment is pre-configured for a user to securely connect to the data platform via SSO. Each user gets their own area in the data platform to discover, work with and develop their own data pipelines or analytics. Sandboxes enable the data engineers to develop locally and concurrently.

Sandboxes have provided the technical team with the following benefits:

- Each user gets their own database, where they can stage their own data or, using containers can develop their own data pipelines and take advantage of the data platform functionality.

- Users can access existing raw and curated data models to supplement their project and bring that data into their own workspace.

- E2E development – Data engineers have an identical production-like environment to test their data models.

- Users can collaborate on data in sandboxes with others in their functional role and importantly data can be surfaced early for dashboarding and getting early feedback from our product owners and customers.

Once the technical work is complete it is committed to Source Control and a Pull Request is raised by the engineer which is peer reviewed by the Internal Insight hub team prior to release to development.

Development

At this stage features of the data model are organised into deployable artifacts called releases, this stage aims to address the following:

- Operationalise the data model and orchestrating the data pipeline end-to-end.

- Performing QA, internal end to end testing of the data pipeline and respective tasks.

- Documentation and lineage generation which should align to the design outlined in Sprint 0.

- Formal peer review and approval of all artifacts is undertaken at this stage before moving to UAT.

UAT / test

At this stage the approved release is deployed to the UAT environment for customer testing, this stage aims to address the following:

- Completion of formal test pack, testing the requirements of the acceptance criteria and requirements outlined in the sprint 0.

- Stakeholder signoff and approval for production based on testing.

- Resolve any identified defects and update release.

- Review by the Change Approval Board (CAB) to approve the production release by operations.

Production / deploy

On approval of the release artifact the release is automatically deployed to production where the release is handed over to the DevOps team for support and monitoring.

Implementation learnings

Following an agile methodology, the retro ceremonies allowed us to identify key areas to improve going into the next sprint cycle. Changes were made to how we work to improve delivery cadence.

One of the changes with the biggest impact was the introduction of weekly regular technical check-ins with the internal and external pod teams. In addition, feature releases were changed to be broken down into smaller changes which enabled us to adopt continuous delivery.

Pre-change: Releases were large, and reviews took longer, and feedback arrived too late putting pressure on the delivery team.

Post change: The internal team were kept appraised of changes in advance, review lead times were significantly reduced, and feedback was timely into the overall build and design and could be clearly communicated to stakeholders going into the User Acceptance Testing (UAT) process.

Execution

Execution of the Site Reporting Project commenced in late November 2020 and continued through to mid-September 2021 at which point, the project transitioned from external execution to internal execution:

- Data Sources – At the point of transition, there were a total of five data sources connected, with another two partial data sources modelled. A key focus throughout the project for each data source was how to enable 'live' flow-through of data. Balancing the capabilities of the system against the wants of business units, we agreed on a refresh interval of 15 mins as an optimal target refresh rate for in-shift data sources while other sources that did not require in-shift focus remained with a refresh rate of once per day.

- A commentary solution was created by way of utilising Sharepoint as the comment entry point which then extracted information via a date reference and joined this data to the required fields in Tableau dashboards that allowed users to apply comments for variance again shift targets.

- Dashboards – At the point of transition, there were a total of 14 dashboard packs created, however many of the dashboard packs had multiple subpages included within them. In total, there were 85 pages created to support the underground mining teams.

- Change Management – To effectively embed the dashboards into the normal way of business, there were several key areas that required change:

  ○ Daily meetings – Agendas were updated for daily meetings to align content with information displayed on the dashboards and use these as the tool to progress through the meeting and review performance (see Figure 8).

  ○ Weekly meetings – Similar to the daily meetings, the weekly meeting agenda was also adjusted, and meeting minutes were modified to include a pdf printout of the dashboards as a snapshot in time for the previous week's performance.

  ○ In-shift triggers – Key examples of in-shift changes included a change to daily servicing strategies for loaders where a dashboard was employed to track engine hours and ensure the machine was being serviced at the right time, every time. Another example was the utilisation of a dashboard which tracked heading turnover in a priority drive and ensured that alerts were created when threshold times were exceeded so that delays could be escalated to appropriate personnel for resolution.

- Project Management – A dedicated project manager was appointed to the project and a steerco committee was set-up to meet on a fortnightly basis to report on progress against time, budget, quality, and variance to schedule:

  ○ Time – The project took 10 months to set-up five data sources and two partially completed data sources against an original target of eight data sources. Many of the data sources were not well known to the vendor, so this added complexity in terms of best extraction methods. Some of the data sources were stored in supplier warehouses and so negotiation was required to ensure fair and reasonable access to this data which impacted timelines. During UAT of the dashboards, access to end users and a dependency to provide quick turnaround times meant that timelines were impacted when feedback was not provided with a quick turnaround the expected time frames.

    There were also periods of extensive rework post deployment of the dashboard to production. This was as a result of coding errors or requests for additional features or enhancements.

  ○ Budget – The initial tendering for the project showed a large variance between proposal costs. The selected vendor had a median price which looked favourable over higher prices proposed by other vendors however during project management, it quickly became apparent the level of work that was required to onboard each data source and develop the dashboards. We were expected to meet budget far quicker than expected and began to recognise that we would need to forego some of the data sources, some of the dashboards or aspects of both. To address this, the project team took a few measures to help prolong the budget:

- Reduced the resourcing size – effectively halving data engineers and data analysts with a preference to keep those with highest work outputs based on progress delivered to date.

- Re-prioritised data sources and dashboards with a higher focus on ingesting data sources.

- Revisited to project charter to establish clear responsibility across the project team to increase levels of accountability.

- Requested additional funds to finalise the key data sources from the original list of eight sources and shift the remainder of dashboard creation to an internal resource.

o Quality – A governance process was in the early stages of being established when the project commenced meaning that templates and expectations were still being trialled and tested to ensure we had the right processes in place to capture relevant project documentation. Initially there was a large focus to help the vendor conform to the documentation requirements. Data modelling and dashboards were reviewed as part of the approval process of pushing from development to test to production environments. At the surface these appeared ok and assumed to be correct as the system was working, however it wasn't until the backend of the project that several key coding issues were flagged and later required re-modelling by the internal resource. Issues flagged involved tables that were duplicated multiple times to achieve similar use cases instead of mapping back to one overarching table to keep the system simple along with other issues of intertwining two data sources at the ingestion period which would cause impacts later on when switching out different source systems.

o Variance to schedule – Variance to schedule was tracked on two different levels:

- Project team meeting – Allowed users to flag key risks/support required during the project and why tasks were not achieved. Actions were captured using an agile methodology and tracked on a sprint board with person responsible and expected completion date.

- Steerco meeting – Project leads used this forum to flag key project risks particularly around time, budget and quality and discussed alternative methods that would help to steer the project in the right direction. A monthly report and regular meeting minutes were distributed as part of the steerco meeting.

o Learnings – The following was learnt during the project management phase:

- Detailed scoping wasn't undertaken at the start of the project so there were surprises encountered during the execution stage.

- Early engagement with data source vendors should be made as part of the scoping study to understand complexity involved with accessing data and expected threats that can be mitigated in planning phase, as opposed to execution phase.

- Strict due dates in UAT testing should be provided to end-users for completion of UAT.

- Variation in UAT testing protocols resulted in a lengthy pipeline of remediation works post-deployment of the dashboards.

- A major data source change was not expected in the original scope of works, however became a key focus later in the project and required a repointing exercise.

- Structuring data from the start of the project to be source system agnostic.

- Benefits associated with internal resourcing over external resourcing to allow greater ownership and alignment with business processes and compliance to governance processes.

- Balancing project requirements and complex problem solving with simple solutions that do not create more work or confusion than required.

- It takes a lot longer to do this work when source systems have already been in use and embedded, particularly the change management and perception of other business units to adopt a new method of reporting.

Mining Operations Daily Review Agenda

Time:- 8am Daily
Venue:- Muster Room 1
Style:- Sit Down Meeting (Informal)
Resources:- Screen to display dashboards
Total Duration:- 35 minutes

Attendees:-
Chair:-
Prerequisites:-

Mining Manager, Project Manager, Senior Foreman, Development Foreman, Production Foreman, Maintenance Foreman, Fixed Plant, Scheduling Engineer, Mine Projects Coordinator, Jetcrete Supervisor, BAPL Safety
Project Manager or Senior Foreman
Variance Commentary & Forecasts to be submitted by Foreman prior to 7:45

DASHBOARD NAME	RESPONSIBLE FOR COMMENT	FACILITATOR QUESTIONS	REQUIRED INFORMATION	Duration (Min)
SAFETY	ALL workgroups	Were there any incidents?	Describe incident. Describe actual outcome and potential outcome. Describe interim controls. State the investigation level and whether incident has been or will be entered in INX	3
		Safety Verifications	Talk to Safety Verifications completed and detail if any learnings were obtained to share with group	
FIRING & RE-ENTRY	Development Foreman, Production Foreman	What was our firing & re-entry time	Comment on the firing and re-entry times and whether there are any improvement opportunities	1
TIMELINE - PRODUCTION	Maintenance/Production Foreman	Was truck and loader availability on target?	Describe the truck and loader availability, major impacts to availability and plan in place to recover	2
HAULAGE SUMMARY	Production Foreman	Did we hit plan for last 24 hours?	Talk to the variance the last 24 hour forecast, inlcuding any actions required to get back on track	2
		What is the forecast for the next 24 hours?	Describe the forecast for the next 24 hours and whether we are on track for the weekly plan	
MHS	Engineering & Maintenance	Any variance to planned downtime last 24 hours?	If no, continue. If yes, describe the variance and any actions to put in place for next time	2
		Top unplanned downtime for the last 24 hours?	Describe the top 2 or 3 unplanned downtime events, and any actions required to mitigate for next time	
		Variance to planned downtime for the next 24hrs?	If no, continue. If yes, describe the variance to the weekly plan	
BROKEN STOCKS	Production Foreman	Is Broken Stocks a contraint?	If no, continue. If yes, talk to number of drawpoints available and planned/actua	
PRODUCTION BOGGING SUMMARY	Production Foreman	Is Production Bogging a constraint?	If no, continue. If yes, talk to any variance for the last 24 hour forecast, next 24 f	

Agenda
continues
(Extract only)

FIG 8 – Extract from Daily Mining Review Agenda.

MAINTENANCE

The following has been implemented to maintain the data and dashboards:

- Internal resources – Internal resources have been appointed to specific roles to maintain progress toward site reporting goals and instil ownership of the following:

  o troubleshooting of data source/dashboard issues

  o ingestion of new data sources

  o development of new dashboards

  o development of a business data model

  o maintain conformance to governance standards.

- Weekly technical check-ins with the Insights Hub central team were set-up to ensure alignment between corporate central data team and site data teams, particularly around approvals and governance (ie consistent layouts, peer reviews and appropriate documentation).

- Weekly Site Priority Meeting – Set-up to ensure site priorities are maintained and the right work is focused on at the right time. This includes a balance between managing change requests on existing dashboards against creation of new data sources and dashboards.

- Established register of issues – Set-up to monitor the incoming stream of new requests that relate to either new dashboards, or enhancement requests against existing dashboards.

Users provide their request via a form which is logged with the site data team and prioritised accordingly against existing work tasks.

ONGOING WORK

Data management

Business Domain Models (BDMs) Whilst delivering on the use-case scope of works a key area of focus is to enable the site data teams with a platform for self-service. The Insights Hub team in collaboration with site-based domain and functional experts are building out domain-based data models to provide a single source of truth of data for consumers (ie dashboards or self-service users).

Data sources

The following data sources remain the current priority for continuation of the site reporting project:

- Fleet Optimisation Tool – A system used to capture operator entered data metrics which required a tailored reporting solution. The snowflake platform has been utilised to house the data created from this system and has been modelled to allow end-user reporting via:
  - tableau pre-defined dashboards
  - tableau pre-defined tables
  - OLEDB connection linked via Excel for remaining Excel-based reports.
- Control System Historian – A system used to capture real-time data from fixed plant equipment across the mine site. An ODBC connection has been established between the Data Historian and snowflake to allow the ingestion of data into the data platform and transformation of this data into existing connections. Further work continues to join this data with other data sources and bring multiple information streams together.
- Mobile Fleet Machine Data system – A system used to capture real-time data from mobile fleet equipment in the underground environment via wi-fi connection back to the data historian. This is then exported via csv files and ingested into the snowflake data platform to allow trending of historical sensors, equipment alarms and current state overview of maintenance related items to inform maintenance teams of emerging issues.

FUTURE WORK

- Future Data Ingestion – Along with the above-mentioned priority data sources to be ingested, there is still a large list of other data sources that the business is looking to ingest into our data platform which had stemmed from the original requirements gathering exercise.
- Site Reporting Tool – As each data source system is ingested, we plan to use the Tableau platform to enable self-service reporting for all end users. To achieve this, we will need to index each data field and align it with an overarching business data model that has clear naming conventions and joins between common fields, for example date/time fields. This will allow the site to quickly and easily overlay data from different systems and analyse the data set for patterns or relationships that will help in address root cause issues.
- Predictive tools – As more data sources are ingested, the site will be well positioned to begin using the data to inform users and create predictive tools. A simple example of where this has already been employed includes a loader servicing dashboards that uses current engine hours and operational time to begin forward estimating the exact time that the next service will be required. This can be assessed against the existing fleet to ensure two loaders are never being sent at the same time. There is more work to be achieved with predictive tools and our internal resource has begun collecting use cases.

BUSINESS RESULTS

The delivery of this project has yielded significant positive results, that are not limited to simply an increase in actual performance. The mining team now operates with a culture of transparency and action (around variances to plan) that previously could not be maintained without the data available in a way that allows the viewer to quickly understand the real-world events.

The understanding of opportunities and threats has increased, yielding quicker and more effective action. Here are some examples of action prompted by the data during daily mining reviews, in relation to trucking only (mine bottleneck) during a three day period in June 2022.

Table 4 contains events and solutions which would be of no surprise to mine operators. The role the data plays in adding value is:

- Transparency that a problem has occurred (everyone can see it on the dashboard).

- Understanding of the impact (eg the dashboard shows how many trucks have stopped and for how long).

- Accountability to fix the problem (it can be easily seen again the next day if the issue has not been rectified).

TABLE 4

Examples of data resulting in increased productivity.

Dashboard observation	Root cause	Derived action
Truck daily service schedule not adhered to (trucks delayed)	Recent change to daily service location restricted number of parallel services	Install additional dome shelter for daily services to occur in new location
Increased re-entry time delaying truck first loads	Ventilation fan setting had been left on local control when should be on SOC control	Implement stricter access control to fan settings so they cannot be adjusted without a pass code
Blocked decline 45 minutes for all trucks on night shift	Vent bag scraped by truck	Install nappies under vent bag to prevent tearing
Trucks waiting refuelling on surface at start of shift	Primary fuel tank empty requiring transferring from secondary	Implement a process to keep primary tank full

These concepts are not just relevant to the daily time frame. The team has seen the same benefits in weekly to monthly performance trends.

The processes which have been implemented in the daily and weekly mining review meetings have become relied upon by those facilitating and presenting at the meeting. For the participants they are now able to come prepared to the meeting knowing what data will be presented and what information they are required to know in advance. For the facilitator they have a clear structure to follow, which frees them up to focus on effectively listening to the participants and understanding the information presented.

Of note is that with the increase in reliance upon the dashboard packages, there is an increased need to maintain data quality. There is now greater emphasis on data validation, with the most variances seen where data capture systems/processes are more manual.

Finally, short interval control using dashboards has been achieved. In particular, the loader servicing dashboard which allows a dynamic servicing schedule to be easily maintained and complied to. This has increased daily service compliance while decreasing daily service delays.

CONCLUSIONS

This projected resulted in five full data sources being ingested to a centralised data lake in a data agnostic way, and paving the way for additional data sources to be added. A self-service reporting tool has been set-up and deployed, however further training and embedment is required to cement this and start seeing value. Fourteen dashboard packages have been created with a total of 85 pages of information and are cemented into business processes so they are used hourly, daily and weekly.

The project has achieved the goals set-out and is achieving the intended benefits. For OZ Minerals and Carrapateena there have been significant learnings around project scoping, time frame, cost, skillset and systems which have been discussed in the paper.

The culture around performance at Carrapateena has been transformed. Transparency and accountability are more easily maintained by leaders who now have the data at their fingertips. Understanding of threats/opportunities, and ability to quickly align on the action required is now built into our management operating system. Meeting agendas set clear expectations for both the facilitator and the participants. This whole process is supported by the right data at the right time.

ACKNOWLEDGEMENTS

The authors would like to thank: Interworks; First Principles Consulting; and OZ Minerals.

The future is electrifying – introduction of a battery electric loader to CSA Mine

M Ireland[1]

1. MAusIMM, Mining Manager, Glencore – CSA, Cobar NSW 2835.
   Email: matthew.ireland@glencore.com.au

ABSTRACT

Extensive use of diesel-powered mobile equipment in modern mines for materials handling, such as load-haul-dump machines, constitute a significant source of heat, exhaust gases and diesel particulate matter. Provision of sufficient quantities of fresh, cool air are required to remove such atmospheric contaminants from the environment for miners, particularly at depths where ventilation constraints and associated operating costs have an increasingly adverse impact. An alternative solution to diesel equipment to combat economic, environmental, social and health challenges, are battery electric vehicles (BEV), with the potential to expand production and substantially reduce ventilation and cooling demand, subsequently reducing operating costs.

In this context, this paper focuses on the introduction of a battery electric loader to a diesel fleet at one of the deepest underground mines in Australia, the CSA Mine, a high-grade copper operation in Cobar, New South Wales. Analysis of a trial battery electric loader compared to existing diesel machines is performed with regards to infrastructure, technical and operational issues. Key benefits of battery electric vehicles over conventional diesel-powered vehicles are discussed, including lower energy consumption and greater efficiency, reduction of emissions and heat, reduced need for ventilation and lower operating costs. Challenges of introducing BEVs into an operation, along with the results from the comparative study are addressed under impacts to energy consumption, equipment selection, effects on ventilation and cooling, mine design and economic viability for future fleet implementation scenarios whilst meeting production targets.

INTRODUCTION

Intense public scrutiny of workers' health and safety, social licence, and environmental issues, are factors driving stricter requirements from governmental institutions regarding air quality and quantity. Furthermore, commitments made to decrease dependency on fossil fuels to meet the world's climate goals and the strategic role of critical minerals in clean energy transitions is of central importance, will result in immense demand growth. One such commitment is that of Glencore, aiming to reduce total emissions (Scope 1, 2 and 3) relative to 2019 levels by 15 per cent 2026, 50 per cent by 2035 and net zero by 2050 (Glencore, 2021).

Diesel-powered mobile equipment dominates modern mines for materials handling, such as load-haul-dump (LHD) machines and trucks. These heavy vehicles constitute a significant source of heat, exhaust gases, diesel particulate matter (DPM), noise and vibration. Ensuring enough fresh, cool air to remove atmospheric contaminants from the underground environment for miners, particularly as mines progressively becoming deeper where ventilation constraints and associated operating costs have an increasingly adverse impact.

A critical step to improve working conditions in underground hard rock mines by reducing diesel-powered machine contaminants is proposed to replace these with electric vehicles which produce none. This alternative solution will combat economic, environmental, social and health challenges, with the potential to reduce mine greenhouse gas emissions, expand production, substantially reduce ventilation and cooling demand quantities, subsequently reducing operating costs and make the mining industry a more attractive workplace.

Despite many advantages, few underground hard rock mines globally use electric vehicles, typically due to market availability, battery life, change out and charging time frames, safety concerns and battery technology. A new generation of reliable BEVs are now manufactured following recent developments in battery technology, offering the same flexibility as diesel machines. To make a business case for a potential transition to a cleaner fleet and prove viability in real mine conditions, field trials were performed at the CSA Mine with the introduction of an Epiroc ST14 battery electric loader.

This paper outlines the results and observations of the field trials comparing the battery electric loader to existing diesel loaders, with respect to improvement of environmental working conditions, specifically temperature, emissions, vibration, and noise. Additionally, survey feedback from mine employees regarding their opinions towards BEVs is also provided.

CSA MINE OVERVIEW

CSA Mine is located 11 km north-west of Cobar or approximately 700 km west-north-west of Sydney in Central Western NSW (Figure 1). Having a long history, mining commenced in 1871 with an erratic production history until 1964, when Broken Hill South Ltd began large scale production, before passing to CRA in 1980, then to Golden Shamrock Mines in 1992 (Glencore, 2022). Closing in 1998 following acquisition by Ashanti Goldfields, Glencore reopened CSA Mine in 1999 under wholly owned subsidiary Cobar Management Pty Ltd (CMPL).

FIG 1 – CSA Mine location in Central Western New South Wales.

Operating 24 hours a day, seven days a week, CSA Mine provides direct employment for over 500 people, the majority of whom are based locally. As an important contributor to the local community and economy, CSA Mine is committed to operating to the highest safety standards and focuses on minimising environmental impacts.

CSA Mine consists of a deep, high-grade underground mine producing 1.2 Mt/a of copper ore, processed on-site to produce more than 165 kt/a of copper concentrate. The concentrate with no deleterious minerals, contains approximately 26 per cent copper metal (~45 kt/a), plus an economic co-product of silver, which is railed to Newcastle for subsequent export.

Mining operations

At a current depth of ~1875 m and ore averaging four per cent mined grade (some as high as 12 per cent), CSA Mine is one of Australia's deepest underground mines and the one of the richest copper deposits in the world. The underground mine is serviced by two hoisting shafts and a decline.

The deposit situated within the Cobar Basin, sees mineralisation at CSA Mine hosted within a steeply dipping sequence of interbedded siltstones and sandstones. Five systems (Eastern, Western, QTS North, QTS Central and QTS South), contain multiple subvertical lenses running north south, are typically 5–30 m wide and have short strike lengths. The orebody with dominant minerals of chalcopyrite ($CuFeS_2$) and Acanthite (Ag_2S), remains open down plunge, with the deepest drill intersection currently at ~2200 m below surface (Figure 2).

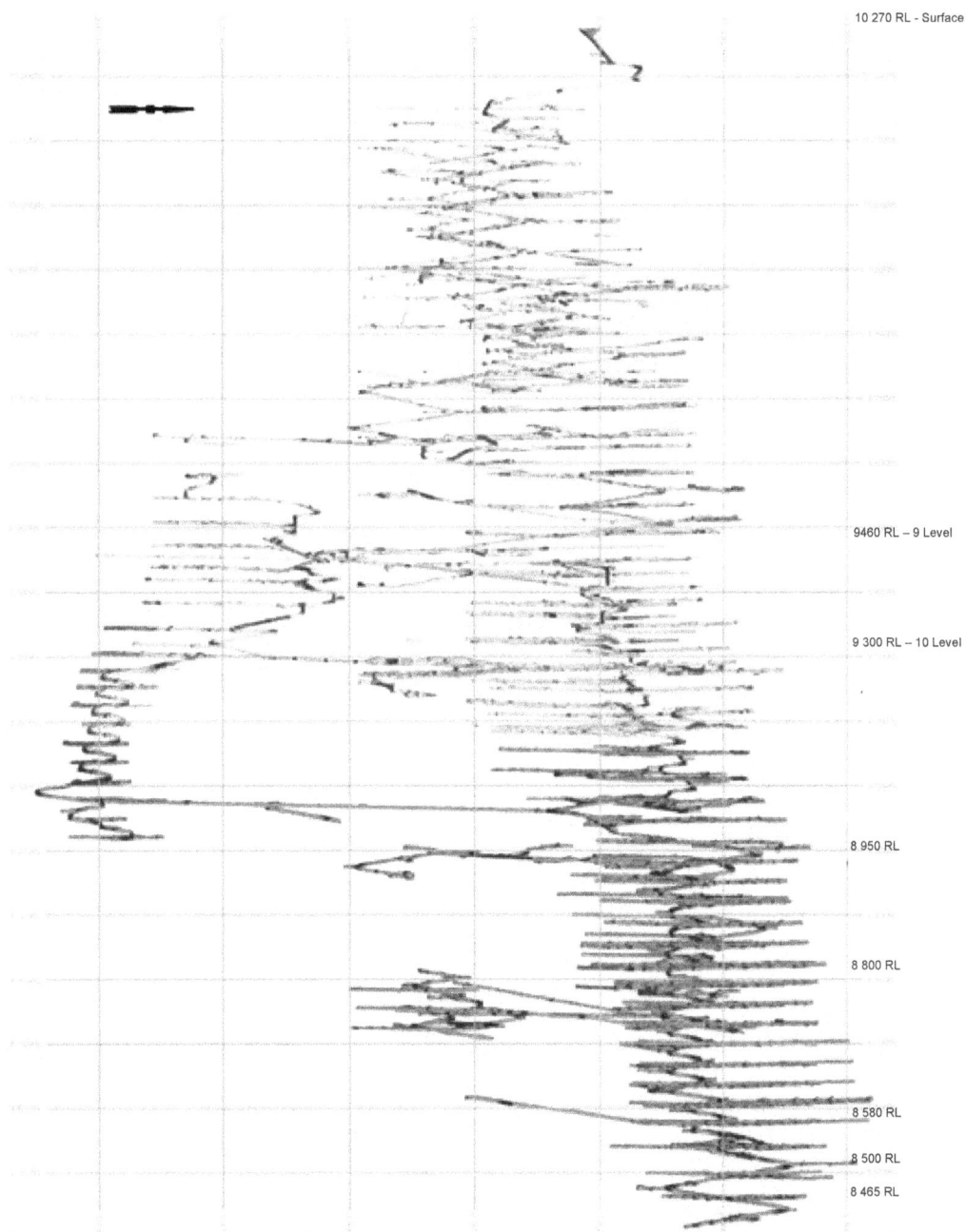

FIG 2 – Annotated long section of CSA Mine.

The current sublevel interval is set at 35 m, however this has varied between 30–40 m in recent years. The bulk of the deposit is extracted using conventional mechanised sublevel longhole open stoping methods with east–west transverse drives and a top-down, centre-out sequence, utilising cemented paste fill to fill voids and mine adjacent areas. A modified Avoca stoping method is used in the narrower lenses, principally in QTS Central, Western and longitudinal extraction in QTS North.

Ore is hauled to underground jaw crushers before being hoisted to surface via the two shafts. The 2019 headframe and double-drum winder for shaft one hoists material from 10 level (9375 RL or 895 m underground) to surface, having a capacity of 700 kt/a. Shaft two, a tower mounted Koepe friction winder, having a capacity of 1.6 Mt/a hoists material from 9 level (9460 RL or 810 m underground) to surface and is also used for personnel transportation.

Power is supplied to the site is via 132 kV transmission line as part of the NSW power grid. The current available capacity of the supply facilities is around 26 MVA, upgrading to 32 MVA in December 2022.

The recently upgraded primary ventilation system consists of four surface exhaust fans providing a combined volumetric flow rate of approximately 1000 m^3/s, to extract atmospheric contaminants out of the mine via a dedicated series of ventilation raises from the bottom of the mine. Primary fresh air intakes include the main decline from the surface, shaft one, shaft two, and fresh air raise one (FAR1). The latter three chilled by ammonia refrigeration plants having a combined cooling capacity of 12 MWBAC, upgrading to 24 MWBAC with the addition of Fresh Air Raise 2 in December 2022. Active working areas are force-ventilated with auxiliary fans moving fresh air from a series of staggered and interconnected fresh air raises. The mine's geothermal gradient is 2°C per 100 m, with virgin rock temperature being approximately 55°C at 8500 level (1770 m below surface).

Underground mobile fleet

Like many underground mines around the world, the CSA Mine mobile fleet consists of rubber tyred diesel equipment. The primary fleet supplied by a variety of equipment manufacturers is outlined in Table 1, with additional ancillary support vehicles omitted.

TABLE 1

Primary mobile fleet in use at CSA Mine.

Equipment type	Model	Quantity	Power rating/ unit (kW)
Jumbo	Sandvik DD421–60C	3	119
Cabolter	Sandvik DS421-C	2	110
Production drill	Epiroc E7C	2	120
Loader	Sandvik LH517i	6	310
Truck	Epiroc MT5020	2	485
Truck	Epiroc MT6020	2	567
Truck	Sandvik TH663i	5	565

CSA Mine is midway through an asset replacement program for underground diesel loaders and trucks to overcome an aging fleet with low utilisation rates, frequent breakdowns, and high costs to maintain. These load and haul assets account for of the majority of the mine's diesel burn and emissions. Future asset replacement cycles consider fewer pieces of equipment with increased utilisation rates, along with battery electric trucks and loaders as a positive alternative to diesel to help reduce ventilation and refrigeration requirements.

Electric alternatives

The exception to the diesel fleet listed above, is the Epiroc ST14 battery electric loader. Battery electric vehicles are an alternative to diesel-powered mobile equipment and an increasingly popular trend in underground mining, largely due to environmental benefits offered within the workplace.

Paraszczak *et al* (2014) also categorises cable-powered and trolley-powered vehicles as electric alternatives to diesel and battery forms of energy supply to machine engines.

Unlike diesel-powered machines, electric vehicles offer benefits of producing no toxic gases, no diesel particulate matter (DPM), reduced vibrations and significantly less noise. Furthermore, electric motors are significantly more efficient at around 95 per cent and therefore produce less heat through energy losses into the surrounding rock, equipment and environment (Macdonald *et al*, 2019). This heat dissipated by electric motors is sensible heat, whereas the heat from a diesel motor reports as a combination of sensible and latent heat due to the combustion process and vapourisation of water. Operating costs are also reported to be lower from a maintenance and fuel perspective, along with reduced ventilation capital as a result of fewer emissions (Moore, 2010; Chadwick, 1992). Moreover, batteries are getting lighter, more efficient and have a more extensive range with chemistries constantly being upgraded.

Conversely, diesel-powered mobile equipment offers advantages over electric equipment including lower capital cost to purchase, fewer infrastructure requirements such as pantograph trolley lines, increased autonomy and flexibility in operation (Varaschin, 2016). Further holistic consideration needs to be taken before making equipment selection including project economics, mine design and ventilation requirements, along with being a greenfield operation commencing or transitioning fleet in a brownfield operation.

Epiroc ST14 battery-electric loader

Numerous changes in key personnel have occurred since 2018 when the project commenced. The battery electric loader purchased to conduct the trial based on those available on the market at the time, was the Epiroc ST14. Designed on the diesel Scooptram ST14, having the same exterior dimensions and appearance, it has a 14 t tramming capacity and 18 t breakout force. Other key mechanical features include a four-speed automatic transmission, fully enclosed wet disc brakes and two electric, liquid cooled motors replacing the diesel engine (Epiroc, 2022).

The ABB 200 kW traction motor and ABB 160 kW auxiliary motor are driven by a 4.5 t lithium ion (Li-ion) battery, situated where the diesel engine would be (Epiroc, 2022). The Li-ion NMC battery pack has four subpacks with liquid cooled cells and an integrated thermal management system. The 360 kW battery, having a usable 300 kW is charged by an 850 VDC source via a CCS2 plug and can be removed for offboard charging or charged onboard in a minimum of two hours.

CLEAN ENERGY TRANSITION

Intense public scrutiny of workers' health and safety, social licence and environmental issues, are factors driving stricter requirements from governmental institutions regarding air quality and quantity. Environmental, social and governance (ESG) is an integral part in ensuring companies act responsibly for accessing mineral endowment. Continued technology investment is important towards achieving sustainable goals maintaining a social license to operate, with Australian resource industry in a position to lead to the world (Phillips, 2022).

Mining is responsible for large quantities of global greenhouse gas emissions and is therefore a large contributor to realising reductions in emissions. Commitments have been made to decrease dependency on fossil fuels to meet the world's climate goals and collaboration is common from companies striving to reach this ambition, accelerating progress towards fully electrified mining operations, renewable energy generation and zero-emissions goals. In this respect, critical minerals play strategic role in clean energy transitions, with immense demand growth expected, coupled with a strong economic outlook for the industry forecast. According to Zakharia (2021), the abundance of critical minerals required for battery systems in Australia allows increased competitiveness, enhancing ESG, transitioning to sustainable energy through responsibly sourced raw materials and a strong regulatory framework.

BATTERY ELECTRIC LOADER TRIALS

Following a thorough safety, mechanical and electrical onboarding to site, a phased approach was planned for the trial of the ST14 at CSA Mine. Operator training and commissioning of the loader took place on surface, prior to going underground for placement at the 9 level transfer (9460 Level), adjacent to the 150 kW fixed charging station (150–850 VDC, 250 A) and shaft two crusher.

Trial 1

The first trial involved rehandling stockpiled ore into the crusher in a localised area with minimal gradient. Observations of tractive effort, breakout lifting capacity and fill factors were made on the loader bucket penetrating the fragmented rock of non-homogeneous nature. Similarly, hauling and dumping cycles were monitored when tipping over the grizzly bin into the crusher. Environmental conditions are relatively benign in this area being only ~810 m below surface and close to downcasting fresh air with temperatures measured being 24°C DB and 26°C WB. Onboard charging times and run times were also assessed to gauge impact to productivity in lieu of a second battery being available to swap-out. Run times ranged from 2–4 hrs depending on the operator and charge times were in line with the original equipment manufacturer guidelines, being up to two hours from a depleted state of charge. Productivity performance metrics for 2022 are outlined in Figure 3, with 185 t per motor hour at 15 t per bucket. Despite the low utilisation, the ST14 loader at CSA Mine is one of the best performing BEV in Epiroc fleet globally based on data analytics from Kenley (February 2022, personal communication).

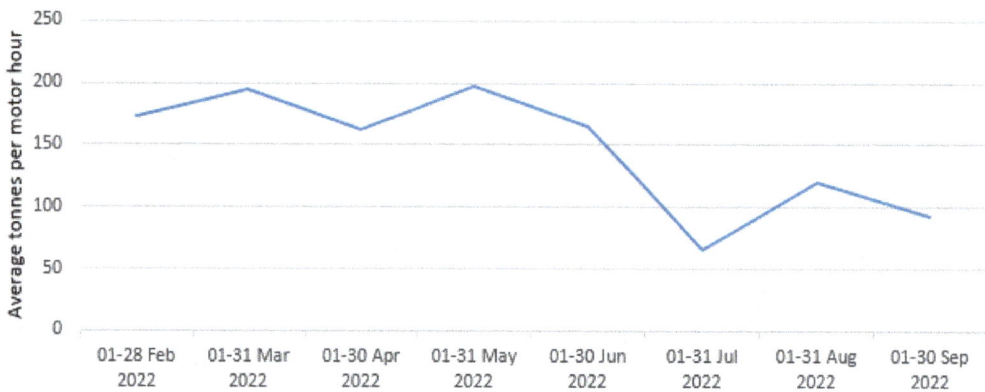

FIG 3 – 2022 productivity performance metrics.

Trial 2

A second trial was performed with the loader driving down to 10 level to feed shaft one crusher with waste. In this trial, regenerative brake functionality was confirmed, along with quantifying the depletion of state of charge when tramming the route back to 9 level charge station as shown in Figure 4. Battery regeneration while tramming down-grade was positive, increasing to full charge over 2.1 km, while 15 to 20 per cent battery depletion occurred travelling the 2.3 km up-grade back from 10 to 9 level.

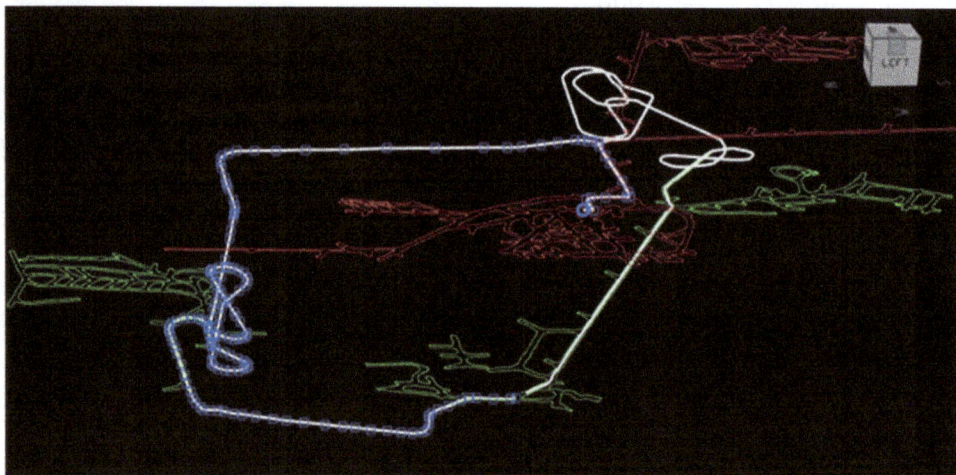

FIG 4 – Loader tramming route from 10 level up to 9 level.

Trial 3

A third trial was conducted to compare the ST14 to the remainder of the fleet being diesel powered, with respect to impact on environmental conditions. In this assessment, measurements were taken in an ore drive at the 8950 level (510 m below 9 level and 1320 m below surface). The repeated tramming path with minimal grade was approximately 65 m from the drive face to the stockpile (Figure 5). Both diesel and battery electric loaders performed simulated loading cycles over a two-day period, consisting of travelling forward to the face of the drive, bogging fragmented ore, reversing to the stockpile and completing a lift-lower cycle. Like trials by Halim *et al* (2021) and MDEC (2018) in Sweden and Canada respectively, variables measured during this trial included:

- heat input to the ventilation airstream (wet bulb and dry bulb temperatures)
- atmospheric contaminants
- internal and external noise levels
- whole body vibration within the cabin.

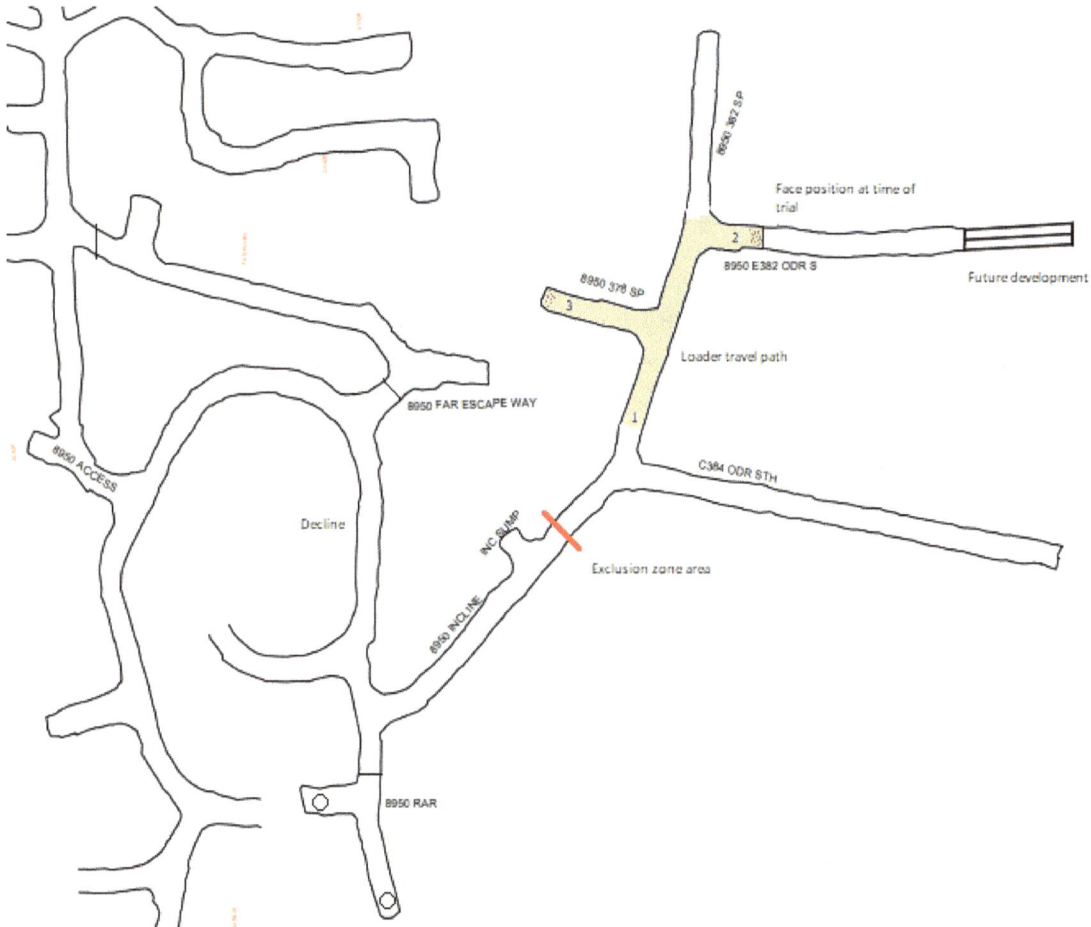

FIG 5 – Plan view of 8950 level showing tramming path from start (1) to face (2) to stockpile (3).

Temperature

Thermal environment measurements were conducted using a Calor HSM II Heat Stress Monitor. Measurements were taken in the local ventilation airstream for dry and wet bulb temperatures, along with globe temperature, relative humidity, and an airflow. The combinations of these measurements as outlined by Macdonald *et al* (2019), impact workplace conditions and perceived conditions felt by employees, when considering latent heat of vapourisation in addition to sensible heat.

The data for the LH517i summarised in Table 2 indicated that the ambient temperature increased in the area during bogging. The Sandvik loader test realised a 14.6°C increase from the baseline in ambient temperature both during and after completion of the test. Conversely, the wet bulb temperature was lower following the test, likely due to the significant reduction in relative humidity (-38 per cent) to which the wet bulb is particularly sensitive.

TABLE 2

Thermal environment monitoring results – Sandvik LH517i.

	Dry bulb (°C)	Wet bulb (°C)	Globe (°C)	Relative humidity (%)
Start	32.3	28.1	33.7	72
End	38.5	25.3	36.7	34
Min	32.3	25.3	33.7	23
Max	46.9	30.3	45.9	72

The ST14 bogged for 50 min, repaired vent ducting for 20 min, then continued bogging for 41 min. The state of charge decreased from 99 to 73 per cent during this time and 17 buckets were bogged before commencing tramming back to 9 level charge station. The state of charge further decreased to 57 per cent during this up-grade tramming cycle.

During sampling the instrument was found to have a volatile memory, resulting in the loss of sampling data collected for the Epiroc loader test. This quantitative data was not able to be recovered and the test was not able to be repeated, nor a direct comparison drawn. Anecdotally, the LH517i produced more total heat than the ST14, in line theoretical calculations for a diesel engine being less efficient.

Atmospheric contaminants

Gas concentrations generated through the trial were measured using Dräger X-am 5000 monitors and compared to occupational exposure standards. Calculations of concentration limits for a 10 hr working shift using the Brief and Scala method were made. Diesel exhaust gas emissions of nitrogen dioxide and carbon monoxide were below the threshold limit value–time weighted average (TLV-TWA) of 3 ppm and 20 ppm respectively. No gases were detected when the ST14 was operating.

Real-time monitoring (RTM) was undertaken using two AM520i SidePak aerosol monitors. One monitor was fitted with a PM_{10} impactor inlet, and the other with a combined Dorr-Oliver cyclone (0.8 μm cyclone inlet). Thoracic dust sampling was undertaken using a 37 mm BGI 2.69 cyclone. The deposit was collected in a 35 mm conductive plastic filter cassette on a 5 μm PVC membrane filter (Rae, 2022a).

Gravimetric and elemental analysis of samples was performed with results outlined in Table 3. Where analytes were detected, the percentage difference between the two loaders has been calculated at approximately 600 per cent. Fugitive dust while bogging with the Epiroc loader was substantially lower than using the Sandvik loader. A plausible contributor is deposited dust made airborne by wind generated from radiator fans and engine exhaust for the LH517i. Furthermore, the difference would likely become more pronounced as the duration of continuous work increases as the background level does not have sufficient time to return to the pre-disturbance level after each cycle.

TABLE 3

Dust, DPM and elemental analysis results.

Variable	Epiroc (mg/m^3)	Sandvik (mg/m^3)	Difference (%)
Thoracic Dust	1.9	12	632
Aluminium	0.098	0.67	684
Barium	<0.0023	<0.0040	174
Copper	0.047	0.27	574
Iron	0.31	2.0	645
Potassium	<0.023	<0.040	174
Elemental Carbon	<0.010	<0.017	170
Total Carbon	0.063	0.12	190

Diesel particulate matter (DPM) monitoring was undertaken using a 0.8 μm cut point cyclone for analysis of both elemental (EC) and total carbon (TC). Laboratory analysis returned results below the laboratory limit of detection for elemental carbon (surrogate for DPM) for both tests. This result supports the effectiveness of the exhaust treatment system used by the Sandvik loader and correct maintenance. The results of total carbon analysis were approximately 190 per cent greater from the Sandvik loader than the Epiroc loader. This result is likely associated with semi-volatile hydrocarbons which pass through the exhaust filter as a vapour and then condense as submicron particles on exiting the tailpipe in the relatively cooler ambient air. Correlations of 0.911 and 0.639 were found for the LH517i and ST14 respectively between measured PM$_{10}$ and submicron size fractions using linear regression analysis. The strong correlation supports that these particulates were re-aerosolised mineral dust, likely generated from radiator fans and engine exhaust of the diesel powered loader.

Noise

Noise emissions were carried out in accordance with Australian Standard AS1269:2005 and the NSW SafeWork Code of Practice 2019: Managing Noise and Preventing Hearing Loss at Work. Handheld noise data was collected with a NATA calibrated Bruel and Kjaer 2250L, Type 1 Sound Level Meter with a type 4950 microphone. A Bruel and Kjaer Type 4448 noise dosimeter was used with exposure levels normalised and adjusted in accordance with AS1269.1:2005.

The ST14 generated noise levels between 69–75 db (A) outside the cab and would not exceed the exposure standard for more than 12 hours continuous exposure. Comparatively, the LH517i produced external sound levels of 85–93 db (A), with higher values returned on either side of the rear of the loader (Figure 6). These values exceed the exposure standard, with allowable exposure decreasing after 1.5 hr. An external difference up to 23.2 db (A) between the loaders equates to approximately 15 times lower sound pressure for the electric loader. This notable difference is in line with anecdotal observations and predictions for the battery electric vehicle being quieter that a diesel unit based on a more efficient drivetrain and motor.

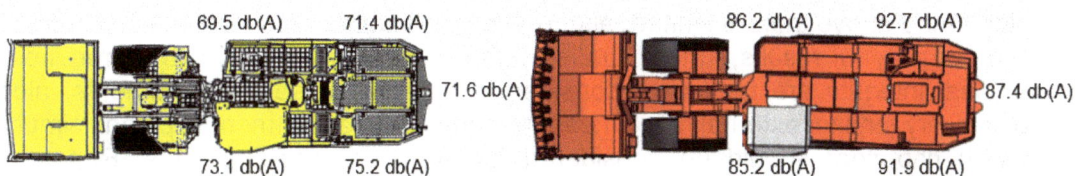

FIG 6 – Measurement locations and results around loaders.

First attempts at idle for the internal personal dosimetry measurements for the ST14 could not be verified due to a mismatch with times recorded. A secondary attempt while tramming recorded

83.4 db (A) in cab for the ST14. Comparatively, the LH517i produced sound levels of 85.3 db (A) inside the cab at idle, translating to a time of approximately six hours to reach the occupational exposure standard of 85 db (A) for an eight hour day. Whilst a direct comparison could not be drawn, it is suggested that a similar variance would be realised to that measured externally with a noise dampened cab.

Vibration

Whole body vibration value was measured inside the cab according to Australian Standard AS 2670.1:2001, using a Svantek SV100A Human Vibration Dosimeter (Rae, 2022b). Features assisting vibration dampening for the operator on both loaders include rubber mounted cabin, air suspended seat, steering and boom soft stop. Exposure duration is the same for both loaders to allow direct comparison of exposure results. In practice, the Sandvik loader is more likely to be used for longer durations during the shift which will result in greater exposure from the Sandvik.

A basic evaluation method was used, involving the measurement of the weighted root mean square (RMS) acceleration. Both loaders were below the exposure action value (EAV), returning results of 0.85 m/s^2 and 0.93 m/s^2 for the ST14 and LH517i respectively. The dominant axis for the basic evaluation was the x-axis, being forward and backwards for the operator, but corresponds to side-to-side movement of the loader. A notable difference in the z-axis (vertical) was also measured, being 0.23 m/s^2 higher in the ST14 (Table 4).

TABLE 4

Primary mobile fleet in use at CSA Mine.

Equipment ID	Model	Weighted RMS (m/s^2)			Exposure categorisation
		x-axis	y-axis	z-axis	
LH25	Epiroc ST14	0.85	0.82	0.82	≤EAV
LH32	Sandvik LH517i	0.93	0.82	0.59	≤EAV

A crest factor greater than nine was noted for the basic evaluation, indicating jolts and jars are likely to have a significant impact on exposures. An alternate evaluation method, maximum vibration dose value (VDV)(T) is a more appropriate evaluation. This showed that results were between action and limit values for both loaders, with the exposure greater across all three axes for the ST14 (Table 5). Further simultaneous measurements of both floor and seat would need to be undertaken to gain further insight, gauging operator's seat adjustment and seat performance to attenuate vibration.

TABLE 5

Primary mobile fleet in use at CSA Mine.

Equipment ID	Model	Weighted VDV (T) (m/s$^{1.75}$)			Exposure categorisation
		x-axis	y-axis	z-axis	
LH25	Epiroc ST14	9.7	9.8	13.2	>EAV, ≤ELV
LH32	Sandvik LH517i	9.3	8.3	8.9	>EAV, ≤ELV

Future trials

Data collected in the above trials is limited, although representative of conditions. To obtain greater accuracy and certainty for results, further studies with reduced variables would be required. Radio volume and seal integrity are factors to consider for internal noise measurements, along with proximity to auxiliary fans for external noise measurements. Seat adjustment, type and simultaneous monitoring with cab floor should be considered for vibration measurements. Repeat thermal measurements and heat modelling with reliable data capturing is also required for comparative results. Other parameters to be considered in tests include measuring machines from the same manufacturer, strata temperatures, cycle times, breakout force, acceleration, and deceleration.

Similar trials to those conducted in upper areas of the mine are planned for 2023 in lower levels of the mine. Limiting factors preventing further work being conducted to date are battery performance and distances to active areas (~500 m vertical to closest heading) relative to the fixed charging station at 9 level. To overcome this constraint, a second charging station is planned to be placed at the 8460 RAW, being 65 m above the lowest part of the decline and closer to the active mining centroid. Following further trials, a battery swap location may also be determined as the current ST14 requires an overhead gantry crane to hoist out the battery, unlike other products on the market.

Operator feedback

Overall sentiment from operators has been mixed from the onset, associated with acceptance of new technology and a different machine. Reluctance to adopt and embrace the change to a different manufacturer as well as a non-diesel-powered loader was soon overcome due to the positive environmental and ergonomic benefits. During the introduction of the ST14 battery electric loader and various trials conducted, employees were surveyed on their opinions on working conditions relative to diesel machines.

All respondents noted improved environmental conditions of reduced noise, heat output, in-cabin vibrations, and improved air quality with zero-emissions. Similarly, operational benefits of speed and manoeuvrability from a high-power traction motor connected to a high efficiency driveline and a separate auxiliary motor that delivers hydraulic power on-demand. Mechanically less complex, easier to maintain than diesel machines with fewer moving parts with battery removal done in minutes, allowing easy access to hydraulics, and performing diagnostics via the battery management system were seen as positive. Considering the improved social and environmental performance, using modern technology, improved public perception from a sustainability perspective it is believed to improve the attractiveness of the mining industry.

Conversely, employees did not like the frequency at which the battery had to be changed, impacting productivity, supply of future batteries due to critical mineral production, the quiet nature being difficult to hear in an underground environment, and potential fire risk from the battery pack. Respondents were indifferent to the physical and mental effort required to operate the machine and work performance when loading.

Mine design and effects on ventilation and cooling

As mining approaches 2000 m below surface, sustained production will depend on developing sufficient mining fronts and having a mining fleet to support production output and subsequent ventilation and cooling to suit. Currently underground ventilation requirements are predominantly driven by the quantity of diesel-powered mobile equipment units operating and strata temperatures. The successful introduction of battery electric vehicles into underground mines provides an opportunity to reduce the quantity of airflow due to the absence of exhaust contaminants and lower heat produced, realising a positive impact on future ventilation and cooling requirements.

A reduction in primary ventilation and cooling may also provide mine design opportunities to reduce drive and airway dimensions along with equipment selection. These factors result in a reduction in capital required, (this may impact productivities from smaller machines), dilution of mine strata gases and extended blasting re-entry times. Furthermore, a significant reduction in operating expenses is possible from reduced electrical power to run the ventilation and cooling network, with fan electrical power being proportional to the cube of airflow quantity.

Key limitations of BEV relative to diesel equipment affecting mine design are their limited range and the time needed to charge. Mine plans need to be equipped to accommodate the BEVs to fully harnessing the benefits, considering materials handling systems, ramp design and layout, maintenance areas, and charging infrastructure (Global Mining Guidelines Group, 2022). Performance of the system can then be measured to determine the effectiveness in comparison to diesel equipment.

CONCLUSIONS

Overall, as a trial unit for CSA Mine to start its journey into battery electric vehicles, the Epiroc ST14 has proven useful, with value added from current trials and to be added from future planned trials.

The progress has been slower than desired with considerable delays since the inception of the initiative, delays in delivery, turnover of key personnel and the size of the mine.

The field trials conducted provide insight into the different environmental effects between the battery electric ST14 and the diesel LH517i when performing similar activities in the same environment. Initial results demonstrate significantly improved conditions with respect to atmospheric contaminants and noise, yet a lesser extent for heat and vibration. Whilst not all results were able to be compared due to equipment faults, data obtained provides a baseline for future comparisons. It is recommended that the installation of a second charging station in lower levels of the mine closer to active areas will be beneficial to gather further data, improving the uptake of battery electric mobile equipment.

Anecdotal feedback from the survey of employees reiterated these observations, with improved social perception and attractiveness to the mining industry. Further economic assessments need to be conducted for life-of-mine planning as the mine expands and exceeds 2000 m depth, specifically on ventilation and cooling upgrades.

Increased collaboration with tertiary institutions, manufacturers, and regulators for sharing learnings about electrified technology, risk management and improved battery performance is needed. A transition to battery electric powered mobile equipment at CSA Mine can be realised and should be considered as part of future asset replacement cycles with minimal changes to the operation.

ACKNOWLEDGEMENTS

The author would like to acknowledge Glencore for their permission to publish the contents of this paper and provision of facilities to undertake this work. Special thanks to all personnel from CSA Mine for their support of this project, from operators to technical staff and leadership – this paper would not have been possible without input from you. Similarly, the author would like to thank Epiroc, ABB and GCG for their collaboration and contribution in respect to loader, charge station and hygiene samples respectively.

REFERENCES

Chadwick, J, 1992. Diesel or Electric? *Mining Magazine*, August p 92.

Epiroc, 2022. Scooptram ST14 Battery technical specification [online], Australia, Epiroc Australia Pty. Available from: https://www.epiroc.com/en-au/products/loaders-and-trucks/electric-loaders/scooptram-st14-battery [Accessed: 15 June 2022].

Glencore, 2021. Pathway to net zero – 2021 progress report [online], Switzerland, Glencore plc. Available from: https://www.glencore.com/dam/jcr:ad341247-c81e-45b4–899d-a7f32a9d69a0/2021-Climate-Change-Report-.pdf [Accessed: 15 June 2022].

Glencore, 2022. CSA Mine [online]. 2022, Switzerland, Glencore plc. Available from: https://www.glencore.com.au/operations-and-projects/csa-mine [Accessed: 15 June 2022].

Global Mining Guidelines Group, 2022. Recommended Practices for Battery Electric Vehicles in Underground Mining – Version 3 [online], GMG Group. Available from: https://gmggroup.org/wp-content/uploads/2022/06/2022–06–23_Recommended-Practices-for-Battery-Electric-Vehicles-in-Underground-Mining.pdf [Accessed: 15 September 2022].

Halim, A, Lööw, J, Johansson, J, Gustafsson, J, van Wageningen, A and Kocsis, K, 2021. Improvement of working conditions and opinions of mine workers when battery electric vehicles (BEVs) are used instead of diesel machines – results of field trial at the Kittilä Mine, Finland, *Mining, Metallurgy and Exploration 39*, pp 203–219.

Macdonald, K, McGuire, C, Harris, W and Witow, D, 2019. Psychrometric evaluation of workplace impacts upon change to battery mobile equipment from diesel, in *Proceedings Australian Mine Ventilation Conference 2019*, pp 244–248 (The Australasian Institute of Mining and Metallurgy: Melbourne).

MDEC, 2018. Battery vs. diesel underground LHDs – direct comparison of heat generation [online], Toronto, Caterpillar. Available from: https://mdec.ca/2018/S7P3_Jay_Armburger.pdf [Accessed: 15 June 2022].

Moore, P, 2010. Plugging the gap underground, *Mining Magazine*, November, pp 40–46.

Paraszczak, J, Svedlund, E, Fytas, K and Laflamme, M, 2014. Electrification of loaders and trucks – a step towards more sustainable underground mining, in *Proceedings International Conference on Renewable Energies and Power Quality*, pp 81–86 (Renewable Energy and Power Quality Journal: Cordoba, Spain).

Phillips, M, 2022. Powering towards a sustainable future, *Australian Mining*, February, pp 36–37.

Rae, H, 2022a. Comparison of environmental effects between an electric and diesel underground loader, GCG Project Rpt No 4714–1.

Rae, H, 2022b. Comparison of noise and vibration exposures between an electric and diesel underground loader, GCG Project Rpt No 4602–1.

Varaschin, J, 2016. The economic case for electric mining equipment and technical considerations relating to their implementation, Master thesis (unpublished), Queen's University, Kingston.

Zakharia, N, 2021. Electric vehicles to drive Australia's mining future, *Australian Resources and Investment*, September 2021, 15(2):28–29.

Intersection reinforcement with self-drilling anchors for improved productivity at Kanmantoo

J Jardine[1], B Roache[2], S Thomas[3] and P Jere[4]

1. MAusIMM, Mining Manager – Kanmantoo Copper, Hillgrove Resources Limited, Kanmantoo SA 5252. Email: jol.jardine@hillgroveresources.com.au
2. Principal Geotechnical Engineer, Neboro, Melbourne Vic 3000. Email: broache@neboro.com.au
3. General Manager, ME Safe, Melbourne Vic 3000. Email: sam@mesafe.com.au
4. Principal Geotechnical Engineer, Master Builders Solutions, Perth WA 6000. Email: precious.jere@mbcc-group.com

ABSTRACT

An improved ground support method for development intersections, wide spans and stope brows was implemented at Kanmantoo mine. The method uses a jumbo to simultaneously drill and install self-drilling anchors (SDAs) to depths from 4.5 m to 7.0 m. Resin is subsequently pumped into each SDA using a specially designed integrated tool carrier basket, with plating occurring within 10 mins due to rapid resin cure. This methodology means that SDAs can be installed rapidly due to reduced process steps and wait times with reduced hazards to personnel.

Cablebolts are used extensively in underground mines to reinforce the ground in wide spans, such as at intersections and stope brows. The flexible cablebolt tendon allows for a long, continuous and high-capacity reinforcing element to be inserted into the rock. A common cablebolt installation practice at Australian underground mines that do not utilise a mechanised cable bolter, is for cablebolt holes to be drilled by a jumbo or longhole rig, hand installation and grouting by a service crew, then plate tensioning following a grout curing wait time of typically 12 to 24 hours. This process involves multiple separate labour-intensive work processes and results in reduced availability of priority headings.

This paper discusses the development history of vertical SDA use in wide spans at Kanmantoo mine. The installation process is fully described, including aspects such as reliable in-hole coupling of SDAs, resin selection and pumping aspects. A cost benefit assessment, and a case for ground stability are presented to compare cablebolt use to SDAs in development intersections.

INTRODUCTION

Deep ground support installation is often required within development intersections at Australian Mines. While many of the larger mines use dedicated cablebolting rigs, there are a greater number of smaller mines that install cablebolts without these machines, primarily due to the capital cost involved and the periodic need. These mines use jumbos (development drill rigs) to drill holes and service crews for the hand installation, grouting and plating of cablebolts. This practice involves multiple labour intensive work processes and causes reduced availability of priority development headings. This is mainly because cablebolts are used as the long reinforcement element with a grout encapsulation medium. The cablebolting process slows down development advance rates at these mines and requires multiple work teams to complete.

Speeding up intersection ground support installation has significant benefits to an underground mine. The benefits of increased efficiency in this area were identified as an implementation priority by Hillgrove Resources Limited at their Kanmantoo copper mine in South Australia.

The use of SDAs at intersections instead of cablebolts at Kanmantoo has resulted in significant improvements such as faster priority heading advance. The SDAs are installed with a jumbo then resin filled by a service crew. Reinforcement of intersections is achieved by installing high tensile strength SDAs to reinforcement depths of 4.5 m or 7 m.

BACKGROUND

The application of long reinforcement for stope support with 'long flexible cables has been under continual development since the early 1970s' according to Fuller (1983). Historically, the wide spans formed at intersections were not always supported with cablebolts as deep support. Risk mitigation has led to this becoming adopted as standard practice. Potvin and Nedin (2003) released *Management of Rockfall Risks in Underground Metalliferous Mines – A Reference Manual* and stated that:

> Intersections, or areas where two or more drives connect, may require special consideration and specific ground control measures. Larger spans are exposed in intersections which can allow large wedges to daylight. Some mines require that all intersections be systematically cable-bolted either during, or immediately after being mined.

Usually, patterns of cablebolts are installed in the backs with the total number designed to reinforce and hold a loosened zone that accounts for the width of the intersection. This method of intersection design, referred to as the parabolic arch methodology, has been practiced by some in the mining industry for well over 20 years and was more recently documented by Potvin and Hadjigeorgiou (2020).

The utilisation of SDAs in mines has been occurring for many years particularly in portal development and at major infrastructure locations. In hard rock mines, SDAs are used for specialised applications such as perimeter spiling bars for development through backfill and in poor ground conditions. Until now, use has not extended to reinforcing in the vertical plane as there was no way to hold the progressive segments up a vertical or inclined hole while drilling. The use of SDAs became more compelling with the development of commercially available pumpable thixotropic resins. The 20 min cure time of these resins eliminated the need for 12 to 24 hour cure times associated with using cementitious grout. Thixotropic resins allow single pass pumping of resin through the SDA and back to the hole collar along the annulus without the need for wadding or breather tubes, thereby simplifying and improving the quality of the installation.

A new mine in early development may only have a development jumbo drill available for use. In this situation jumbo operators use a range of techniques to drill holes of suitable length for hand installed cablebolts. These techniques vary from stabbing extension steels with the second boom, to the use of pitot jaws on booms (generally without the equipment suppliers' recommended rod handler) and other iterations of these with varying levels of practicality and safety. Changing a jumbo that is set-up to drill and support a face so it is ready to drill cablebolt holes includes several steps, such as using different drill rods, adding pitot jaws and a rod handler, taking several hours.

Small mines in their production phase generally migrate from using the jumbos to drill intersection support to using production drills. This solves the inefficiency and safety issues of using a jumbo but it reduces the efficiency of the production drills. Production drills are required to tram to the intersection for a relatively small amount of drilling, often significant distances from the areas where they are undertaking production drilling.

Drilling these deeper holes with either of these drill types then requires cables to be installed, grouted, plated and tensioned manually, typically from tool carrier baskets. There is significant evidence in the industry of ongoing issues with this process, which has typically resulted in impact and strain injuries, grout burns and eye injury. In addition, logistics associated with each of these processes and the associated minimum curing time of the cementitious grouts used, typically leads to delays in critical development headings of several days. Miners accordingly schedule their mine development allowing for the impact of these delays on the critical path mine access development.

Larger mines mitigate the above requirements with the operation of cable bolters, however this has the associated costs and requirements of dedicated capital machines and specially trained operators. Ideally, operators would like to install deep support elements in intersections and wide spans with development drills without the issues and delays of cablebolt installation.

KANMANTOO MINE DECISION-MAKING PROCESS

The Kanmantoo copper mine in South Australia, operated by Hillgrove Resources, commenced underground development during 2021 following a successful period of mining within the Kanmantoo open pit. Hillgrove Resources decided to install SDAs in development intersections instead of following the standard industry practice of using cablebolts. It was believed this change was worth pursuing due to the improved cycle times related to reduced wait times and reduced work tasks for each intersection support cycle.

Process simplification and time gains

A development intersection that is reinforced by hand installed cablebolts requires several separate work processes, including drilling of the cablebolt holes, installing the cablebolts and grouting, a wait time for grout cure and plating/tensioning. This intensive ground support cycle is often repeated a second time when the designer has determined that second pass cablebolts are required for a wider spread of cablebolts throughout the intersection, as shown in Figure 1.

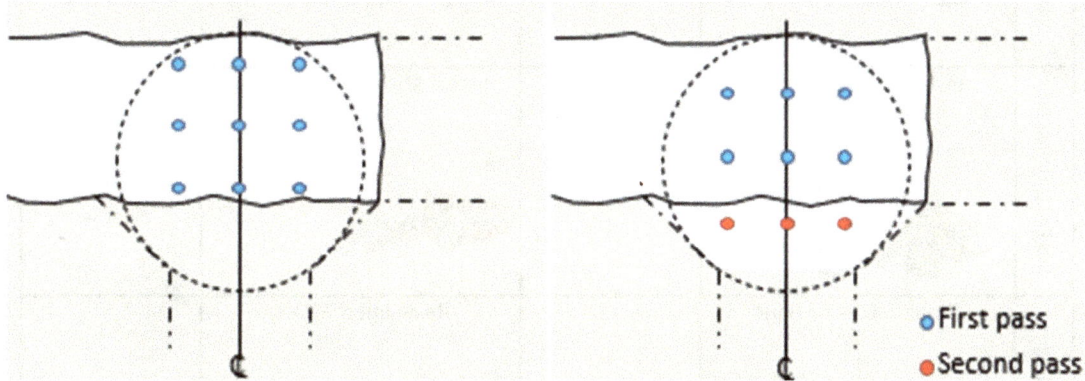

FIG 1 – First pass and second pass intersection cablebolting.

When using SDAs instead of cablebolts the number of work processes reduces to drilling the SDAs and pumping the resin. Wait time is not required due to the speed of resin strength gain, and plating/tensioning the SDAs is completed by the jumbo when drilling the next face. This reduces the number of work steps from four to two during single pass intersection support cycles and decreases the time required for long support installation from as much as 24 or 36 hours to 9 hours. This is summarised in Figure 2. Cablebolts are prone to unravelling prior to plating when close to blasting, making the subsequent plating and tensioning unpractical. The minimal protrusion required for the SDAs make them far less susceptible to damage from adjacent blasts compared to cablebolts.

DEEP GROUND SUPPORT				
	Current Process in Small Mines	**Typical Duration** (Hours)	**Planned Process at KCM**	**Typical Duration** (Hours)
DRILL HOLE	Drill Holes with Jumbo 	6	Drill Holes with Jumbo 	4
INSTALL ELEMENT	Install Cables by Hand 	4	Install SDA's While Drilling 	0
SECURE	Grout Cables by Hand 	5	Resin Inject SDA's 	3
CURE		12		0
PLATE	Plate and tension after Cure 	2	Plate SDA's 	2
	Total Time	**29**	Total Time	**9**

FIG 2 – Intersection long reinforcement work steps and timings (KCM – Kanmantoo Copper Mine).

Costing considerations

A costing scenario of reinforcing an intersection with cablebolts verses SDAs is provided in Table 1. This information breaks down the costs for each of the work steps described in Figure 2 for a standard intersection with eight 6 m long twin strand cablebolts compared to eight 4.8 m long 38 mm SDAs. This is based on the mines owner operator costs and does not allow for fixed overheads or contractor profit margins. Each element in the cost comparison considers the ownership capital cost of equipment, the operating cost of that equipment, the consumables costs and labour required.

TABLE 1

Cablebolt and SDA costing comparison for intersections.

Cost comparison	Conventional cablebolts	SDAs
Logistics	$520	$250
Drilling	$4250	$3010
Element supply	$2530	$1927
Element installation	$1700	-
Grouting/resin injection	$3682	$3270
Plating/tensioning	$840	$330
Total cost	**$13 522**	**$8787**
Unit cost (per m of installed element)	**$282**	**$229**

The costing comparison shows that an intersection can be reinforced using SDAs for significantly less than using cablebolts.

A further significant saving that is not captured in the direct cost comparison is the reduced time required to fully support an intersection when using SDAs. The SDAs can be fully installed in less than a shift compared to a minimum of two shifts required for conventional cablebolts. This is a material difference for a rapid development decline which pulls forward the development completion date and subsequent mining. This approach can be fully appreciated when managing a single heading decline which has an entire fleet of gear and team of personnel waiting for 12 to 24 hours for grout to cure in their only development heading.

SDA and resin selection

SDAs are a hollow steel bar with a threaded outer surface. The sacrificial drill bit uses the flushing holes to allow resin flow-through the bit, once installed. Sections of SDA can be joined with a high strength internally threaded coupling. The SDA is filled with resin by connecting onto the installed SDA at the collar and pumping resin through the SDA's hollow core. The resin flows through the bit and back down the drill hole annulus to the collar, as shown in Figure 3. Full encapsulation is achieved, and this results in a stiff bolt suited to static ground conditions.

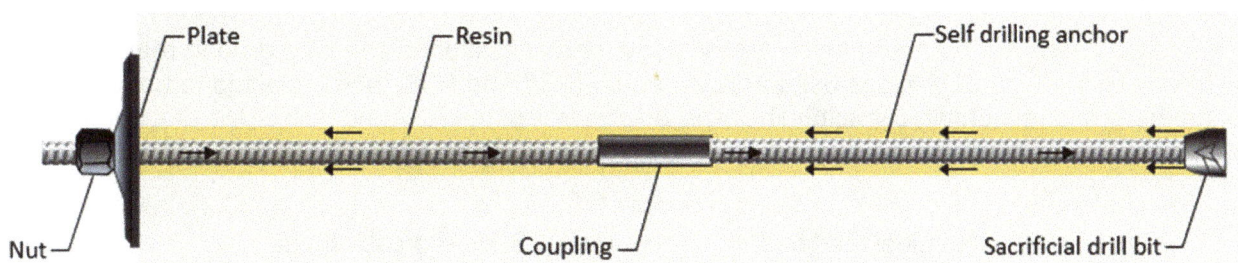

FIG 3 – Conventional SDA, with arrows showing resin flow direction.

SDAs used for supporting intersections are available from ground support or drilling consumable supply companies. At intersections, Kanmantoo mine use 2.4 m lengths of the R32 or R38 SDA, which are coupled together during installation for 4.8 m long SDAs (4.7 m effective length, with 100 mm extruding from the hole). Specifications for the SDAs are shown in Table 2.

TABLE 2
SDA specifications.

SDA description	R32	R38
External diameter (mm)	31.1	37.8
Internal diameter (mm)	18.5	19
Yield load (kN)	230	400
Ultimate tensile load (kN)	280	500
Ultimate load strain (Agt%)	≥5%	≥5%
Weight (kg/m)	3.2	5.85

The SDA steel stretches under load, as shown by having an ultimate load strain (Agt%) value of over 5 per cent. This is a measure of the total elongation at maximum load, excluding the plastic deformation of the steel. It provides a better indication of rock bolt stretch under normal working loads, rather than considering the total elongation which includes consideration of the plastic deformation prior to failure. Cablebolts are often described in terms of their total elongation to failure, with typical values of 6.5 per cent strain for 600 mm test lengths of 15.2 mm cable. SDA bar compares favourably to this and was concluded by Kanmantoo to have suitable steel properties for use as long reinforcement.

Site trials were conducted on five different brands of two-part thixotropic resin. All showed suitable strength and pumpability characteristics during trials. Some were temperature sensitive for pumpability and some had fast cure times which was a challenge when pumping SDAs over 9 m long. Master Builders Solutions provided MasterRoc® RBA380 an adaptive curing (3–5 mins) resin and worked with Kanmantoo to optimise the static mixer matched for their resin product. Samples of resin were taken during trials which enabled cure times to be measured and later testing for strength using UCS testing.

Ground stability considerations using SDAs

The local mine ground conditions at Kanmantoo are considered static with competent rock and low to medium stress conditions. Static conditions for ground support design are essentially the same as 'normal conditions' which were defined by Potvin and Hadjigeorgiou (2020). They described normal conditions as strong rock, with 'good' rock mass quality and mining induced stresses that are lower than the amounts required to cause initial visual signs of stress damage, such as face spalling or shotcrete cracking.

Kanmantoo uses the parabolic arch methodology for justification of the long ground reinforcement at intersections. The SDA types used at Kanmantoo (R32 and R38) were selected to balance ease of handling and reinforcing capacity. The tensile strengths of the R32 and R38 SDAs match relatively well with single strand and twin strand cablebolts respectively as shown in Table 3.

TABLE 3
Nominal load bearing properties for SDA and cablebolts.

Load bearing properties	Unit	6 m × 15.2 mm bulbed CB	R32 SDA	6 m × 15.2 mm twin strand bulbed CB	R38 SDA
Nominal yield load	[kN]	250	230	500	400
Nominal ultimate load	[kN]	265	280	530	500

The elongation properties of the SDA bars describe that the bar steel will stretch, but the way the bolt behaves is different to a cablebolt, mainly due to stiffness of a fully installed SDA. The coarse thread on the outside of the SDA locks the bar in place within the resin and following installation there is limited free length of SDA bar, except for a very minor amount at the collar depending on the seat of the plate against the excavation surface. A hand installed cablebolt has free length at the collar due to cotton wadding, and a loaded cablebolt is more likely than an SDA to exhibit some amount of debonding at the grout to bolt interface when under load. Both SDAs and cablebolts can expect to have some system elongation due to movement at the plate. Cablebolts are more capable of absorbing ground movement prior to failure than a fully encapsulated SDA. This concept is described in Figure 4.

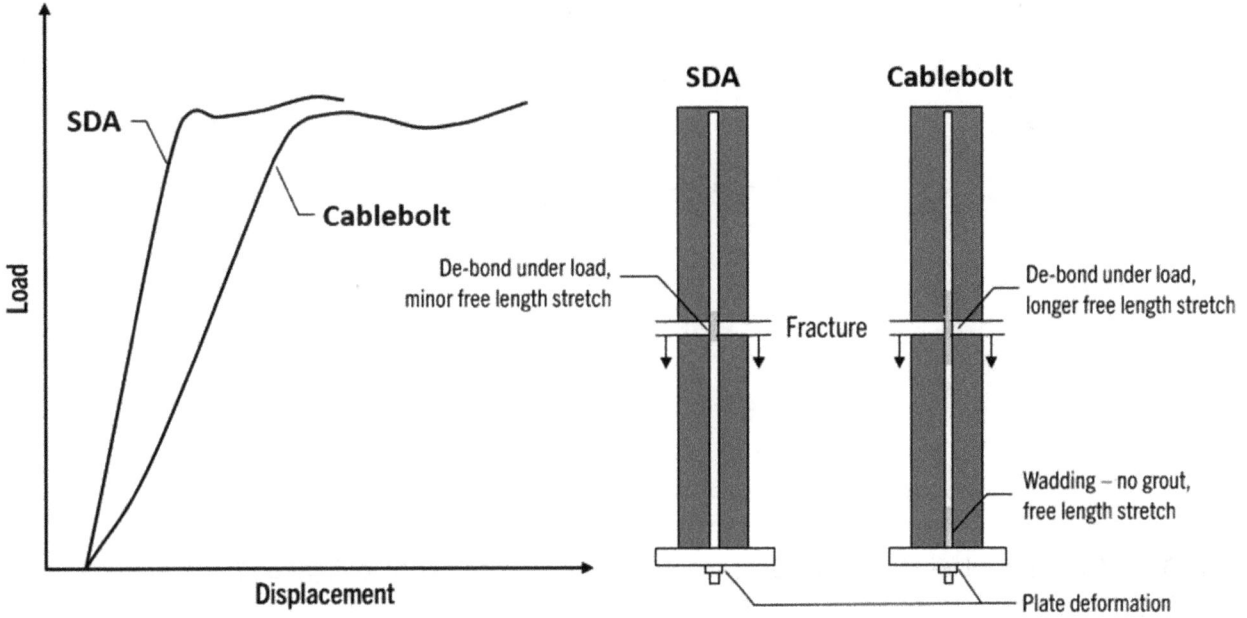

FIG 4 – Conceptual diagram of SDA and cablebolt displacement under load.

When the SDA ground support system is required to manage high stress conditions and exhibit greater system ductility, there are commercially available SDAs with smooth, annealed tube sections, intended to absorb energy by allowing the steel to stretch.

Intersection spans reinforced with cablebolts often use 6 m cable lengths at mines with standard intersection dimensions. The 6 m cable lengths are required due to the anchorage needed to lock the cable into the ground beyond the parabolic arch loosened zone, as shown in Figure 5. The required anchorage length for bulbed cablebolts, or the critical embedment length, is usually thought to be about 2 m. A 2 m cable anchorage length allows for two bulbs on each cable of a twin strand to be locked in place beyond the loosened zone. It also adds extra conservatism for variable grout mix strengths and poor grout encapsulation at the end of the cable where air pockets can form while grouting. The SDA is a more reliable system for full encapsulation at the end of the bolt in the anchorage zone. For typical intersection sizes at Kanmantoo mine such as a 11 m intersection span (and a 3.7 m high parabolic arch loosened zone), 4.8 m long SDAs (two 2.4 m long bars coupled together) provides 4.7 m of SDA embedment, with 1 m of SDA in the anchorage zone at the highest part of the parabolic arch.

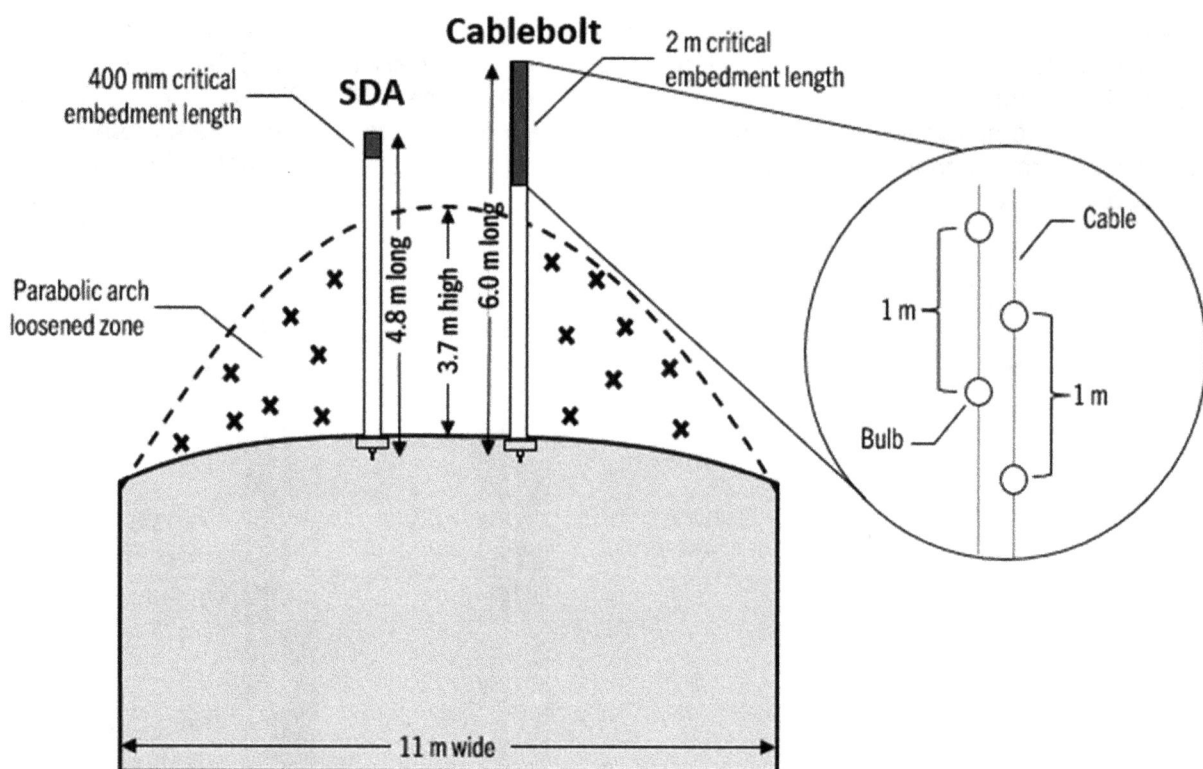

FIG 5 – Conceptual wide span cross-section of SDA and cablebolt critical embedment lengths.

Field tests

SDA short encapsulation pull tests to determine critical embedment length were undertaken at Kanmantoo with results shown in Table 4. During the testing it was evident that SDAs achieved high strengths up to the SDA steel tensile strength with very short encapsulation lengths. SDA steel tensile capacity was reached in 300 mm to 400 mm of resin encapsulation.

TABLE 4

Minimum encapsulation test results for R32N SDA.

Bolt #	Encapsulation (mm)	Max load		Residual load		Failure mechanism
		(kN)	(t)	(kN)	(t)	
1	300	220	22	80	8	Resin/bolt Interface – pulling bit through
2	300	310	31	260	26	Resin/bolt interface – pulling bit through
3	400	380+	38+	-	-	Bolt tensile failure
4	400	260	26	250	25	Resin bolt interface – pulling bit through
5	500	380+	38+	-	-	Bolt tensile failure
6	500	350	35	230	23	Resin bolt Interface – pulling bit through
7	600	360	35	210	21	Resin bolt Interface – pulling bit through
8	600	380+	38+	210	21	As 7 plus Nut failure over SDA thread

Annulus width is the distance from the outside diameter of the SDA to the hole wall. Trials completed with vertical clear Perspex tubes by PYBAR at Dargues Gold Mine showed that at large annuluses there is potential for resin to free fall and not fully encapsulate the SDA. The testing set-up is shown in Figure 6. It is recommended that 51 mm holes for R32 SDA's and 64 mm holes for R38 SDAs be the maximum size used to prevent the risk of incomplete encapsulation. At these dimensions the thixotropic resin sticks to the drill hole walls while returning down the annulus, slowing the resin flow and preventing resin free fall.

FIG 6 – Resin encapsulation tests using Perspex tubes.

INSTALLATION OF INTERSECTION SDAS

SDAs can be manufactured to any length, typically 2.4 m long for development drills or 1.8 m long for production drills. They can be manufactured with prefabricated bushes and couplings to assist with ease of installation. Kanmantoo use the SDA LOCK™ manufactured by ME Safe to suspend the SDA in the hole once drilled prior to resin encapsulation. The SDA LOCK™ is a retainer device fitted to the SDA between the drill bit and a shoulder member (either a swage element or a coupling as shown in Figure 7) enabling the SDA to be installed as per the standard procedure.

During drilling, the SDA LOCK™ abuts against the shoulder member. The SDA can rotate independently within the SDA LOCK™ (Figure 7, Diagrams 1 and 2), and the drill bit rests on top of the SDA LOCK™ retaining the SDA in the hole with the folded down tabs providing the retaining force (Figure 7, Diagram 3).

During the development of the SDA LOCK™, significant trialling was undertaken applying different loads in varying rock types to build a database of their reliability. Installation was completed with a Sandvik DD 421 jumbo using the second boom to stab and hold the installed SDA while breaking the coupling at the shank to add subsequent lengths of SDAs. A combination of 4.8 m, (two lengths) and 7.2 m, (three lengths) SDAs have been installed to date depending on intersection span and design.

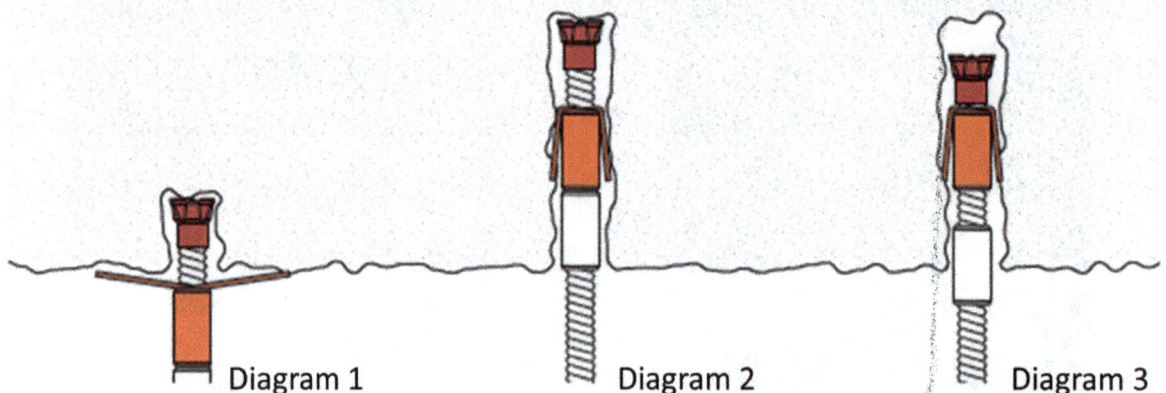

FIG 7 – SDA LOCK™ diagrams of drilling installation.

Best practice SDA installations commence from the furthest to nearest SDA in the drive so no person travels under them, reducing the likelihood of exposure to a failed spring or steel element. If for some reason the SDAs cannot immediately be resin encapsulated, they can be secured via chain to the mesh (or rock bolt plate flanges, if available) with captive caps as is often practiced with cablebolts during installation and grout curing.

Once the SDAs have been installed, they can be resin encapsulated. Kanmantoo's approach is to keep this process independent of the drill, allowing it to return to its primary role of drilling and supporting development headings. A gear pump is used to pump a two-part thixotropic resin up the centre hole of the SDA and back down the hole via the annulus. The two parts of the resin are pumped separately and only mixed in a static mixer at the head that is attached to the SDA. This means that there is minimal resin wastage and clean-up. Some resins can be temperature sensitive, and manufacturers should be consulted when they are used in cold temperatures or conditions of cyclic temperature.

The use of resin in this method significantly reduces the likelihood of cement dust or grout entering operators' eyes which is common in existing practices and has led to most sites having diphoderine or eyewashes on the mixing gear and mandatory goggle use. Grout is commonly pumped into holes under pressure using poly fittings and low-quality hose which can easily blow out when poorly mixed grout is used or obstructions in filler or breather pipes occur. These blow outs can lead to eye injury, ingestion, and skin rashes. Resin pumping is done using high quality, fit for purpose designed hoses and pressure relief valve systems to prevent such occurrences.

The small tanks for part A and part B of the resin, the gear pump and associated hosing fit into a tool carrier basket. Kanmantoo mine have fabricated a high lift basket with the tanks and pump on the lower deck with the controls on the upper deck, allowing resin injection to be completed without the trip hazards in the basket of the pump and tanks. This same set-up can be used for Kanmantoo's thin spray on liner (TSL) application, making the whole set-up multipurpose and pumps interchangeable across both operations, as shown in Figure 8.

FIG 8 – SDA and TSL basket used at Kanmantoo Copper Mine.

CONCLUSIONS

The SDA and resin solution implemented at Kanmantoo mine provides a dependable deep support system for wide spans. Flow on improvements to safety, cost and efficiency have resulted in the use of resin filled SDAs at Kanmantoo in intersections and wide span areas of the underground mine.

ACKNOWLEDGEMENTS

PYBAR Mining services should be recognised for initiating this project and for supporting SDA trials at multiple mines including; Carrapateena, Henty, Dargues and Thalanga. The miners at these sites along with Warren Attwell and Sam Lennon at Kanmantoo are acknowledged for their input and guidance and having the patience in the small hours of the night to get the system just right. Strata Consolidation have had a significant impact on the development of this system with the use of their resin pumps, their pumping experience and provision of professional pumping services over several years to get to this point. Quarry Mining Services and the teams at DSI and Split Set Mining Systems have also greatly helped in providing SDAs for trials and input into the process.

REFERENCES

Fuller, P, 1983. Cable support in mining A keynote lecture, in *Proceedings of the International Symposium on Rock Bolting*, pp 511–522.

Potvin, Y and Hadjigeorgiou, J, 2020. *Ground Support for Underground Mines*, 520 p (Minerals Council of Australia).

Potvin, Y and Nedin, P, 2003. *Management of Rockfall Risks in Underground Metalliferous Mines – A Reference Manual*, 160 p (Australian Centre for Geomechanics: Perth).

No entry mechanised presink of the Wira Hoisting Shaft

D Kilkenny[1], J Clark[2] and C Hill[3]

1. General Manager, Raising Australia, Perth WA 6105. Email: dave.kilkenny@byrnecut.com.au
2. Superintendent – PHOX Execution, OZ Minerals, Adelaide SA 5950.
   Email: jesse.clark@ozminerals.com
3. Project Manager – Shafts, Byrnecut Australia, Brisbane Qld 4000.
   Email: christopher.hill@byrnecut.com.au

ABSTRACT

This paper discusses the use of a mechanised shaft sinking machine Herrenknecht Vertical Shaft Sinking Machine (VSM) – to conduct the 93 m presink for the Wira Hoisting Shaft at the Prominent Hill Mine in South Australia. This technology has been widely used in the global construction industry, however, this was the first time it had been used in Australia and the second time in a mining application. The non-entry shaft sinking system eliminates risks associated with a traditional manual method and allows all shaft construction operation to take place from surface.

The geology in the upper 100 m at the Prominent Hill mine is conducive to mechanical excavation, with expected rock strengths below 130 MPa. This enabled an alternative shaft construction methodology to be considered. The Herrenknecht VSM system was selected due to its extensive work history and demonstrated performance in similar ground conditions.

In collaboration with the operations and design team, the shaft collar and surface configuration were modified to suit the requirements of the VSM, additional equipment, and loading jacks that support the pre-cast lining. Dedicated concrete moulds were designed in Germany and built in Italy. The concrete segments were cast in Adelaide and road freighted to site. This paper outlines each stage of the presink – refurbishment, transportation, logistics, establishment, operation and demobilisation.

Selecting this shaft presink excavation methodology, delivered through positive collaboration and communication despite some challenges, shows the benefit to the mining industry of applying new technology and innovation to improve safety and productivity.

INTRODUCTION

Shaft construction is seeing a renaissance in Australia. At the time of writing, three large hoisting shafts are underway (Tanami, Odysseus and Prominent Hill) and more are currently in planning stages (two shafts at Appin). Several pumped hydro power projects will also require shafts for water storage and transfer.

The initial phase of shaft construction is called the 'presink'. During the presink phase, an initial excavation is established to a predetermined depth to allow installation of the equipment required to sink the shaft to full depth. A multi deck work platform, known as a stage or Galloway, is used to provide a stable work area from which to conduct the shaft sinking operations of drill and blast, excavation, ground support, lining and services installation.

The presink depth is dictated by the geological conditions on-site along with the proposed sinking methodology and stage design. Generally, this depth varies between 30 m and 100 m deep. In Australia, presinks are often limited to 50 m as mining regulations require all hoisting operations over this depth to have a guidance system, which increases the complexity of the presink, design registration and lead time on equipment.

Whilst the main sink phase of the operation is becoming increasingly mechanised and sophisticated, the presink phase is often constructed using manual methods. A focus on workplace safety and reduction in manual operations across the mining industry has resulted in significantly larger and heavier stages. For example, the stage used to sink the 1126 m deep Telfer Hoisting Shaft in 2004 had an operating weight of 55 t, in comparison with the sinking stage proposed for the 1330 m Prominent Hill Wira Shaft which has an operating weight of 110 t.

TRADITIONAL PRESINK

The methodology used to construct a traditional presink varies with site geology and the target presink depth. Weak and unconsolidated ground can be directly excavated using small 'mini-excavators' and cranes that can hoist out material. Harder and more competent ground will require drilling and blasting prior to excavation. The completed presink will require sufficient depth to house the sinking stage and provide sufficient clearance for blasting below the stage.

Modern sinking stages are commonly between 20 m and 30 m in total length. Presink depths of 50 m have been acceptable when combined with a minimum blasting distance of 20 m for benching, or strip and line methods, to prevent stage damage. However, larger stages and alternative blast methodologies (ie full face) require deeper presinks.

Whilst the main sink has the benefit of a full-scale hoisting system with high capacity winders and headframes, the presink must be constructed prior to this infrastructure being available. Methodologies can include a basic mobile crane hoisting system with manual drill, blast and excavation, or temporary headframe systems with light duty hoisting systems.

Regardless of the hoisting system used, the presink must excavate through any oxidised surface material prior to the installation of any stage system. Where ground conditions allow, this can be 'free dig' excavated. However, drilling and blasting is often required. There is a point where mechanised surface drill and blast equipment can no longer be used and the shaft is still too shallow to allow heavier duty shaft sinking equipment.

Traditionally, hand-held drilling equipment such as airleg/sinker and rockdrill, has been used across the industry as shown in Figure 1. However, this can be contrary to other work practices and procedures in place on the mine site, as it reintroduces operational hazards that may have been previously removed. More recently, mini-excavator based drilling systems have been trialled across the industry with varying degrees of success.

FIG 1 – Handheld drilling.

Similarly, excavation of the shaft has traditionally relied on kibbles being filled on the shaft bench and hoisted to surface for disposal. Methods of filling kibbles vary from manual shovel to air powered 'super-suckers' and mini-excavators as shown in Figure 2. The kibbles are then hoisted via crane or winder to surface for disposal.

FIG 2 – Presink excavation.

Main sink operations have been modified over time to remove personnel from operational hazards such as working directly on the bench, manual handling of drills, and provision of overhead protection. However, it is often impractical to apply this approach to the presink.

During a risk assessment, the presink activities generally rank higher on the overall risk profile than the main sink activities. However, the presink parts of any shaft sink project are not well documented or published. A search of the OneMine database for 'pre-sink' or 'presink' returned only ten results, of which only one detailed the presink operation (Heever, 1981). Of these papers, one details use of a crane mounted Auger, and the other describes the surface headgear arrangement used to hoist material rather than outline shaft activities.

MECHANISED PRESINK

A mechanised non-entry presink has been a long-term aspiration for the shaft sinking and risk management industry. Herrenknecht AG, a German tunnelling firm, developed a mechanised shaft sinking system 'Herrenknecht VSM' in 2004 for a series of utility tunnels in Kuwait (Schmäh, 2007).

The VSM concept (as shown in Figure 3) has been used in numerous civil construction applications worldwide. However, it had only been used once in a mining application, at the Woodsmith Mine in the United Kingdom. Two key challenges were ground conditions and machine availability. The VSM cutter drum can cut rock up to approximately 125 MPa uniaxial compressive strength (UCS), which has not been compatible with most Australian mine sites. However, the geology at Prominent Hill was expected to not exceed this rock strength in the first 100 m of shaft, making it suited to mechanical cutting. Additionally, Herrenknecht had recently established a rental option for the VSM rather than outright purchase, which increased machine availability for one off projects. A major advantage of this mechanised presink system is the ability to commence early works ahead of the traditional critical path schedule and ensure project delivery.

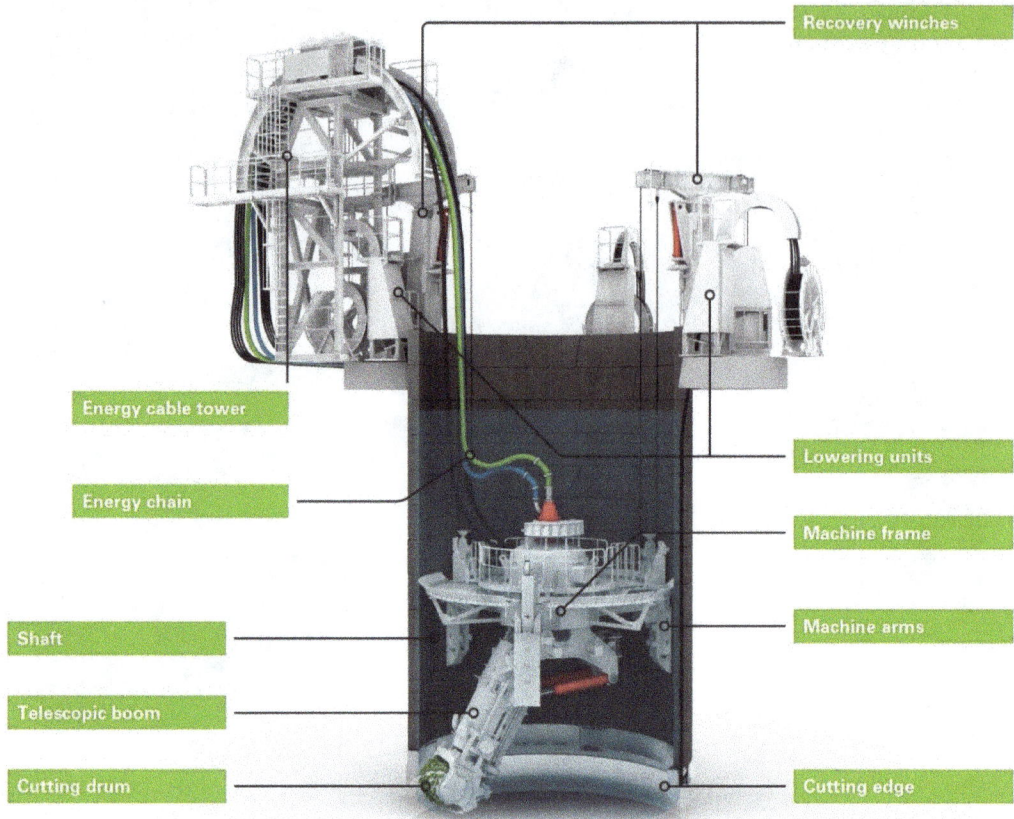

FIG 3 – Herrenknecht VSM.

Since the initial project in 2006 (Kuwait), over 100 shafts have been sunk using this system, with diameters ranging from 4.5 m to 12.0 m and depths up to 115.2 m (Frey and Schmäh, 2019).

PROMINENT HILL

Located in South Australia on the north-eastern margin of the Gawler Craton, approximately 130 km south-east of Coober Pedy (as shown in Figure 4), the Prominent Hill Mine is a significant global copper resource.

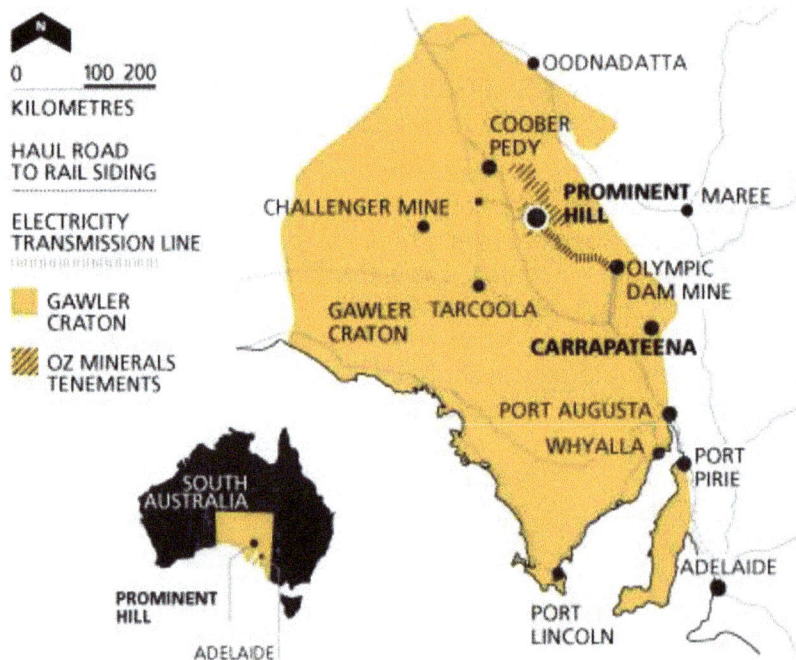

FIG 4 – Prominent Hill Mine.

The cover sequence (0–100 m below surface) consists of the following layers from the top down, (refer Figure 5 and Table 1), Silcrete, Oxidised Bulldog Shale, Fresh Bulldog Shale, Cadna-Owie Formation and Permian Boorthanna Formation.

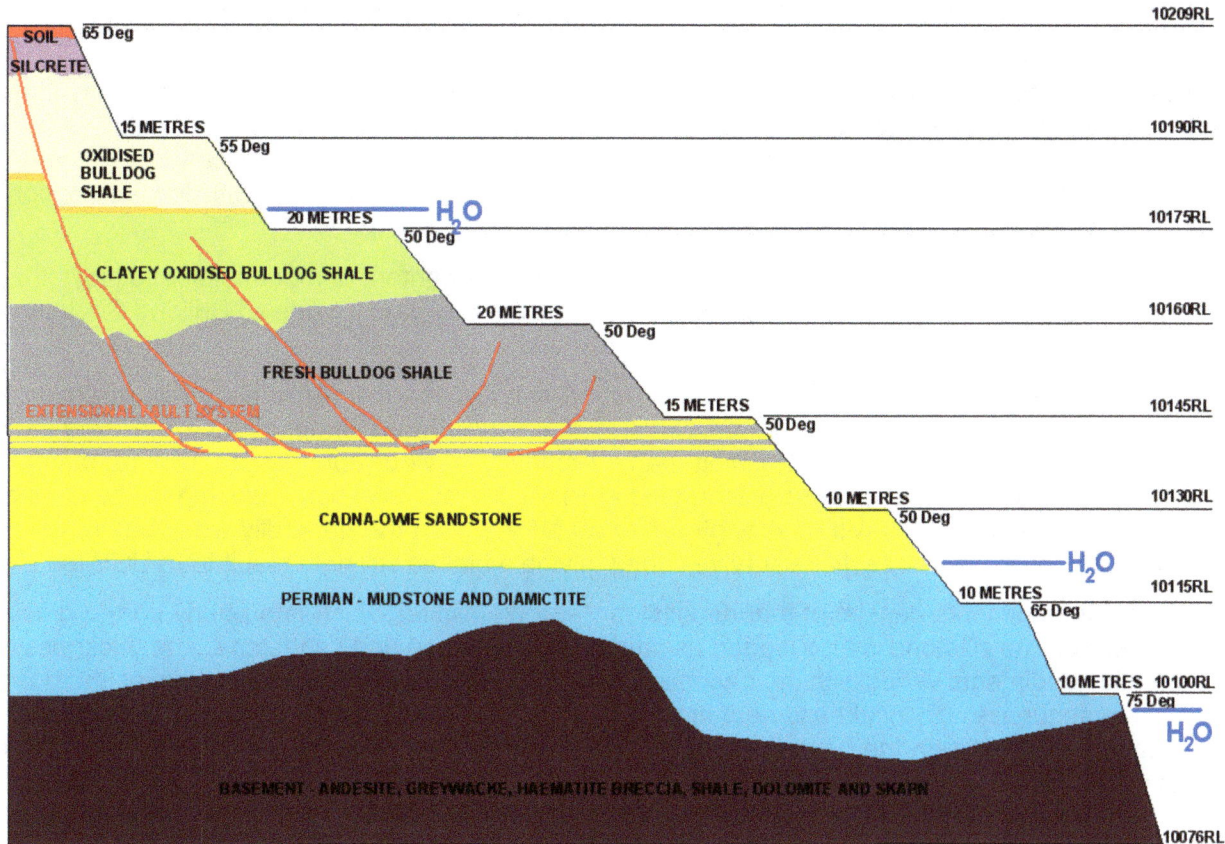

FIG 5 – Prominent Hill geology.

TABLE 1

Prominent Hill ground conditions.

Unit		Nominal depth		Nominal thickness	UCS			Density	Rock mass quality[1]	
		From	To		Tests	Mean	Standard deviation			
		(m)	(m)	(m)		(MPa)	(MPa)	(kg/m³)		
Silcrete		0	5	5	2	8.1	6.9	1,940	4.0	Fair
Bulldog Shale	Oxidised Bulldog Shale	5	28	23	8	5.8	2.2	1,660	3.3	Poor
	Sandstone bed	28	30	~2	–	10[2]	–	2,500	10	Fair
	Clayey oxidised Bulldog Shale	30	45	15	5	1.8	1.6	1,660	3.3	Poor
	Fresh Bulldog Shale	45	75	30	9	2.4	1.5	1,780	3.4[3]	Poor
Cadna-owie Sand		75	90	15	1	1.9	–	2,080	20	Good
Permian sediments	Siltstone, sandstone	90	150	0 to 60[4]	8	5.1	2.5	2,470	7.9	Fair
	Diamictite				2	3.7	4.3	2,470	7.9	Fair
Footwall volcanics		150	–	–	21	102	49	2,860[5]	4 to 100	Fair to very good

The silcrete unit is a near-surface siliceous layer with variable strength. The thickness varies from 4 to 12 metres. In places, relict bedding dipping at 5° to 20° and two possible relict joint sets dipping at 20° to 50° and 70° to 90° occur. The silcrete has a very high slake durability.

The Bulldog Shale is a weak, carbonaceous and pyritic marine mudstone with fine-grained sand intervals increasing in frequency towards the base of the unit. Weathering divides the Bulldog Shale into oxidised and fresh units which are visually and geotechnically distinct. The base of weathering is nominally 45 m deep, but varies locally. A sandstone bed at about 28 m deep further divides the oxidised material, with the oxidised material below the sandstone bed having a higher swelling clay content.

The Bulldog Shale above the sandstone bed is weathered to form a generally competent rock mass, where structures are usually re-cemented. Little deterioration has been observed in this unit over time.

The sandstone bed, which varies in thickness up to about 2 m, separates the oxidised Bulldog Shale and the underlying clayey oxidised Bulldog Shale. The bed is variably weathered with variable strength. No UCS tests were performed in this unit but pit experience and field strength estimates from geotechnical logging based the typical UCS ranges from 5 to 20 MPa with limited sections where the estimated strength was up to 100 MPa.

The clayey oxidised Bulldog Shale appeared to have a higher content of montmorillonite clays than the overlying oxidised Bulldog Shale. Some deterioration of the clayey oxidised Bulldog Shale was seen over time.

The Cadna-owie Sand is a weakly cemented marine sandstone, comprised mainly of quartz with minor carbonate cement, clay coatings on particles and pyrite. Hand samples of Cadna-owie sand can be easily crushed by hand and exhibit marked brittle behaviour upon failure, disintegrating to free-running sand. The field strength is estimated at ~1 MPa. Near local basement highs, the Cadna-owie Sand is well cemented with a strength up to 50 MPa. Given the generally weak nature of the material, only the strongest sections survive coring, giving a biased strength result from UCS testing.

The Permian sediments consist of marine siltstone and diamictite, which is a poorly sorted glacial conglomerate. The siltstone is also logged as mudstone, claystone and sandstone. The thickness of the Permian sediments varies with the basement topography. Diamictite typically occupies the basal third of the sequence. On local basement highs, the Permian sediments are absent and the Cadna-owie Sand lies directly on the basement rocks. The basement sections of the proposed ventilation shafts will lie within the footwall volcanics sequence, which is comprised primarily of andesite, with minor mafic volcanics and dolerite dykes. Based on geotechnical drilling and experience in the pit, up to three joint sets are expected in the volcanics, with potentially blocky ground conditions (Lilley, 2009).

The Bulldog Shale, Cadna-owie Sands and Permian sediments are not suitable for raiseboring and required a presink to be conducted prior to drilling the raise bore pilot hole for the shaft stirp and line operation.

WIRA SHAFT PRESINK

Phases of presink

This paper will consider the Wira Shaft presink as five distinct phases. Each phase is discussed in further detail.

Design

The VSM system has previously been used to construct shafts directly from surface, whereas the Wira Shaft has a brace and sub-brace arrangement. A standard VSM presink consists of a ring beam constructed around the shaft perimeter to support the VSM strand jacks and lowering units. This ring beam is a formed and cast concrete structure where the shaft collar allows, alternatively, a temporary steel ring beam has been installed where a cast concrete foundation is not practical (see Figure 6).

FIG 6 – Ring Beam foundation.

However, due to the ventilation plenum and sub-brace structure required for Prominent Hill, the cast concrete ring beam was not suitable and a temporary steel ring beam was required. The depth of the shaft and the total weight of the segments dictates the loading conditions for the strand jacks, which needed to be incorporated in the collar design. For Prominent Hill, with a nominal presink depth of 93 m, this resulted in individual strand jack loads of 4500 kN, with a required minimum bending radius of 5.0 m for the individual strands (refer to Figure 7).

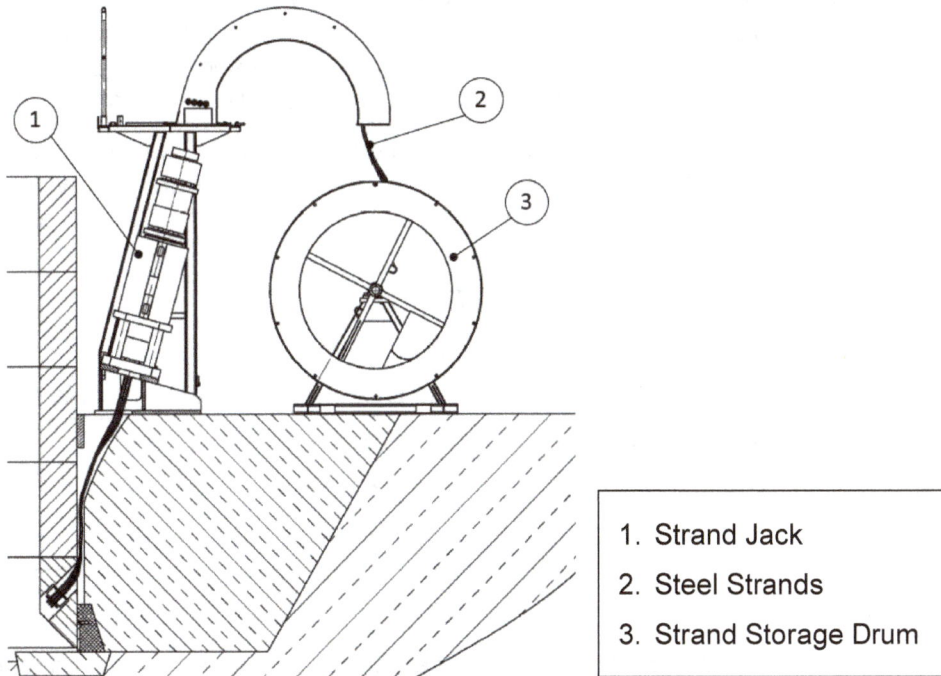

1. Strand Jack
2. Steel Strands
3. Strand Storage Drum

FIG 7 – Strand Jack arrangement.

Mobilisation

As the VSM is an existing rental unit and had been used on other projects, it required a refurbishment and extensive cleaning in line with Australian Quality Inspection Solutions (AQIS) entry requirements prior to leaving Germany.

The original schedule anticipated the equipment would be available for dispatch four months following the finalisation of the rental agreement (Figure 8).

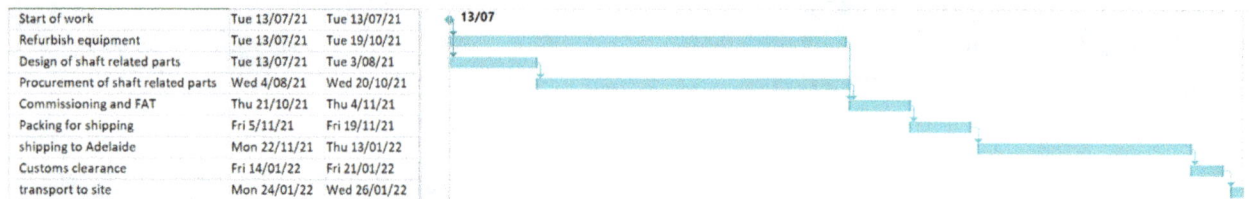

Start of work	Tue 13/07/21	Tue 13/07/21
Refurbish equipment	Tue 13/07/21	Tue 19/10/21
Design of shaft related parts	Tue 13/07/21	Tue 3/08/21
Procurement of shaft related parts	Wed 4/08/21	Wed 20/10/21
Commissioning and FAT	Thu 21/10/21	Thu 4/11/21
Packing for shipping	Fri 5/11/21	Fri 19/11/21
shipping to Adelaide	Mon 22/11/21	Thu 13/01/22
Customs clearance	Fri 14/01/22	Fri 21/01/22
transport to site	Mon 24/01/22	Wed 26/01/22

FIG 8 – VSM Mobilisation schedule.

The total movement of plant required for the VSM presink is outlined in Table 2 and reflects the original freight booking.

TABLE 2
VSM plant freight schedule.

Shipment	Parts	Port of loading	Port of arrival	ETS	ETA	Loose	40' OT
1	Segment Moulds	Genoa	Adelaide	29.11.2021	24.01.2022	-	6
2	Separation Plant	Antwerp	Adelaide	30.11.2021	19.01.2022	9	1
3	VSM	Antwerp	Adelaide	04.01.2022	23.02.2022	15	15

In total, there were 22 sea containers (40') and 24 loose ('break bulk') loads required to transport the VSM equipment to site.

The original project schedule was based around the separation plant being delivered to site first, giving time to set it up before the remaining equipment arrived. Next, the segment moulds were to be delivered to Adelaide late January 2022 to commence concrete segment manufacture. Finally, the VSM itself was to arrive late February 2022 to commence site establishment in March.

This schedule was impacted by delays in global shipping schedules, caused by the COVID-19 pandemic, ship changes and port and quarantine issues. The final schedule is provided in Table 3. In total, the VSM and associated equipment were delayed by three months, however, this was still within the project timeframe.

TABLE 3
Actual delivery dates.

Shipment	Parts	Original ETA	Actual delivery dates
1	Segment Moulds	24.01.2022	25.02.2022
2	Separation Plant	19.01.2022	3.03.2022
3	VSM	23.02.2022	30.05.2022

Concrete segments

The VSM presink system uses pre-cast concrete segments to line the shaft concurrently with excavation. Following a competitive tender process for both local and international suppliers, it was decided to award the concrete segment manufacture to Bianco Precast in Adelaide. Once the pre-fabricated steel moulds arrived in Australia, they were delivered to Bianco to commence manufacture of the concrete segments (refer Figure 9 for mould schedule; Figures 10 and 11 show the moulds and segments).

The lining design resulted in four segments per ring. Each ring contained two of each panel, A and B. The segment dimensions were 1250 mm high, 6279 mm arc length, 350 mm thick and each panel weighed 7.8 t. In addition, four segments were cast to form the cutting ring (ring 1). These included reception boxes to which the strands from the strand jacks were attached and significant cast-in steel for the cutting edge itself. A total of 292 segments were cast for the project over a period of four months.

start of work	Tue 1/06/21	Tue 1/06/21
Segment Design	Tue 1/06/21	Tue 27/07/21
manufacturing segment moulds	Thu 1/07/21	Tue 19/10/21
manufacturing of cast-ins and connection parts for the segments	Wed 28/07/21	Tue 19/10/21
shipping to Adelaide	Tue 19/10/21	Fri 10/12/21
customs clearance	Wed 15/12/21	Wed 22/12/21
transport to precast company	Thu 23/12/21	Mon 27/12/21

FIG 9 – Mould manufacture and delivery schedule.

FIG 10 – Moulds at Bianco for casting.

FIG 11 – Final Cast segment.

Establishment

Site establishment commenced with installation of the separation plant (refer to Figure 12), which is used to process the slurry pumped from the shaft bottom and separate out the solids whilst returning water to the shaft. The HKS 500 separation plant is a proprietary system designed and manufactured by Herrenknecht with a throughput capability of up to 500 m^3/hr and requires a 190 kW 400 V power supply.

FIG 12 – Separation plant.

Once the segments were delivered to site, the first ring, with integrated 'cutting edge', was installed in the shaft (Figure 13). A temporary support structure was installed in the shaft to allow the first segment to be levelled and bolted together in the sub-brace.

FIG 13 – Shaft segment with cutting ring.

As previously mentioned, a separate steel ring beam foundation was required to support the surface infrastructure on the shaft collar before shaft sinking could commence. This resulted in a delay to the excavation commencement whilst the support structure was designed and fabricated.

The ring beam was delivered to site in August and installed on the shaft collar over three days (Figure 14).

FIG 14 – Steel ring beam installation.

Following installation of the ring beam, the VSM surface infrastructure was installed, calibrated and commissioned to commence cutting.

This phase of the operation took one month to complete and involved the following steps:

- Grouting and curing for the ring beam.
- Installation of strand jacks and connection to Ring 1.
- Installation of Rings 2–5 to bring segmented lining to shaft collar.
- Set-up dummy frame to weld VSM attachment shoes to the bottom ring.
- Install VSM in shaft (Figure 15).
- Calibrate and test equipment.
- Machine repairs due to hydraulic fault.
- Fill shaft with water ready for excavation to commence.

Assembly and dry-commissioning of the VSM itself was completed in an assembly stand prior to the construction of the ring beam.

FIG 15 – VSM into shaft.

Excavation and lining

Mechanical cutting of the Wira Shaft commenced in September 2022.

Excavation of the presink uses a cutting drum fixed to a telescopic boom and slewing arrangement which allows for an overcut underneath the cutting ring to a predetermined distance. Material is cut in cycles with the slurry being pumped through to the separation plant for processing. After successive cutting cycles the shaft lining, which the VSM is fixed to, is lowered and set-up for further cutting cycles. When the lining is lowered sufficiently a new ring is built on top (2 A segments and 2 B segments) which is then tied to the lower ring using Z bars and horizontal bolts. A combination of mobile crane (Franna) and slewing hydraulic crane (80 t) are used to bring the segments to the shaft before hoisting into position.

Material from the separation plant is transferred to a temporary waste dump using a small wheel loader.

As required by material output volume, the surface mining contractor (Thiess) rehandled the waste to the tailings storage facility for disposal.

The initial results were positive, with 1 m cut and lined on the first shift. A blockage in the suction line during the next shift reduced progress to 0.6 m, however this was quickly overcome. Over the next week, shaft progress increased up to 4 m cut and lined in a 24 hr period.

Each shift comprised of the following activities:

- Excavate in two passes of 100 mm increments.

- Surface crew transfers the segments to the shaft collar and builds the next ring *in situ.*

- Lower the shaft lining 200 m then commence the next cut.

- Surface crew transfers the segments to the shaft collar and build the next ring *in situ* (when the lining had lowered to a sufficient point).

The specialist personnel required to operate the VSM are outlined in Table 4.

TABLE 4

Specialist labour for VSM.

	Day shift	Night shift
Project Manager	1	
Supervisor	1	
Operator	1	1
Mechanic	1	1
PLC Electrician	1	1
Mud Man	1	1

Challenge #1

In October, at shaft depth of 42 m below collar, a survey found the shaft had deviated from centre and was 149 mm offline (Figure 16).

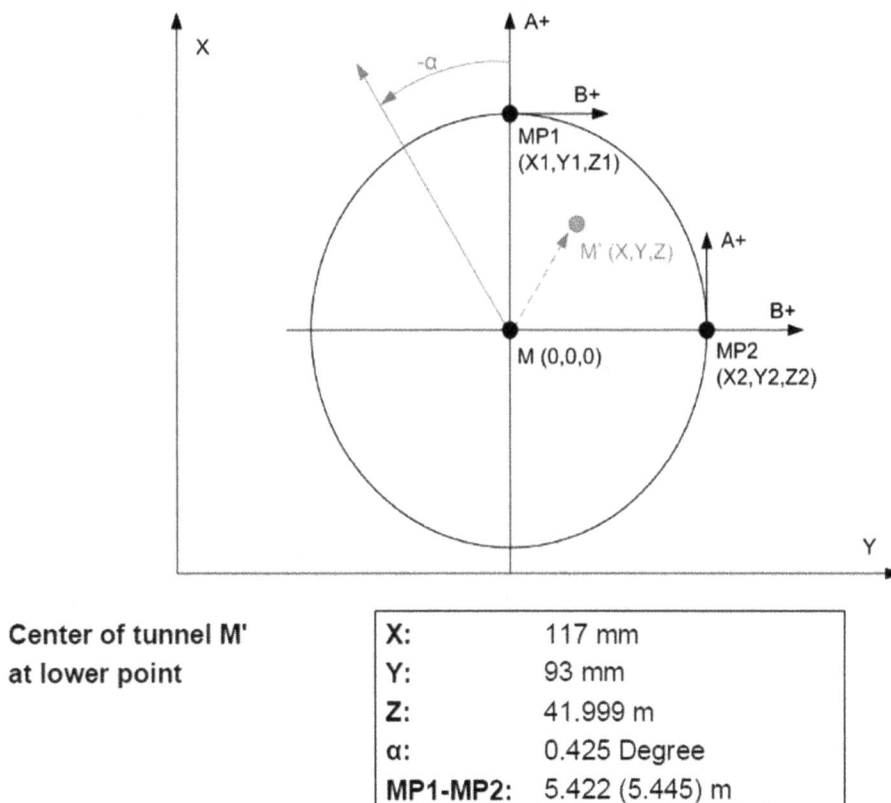

Center of tunnel M' at lower point	**X:**	117 mm
	Y:	93 mm
	Z:	41.999 m
	α:	0.425 Degree
	MP1-MP2:	5.422 (5.445) m

FIG 16 – Shaft survey.

The VSM uses two survey tools during operation to maintain alignment. Firstly, an inclinometer in the machine itself which allows the operator to keep the machine level, while also acting as a reference point for the cutting boom.

The second system uses two survey pipes installed along the length of the shaft at 90° to each other. On regular intervals, a manual survey is conducted using a giro survey tool to survey the entire length of the shaft wall. The results for both survey pipes are then combined to give an overall shaft alignment. The operators use two methods to control the shaft position. Firstly, by proportioning the tension on the strand jacks combined with asymmetrical lining lowering distances, the operator can steer the shaft. Secondly, adjusting the overcut on any quadrant of the shaft allows the lining to drift in that direction.

As can be seen from Figure 17, the shaft was offline, heading away from the centreline and had potential to exceed allowable shaft tolerances.

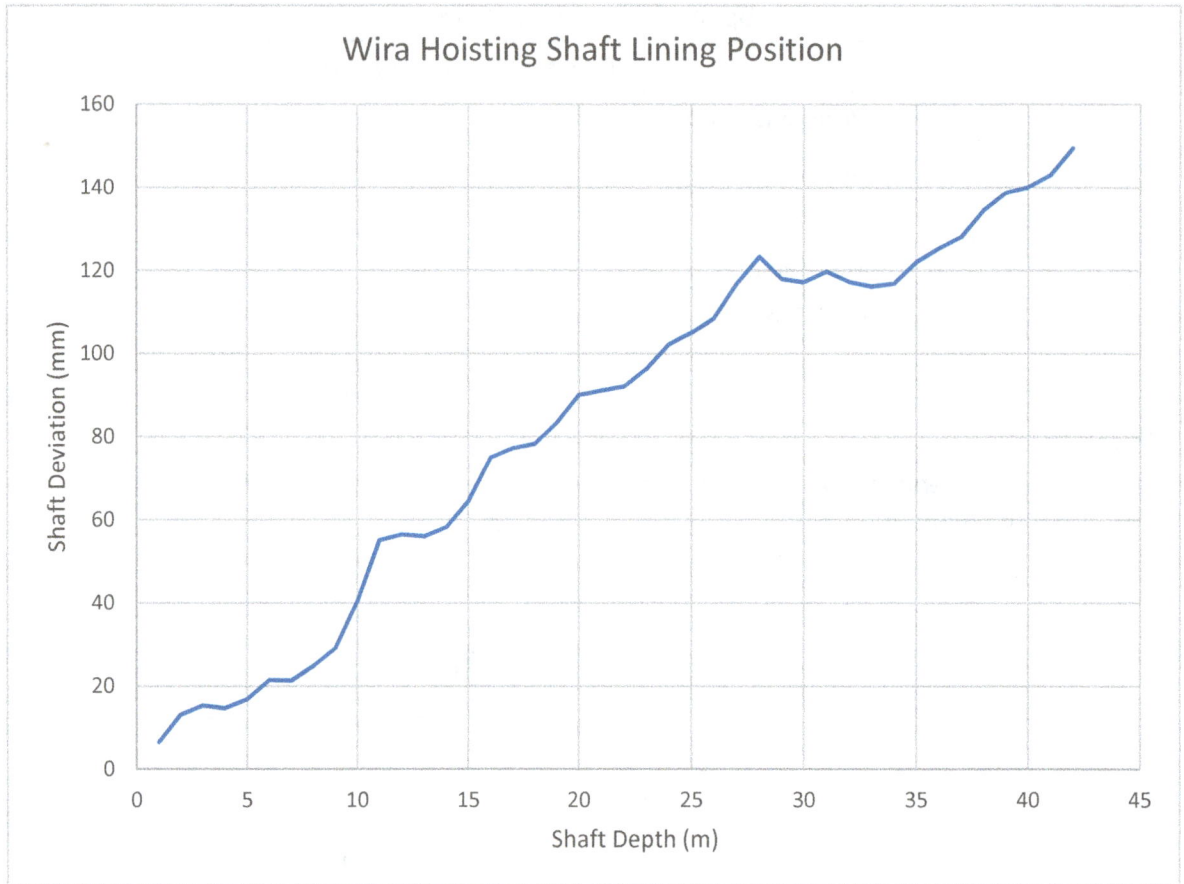

FIG 17 – Lining position.

Once an explanation of the way the lining behaves, and an action plan to correct the issue, was provided from Herrenknecht, we were able to address this challenge.

The lining itself behaves as a stiff beam, which floats in the excavation on a lining of bentonite between the concrete segments and the excavated perimeter.

The first stage to correct the issue was to over-excavate the opposite quadrant to allow the lining to return to centre. Excavation advance was also restricted to 1 m increments followed by a survey. Initially, the shaft bottom deviation continued to increase, but began to pull back on centre at 56 m in depth (14 m deeper). Over the next 13 m, the shaft bottom had been brought back to 87 mm offline and was continuing to swing back to centre.

The shaft lining had behaved as intended and now flexed around a point of maximum deviation approximately 52 m below collar. With further excavation, the curve in the shaft continued to flatten out (Figure 18) and resulted in the lining returning back towards shaft centre.

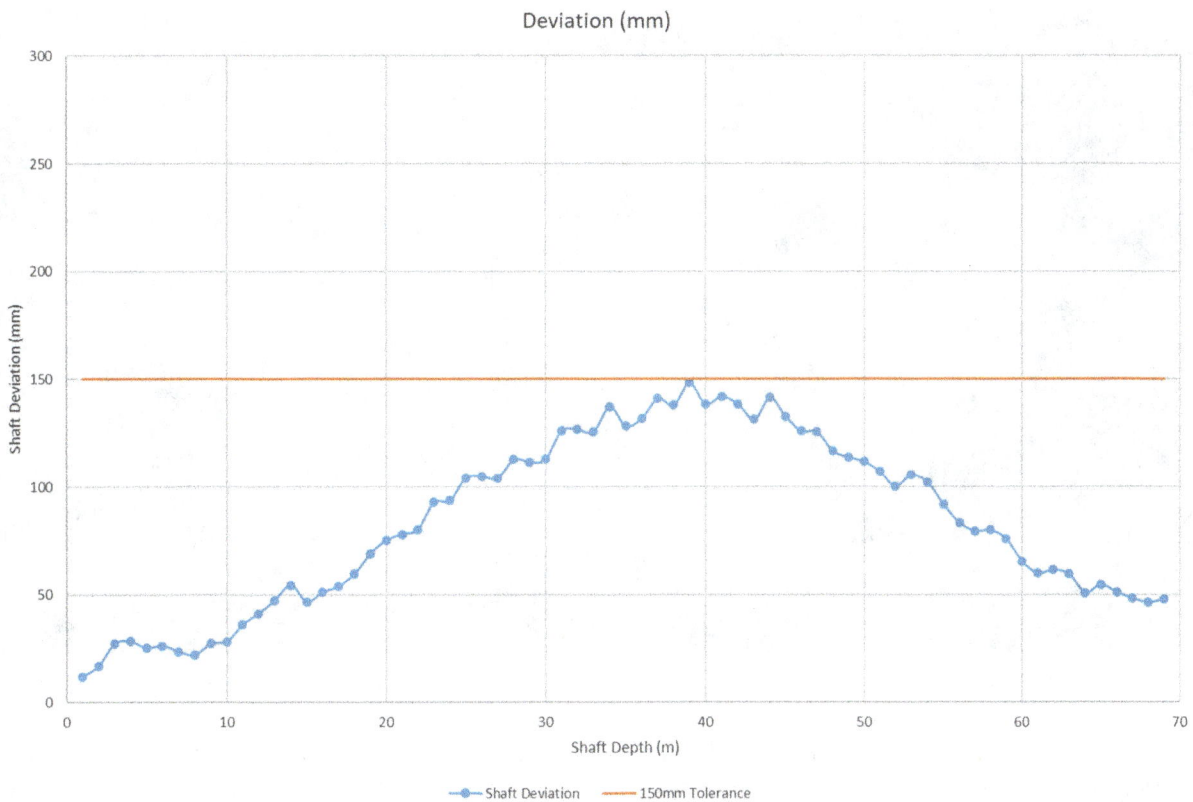

Deviation (mm)

FIG 18 – Lining recover.

Challenge #2

At the end of October 2022, with the shaft now back in tolerance, 71 m deep and advancing at an average 1.7 m per day for the previous week, one of the survey tubes was damaged and a survey tool lost, restricting the ability to measure shaft verticality.

Shaft excavation continued at a reduced advance rate whilst the new survey tool was personally express delivered by Herrenknecht. The replacement tool was received and the shaft surveyed mid-November. The cumulative advance rate during this time was between 1.82 m/day and 1.42 m/day.

Due to the confidence obtained in the behaviour of the shaft lining during the previous challenge, potential delays were mitigated and shaft progress continued.

Challenge #3

The final challenge was the presence of hard ground. Initially, resistance on the cutterhead (increased torque and thrust pressures) was identified as being related to clays which were present in the separation plant discharge pile. The VSM was raised so the cutterhead could be inspected and the clays washed off. However, inspection of the cutterhead showed significant wear in the centre section of both picks and pick holders (Figure 19).

FIG 19 – Cutterhead and Pick wear (new pick on left, worn pick on right).

The cutterhead was repaired, picks replaced and the machine reinstated. It immediately encountered high pressures indicating hard ground. After 200 mm advance the machine was raised for inspection, which revealed further damage to the cutterhead and heavily worn picks.

A band of material that had not been present on the geotechnical logs was in the shaft and causing significant difficulty to the cutterhead. Investigations of drill core and the adjacent open pit led to a conservative estimate of a 1.5 m band of up to 150 MPa and approaching CERCHAR abrasivity 5 material potentially covering the entire shaft.

As the shaft was now below 70 m in depth, raising and lowering the machine took approximately 24 hours to complete. A revised plan was instigated to continue cutting in a maximum of 200 mm in two successive 100 mm passes then raise and inspect the head.

Due to the hardness of the ground, the machine was limited to cut 100 mm (in 50 mm passes) before requiring new picks. During this period the VSM was raised nine times for pick changes and cutting drum repairs.

This band of very hard rock was 1200 mm thick. Once traversed, efficient cutting recommenced.

RESULTS

The initial schedule had allowed for 125 days to complete the presink from when the machine arrived on-site. Construction challenges resulted in this phase of the project taking 210 days, after the main VSM components arrived on-site.

The shaft was forecast to be excavated at an average of 1.5 m per day. The final excavation component averaged 1.43 m, and achieved a peak advance of 4.0 m in a day. Figure 20 shows the overall shaft excavation with the extended disruption due to breakdown and hard ground between 30 October and 28 November.

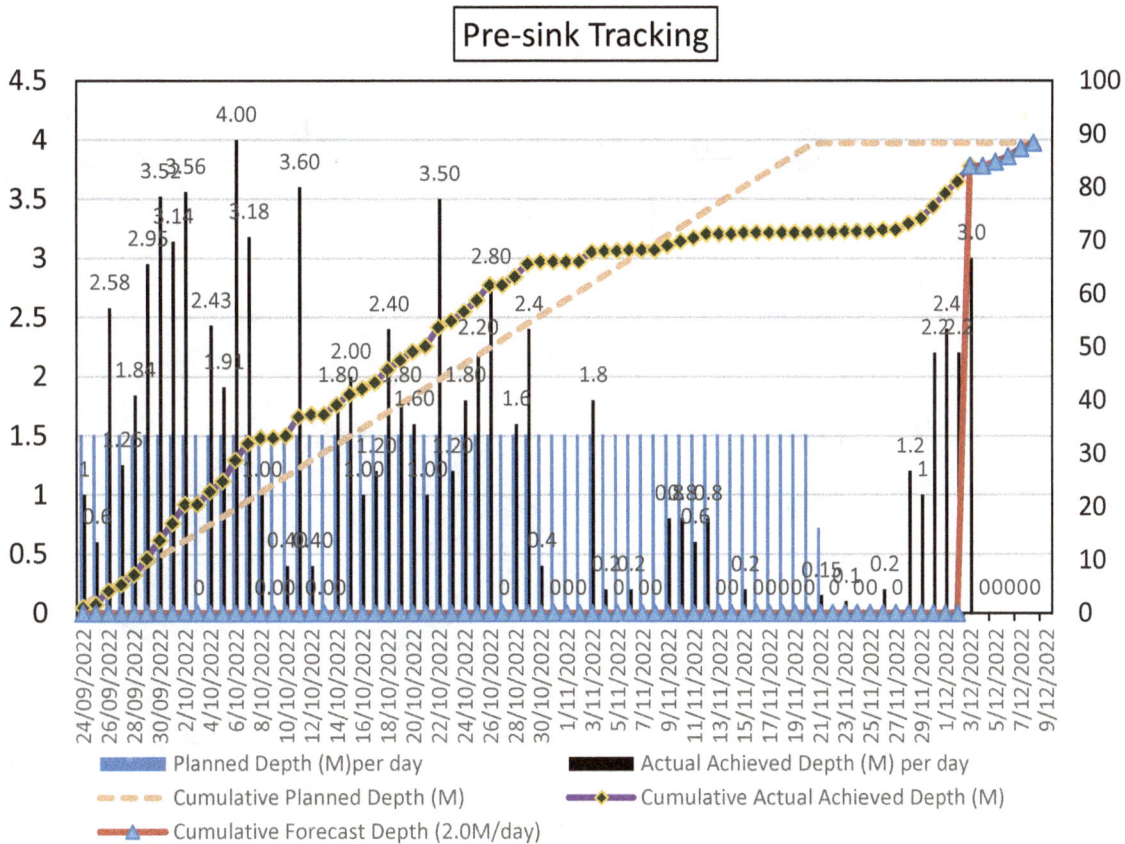

FIG 20 – Shaft excavation schedule.

RECOMMENDATIONS

A summary of known lessons (to date) is provided:

- Collar design to incorporate VSM loading requirements.

- Use multiple geotechnical holes to cover the shaft area. A single hole is limited in its effectiveness in identifying structures that can cause significant impact to excavation.

- Redesign VSM steel consumables in line with local (Australian) steel specifications. Plate, rebar, and fittings were not available in Australia and had to be sourced from Europe.

- Utilise existing segment moulds to minimise set-up time and cost.

- Allocate sufficient real estate ('the runway') adjacent to the shaft to pull the entire energy chain out of the shaft and keep it intact to minimise delays due to raising or lowering the VSM (Figure 21). This improvement reduced a nine hour operation to 1.5 hours for 70 m of energy chain.

- Install backup survey pipes on the shaft lining.

- Undertake regular shaft surveys from day 1.

- Consider post excavation final discharge and treatment of slurry – Prominent Hill had an adjacent paste plant that was able to process the material.

FIG 21 – Energy chain runway.

CONCLUSIONS

Shaft presink excavation was completed in December 2022. This was the first time this method had been applied in Australia, and the second time it had been applied in the mining industry.

Positive communication, collaboration, early intervention and competent support allowed each challenge to be navigated and overcome. The outcome was a safely constructed, mechanised, non-entry concrete lined shaft (refer Figure 22). The 93 m presink was excavated and lined quickly and efficiently – despite shipping and construction delays.

The mobilisation issues encountered were exacerbated by the interruption and delays to global shipping caused by the pandemic and could not have been foreseen or prevented. However, selecting this shaft presink excavation methodology meant works could be carried out more safely and quickly compared to a traditional construction method, allowing the overall project to remain on schedule, and recognises the benefit to the mining industry of applying new technology and innovation.

FIG 22 – Concrete lined presink.

ACKNOWLEDGEMENTS

The authors acknowledge OZ Minerals for allowing us to present this paper and the assistance provided by Herrenknecht AG.

REFERENCES

Heever, C, 1981. New Techniques In Deep Presinking Of Mine Shafts, Society for Mining, Metallurgy & Exploration 38822 SME.

Frey, S and Schmäh, P, 2019. The role of mechanized shaft sinking in international tunnelling projects, Proceedings of the WTC 2019 ITA-AITES World Tunnel Congress (WTC 2019), 1st ed (CRC Press).

Lilley, C, 2009. Geotechnical conditions for ventilation shafts at Prominent Hill. OZ Minerals.

Schmäh, P, 2007. Vertical shaft machines, State of the art and vision, Acta Montanistica Slovaca.

Chamfered backs to improve open stope stability in poor ground conditions

G Capes[1], R Lachenicht[2] and N Mitkas[3]

1. Manager Mine Planning, BHP, Adelaide SA 5000. Email: geoff.capes@bhp,com
2. Principal Geotechnical Engineer, Itasca Australia Pty Ltd, Perth WA 6121. Email: rlachenicht@itasca.com.au
3. Lead Mine Design Engineer, BHP, Adelaide SA 5000. Email: nicholas.mitkas@bhp.com

ABSTRACT

A novel open stope design approach is presented that delivered improved performance in a mining area with poor ground conditions at the Olympic Dam mine. During 2019, a large open stope had an unplanned caving event resulting in the inability to extract the planned tonnes and subsequent sterilisation of nearby stopes. A design limitation was the lack of data and understanding around the alteration domain extents, the rock mass strength controls and limits, the failure modes controlling stope stability in the local area, and the use of conventional stope shapes. From the available data and failure observations, it was inferred that the dominant caving failure mode resulted from a blocky rock mass impacted by structure and alteration causing an unravelling failure mechanism. Numerical and empirical analyses were conducted to test this hypothesis and establish strength limits for design. The strength limits for the design were then used to provide a range of outcomes from which future stope design shapes such as chamfered backs could be selected. Three alternate chamfered geometries were selected to trial the innovative designs. The application of this methodology in 2020 resulted in overbreak reduction from >100 per cent in the initial caved stope to <20 per cent in the trial stopes.

INTRODUCTION

Olympic Dam (OD) mine is situated approximately 570 km NNW of Adelaide, South Australia. OD has historically operated in the Northern Mine Area (NMA) since operations commenced in the 1980s. The OD mine has been extending into the Southern Mine Area (SMA) and peripheral areas within the NMA. This extension is being supported by key geoscience and resource engineering studies. As the main NMA orebody transitions to remnant tertiary stope and pillar extraction, there is an increased focus on extending into the SMA and into deeper areas of the NMA. New NMA blocks have extended deeper with the Deep Greens, Deep Purples and Deep Scarlets while also developing further to the SW into the Limes mining area, which is the focus of this paper (Figure 1).

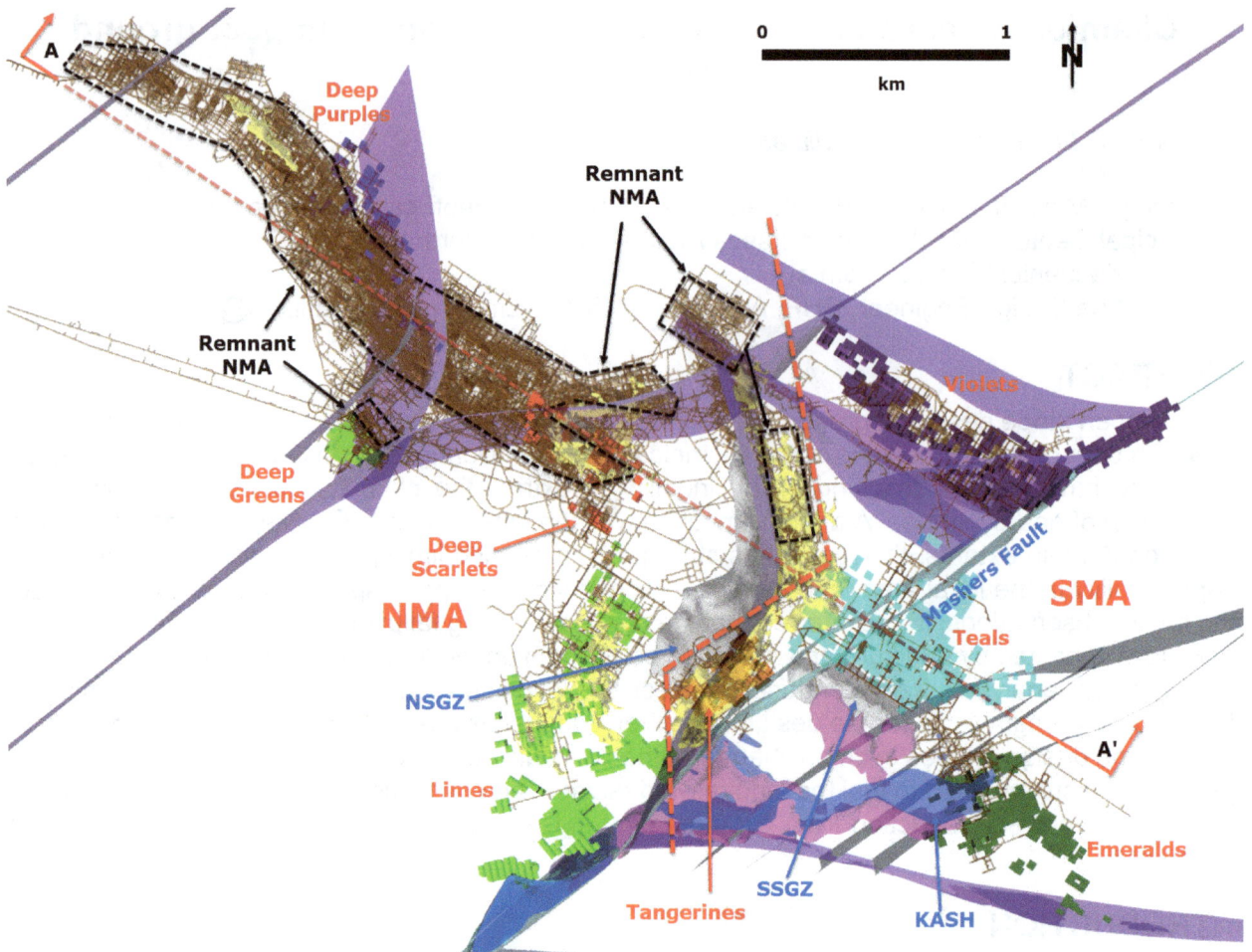

FIG 1 – Mining schematic showing remnant NMA mining blocks, new NMA mining blocks and initial SMA mining blocks, as well as lithological/structural complexity comparison between NMA and SMA (regional faults in purple).

The standard mining method in all of the mining blocks is mechanised sublevel longhole open stoping (SLOS) with cemented aggregate backfill. Each of the new mining blocks present unique geotechnical conditions, including new rock mass characteristics and domains, higher stresses and structures. Many different failure modes have been observed in the walls and backs of stopes in the catalogue of over 600 case histories (Jones *et al*, 2020). A failure of interest in 2019 was a large stope caving event in which a stope in the Limes mining area (Stope Back Hydraulic Radius [HR] = 7.5) caved ~100 m above the planned stope back. The final stope position was ascertained using probe holes from nearby drives (Figure 2). On this basis a forward-looking rock mechanics program involving back analysis, data collection, and instrumentation has been underway since 2019 to improve stope performance. The back analysis of the Lime 290 (LIM290) stope and resulting stope design improvements is the subject of this discussion.

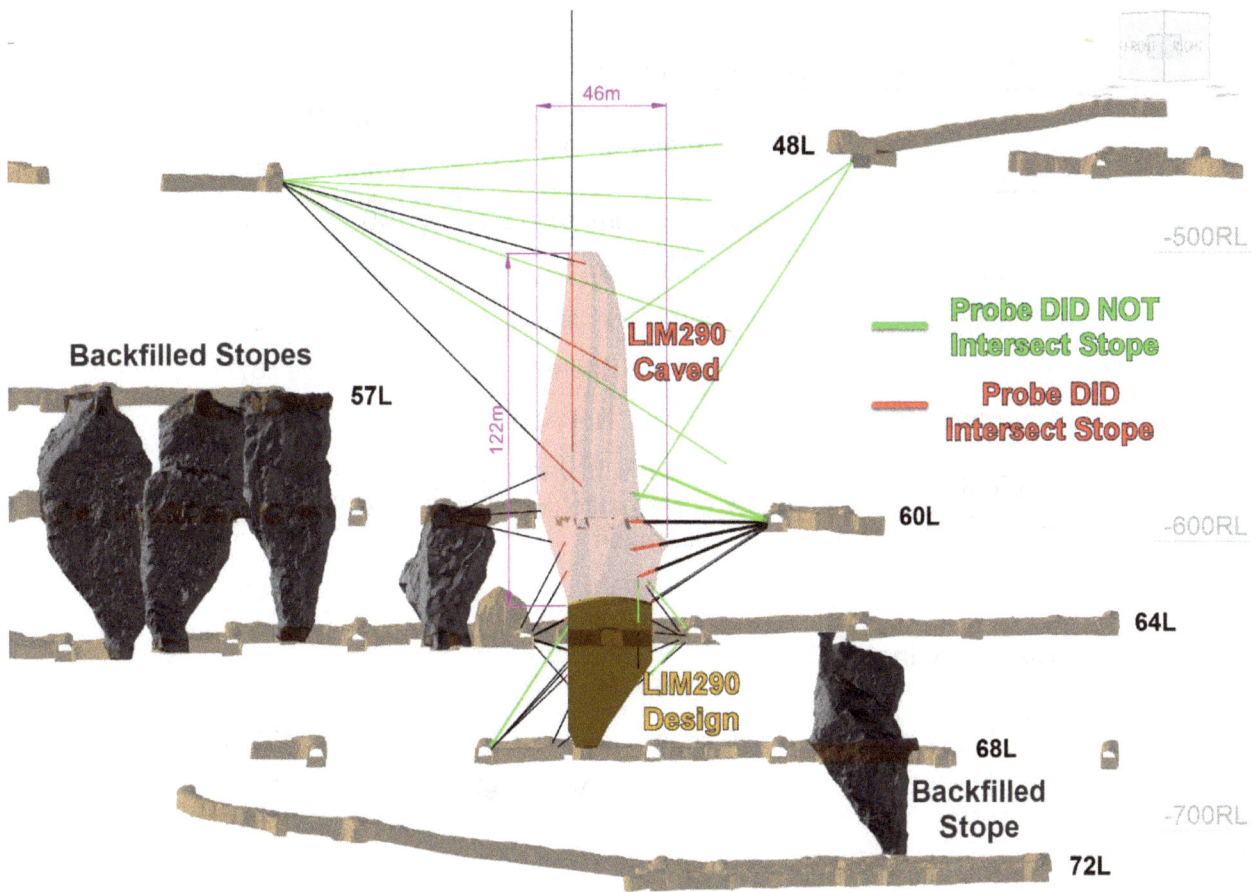

FIG 2 – Stope LIM290 in the Limes mining area which caved in an uncontrolled manner.

GEOLOGICAL AND GEOTECHNICAL BACKGROUND

The *in situ* stress field at Olympic Dam has been determined from 17 Hollow Inclusion (HI) cells measurements. Other methods, including Acoustic Emission (AE) and borehole breakout have confirmed this orientation. The mean results of the basement stress fields are shown in Table 1. The major stresses run approximately NW-SE (Mine Planning Grid – or W-E True North) against the orebody. The overburden stresses do not have a major impact on the competent unit and instability is mostly structurally controlled.

TABLE 1

Summary of the basement stress field.

Basement stress field	Bearing/dip	Depth relationship (D = depth below surface)
σ_1	133°/02°	$6 + 0.041 \times D$ MPa
σ_2	224°/18°	$2 + 0.036 \times D$ MPa
σ_3	037°/72°	$0.033 \times D$ MPa

The orebody has been described by (Reeve *et al*, 1990; Ehrig, McPhie and Kamenetsky, 2012; Clark, Passmore and Poznik, 2017). It exhibits intense, structurally destructive hematite and sericite alteration which can result in both mineralisation and a reduction in rock mass strength. The general rock mass properties used at OD at the time of the back analysis of the LIM290 are shown in Table 2. Of note is the variability in rock mass condition that exist within the mine. The NMA rock mass characteristics described as the typical basement rock have been domained by mineralogy – level of hematite alteration. Increasing levels of sericite and chlorite alteration of the rock mass correlate with a gradual decrease in rock strength as the rock matrix degrades to almost clay-like conditions through highly altered zones.

TABLE 2

Summary of cover sequence, NMA and SMA rock mass characteristics.

Rock mass domain		UCS [MPa]	
		Mean	Std
Cover sequence	Andamooka Limestone	49	20
	Arcoona Quartzite	116	40
	Corraberra Sandstone	92	26
	Tregolana Shale	104	67
Unconformity	Not Applicable		
Typical basement	Granite (GRN) with varying degrees of Hematite (HEM) alteration	135–239	~100
New domains, structure, and alteration	HEMQ	149	90
	KASH	35	14
	KHEMQ	112	50
	Gouge Zones	<20	TBC
	Sericite Alteration (Variable)	~2–50	TBC

ANALYSIS

At the time of back analysis, a design limitation was the lack of local area data and understanding around the alteration domain extents, the rock mass strength controls and limits, and the SMA failure modes controlling stope stability.

Towards investigating the stope failure modes and establishing a stope design assessment tool, numerical and empirical sensitivity analyses were considered.

Numerical analysis

The numerical analysis methodology was tested based around the following stability hypotheses:

- Stope stability in the SMA is a function of stress, stope dimensions, rock quality, rock strength, geological structure interaction and support/reinforcement design.

- Rock mass failure can occur based on stress induced damage of the rock mass or from unravelling, tensile failure of the rock mass. Failure can initiate and localise within weak zones controlled by structure and alteration with the final failure state a function of *in situ* stress, extraction geometry and support/reinforcement.

- Rock strength is a function of the base lithology and the degree or intensity of alteration including both sericite alteration and a newly interpreted 'indicator' alteration wireframe.

- Controls on alteration intensity impacting on rock mass strength include structure with limited current understanding.

- Rock quality is interpreted to be controlled by structure interaction (specifically influencing how available rock quality data is interpreted and modelled).

- From the available data and failure observations of the large-scale stope failures, it was inferred that the dominant caving failure mode results from a blocky rock mass impacted by structure and alteration causing an unravelling failure mechanism. The alteration results in a strength reduction of the discontinuity infill and the intact rock strength. The purpose of the numerical analyses was to test this hypothesis and establish strength limits for design.

To implement the numerical analysis methodology, the following approach was adopted:

- Flac3D, was selected as it allows for rapid set-up and testing of stope stability analyses.
- The methodology captures stress through the *in situ* stress assumptions and the incorporation of approximately 150 m of the surrounding extraction layout (a total of a 300 m modelled span). The surrounding layout that is modelled is limited in extent to ensure the analyses run rapidly allowing for multiple sensitivity studies to be undertaken. The design approach is termed 'Stope-Lab'.
- Local stope rock quality was interpolated into a three-dimensional model and mapped into the Flac3D analyses. In the initial analyses, RQD was interpolated using a spherical search radius.
- The modelling methodology captures the stope extraction sequence (path dependency), limited to the modelled surrounding extraction layout extent (300 m span).
- In the initial approach structure was incorporated as ubiquitous joints with all structure assigned the same strength parameters. Ubiquitous joints were selected to ensure speed of analysis.

Due to unknown local ground conditions, a sensitivity assessment was completed to estimate the conditions at which caving would occur relative to the level of alteration present. The back analysis relied on a newly developed geological alteration wireframe referred locally as the 'indicator', having a significant presence in the Limes mining area. The rock mass properties for the modelled runs were weakened as follows to account for local alteration:

- Area of altered ground known locally as the 'Indicator' is weakened to a $UCS_{intact} = 20$ MPa and a Ja (Joint Alteration) = 6 as an estimate for the sericite/clay alteration.
- Area surrounding the 'Indicator' ±15 m is weakened to a $UCS_{intact} = 70$ MPa and a Ja = 6 rock quality assumption.

From the strength sensitivities, the analysis noted that the back analysed LIM290 design was vulnerable to large scale crown collapse concurring to the observed stope performance. The 'Indicator' ±15 m scenario resulted in an unstable model indicating potential for ongoing continuous caving matching the observed stope performance.

Empirical analysis

From an empirical design perspective, rock mass conditions are implicitly represented by the modified stability number (N') as per Potvin (1988) which has a rock mass classification index input Q' (Barton, Lien and Lunde, 1974). The stability number is compared against a shape factor known as the Hydraulic Radius (HR) to estimate if a stope will perform as a stable/unstable/caved outcome. Understanding the rock mass factors influencing Q' is key to estimating open stope performance. One of the factors – joint alteration (Ja) can range considerably relative to the altered condition of the rock mass joints and type of clay infilling. At the time of stope design, rock mass weakening due to alteration in the area had not been considered and Ja was assumed as 1 and a stable outcome was estimated. In reviewing the stope performance and the new alteration wireframe, the following adjustments were made to represent local conditions which identified a caved outcome:

- Reducing RQD assumption, a rock quality estimate from drill core, from 85 (original) to 50.
- Increasing Ja (joint alteration) from 1 (original) to 6.

Forward analysis of stope with chamfered back

Based on the insights from the numerical and empirical analyses, stope geometry modifications were proposed by the technical team members as a series of trial stopes. They hypothesised that a chamfered stope crown would potentially minimise ore loss from the planned stope shape and prevent significant overbreak. For the forward analysis, numerical analyses were chosen as it captured the unique geometry considerations of the stope. The outcome of the analysis indicated it to be less vulnerable, comparatively to the LIM290, to large scale collapse. The modelled outcomes provided the following insights from the range of scenarios considered:

- The chamfered design presents a low risk of extending past 1 m depth of failure with the explicit 'Indicator' and sericite alteration interpretations with an intact UCS strength reduction to 20 MPa (Figure 3).

- A worst case 'Indicator' interaction scenario (±15 m) indicates the potential for more severe stope failure and depth of damage extending to 7 m (but not uncontrolled caving failure as per the equivalent LIM290 models) (Figure 4).

- The chamfered design incorporates a small crown span (HR 2.1) which is not vulnerable to collapse from a gravity driven rock mass unravelling, ie a drive scale exposure with planned deep anchorage reinforcement seated back to surface support. This is a significant reduction from the full stope span of HR 7.5. Comparing the modelling to the stability graph Q' modifications (RQD = 50; Ja = 6; Jr = 1) indicated the crown as stable with cable support (drive scale) as an additional design check.

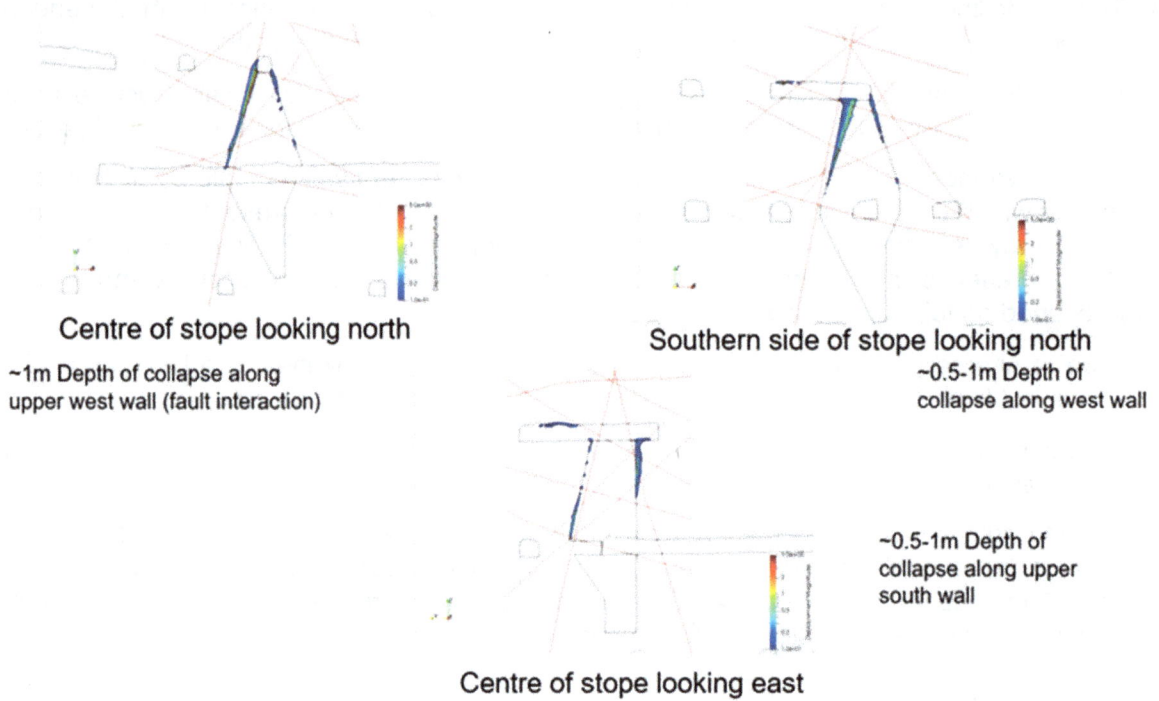

Centre of stope looking north

~1m Depth of collapse along
upper west wall (fault interaction)

Southern side of stope looking north

~0.5-1m Depth of
collapse along west wall

~0.5-1m Depth of
collapse along upper
south wall

Centre of stope looking east

FIG 3 – Model Outcomes (indicator weakened). Modelled displacement colour scale shown ranges from 0.1 m (blue) to 5 m (red).

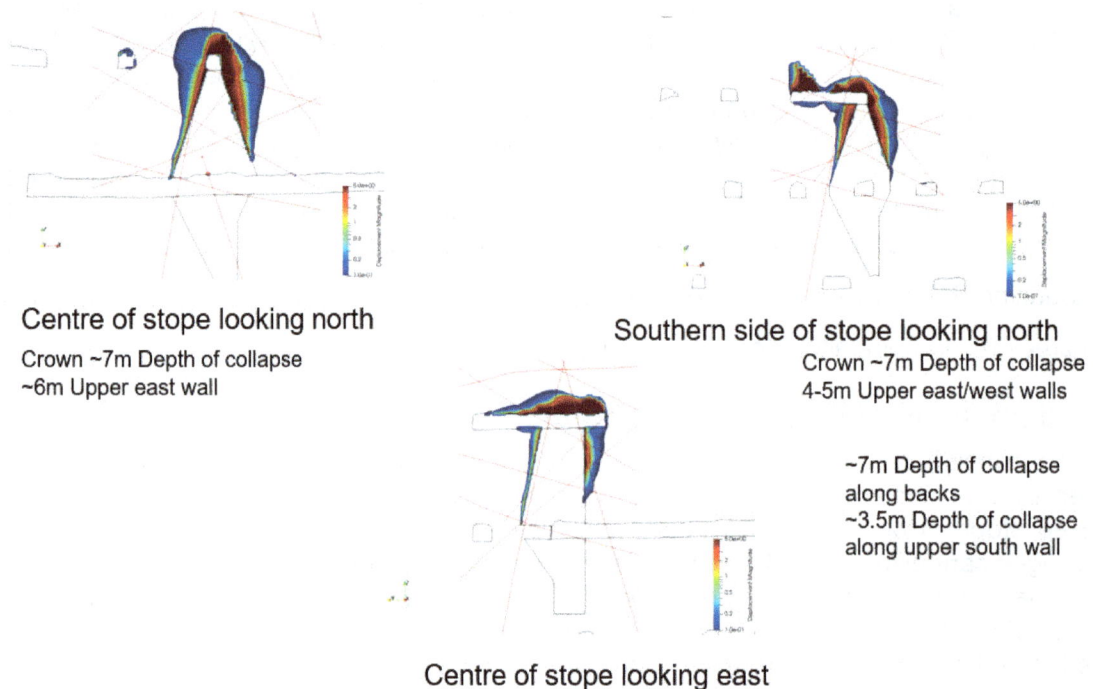

Centre of stope looking north

Crown ~7m Depth of collapse
~6m Upper east wall

Southern side of stope looking north

Crown ~7m Depth of collapse
4-5m Upper east/west walls

~7m Depth of collapse
along backs
~3.5m Depth of collapse
along upper south wall

Centre of stope looking east

FIG 4 – Model Outcomes (indicator ±15 m weakened). Modelled displacement colour scale shown ranges from 0.1 m (blue) to 5 m (red).

Underground Operators Conference 2023 | Brisbane, Qld | 27–29 March 2023

STOPE TRIAL – CASE HISTORIES

Three separate stopes with similar ground conditions as the previous uncontrolled failure (LIM290 stope) were targeted to serve as trial environments for applying a variety of chamfered back open stope designs. The design teams brainstormed options to mitigate expected risk, deciding on the experimental chamfered stope backs at various angles resulting in various stope sizes. The three stopes are summarised below:

- Test stope 1 (068_LIM_468_489_STP) was designed and production started in August 2020. The stope chamfer was tight, and size was reduced to 100 Kt.

- Test stope 2 (072_LIM_456_495_STP) was designed and production started in June 2020. The stope chamfer was moderate, and size was increased to 160 Kt.

- Test stope 3 (065_LIM_459_474_STP) was designed and production started in February 2020. The stope chamfer was moderate, and size was increased to 190 Kt.

Stope performance (overbreak and underbreak) was reviewed at the stope close-out step to determine if the stopes were in planned tolerance. All of the stope experience a degree of instability, however, none of the stopes caved which made the trial a success. The smallest stope with the tightest chamfer had the best overbreak performance of 8 per cent (within a planned tolerance of <15 per cent) whilst the larger stopes had ~20 per cent overbreak which in this instant occurred into ore existing outside the stope shape.

Specific stope results include (Figure 5):

- Test stope 1 (068_LIM_468_489_STP) had overbreak of 8 per cent (ie within planned tolerance).

- Test stope 2 (072_LIM_456_495_STP) had overbreak of 20 per cent (ie significantly exceeds planned tolerance).

- Test stope 3 (065_LIM_459_474_STP) had overbreak of 17 per cent (ie exceeds planned tolerance).

The chamfered stopes proved to be less vulnerable to large scale collapse. The design method developed presents a low risk of extending past 5 m depth of failure even with very low rock quality.

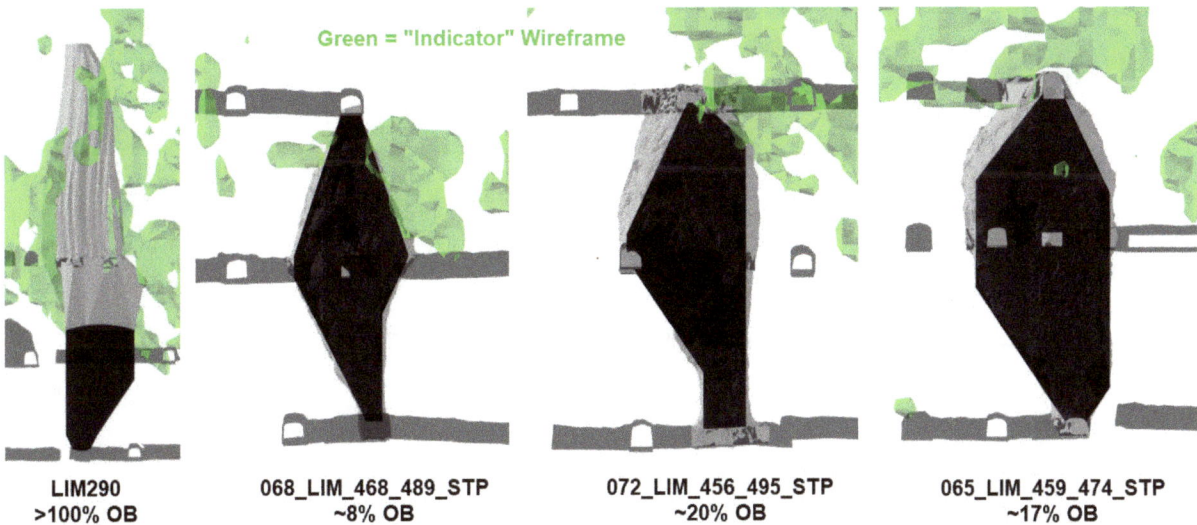

| LIM290
>100% OB | 068_LIM_468_489_STP
~8% OB | 072_LIM_456_495_STP
~20% OB | 065_LIM_459_474_STP
~17% OB |

FIG 5 – Caved stope and outcomes of three trial stope outcomes with actual performance shown in light grey.

CONCLUSIONS

A large stope failure required a novel approach to enable recovery of open stopes in an area of poor ground conditions. Due to unknown local performance outcomes due to a lack of available data, a sensitivity analysis was conducted to determine a range of outcomes that could occur from a revised stope shape. Insights from numerical and empirical back analysis led to the hypothesis of a stope shape that would result in a stable outcome for stope extraction. A numerical simulation was conducted to determine if the stope would be stable or fail continuously based on the parameters developed from the back analysis. The simulation showed a degree of instability would occur but that stope caving would not occur. The mine completed three trial stopes to test the stability hypothesis. The outcomes of the trial demonstrated that chamfered back design is an effective method to mitigate stope caving in highly altered, poor ground conditions at Olympic Dam. Ongoing data collection and domaining has continued to identify areas of high alteration and other unique conditions which require novel mine design approaches and reconciliation as part of the mine's Plan-Do-Check-Act processes.

REFERENCES

Barton, N, Lien, R and Lunde, J, 1974. Engineering classifications of rock masses for the design of tunnel support, *Rock Mech*, 6:97–106.

Clark, J M, Passmore, M and Poznik, N, 2017. Olympic Dam rock quality designation model – an integrated approach, *Tenth International Mining Geology Conference* (The Australasian Institute of Mining and Metallurgy: Melbourne).

Ehrig, K J, McPhie, J and Kamenetsky, V, 2012. Geology and mineralogical zonation of the Olympic Dam iron oxide Cu-U-Au-Ag deposit, South Australia, in *Geology and Genesis of Major Copper Deposits and Districts of the World: A Tribute to Richard H Sillitoe* (eds: M Harris, F Camus J W Hedenquist), Special Publication 16:237–268 (Society of Economic Geologists: Littleton).

Jones, E, Mitkas, N, Capes, G, Morizzi, L and Grant, D, 2020. Olympic Dam open stope observations, in *Proceedings of the Underground Operators Conference 2020* (The Australasian Institute of Mining and Metallurgy: Melbourne).

Potvin, Y, 1988. Empirical open stope design in Canada, PhD thesis, The University of British Columbia, 350 p.

Reeve, J S, Cross, K C, Smith, R N and Oreskes, N, 1990. Olympic Dam copper-uranium-gold-silver deposit, *Geology of the Mineral Deposits of Australia and Papua New Guinea* (ed: F E Hughes), pp 1009–1035 (The Australasian Institute of Mining and Metallurgy: Melbourne).

Innovative digital tools shaping the future of sustainable sprayed concrete operations and operator training

P Oikkonen[1] and B Aga[2]

1. Director Underground Process Excellence, Normet Group, Espoo 02920 Finland.
   Email: panu.oikkonen@normet.com
2. Equipment Manager, Hæhre Entreprenør AS, 1301 Sandvika, Norway.
   Email: bjarte.aga@akh.no

ABSTRACT

Sprayed concrete has played an integral role in underground rock support for nearly a century. However, concrete is also considered to be one of the biggest environmental problems in construction and mining, largely because of the high carbon footprint of cement and aggregates production. This is driving the industry to find ways to cut concrete usage where possible. Improving sustainability by cutting concrete material consumption also reduces application time and costs.

Improving concrete properties can make it possible to design thinner linings, but as awareness of safety and longer design lifetimes are driving developments in the opposite direction, it is not realistic to see this as a major factor in cutting concrete consumption. More accessible solutions are focusing on layer thickness control and minimising rebound and wastage.

Digital technology and innovation offer excellent tools for improving sustainability while optimising other aspects of the spraying process at the same time. A digital layer thickness control system gives near-real-time information to the sprayer operator with full coverage of the sprayed surface. This enables the operator to achieve consistent layer thickness on the first pass, in comparison to the traditional methods of overspraying with the aim of achieving sufficient layer thickness, probe drilling after the concrete has set, and respraying at a later stage if needed.

Operator competence is the single biggest variable in a successful sprayed concrete process, affecting rebound and wastage levels. Unfortunately, the learning curve to become a master sprayer is quite steep and training on a real sprayer and real concrete requires a significant amount of time and money. Training in a virtual reality (VR) environment enables operator training to be carried out in a low-risk environment and provides a way of measuring performance and assessing and certifying the operators. After the VR training period, a novice nozzle operator has the basic understanding to start working on a real sprayer, and an experienced nozzle operator can benefit from the possibility of assessing their performance.

INTRODUCTION

This paper discusses the most common factors affecting the environmental impact of the concrete spraying process. Special attention is paid to concrete consumption exceeding the theoretical or optimal requirement in order to reach the desired sprayed concrete lining.

New digital technologies that can bring further improvement to narrow the gap between theoretical and actual concrete consumptions are discussed through presentation of the underlying technology as well as case studies.

The paper discusses digital layer thickness control systems including a case study from a tunnelling project. When automatic georeferencing is added, these systems are also capable of readily producing near-real-time as-built layer thickness data and surface mapping in the project or mine coordinate system.

The effect of operator competence is also discussed. Concrete spraying is considered to be one of the most demanding tasks underground, and the operator has a significant effect on quality, safety and material consumption.

IN SPRAYED CONCRETE, SUSTAINABILITY AND SAVINGS GO HAND IN HAND

Meeting expectations regarding sustainability in all fields of the underground industry is part of the licence to operate.

In concrete spraying there are ways to influence quantities such as material consumption and improve sustainability in this way. The carbon footprint of traditional concrete mix is known to be extremely high, and because of the nature of the concrete spraying process, there is a significant difference between the theoretical quantity of concrete to be used to produce a certain sprayed concrete structure, and the actual consumption of concrete to produce the same structure. Given that, in concrete spraying, the total process cost, time and carbon emissions of the process are all directly dependent on the concrete consumption, narrowing the aforementioned gap will improve all three factors – process cost, time and carbon emissions – instead of sustainability improvements just increasing the costs. Lowering the carbon footprint of the concrete mix itself can bring more sustainability improvements, and the possible increase in unit cost can still be covered in terms of the total process cost by lowering the consumption.

SAVING POTENTIAL IN SPRAYED CONCRETE CONSUMPTION

The areas where waste and excess concrete consumption is generated can be divided into six categories.

Design

Sprayed concrete structures and linings are usually designed for rock support and thus to guarantee the safety of people using the underground facility. This means that linings will need to be sufficient in all situations and the developing awareness of safety needs to be considered. In some cases, this might even mean that there may be pressure to design thicker or otherwise stronger linings. More often though, there can be scope to design thinner linings using modern high-performance concrete mixes and fibres. Put simply, the same support effect can be achieved with a thinner lining. This is especially the case if primary rock support can be considered as part of the permanent lining.

Especially in tunnelling, moving away from cast or double-shell linings to single-shell sprayed concrete linings has a significant positive effect on concrete consumption, even before the actual project has started.

The reduction in embodied CO_2 when moving from double-shell tunnel lining consisting of a sprayed primary lining and cast-*in situ* permanent lining to sprayed concrete lining has been estimated by International Tunnelling Association (ITA) (2020) to vary from about 20 per cent in soft grounds, to about 50 per cent in hard rock.

Excavation quality

The effect of excavation quality on sprayed concrete consumption is very easily neglected. It may be true that, especially in hard rock drill and blast, saving some holes in the drill pattern and cutting some corners on smooth blasting will accelerate the pace of excavation slightly, but it will ultimately result in an increased need for spraying. This can be divided into three different effects:

1. With a rough surface finish and wide damage zone, the need to fill cavities and spray weakness zones excessively is greater.

2. Over a rough and feature-rich surface with a lot of overbreaks, nozzle control is more difficult and both rebound and excess spraying increase – on a smooth surface, spraying is more systematic.

3. The surface area to be sprayed is simply larger (Figure 1).

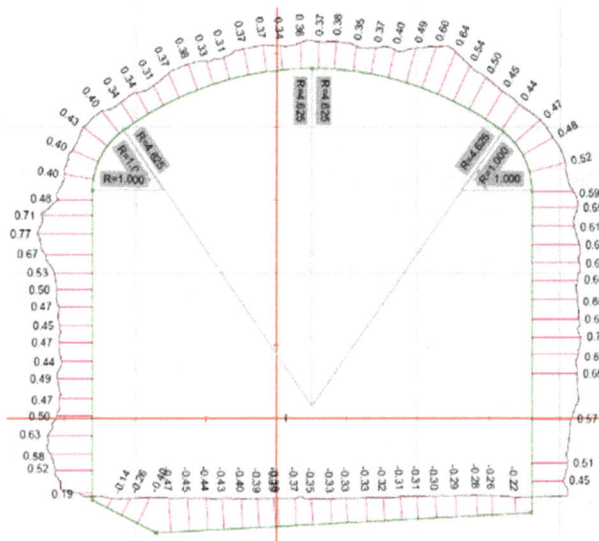

FIG 1 – As-built record of excavated profile in comparison to theoretical.

It is also to be noted that rough excavation quality increases the time for mucking as the amount of muck is simply greater, and the need for scaling as the surface quality is rougher. Safety risks increase as the rough surface finish and wider damage zone increase the risk of ground fall.

Smooth surface quality and narrow excavation tolerances bring improvements to more or less all the subsequent processes through lower costs and environmental impact, increased safety and better control over the excavation cycle.

In hard rock, excavation surface quality can be improved through careful drill pattern and charging pattern design, accurate and careful drilling using the correct parameters, and smooth blasting procedures using accurate charging methods and explosives, as well as taking into account the rock quality at hand and adjusting the process as needed.

In soft ground mechanical excavation methods such as road header, maintaining control over excavation surface quality is usually easier, depending on the ground type. On the other hand, the effect of over-excavation is direct if the spraying is carried out based on design profile instead of design layer thickness.

The estimation of wastage amount and material saving potential is highly dependent on the rock or ground type and excavation method.

Overspraying

It is just as important to avoid overspraying as it is to avoid underspraying.

Overspraying the design layer thickness has a direct effect on concrete consumption – the amount oversprayed is all excess. However, because of the nature of the sprayed concrete application process, precisely spraying the design layer thickness is extremely challenging, and aiming for this increases the risk of underspraying.

The effect of underspraying is slightly more complex. It is clear that systematic underspraying cannot be allowed for both short- and long-term safety reasons. Given the nature of concrete spraying application and the often uneven receiving surfaces, a certain amount of local underspraying is likely to occur when aiming for a design layer thickness. Depending on the local standards or project requirements, a certain amount of undersprayed areas of a certain size is usually allowed. This is elaborated in more detail later in the paper. Any other underspraying needs to be resprayed. Having to respray an area will result in increased concrete consumption, as controlling the usually very thin layer thickness – especially on an irregularly shaped area – is even more challenging than normal. Furthermore, starting and ending the small patch work leads to more waste than systematic spraying.

The estimation of wastage amount is very much dependent on roughness of the receiving surface, and operator skill and motivation level.

Overlap

In cases where the surface of every blasted round needs to be sprayed before excavating the next one, some of the concrete will be removed in blasts or in mechanical excavation. This is highly dependent on the rock quality. In the case of poor-quality rock, the rounds are shorter and thus the quantity of overlaps is higher. In addition, the layer thickness is often greater, and if the rock quality is poor enough the face wall will need to be partly or completely sprayed.

This type of overlap is difficult to avoid if the reason behind it is poor rock quality and thus safety. Careful layer thickness control and spraying only what is needed is important on the areas that will be excavated away as well.

Some overlap also occurs when continuing spraying after the next round is excavated or between sprayer positions. Between excavated rounds, the border of the previously sprayed concrete can be rough and may need some overlap to achieve a smooth finish and continuous rock support. The concrete with a weakened bond to the rock surface due to blasting needs to be scaled off and the area resprayed. Some overlap usually also occurs between sprayer positions to ensure smooth and continuous layer thickness.

In these cases, it is important to spray systematically and avoid spraying full layer thicknesses over areas that are already acceptable. Careful layer thickness control helps.

Rebound

The effect of rebound on concrete consumption is also clear and direct – everything that ends up anywhere other than on the wall is excess.

However, the cost and environmental impact of rebound is higher than that of excess concrete on the wall. In addition to the cost and impact of the concrete that is manufactured and sprayed, there is also the cost and impact of disposing of the rebound to be considered. Rebound is not collected and disposed of in all areas but, as environmental awareness picks up, this is likely to increase.

The amount of rebound is affected by several factors:

- mix design
- mix quality (when batched)
- mix quality when at nozzle (transportation, unloading, pumping)
- surface quality
- surface preparation
- accelerator dosage
- pulsation in pumping
- application (stand-off distance, angle, spraying direction, spraying sequence)
- spraying and equipment parameters
- conditions (temperature etc).

Optimising these factors will bring improvement to rebound control and rebound percentage, but they all need to be taken into account to achieve excellent results. In another words, rebound is also an excellent indicator of the quality level of the whole sprayed concrete process – high rebound percentage always indicates that there is something to be improved in the process, whereas a very low rebound percentage indicates that all the factors in the process are optimised to a very high level. A sudden increase in the rebound percentage always indicates a change or a fault in the process.

One of the unfortunate things about rebound is that it is extremely difficult to measure accurately, continuously and in real time. Although it is possible to roughly observe the level of rebound visually, a proper reading based on an accurate measurement has traditionally required a special test arrangement where an accurately measured amount of concrete is sprayed over a sheet from which the rebound concrete can be collected and weighed. This test procedure is described for example by the Norwegian Concrete Institute (2011). In future, modern technology can bring improvement in fast and accurate rebound control, as volumes of concrete sprayed and volumes of concrete on the wall can be measured and compared accurately already during application.

It is important to note that all the overspraying also causes rebound so reducing the overspraying will also reduce the absolute amount of rebound, which is the most expensive type of sprayed concrete as it does not contribute to rock support but often needs to be disposed of with added cost, time and environmental impact.

A good rebound level is often considered to be 5–10 per cent or less, but depending on the conditions, a poor level can be almost anything above.

Waste and spill

In this study we consider waste to be the concrete or ingredients of concrete that are never actually sprayed through the nozzle. Spill, as part of waste, would be the concrete or ingredients that either accidently or because of negligence end up on the floor instead of proceeding through the process and being sprayed. Waste and spill can be generated anywhere in the process from the transportation of the ingredients to the nozzle. Typical sources of waste include:

- Ingredients that leave their origins of supply but do not end up being used at the batching plant.

- Ingredients that are wasted or overdosed during the batching process.

- Concrete that is mixed but does not end up in the transportation unit (this includes concrete that stays in the mixer drum and batches that are rejected for quality reasons).

- Concrete that gets delivered to the sprayer but does not end up in the hopper (this includes spill and concrete that stays in the transportation unit drum).

- Concrete that is transported but not unloaded to the hopper (this includes batches that are rejected for quality reasons or because the sprayer is not ready for unloading and the batch needs to be disposed of).

- Concrete that ends up in the hopper but does not get sprayed (this includes concrete that stays in the hopper, pump and concrete lines when finishing the spraying, and concrete that is released from the pumping process line because of blockages etc).

- Concrete that comes through the nozzle but is not sprayed on the surface and ends up on the floor (this includes, for example, concrete that is pumped through the line and onto the floor before starting to spray the surface, and concrete that gets sprayed on the floor).

A lot of this kind of waste is generated because there are quality control issues or equipment availability issues in the process cycle.

Combining these six areas where waste and excess concrete consumption is generated gives us an idea of the saving potential in terms of CO_2, time and cost, but a detailed estimation would require more study on each root cause. According to International Tunnelling Association (2022), the amount of embodied CO_2 can be as high as 450–500 kg/m^3 in steel fibre reinforced sprayed concrete mix. Cost and time per m^3 of concrete sprayed are dependent on the area and project.

As explained in the previous section, when it comes to sprayed concrete the time, cost and carbon footprint are all directly dependent on the concrete consumption (assuming no changes in the concrete mix). This means that everything that reduces the consumption will result in direct savings for all three factors – time, cost and carbon footprint. In addition, focusing on holistically improving the process will bring benefits to the whole excavation cycle through better predictability and control over the cycle times. Improved and controlled quality will benefit the life cycle management of the underground facility and can also bring more cost and CO_2 savings through not having to overoptimise the concrete mix.

DIGITAL TOOLS AS PART OF SUSTAINABLE SPRAYED CONCRETE OPERATIONS

Underground processes generally and concrete spraying in particular have not exactly been at the forefront of development when it comes to digitalisation. In mobile machinery, many of the digital tools have focused on collecting information on the equipment operating parameters and presenting this to the users for the purposes of reporting, preventive maintenance planning, process optimisation, etc.

In the context of improving the sustainability of sprayed concrete operations, digital tools should focus on improving the topics introduced in the previous chapter to limit the concrete consumption to a level as close to optimal as possible, to reduce the initial carbon footprint of the operations, and to improve the sprayed concrete lining quality to enable better life cycle management and reduce the need to overoptimise the lining structure. Out of the six factors generating excess concrete consumption, we can define four improvement targets:

1. better layer thickness control

2. lower rebound percentage

3. lower amount of waste

4. better application quality for optimal compaction and bonding.

Modern digital tools can help bring the sprayed concrete process improvements to a higher level. They also help assess and evaluate the process and capabilities, and make it possible to report and document the results comprehensively, quickly and efficiently.

Digital layer thickness control systems

Traditionally, monitoring the sprayed concrete layer thicknesses has meant visual inspections and probe drilling or the use of pre-installed probes or wires. Visual inspection is inaccurate and is affected by the viewer's experience and ability to read layer thicknesses. It also lacks the means to validate and document the results without accepting the effect of human factors. Probing by drilling can only be done after the concrete has set, which means either patching the undersprayed areas separately later or neglecting them. Probe drilling and using pre-installed probes only provides information about those exact locations and is subject to the viewer's reading of the result. In addition, the results can only be documented manually in this case.

Standards and guidelines on layer thickness

How sprayed concrete layer thicknesses are controlled varies globally. Many countries in Europe generally follow European standard EN 14488–6 (2006) but may have more detailed national specifications to stipulate how the requirement in the standard is fulfilled and what other methodology can be used. EN 14488–6 describes a specific probe drilling pattern and method.

In Finland, for example, the national guidelines for sprayed concrete (Concrete Association of Finland, 2015) refer to the European standard but also allow alternative methods such as laser scanning. For scanning, the guidelines state that the required layer thickness needs to be met when taken as an average value, and that on 20 per cent of the scanned area, the layer thickness can be 25 per cent thinner than the required thickness. This follows the spirit of the EN standard which states that the thickness reading in one of the five holes in the probe drilling pattern is allowed to be below the required thickness.

In Norway, the EN standard is not directly referred to, but the country follows the National Publication no. 7 (2011) of the Norwegian Concrete Association, and the updated version of that publication will recognise scanning as an acceptable layer thickness control method. The Norwegian guideline and method are presented in more detail in a case study later in the paper.

The Concrete Institute of Australia (2020) does not yet recognise scanning as a layer thickness control system, and one random probe or hole per 50 m² of sprayed surface is generally required. Some local guidelines, for example in Tasmania (Tasmanian Department of State Growth, 2021) already states, that the contractor may propose alternative methods for the monitoring of sprayed concrete thickness during application.

Laser scanning in layer thickness control

Monitoring the layer thicknesses with laser scanning has been possible since laser scanning was introduced to the underground as-built documentation, and software for comparing point clouds was developed. At its simplest, the rock surface is scanned before and after spraying using a terrestrial laser scanner for surface mapping and a total station for georeferencing, and layer thickness is calculated between these two surfaces afterwards using point cloud analysis software. The method can provide layer thickness information with full coverage on the surface, but the information is not ready to be utilised until long after the sprayed concrete work on the area is completed. In addition, there is a lot of manual work involved in getting the point cloud files processed and aligned with each other before the thickness calculation can be done. This may be acceptable for documenting the results but does not help to improve the process in real time. Presumably for these reasons, utilising terrestrial laser scanners in layer thickness control has not gained a major foothold as part of the spraying process.

With the development of laser scanning hardware, with careful design of process workflows and with proper software and algorithm development, the method can be modified to fit the sprayed concrete process to help the operator achieve and document the correct layer thickness during the process. A digital layer thickness control system can have the benefit of reducing excess concrete consumption, directly influencing the aforementioned root cause categories of overspraying and overlap.

For a scanning system to be able to work efficiently for layer thickness control during the sprayed concrete application and bring about improvements, the following points need to be fulfilled.

Integrated into spraying process

The system needs to be physically integrated into the spraying process so that no extra hardware needs to be carried and set-up at the spraying location. The system also needs to be straightforward and simple so that the sprayer crew can use it during spraying, without unnecessary interruption to their work.

Robust design

The conditions at the spraying location can often be unfavourable for fine electronics. Although the working conditions have improved a lot since the days of dry spraying and water glass accelerators, we have to deal with factors such as humidity, rebound, form oil and a certain amount of dust. The scanner sensor and other hardware must be protected against these elements to ensure trouble-free operation.

Fast operation

The system needs to be able to scan the surfaces and process the results fast enough to cause little or no disturbance in the spraying process. Generally speaking, the process times for the system need to be kept to a minimum, and should be noticeably shorter than those of the aforementioned traditional methods for layer thickness control.

Automatic alignment of point clouds

The thickness calculation cannot be affected by spray rig or scanning sensor movements during spraying. Due to the nature of the spraying process, minor movement of the rig, and thus the mapping sensor mounted on it, is expected and this movement will be transferred to the thickness results without a means of aligning the point clouds. This could be achieved manually by using targets, but installing targets on the area to be scanned would involve extra work and removing them would be a safety hazard as the area would be freshly sprayed. In addition, the thickness calculation needs to be accurate directly after spraying so the operator can assess it without any post-processing.

Relevant thickness

The system needs to be able to calculate and show the relevant layer thickness perpendicular to the sprayed surface, instead of simply calculating the shortest distance between the two point clouds or distance along the laser beam direction of travel.

Clear and visual results

The initial thicknesses of sprayed layers need to be visualised for the operator clearly and intuitively so that it is quick and easy for them to understand where patching may be needed, where layer thickness is at the desirable level and – for learning purposes – where the layer is too thick.

Compatible results and reporting

The layer thickness and profile information need to be in formats that are compatible with the documentation methods used on-site. This makes it possible to harness the full potential of the system as a documentation and quality assessment tool, especially when the scan results are georeferenced in the project coordinate system.

If these points are fulfilled, modern surface mapping technology can be used to present layer thickness information to the operator in near real time, bringing the layer thickness control to a completely different level when compared to traditional methods. The operator can check the layer thickness with full coverage of the sprayed surface as many times as needed during spraying without having to overspray to be sure the result will be acceptable when reporting the results, or when probing later if necessary.

The benefits of the method are not restricted to monitoring the layer thickness – it also helps the operator develop their own ability to control the layer thickness faster. If the operator receives instant feedback on the layer thickness after spraying a layer, they can understand where and why the layer thickness failed to reach the specified value or exceeded it. If they only get feedback from the probe locations and long after spraying a certain location, this will limit the operator's ability to understand which features in the rock and which aspects of the application technique tend to lead to overspraying or underspraying.

Georeferenced surface mapping and layer thickness information

When georeferencing is added to the workflow of the digital layer thickness control system, the usability of the system is expanded to building information modelling (BIM) and as-built documentation. This is something completely new to the underground industry. Georeferenced profile mapping of the sprayed concrete lining with layer thickness information (Figure 2) can be available and uploaded to a project data model only minutes after the spraying task is finished.

There are tried-and-tested methods for georeferencing 3D laser scanning point clouds, but when integrated into the concrete spraying process workflow, acquiring the georeferencing needs to be fast and highly automated for the sprayer crew or operator.

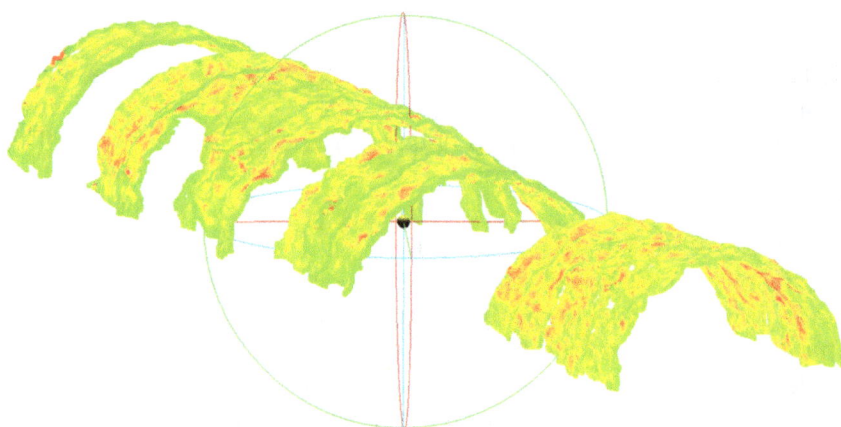

FIG 2 – Georeferenced layer thickness mapping in a point cloud analysis software.

Case study – Rogfast in Norway

E39 Rogfast highway tunnel project

The E39 Rogfast (Rogaland Fixed Link) project is aiming to connect the Norwegian coastline across the Boknafjorden and Kvitsoyfjorden fjords with a subsea highway tunnel. On a larger scale, it is part of the plan to create a ferry-free highway (E39) between Kristiansand and Trondheim. When finished, the Rogfast tunnel will be the world's longest and deepest subsea road tunnel connection with a continuous length of 27 km and a maximum depth of 390 m below sea level. The constructor for the project is Statens Vegvesen, the Norwegian public roads administration.

Regulations for sprayed concrete layer thickness in Norway

Using sprayed concrete for rock support is regulated in Norway by the Norwegian Concrete Association (2011) as follows.

Where sprayed concrete is used for permanent rock support, the layer thickness needs to fulfil the following criteria:

- average measured layer thickness must be at least equal to the prescribed layer thickness
- minimum measured layer thickness must be at least 50 per cent of the prescribed layer thickness.

For controlling and documenting the sprayed concrete layer thickness, it states that one of the following methods can be used:

- measuring the thickness in holes for rock bolts
- probe drilling through hardened concrete
- utilising probe sticks or wires, and spraying until these are covered
- with laser scanning before and after spraying.

At the time of this study, the regulating document Norwegian Publication no. 7 is being revised, and the updated draft (Norwegian Concrete Association, 2022) elaborates on scanning for layer thickness control in more detail. The document states that, with scanning, a finely meshed map of the scanned surface can be obtained, and layer thicknesses can be shown as colour maps where different colours represent different thicknesses. Furthermore, the document recognises that, with sprayed surfaces often being uneven and scanning being a very high-resolution method of measurement, the scan result can show small local areas where the layer thickness is too thin even when the average layer thickness around the area is sufficient. The document suggests that, when utilising scanning for layer thickness control, the average thicknesses should be calculated and shown on areas no smaller than eg from 10 × 10 cm to 25 × 25 cm. The document also states that another major advantage of scanning is that the operators can display and review the scan colour map before they finish spraying and move the spray rig away. They are able to identify areas where the layer thickness is too thin and correct these immediately by spraying an extra layer before moving the rig away.

Project documentation in E39 Rogfast

The project documentation (not publicly available) in the E39 Rogfast tunnelling project states that, in addition to what is prescribed in the Norwegian Publication no. 7, sprayed concrete layer thickness must be controlled with a laser scanner mounted on the spray rig. For the scan resolution it states that a minimum of 16 points per square metre will be measured. Reporting must be carried out both as a PDF file showing the layer thickness as a topological surface map (colour map), and as original scan data in the project coordinate system uploaded to the client's database no later than two days after the application.

Utilising a digital layer thickness control system at the E39 Rogfast Kvitsøy site

Contracting company Hæhre Entreprenør AS is excavating for the Kvitsøy contract package of the E39 Rogfast project. At the time of this study (November 2022), excavation of the 4.1 km access tunnel to the highway tunnel line is under way. The access tunnel profile is 10.5 m wide, and the prescribed average sprayed concrete layer thickness is 80 mm. The geology is mostly black shale and greenstone, which makes the drill and blast works challenging but provides a relatively strong self-supporting rock arch effect as is typical in the Nordic hard rock geologies. A georeferenced laser scanning-based sprayed concrete layer thickness system is in use on the site.

System set-up on sprayer

The system is mounted on a truck-based spray rig. A scanner unit for capturing the profile information is mounted in the back of the truck facing the application area, and an additional LiDAR sensor for georeferencing is mounted on top of the truck behind the cabin, with the open tunnel behind the application area in its field of view. A tablet computer for controlling the system and displaying the scan results is mounted on the side of the truck, near the rear end. The system is tapped to the 24-volt DC electric system of the truck carrier, and all the data communication is hardwired.

Scanning and spraying workflow

When starting the application, the spray truck is taken to the tunnel with the back of the truck (boom) facing the application area and the front of the truck (concrete pump) facing the open tunnel behind. Once the truck has been navigated to the correct position for the application area and stabilised with support legs, the scanning system can be started and the reference scan of the application area can be captured. When initiating the reference scan, the system will first scan for targets on tunnel walls for georeferencing, using the additional LiDAR sensor. After a successful scan of the targets, the system will automatically move on to capturing the reference scan with the scanning unit mounted in the back of the truck. The captured profile will be displayed on the tablet computer screen following a successful scan. This scan workflow for the reference scan takes approximately three minutes.

Now the profile is ready to be sprayed. The operator will spray a layer aiming for the prescribed layer thickness, which is 80 mm in standard rock conditions on this site. After finishing the spraying, the operator will initiate the comparison scan. This time only the scanning unit for the application area is used, and it will capture the profile information over the sprayed area. The calculation of the layer thickness is automated in the workflow following the comparison scan, and after this a change map showing the layer thickness over the whole captured profile is shown on the tablet computer screen. The operator can pan and tilt the 3D point cloud colour map (Figure 3) on the touch screen and review the layer thicknesses. If satisfied with the layer thickness, they can then move the rig to the next spraying position or away from the area. If additional layers are required on some areas of the profile, the operator can spray over these and scan for a second comparison scan. This scan will again be automatically calculated against the original reference profile, and a new change map will be shown on the tablet computer screen. All the scan data from a specific sprayer position is saved to a folder that is automatically created in the system database when the scan project is started with the reference scan. The operator can input relevant information such as application area chainage metres and this will be shown in the project folder name.

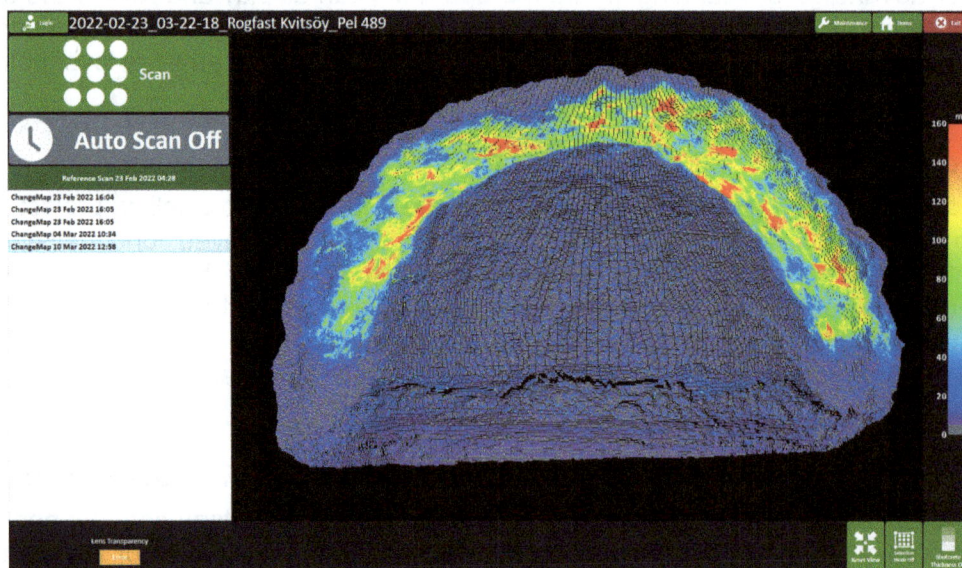

FIG 3 – A layer thickness change map in onboard software.

Georeferencing system description

Georeferencing the scan results is based on picking up surveyed targets on tunnel walls with an additional LiDAR sensor mounted on the spray rig. The targets are mounted on the tunnel walls and surveyed with a standard total station, and target coordinates are uploaded to the scanning system software on the tablet computer prior to operating the system. Creating and expanding the target network is carried out as a separate task, making sure that there are targets within the field of view of the LiDAR sensor on all upcoming application areas. The items of scanning hardware on the rig (LiDAR sensors for both picking up the targets and scanning for profile information) are calibrated in relation to each other, and the system will automatically calculate coordinate locations for each point in the scan point clouds during a successful scan project. The software of the scanning system itself still only shows the scan results and change maps in the local coordinate system, but point clouds can be exported to the global or project coordinate system (in the system in which the targets on tunnel walls were surveyed).

Data processing and reporting

The scan results are used in this project for both fulfilling the contract requirement for documentation and improving the control over sprayed concrete layer thicknesses. After finishing the spray application in the tunnel, the scan project files are transferred to the survey engineer's desktop computer using a USB stick and uploaded to a desktop software program of the scanning system. Here, the individual change map point clouds are cropped to show only the area that was sprayed, and a PDF report is generated. A cropped thickness map is shown in Figure 4.

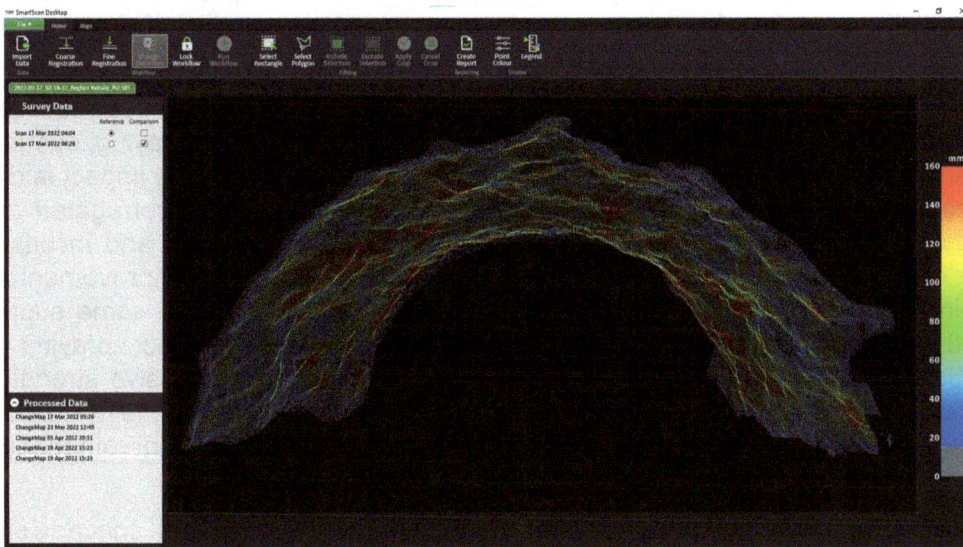

FIG 4 – A cropped layer thickness change map in desktop software.

The PDF report shows the colour map of the thicknesses, calculated average thickness over the measured area and sprayed concrete volume on the wall. This PDF report is delivered to the client as documentation and compliance of the sprayed concrete work. The georeferenced point clouds of the captured profile surface are exported from the desktop software and uploaded to the project 3D model, overlaid over the theoretical tunnel model.

Benefits and challenges of using the system

Generally, the experienced operators in the project are capable of achieving the correct layer thickness even without the system, but the system improves the checking and documentation of the layer thicknesses during the application process and reduces the time required for these tasks later in the tunnelling cycle. With inexperienced operators, the control over layer thicknesses plays a greater role. The biggest impact on the operator's work is that they can be sure of the layer thickness being acceptable and documented before leaving the application area, as they see the same colour map that will later be delivered to the client. Probe drilling through hardened concrete can still be carried out to double-check.

Using the system requires some extra work from the operator, but at the same time it saves effort on the part of the survey engineer. The closest substitute as a means of obtaining the required scan results would be to have the site survey engineer go and scan with a terrestrial laser scanner prior to and after spraying. This would not precisely fulfil the contract requirement and would mean quite a significant amount of extra work as the site operates seven days a week and 24 hours a day, and concrete spraying is often done during the night shift. In addition, the benefit of checking the layer thickness before leaving the application area would be lost.

Digital layer thickness control in Australia

Australia is one of the main markets utilising laser scanning in sprayed concrete layer thickness control. Most of the specialised systems are in use in mining development tunnelling, both with mining companies and contractors. Mining contractors in Australia have also been at the forefront of developing specialised laser scanning-based systems for layer thickness control, with a focus on giving the nozzle operator a tool to monitor and report the layer thicknesses quickly and efficiently using automated PDF-reports.

To the author's knowledge there is at least about 100 specialised sprayer-mounted laser scanning-based layer thickness control systems in use on the Australian mining and tunnelling markets. The number can be bigger.

The role of operator competence

The role of operator competence and motivation cannot be stressed enough in this context. Because of the nature of the sprayed concrete process, the quality of the application process has a huge impact on both the amount of excess concrete usage, and the quality of the finished sprayed concrete lining.

Correct application techniques and careful working practices can reduce excess concrete consumption in the aforementioned categories of excavation quality, overspraying, overlap, rebound and waste and spill. In the case of overlap, rebound and waste and spill the impact is direct; in the case of excavation quality, a competent operator can cope better with corrugated and uneven surfaces, whereas an inexperienced operator will produce more rebound and inconsistent layer thicknesses over the challenging surfaces. In the long run, a general improvement in sprayed concrete lining qualities can even affect the first category, design, as the same support can be achieved with thinner linings. In addition, as carefully taking account of correct spraying parameters and application techniques produces higher and more consistent compressive strength results in final structures, the amount of cement in mix designs can be reduced without the risk of not achieving the desired strength values. The effect of reducing cement content is significant with regard to carbon reduction.

Despite the developments in computer-assisted and automated boom control, concrete spraying can still be considered a highly manual task, where skill development is slow and can only be achieved by acquiring spraying experience. New, inexperienced operators have the highest risk of safety hazards and more challenges in manoeuvring the spray rig and boom when developing their spraying skills. This affects their ability to produce high-quality sprayed concrete linings safely, efficiently and with low excess concrete consumption.

VR simulators in operator training

Modern virtual reality (VR) and computer technology has reached a level where highly realistic simulations can be run with a light and portable simulator set-up. Digital control systems and remote control over controller area network (CAN bus) in real sprayers make it possible to use actual sprayer remote controls and control system software in simulators (Figure 5). These so-called hardware-in-the-loop (HIL) simulators have the benefit of offering an extremely realistic experience for the users and familiarising them with the exact same controls and settings as the real equipment. The machine models are based on real mechanical and hydraulic designs and the parameters can be adjusted. The same digital twins can be used in R&D for the equipment models.

The VR simulator can be used either in beginner operator training and assessment, or to provide refresher training for experienced nozzle operators. VR simulator assessment can also be part of operator certification schemes such as the EFNARC C2 operator training and certification scheme (EFNARC, 2020).

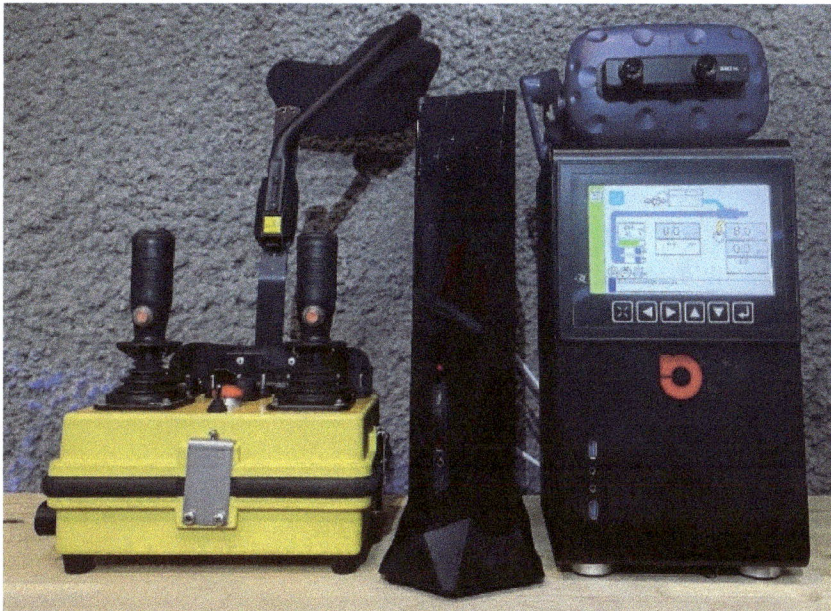

FIG 5 – VR concrete spraying training simulator hardware.

The goal of operator training should not only be to give the candidate basic skills regarding boom and nozzle controls and setting up spraying parameters. Building a comprehensive understanding of the importance of sprayed concrete as rock support for safety or as the final lining in an underground facility plays a key role in motivating the operator to produce high-quality linings with great pride in their craftmanship. Having a good understanding of the cost and environmental impact of choices made during spraying can help strengthen this motivation to strive for the lowest possible concrete consumption and best possible quality. However, the ultimate goal and greatest pride for the nozzle operator should be the fact that the next person entering the underground space, and everyone coming after them, can feel confident that they are completely safe.

DIGITAL ECOSYSTEM SHAPING THE FUTURE OF CONCRETE SPRAYING

The aforementioned digital tools and new technologies already have their place in improving the sprayed concrete process for a more sustainable underground excavation cycle.

What we are likely to see in the near future is that, as well as taking leaps of their own, these technologies will also be integrated into a digital ecosystem revolving around the concrete spraying process and other tunnelling cycle processes. We are already seeing data being fed from different digital systems to common data platforms for analysis, reporting and optimisation, but we are not yet seeing many of the separate systems truly interacting with each other in real time for direct process control and improvement. This type of digital data integration to spraying process control opens completely new possibilities to develop process optimisation tools.

CONCLUSIONS

It is easy to state that there is significant potential to reduce concrete consumption in sprayed concrete operations if the whole process is not at an optimal level. But, as the optimal or minimal levels of excess concrete consumption in different root cause categories are highly affected by characteristic conditions on each site and in every project, it is difficult to define specific levels that can be achieved by optimising layer thickness control and the application process. Conclusions can be drawn from empirical experience gained over the years, but they often apply only in specific conditions. It can also be assumed that, so far, these optimal levels have been achieved using mostly traditional methods, so the potential for further improvement is still there and could even be expanded by utilising new technologies. Sprayed concrete is yet to achieve a factor of 1.0 between theoretical and actual concrete consumption.

More or less all standards and guidelines globally require some level of reporting of sprayed concrete works, usually including at least the most important quality information and location of said works. Based on the Rogfast case study information, a scanning-based layer thickness control system can automate at least part of the work needed for as-built documentation and reporting. Recorded quality data can also be more accurate and more comprehensive. In the case study learnings, this is regarded as a remarkable improvement in process control and optimisation even without the requirement of digital reporting and as-built documentation from the client. However, more study and on-site measurements should be done to quantify the realised savings in concrete consumption.

Further study should also be focused on determining how much wastage each root cause introduced in the chapter 'Saving potential in sprayed concrete consumption' generates in different conditions. This would make it possible to estimate the saving potentials in order to improve methods and technologies more accurately, so that the wastage can be reduced and the improvement in terms of cost and environmental impact can be quantified.

ACKNOWLEDGEMENTS

The presenting author would like to acknowledge the support received from the whole personnel of Hæhre Entreprenør AS working on the Kvitsøy tunnelling site, in both driving for innovation and development for improvement, and in putting together this paper.

REFERENCES

Concrete Association of Finland, 2015. BY63 Guidelines for sprayed concrete 2015 (Suomen Betoniyhdistys, 2015. BY63 Ruiskubetoniohjeet 2015) chapters 7.2.5. and 9.6.8. Available only in Finnish.

Concrete Institute of Australia, 2020. *Shotcreting in Australia - Recommended practice*, third edition.

EFNARC, 2020. EFNARC C2 training and certification plan. Available online: <https://efnarc.org/publications>

European standard EN 14488–6, 2006. Testing sprayed concrete part 6: Thickness of concrete on a substrate.

International Tunnelling Association (ITA), 2020. ITA report n:o 24, Permanent sprayed concrete linings, International Tunnelling Association Working group no. 12 and ITAtech, October 2020, pp 21–22. Available online: <https://about.ita-aites.org/publications/wg-publications/content/17-working-group-12-sprayed-concrete-use>

International Tunnelling Association (ITA), 2022. BIM in tunnelling, guideline for bored tunnels – vol 1, International Tunnelling Association Working group no. 22, Information modelling in tunnelling, p 18. Available online: <https://about.ita-aites.org/publications/wg-publications/content/208-working-group-22-information-modelling-in-tunnelling>

Norwegian Concrete Association, 2011. Publication no. 7 – Sprayed concrete for rock support, August 2011 (Norsk Betongforening, 2011. Publikasjon nr. 7 – Sprøytebetong til bergsikring, August 2011). Chapters 1.5.2.4, 2.2.2, 2.2.3, Available only in Norwegian.

Norwegian Concrete Association, 2022. Publication no. 7 – Sprayed concrete for rock support, revision draft, September 2022 (Norsk Betongforening, 2022. Publikasjon nr. 7 – Sprøytebetong til bergsikring, Høringsutkast kun til høringsformål, September 2022). Chapters 1.5.3, 2.3.1, 4.5.2. Available only in Norwegian.

Tasmanian Department of State Growth, 2021. *Section 684 – Sprayed concrete*. January 2021. Chapter 684.13. Available online: <https://www.transport.tas.gov.au/roads_and_traffic_management/contractor_and_industry_information/specification_listings_-_standard_sections>

Mechanised, wireless development charging – Orica's and Epiroc's Avatel™ charging solution

B Taylor[1] and M Adam[2]

1. Manager, New UG Technology Commercialisation, Orica, Melbourne Vic 3000.
   Email: ben.taylor@orica.com
2. Manager – Domain, New Technology Commercialisation, Orica, Melbourne Vic 3000.
   Email: martin.adam@orica.com

ABSTRACT

Although the underground mining environment is arguably safer now than at any time in history, as underground mining operations across the globe increase in-depth, magnitude, and complexity, protecting workers from harm becomes increasingly challenging. In deep, stressed, seismic mines the development face is one of the most hazardous work areas. The active face is usually the furthest extent of the underground workings, so ventilation can be difficult to maintain, exposing workers to high temperature and humidity while working on foot and doing manual labour. While mechanisation has made many tasks easier in modern times, the reality of underground mining is that development charge-up workers still spend long periods directly exposed hazards at the face. In 2019, Orica and Epiroc partnered to co-develop the world's first fully mechanised development charging unit using fully wireless initiation, called Avatel™. Built on Epiroc's M2 Boomer carrier platform, the system is designed to allow one person to charge a development face from the safety of a fully enclosed cabin, thereby reducing exposure to hazards at the face. The system was prototyped in Europe before entering commercial introduction in 2022. This paper reviews earlier attempts to mechanise underground charge-up and describes the development journey and engineering challenges of building a mechanised development charging machine.

INTRODUCTION – THE MECHANISATION IMPERATIVE

In underground mines, working at the development face exposes people to hazards including rockfall, heat, noise, dust, inrush, trip hazards, heights, and interaction with machines. As easier-to-mine orebodies are progressively exhausted, producers look for ways to safely mine deeper and more difficult deposits. Although most underground mining activities such as scaling, mucking, development drilling, production drilling and installing ground support are already mechanised, and some are automated, production and development charging still have great potential for mechanisation and automation.

Development charge-up involves cleaning and charging horizontal blastholes in a tunnel face. Faces in large, mechanised mines are usually five or six metres high, and a similar width. Typically, a crew of two use an elevating work platform to access blastholes, with one person in the basket and one on the ground. The person in the basket is exposed to hazards including pinching or crushing between the basket and the wall or backs, falls from height, heat, and geotechnical hazards including falling or ejecting rocks. Workers at the face are exposed to noise, dust and debris when cleaning holes with compressed air. The person on the ground is exposed to hazards including being struck by the basket, falling or ejecting rocks or equipment, manual handling injury, heat, tripping and falling. Both are exposed to the hazards associated with handling explosives, including premature initiation due to friction, impact, static and heat. The task of charge-up is not only hazardous – it requires intuition, intelligence and improvisation. Sometimes the pattern drilled does not match the design, and there may be extra holes or missing holes. Remnants of holes from the previous round (butts, bootlegs or sockets) look like blastholes to be charged, but they are not. These must be identified and marked.

The holes in the floor (lifters) are often concealed by broken rock and water. If the lifters are not kept open with HDPE pipe (commonly called lifter tube), a worker must pump the water away and manually clear the buried holes using a pick or shovel. This is hard physical labour that exposes the worker to many hazards for long periods. Data from the mining regulator in Queensland, Australia shows that more people are injured while digging out lifters and charging development blastholes than during any other activity at the face (Figure 1).

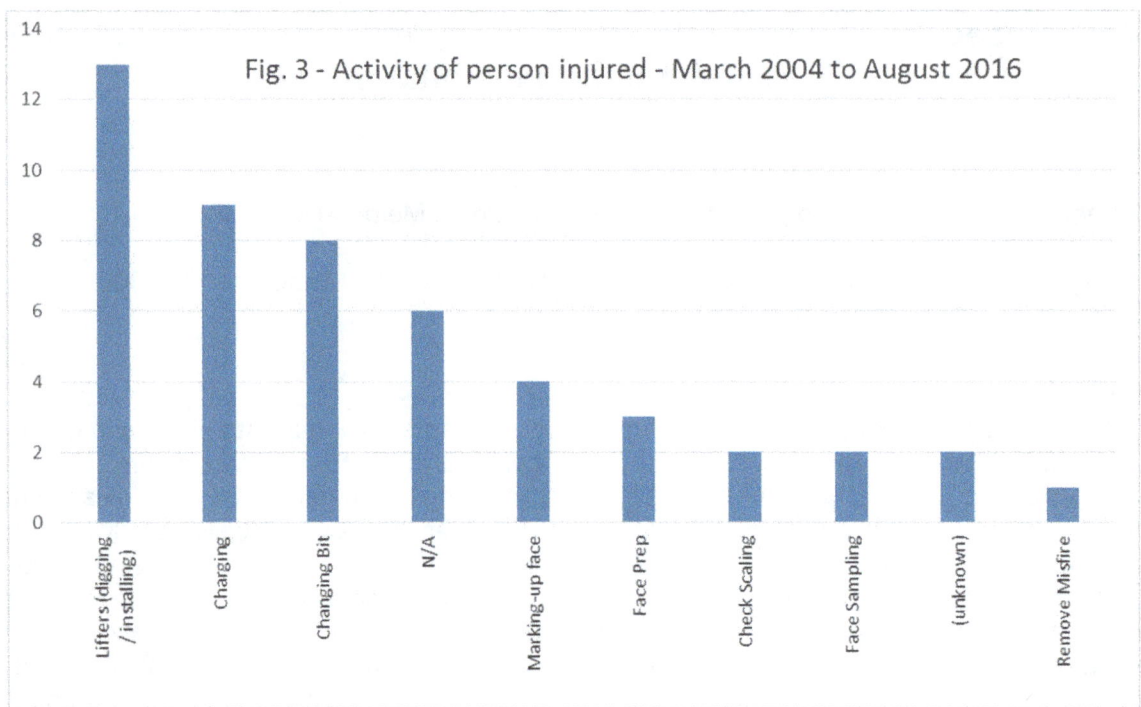

FIG 1 – The highest number of injuries in development headings occurs when digging out the lifters and charging holes. Source: Queensland Department of Natural Resources and Mines, 2016.

In the last 30 years others have made many attempts to mechanise and automate development and production charging. The only notable improvements are the introduction of mechanical hose pushers for charging long upholes, integrated elevated work platforms on charge-up machines, and emulsion pumps. These changes have reduced components of manual handling, such as lifting ANFO bags and pushing long hoses up holes. They have also reduced the duration of the charge-up task, thus reducing exposure time at the face. However, both production and development charge-up still require people at the face or in the drawpoint, either on-foot or in a basket.

Today, the most widely used system of charge-up uses bulk emulsion explosive initiated by a non-electric (signal tube) detonator. Each hole is designed to contain a specific delay detonator that matches the sequence of the blast. The charge-up worker inserts the primer on the end of the charging hose and pumps emulsion into the hole. After all the blastholes are charged, the operator connects the signal tubes hanging from the holes to a loop of detonating cord that spans the face. Later, the detonating cord is connected to a starter detonator, which is in turn connected to a firing cable or remote firing system.

Selecting the correct delay for each hole and connecting the signal tubes to the detonating cord are manual processes that require care and attention to detail to avoid misfires and unplanned initiation. There are hazards associated with handling signal tube and detonating cord around machines that can lead to unplanned initiation. The authors are aware of at least two documented incidents of signal tube initiating after entanglement with a machine during charging underground since the year 2000. The US MSHA published a report (see Appendix 1) of a suspected snap-slap and shoot at an underground quarry owned by Fred Weber Inc on 24 March 2018. Two men were injured, one seriously. In February 2019 a snap-slap and shoot event occurred during production hole charging in the Apatit mine in Russia. The event was recorded on video from the cab of the charge-up machine. Both miners survived because the blasthole only contained a small toe charge. (Details of this event are not in the public domain, but the authors have personal knowledge of it).

To reduce the hazards to charge-up workers in development, many mines now mesh and shotcrete each face after scaling. This extra effort significantly decreases the advance rate and increases development costs. Further, steel contaminates the muck from every round causing blockages in crushers and damage to conveyors. Moving the charge-up worker away from the face offers the potential to eliminate face support, with significant cost savings and productivity improvements.

THE MECHANISATION CHALLENGE

There are reasons why development charge-up is one of the last activities to be mechanised – it is a complex process in an extreme environment. Locating and identifying blastholes in a face, especially lifters, requires strength, dexterity, and intuition. Blastholes must be distinguished from butts (bootlegs, sockets), that are easy to misidentify as holes. Lining up and inserting the hose into a hole requires vision, pattern recognition, positioning and precise multi-axis tool control that is trivial for a human or factory robot, but hard for a hydraulic underground mining machine. Handling and connecting wired or signal tube detonators requires human dexterity and intelligence, especially if the wire or tube tangles. Primers and detonators must be kept separate until they are put in the blasthole, so they are usually assembled by hand.

Mechanisation, remote operation and automation are steps on the pathway to fully autonomous operation. Full autonomy means the machine will travel to and charge a face without human intervention. Partial automation means some parts of the cycle occur without human intervention – for example, the machine locates blastholes and inserts the hose to clean the hole. Remote operation refers to a human controlling the machine from somewhere other than on the machine. Remote operation doesn't require automation, but partial automation makes remote operation more efficient, and may enable one operator to control many machines. A modern example is a remotely operated bogger (scoop) that is controlled by a human while filling the bucket and dumping into the truck, but which trams between the stockpile and truck autonomously.

Earlier attempts to solve charge-up mechanisation

Australia's CSIRO had a project called Blasthole Charging Automation in the early 2000s. The aim was to automate the uphole charging process in underground mining. CSIRO's role was to develop a hydraulic arm to insert the emulsion charging hose into the pre-drilled hole. Bonchis *et al* (2014) describe developing a six degree-of-freedom robot arm with a laser scanning system to identify blasthole collars and insert a charge hose with a primer assembly into a production uphole. The system required a human in the loop to verify blasthole identification. Bonchis *et al* note key learnings from the project included the limitation of cameras in the underground environment, detecting holes, matching holes to plan, identifying collars of holes that are not to be charged, and calibration and precise control of a robust industrial hydraulic multi-axis arm.

Hermann and Elvøy (2004) describe a joint venture to develop an automatic charging system for emulsion explosives suitable for tunnelling. The aims were to eliminate human activity in front of the jumbo drill during drilling and reduce the time for drilling and charging by 20 per cent by automating the charging process. The joint venture completed a feasibility study in 1997 and a prototype in 2003. The project attempted to combine the drilling and charging system on a single machine. It used a vision system to recognise drilled holes and position the charging tube. As the system used non-electric delay detonators, an operator working from behind the jacks and outside the cab had to select each delay and manually feed it into the system. The system used string loading to achieve different energy density from blasthole to blasthole across the face. In the conclusion of their paper, Hermann and Elvøy note the challenge presented handling wired or non-electric detonators:

> *The greatest challenge in achieving a complete automated system is connected to the selection of the numbered detonators for the individual holes and the coupling of the detonators. Electronic detonators will be advantageous. When the explosives industry someday can supply cordless electronic detonators, all-automated charging at face may become a reality.*

The Tesman Remote Loading Arm (Kovatera, 2023) is a machine attachment for charging bulk emulsion, ANFO and cartridge explosives in development rounds. It attaches to an underground carrier. The operator stands in a basket to handle initiating explosives, cartridges and change fittings. The machine keeps the operator at least five metres back from the face. The remote loader partly mechanises the development charging process.

In May 2020 im-mining.com published news of a successful test of a remote charging robot at Boliden's Garpenberg underground zinc mining operation in Sweden (Moore, 2020). The machine, developed by ABB, is described as an autonomous robot that automatically detects boreholes and fills them with explosives and detonators without the presence of humans. The machine uses a vision system, detonator cassette, and charging hose.

In an interview published by GBR (2022), Marco Ruiz Hernandez (Robotics Director, Enaex) stated that 'in April 2022 Enaex performed the first remote blast in an underground productive mine .....without any human interaction in the process at the mine face'. On their website, Enaex (2022) claims the UG-iTruck has an autonomous system designed to remotely perform all the tasks associated with loading explosives for development work, including a mechanised priming system and variable density emulsion, with the entire process being monitored from a remote control station or a centralised operations station. In an article published by Moore (2022), Hernandez further states the system can 'detonate without the need for tie-up since wireless electronic detonators are used'. In a presentation by Enaex to *Mineria Digital 2022*, a video shows the machine inserting a primer with a wire into a blasthole. The authors believe this system uses a wire-to-the-collar electronic detonator.

On their website, Olitek Mining Robotics (2023) describe a Remote Charge Up machine that prepares pre-drilled holes for charge-up and charges emulsion. An article by Weatherell (2021) says the Remote Charge Up unit is enabled by Olitek's unique 'Trigger Assembly' that enables lower cost conventional detonators to be mechanically, safely and efficiently installed.

An operational mechanised charge-up system

Avatel™ is a charge-up machine for development mining based on Epiroc's M2 twin boom development drill carrier (Figure 2). It allows a single operator to safely charge a development round from several metres away from the face. It dewaters, clears debris from lifters, allows an operator to position one of two booms over a blasthole, cleans the hole, inserts a primer and charges bulk emulsion explosive. The operator can manually allocate the charging and delay parameters of each hole or load a blast plan from blast design software. A digital control interface manages emulsion delivery and hose retraction of the emulsion process unit. By controlling the hose retraction rate, the unit is capable of string loading to decouple the explosive and lower the energy in the blasthole.

FIG 2 – Avatel™ is a mechanised, semi-automated development charging solution that uses Orica's WebGen™ wireless primers.

Handling detonator signal tube lead wires presented one of the biggest challenges to early efforts of charge-up mechanisation. Although some groups report solutions and work-arounds to this problem, the approach used in this system is to eliminate the wire and signal tube by adapting a truly wireless blasting system to mechanised applications. The diameter of the first-generation of wireless primers was too large for development mining, so a smaller version that can be assembled, encoded and loaded by this system was developed. A purpose-built onboard magazine stores the wireless primers.

Fitting the primer in the blasthole wasn't the only challenge faced during design. In small diameter blastholes the receiver of the wireless primer creates an inert void space within the explosive column. It is therefore critically important that the booster reliably initiates the surrounding emulsion at the toe of the hole to overcome the void space and initiate the rest of the charge. Many design iterations of the booster were required before a reliable design was found. The first version of the booster contained 50 g of pentolite. The detonation of the booster was filmed using an ultra-high-speed camera at two million frames per second. Testing revealed weaknesses in achieving a complete detonation of the booster's pentolite due to critically small internal geometries. The booster design team went back to the drawing board and worked collaboratively with the mechanical design team to develop a new version based on numerical modelling. The final version of the booster has a diameter of 34 mm and contains 92 g of pentolite. A demanding range of physical testing validated the design.

Clearing the blastholes drilled at floor level (the lifters) is one of the most arduous and manual tasks in development charge-up. As lifters are often under rock, mud and water it is common practice for the driller to insert a one to two metre length of polyethylene pipe in the collar of lifters as they are drilled to protect them and make them easier to find. Clearing the lifters usually involves pumping out the water and manually shovelling and scraping away rocks and mud. Lifters are often charged through the pipe, using either pumped emulsion or cartridges. Either way, finding and charging the lifters requires human dexterity and intuition.

A fundamental objective of the system is to reduce and then eliminate the need for personnel to spend time at the face. Hence, the intent is that the lifter tube will be eliminated from the process entirely. This adds complexity as the lifter holes are less protected and less visible to the operator. The system carries an on-board dewatering pump to remove water from the face, however the main challenge is moving debris away from the collars of the lifter holes to gain access for charging. Several mechanical options will be available depending on work environment, including steel brushes, picks or shovels, and an onboard compressed air system. It is likely some change to drill and blast practices will be necessary for productive use of the system. This may include modifications to drill and blast design or the supporting activities to maintain blasthole access, such as loader clean-up tasks.

Manually controlling the heavy, hydraulically driven booms with the required precision to align them with a blasthole on a development face several metres away is a significant challenge without positioning assistance. The system has aids to support the operator to efficiently position the booms. It will accept an IREDES file containing as-drilled coordinates from most smart drills. This file can be loaded into the machine to provide navigation to a hole's collar coordinates using three-point positional information. If as-drilled data is not available, a blast design can be loaded into the control system, helping the operator navigate to the design collar position and adjust manually if needed. Industrial video cameras on each boom send a live video feed to monitors in the operator's cabin. The combination of these systems allows for intuitive positioning of the booms. The future state of this system may include automated boom movements based on as-drilled data or 3D scans to locate blastholes and position the booms accurately and quickly.

The most complicated aspect of the development was the assembly and integration of the wireless primer. This presented many challenges, primarily around storing the components, assembling the explosive device mechanically and delivering it to the hole. A shuttling system brings primers from the magazine to the arms. Multiple design iterations were used to learn and evolve to the current design. The type of hose, and placement of the hose were changed multiple times before finding a reliable, functional combination. The primer storage must adhere to explosive design standards while also including moving parts for assembly. A mechanised, mobile magazine is not covered by any current known standard, so the design aims to adhere to widely accepted safe explosive storage and handling principles.

The system uses a proven bulk emulsion delivery system with a digital control system to meter delivery of emulsion to each blasthole. This method is commonly referred to as string loading. String loading is an effective method of decoupling energy to reduce undesirable damage to the rock mass surrounding the planned excavation. It is enabled by closed loop control between the progressive cavity pump of the delivery system and the hose retraction system.

During verification and validation of the machine in controlled surface and underground environments, many thousands of individual hole loading simulations were completed. In an underground test mine, the machine has loaded approximately 1200 blastholes in simulated full charging cycles. The test face had 114 × 45 mm diameter × 5 m length drill holes, with between 55 and 80 holes loaded in each simulated cycle. In total 16 simulated faces were loaded with an average cycle time of 2 hours and 10 minutes, and a fastest time of 1 hour and 30 minutes for a face of 71 holes.

The first live blast using the machine took place at an underground mine in Finland on 22 November 2022, where 62 wireless primers and 295 kg of ammonium nitrate emulsion was used to blast a development heading from a surface control room. Since that time a number of blasts have taken place as part of an extended trial campaign which, at the time of writing, is ongoing.

The system is not designed to completely replace a conventional development charge-up crew – it is intended to provide a solution for development charging in locations that are too hazardous for productive, conventional methods. The holistic view of this application is understanding where shifting the risk profile of the work at the face can impact efficiencies in other upstream or downstream areas of the mining cycle.

FUTURE DEVELOPMENTS

The first two production units of the system will be deployed in operating gold mines in Australia and Finland shortly after the date of submission of this paper. Future improvements include using machine vision to navigate around the face, identify each blasthole by name and automatically insert the charging hose. This will enable tele-remote operation, such that a single operator may be able to supervise more than one machine. Conceivably, the unit will be able to autonomously tram between headings and return to the magazine for reloading. As the technology evolves and the loading rate improves, versions will be developed for the civil tunnelling market where there is a higher focus on precision and cycle time. This may result in larger machines that better suit the tunnelling application. The trend toward battery electric vehicles (BEV) has also been considered in the design, with the carrier configuration largely compatible with a BEV driveline in its current form.

Solving the problem of autonomous development charging will eventually lead to fully autonomous production charging. In some ways, the case for autonomous production charging is more compelling and easier to solve than development. The drawpoint or edge of an open stope is an equally dangerous place to work, however wireless detonators have already retired much of this risk by enabling pre-charging from safer locations, further from the brow. Presumably it will be easier to automate location and insertion of the charging hose because production blastholes are larger. Production blasthole charging is sometimes more complex when decking is involved, and when holes are lost or intersect, and it is hard to envisage how a machine will intelligently identify these conditions and make decisions on-the-run without human intervention.

REFERENCES

Bonchis, A, Duff, E, Roberts, J and Bosse, M, 2014. Robotic Explosive Charging in Mining and Construction Applications, *IEEE Transactions on Automation Science and Engineering*, 11(1):245–250, Jan 2014.

Enaex, 2022. Underground Robotics [online]. Enaex ERA. Available from: <https://www.enaex.com/era/en/underground-robotics/> [Accessed: 02/02/2023].

Global Business Reports (GBR), 2022. Enaex Interview [online]. Global Business Reports (GBR). Available from: <https://projects.gbreports.com/chile-mining-2022/enaex-interview> [Accessed: 02/02/2023].

Hermann, R and Elvøy, J, 2004. Automatic charging of emulsion explosives to increase safety, productivity and quality, *Norwegian Tunnelling Society*, Publication No 13.

Kovatera, 2023. Tesman Remote Loading Arm, Kovatera. Available at: <https://www.minecat.com/products/attachments/tesman-remote-loading-arm.php>

Moore, P, 2020. ABB's autonomous remote charger robot shows potential in its SIMS demo at Boliden Garpenberg [online]. International Mining. Available from: <https://im-mining.com/2020/05/06/abbs-autonomous-remote-charger-robot-gets-sims-demo-boliden-garpenberg/> [Accessed: 02/02/2023].

Moore, P, 2022. Enaex claims world first completely robotic and autonomous underground explosives loading operation using its UG-iTruck [online]. International Mining. Available from: <https://im-mining.com/2022/04/22/enaex-claims-worlds-first-completely-robotic-underground-blast-using-its-ug-itruck/> [Accessed: 02/02/2023].

Olitek Mining Robotics (OMR), 2023. Remote Charge Up – Mechanised underground charge-up [online]. Olitek Mining Robotics. Available from: <https://www.olitekminingrobotics.com/charge/> [Accessed: 02/02/2023].

Queensland Department of Natural Resources and Mines, 2016. Managing rockfall hazards at development headings, Mines safety bulletin no. 159, 15 December 2016, version 1.

Weatherell, C, 2021. Olitek Mining Robotics and CMIC Announce collaboration on remote charge-up unit [online]. ReThink Mining. Available from: <https://www.rethinkmining.org/olitek-mining-robotics-and-cmic-announce-collaboration-on-remote-charge-up-unit/> [Accessed: 02/02/2023].

APPENDIX

APPENDIX 1 – This MSHA bulletin was forwarded to the author by a colleague.

Dilution optimisation in sublevel open stoping operations

A Vakili[1,2]

1. Principal Geotechnical Engineer, Mining One Pty Ltd, Melbourne Vic 3000.
   Email: avakili@miningone.com.au
2. Founder and Managing Director, Cavroc Pty Ltd, Melbourne Vic 3000.

INTRODUCTION

Dilution is associated with considerable direct and indirect costs in underground open stoping operations. As shown in previous works, an improved ability to predict dilution enables the economic risks associated with unplanned dilution to be reduced. Depending on the width of the orebody and depth of mining, the economic performance of an open stoping operation can become even more sensitive to dilution.

It is shown in several well-documented studies that as little as 1 per cent of additional dilution can impose millions of dollars of lost profits per annum. Optimisation of a stoping operation is therefore largely dependent on the ability of the operation to minimise dilution. This can only be achieved through sensible analysis of past and future performance of mined stopes using the most appropriate recent available technologies and tools.

This article presents a new design process that utilises advanced geotechnical numerical modelling and various other analytical tools to assist with a more robust mine design of open stoping operations. The focus of this design process is on geotechnical factors affecting dilution during the life-of-mine.

A recent application of this optimisation method for a mining operation is presented in this paper as a case study. Through the application of this method and various other operational improvements, the mine was able to deliver more consistent production and lower operating costs over its life-of-mine outlook and was able to reduce dilution by up to 50 per cent and achieve sustainable record high production rates.

Previous works on narrow vein orebodies completed by Stewart and Trueman (2008) shows the increasing economic effect of unplanned dilution for narrower ore widths (see Figure 1).

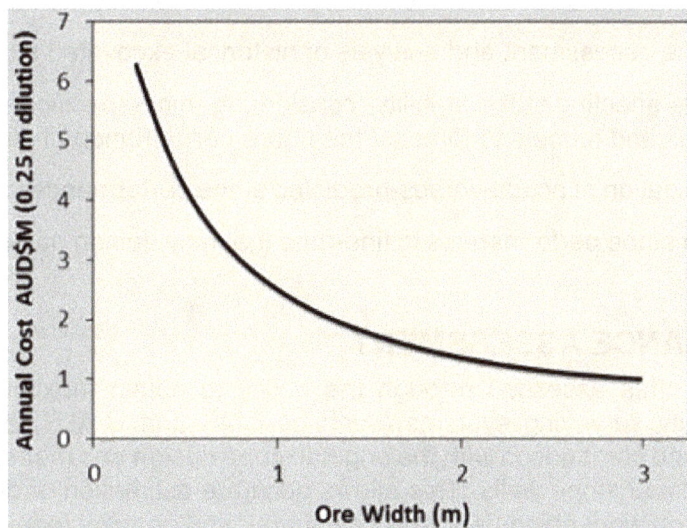

FIG 1 – Annual operating cost for 0.25 m of dilution for typical narrow vein mine (after Stewart and Trueman, 2008).

The relationship between dilution and direct operating cost at varying depths for a mining operation in Australia is also shown in Figure 2.

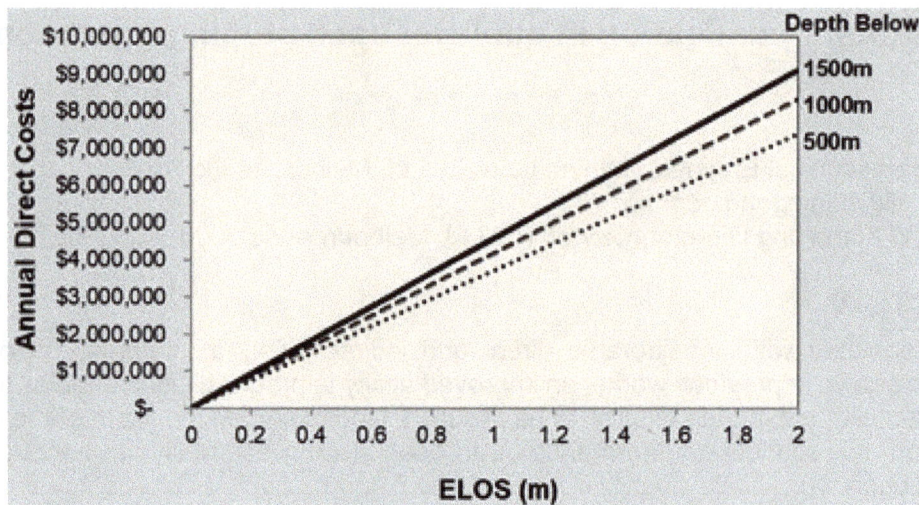

FIG 2 – Annual direct cost associated with dilution at various mining depths for a mining operation in Australia.

Despite the significant impact that dilution can have on the economic performance of underground operations, the standard practice for prediction of dilution is still based on outdated empirical and benchmarking techniques. These techniques have several limitations which are often ignored by many practitioners. As a result, many stoping operations have ongoing unresolvable dilution issues. In many instances, dilution issues are considered unpredictable or unmanageable, or they may be associated with localised rock defects that are not predictable.

Detailed geotechnical reconciliation and dilution optimisation are rarely completed. This translates to a missed opportunity for measurable improvements to overall mine economics.

The approach that is developed by the authors follows a structured process that can significantly improve the reliability of dilution prediction and can assist in finding optimum strategies to minimise dilution. This approach makes use of the most recent computing hardware and software advances in geotechnical engineering and computer simulations to replicate the geotechnical mechanisms that cause dilution.

The steps required for this approach are not complicated, and typically include:

- Stope performance assessment and analysis of historical excavated stopes.

- Analysis of factors affecting stope stability, considering mine-specific factors, and developing predictive methods and models to forecast the future performance of planned stopes.

- Calibration, or validation of actual-versus-predicted stope performances through back analysis.

- Forecast of future stope performances to fine-tune the mine design parameters and assist with planning.

STOPE PERFORMANCE ASSESSMENT

Stope performance is often assessed through the ability to obtain maximum ore recovery with minimum dilution. Cavity surveying systems (such as CMS and C-ALS) are critical in allowing detailed reconciliation and comparison with the original stope design and measuring stope overbreak or underbreak for individual stope walls. This allows accurate estimation of dilution and ore loss to be completed, building on work being performed by survey and geology teams.

For 'geotechnical' reconciliation assessments, however, this task is usually conducted manually. It is labour-intensive and time-consuming and is therefore limited to a few sampled stopes. As a result, there is usually an insufficient number of stopes being reconciled to enable a representative statistical analysis to be performed.

Authors make use of fully-automated computer programs and macros that can complete stope performance assessment from cavity survey data significantly faster and more accurately than other available techniques.

As shown in Figure 3, this technique uses a 3D meshing routine (similar to a block model) to fill the cavity survey and design wireframes, then overlay the two meshes on each other. A computer engine then calculates the differences between the two meshes to output the overbreak and underbreak values for each stope wall. The block models, geological wireframes, and orebody wireframes can also be incorporated into these calculations.

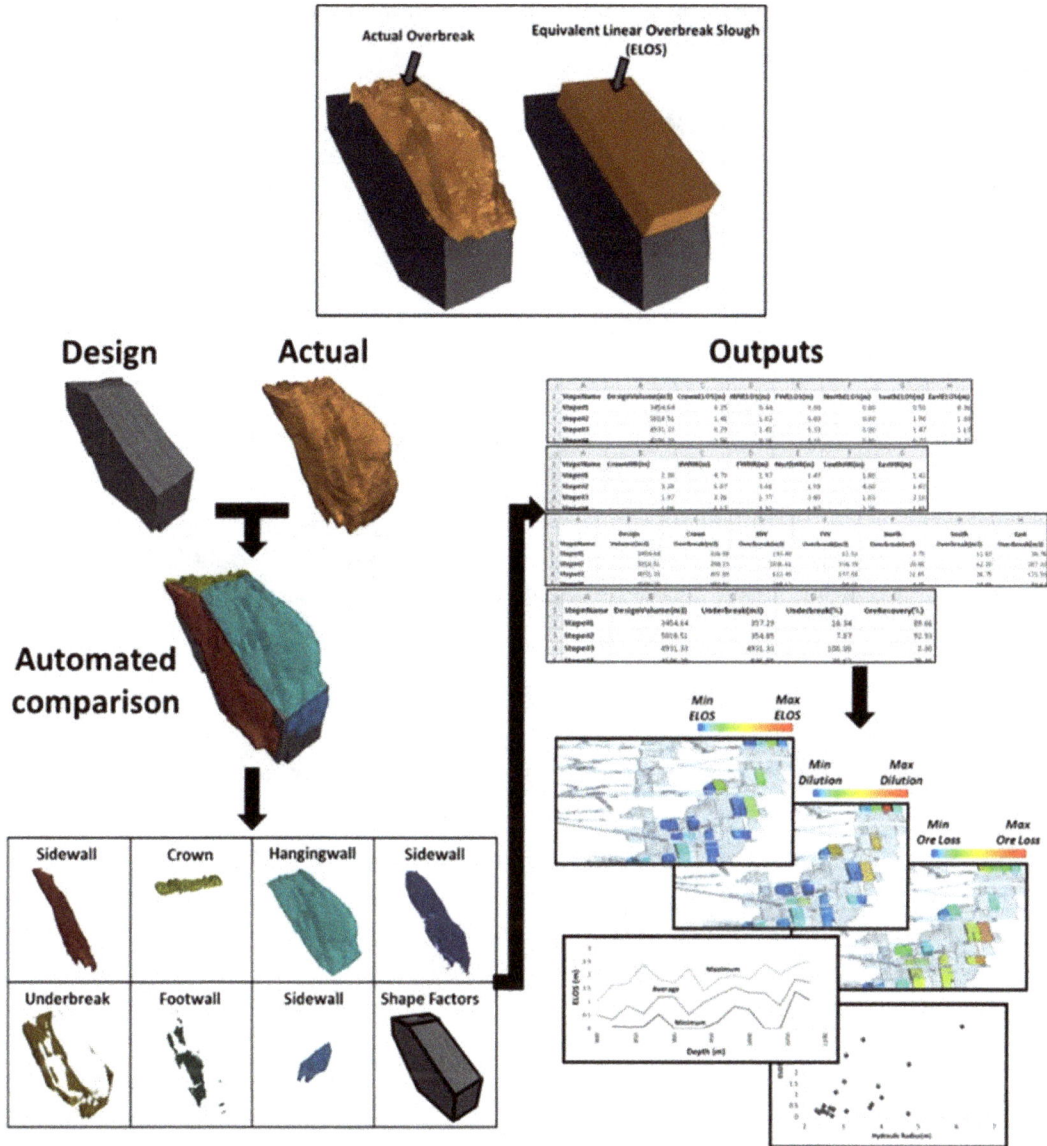

FIG 3 – Automated stope reconciliation assessment.

The outputs from this engine are equivalent linear overbreak slough (or ELOS as introduced by Clark and Pakalnis, 1997), overbreak or underbreak volumes, overbreak or underbreak percentages, and breakdown of material types (such as rock type, ore/waste, and fill) for all assessed stopes, with a breakdown of values for each stope wall.

This refines and enhances work being done for Mine to Mill reconciliation to identify geotechnical characteristics responsible for dilution. This method can complete the performance assessment for hundreds of mined stopes in a short period, generating a complete geotechnical reconciliation database.

ANALYSIS OF FACTORS AFFECTING STOPE STABILITY

Geotechnical factors are amongst the most important parameters affecting dilution. They include rock mass quality, mining sequence, induced stresses, mining depth, stope dimensions, stope geometry, orebody geometry, and proximity to major geological structures. These factors are very site-specific and vary largely from one mine to another. Therefore, for any predictive model or system to be effective, it needs to account for these factors on a site-by-site basis.

Therefore, for site-specific variables to be well understood, empirical and benchmarking systems may not provide reliable insight into the likely mechanisms affecting a particular operation. These empirical systems can only be used when all the limitations for specific mine sites are fully appreciated, and the possible sources of error are adequately quantified and accounted.

BACK ANALYSIS

For computer analysis, the proposed method makes use of state-of-the-art technologies in computational geomechanics and computer simulations to develop predictive models for different mining scenarios. In the proposed framework, FLAC3D (Itasca Consulting Group Inc., 2012) solver is used together with StopeX (Cavroc Pty Ltd, 2017). StopeX is a plug-in to FLAC3D which significantly simplifies the application of advanced numerical modelling and brings a much more advanced material model called Improved Unified Constitutive Model (IUCM). This makes it possible for all geotechnical practitioners, and not just specialist consultants, to learn and apply a more rigorous analysis than the current empirical and numerical tools.

Sufficient and high-quality geotechnical data is the critical input to these models to ensure reliable predictions. However, for any predictive model, back-analysis and calibration are necessary to ensure that models are validated against known past performance.

For back-analysis, the geotechnical stope reconciliation database is often the primary source that is used for comparison with the model outputs. However, often, other data such as development damage mapping, monitoring data and high-quality seismic databases are also used to compare the actual versus model performance.

FORECASTING FUTURE STOPE PERFORMANCE

The validated predictive dilution models are valuable tools for mine-design and to assist with short to long-term mine planning. These models provide vital input to mine planners and geotechnical engineers to:

- foresee operational bottlenecks resulting from unplanned stope dilution
- identify optimised strategies to minimise dilution
- optimise stope dimensions and geometries
- optimise stoping sequence
- compare and analyse various mining scenarios to identify trade-offs between production rates and expected dilution.

CASE STUDY

A recent application of this dilution optimisation method for a mine in Australia showed a great success.

Despite the fact that the mine is located in a challenging geotechnical environment, it was able to deliver more consistent production and lower operating costs over its life-of-mine outlook. Through systematic dilution optimisation methods, the mine was able to reduce dilution by up to 50 per cent and achieve sustainable record high production rates.

In the first and second year, before this method being implemented, the mine experienced increased stope hanging wall dilution. Individual stopes experienced ELOS exceeding 7 m (up to 40 per cent dilution). The total annual dilution reached 11 per cent in year 1 and 16 per cent in year 2.

Together with the site personnel a series of studies were completed over three years to address the dilution issue. The key milestones for this project were to:

- Identify the main contributing factors affecting the stope hanging wall dilution.

- Develop a predictive model to assist the mine in foreseeing the expected dilution resulting from planned stoping sequences.

- Evaluate the future long-term mining scenarios and assist with finding the optimum mining strategies that can maximise production and minimise the stope hanging wall dilution.

HISTORICAL STOPE PERFORMANCE ASSESSMENT AND IDENTIFYING KEY FACTORS AFFECTING DILUTION

The first step, to find the main controlling factors, involved a detailed analysis of historical stope performances. For this purpose, various information was analysed, including stope design and as-built shapes, stope-strike-length, mining sequence, rock mass quality and backfill sequence. Examples of these analyses are shown in Figure 4.

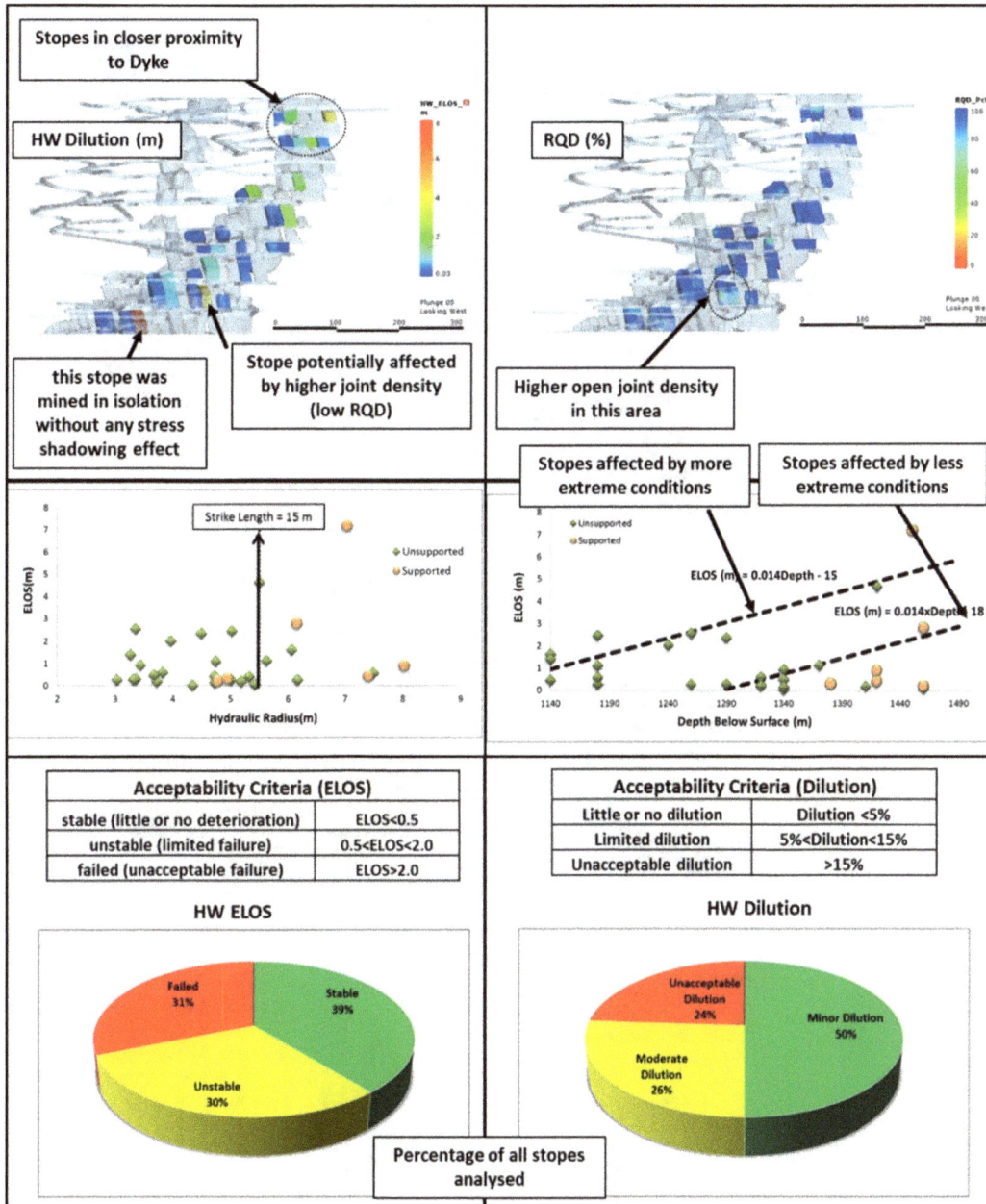

FIG 4 – Historical stope performance assessment (for year 1 and year 2).

The historical stope performance analyses identified some key factors and conclusions relating to the hanging wall dilution, including:

- The first stope on each level had increased dilution while subsequent stopes showed considerably lower dilution. Stress shadowing effect was found to be the leading cause.

- South to north stope retreat showed more dilution than north to south retreat. This was mainly associated with pre-mining stress orientation and geotechnical characteristics of the hanging wall.

- Stopes with a strike-length shorter than 15 m showed considerably lower dilution.

DEVELOPMENT OF A PREDICTIVE DILUTION MODEL

Once a good understanding of the previous stoping performance was attained, predictive models were constructed and validated against the past performance data. Figure 5 shows some of the outputs from the back analysis study and the correlation with the actual stope performance.

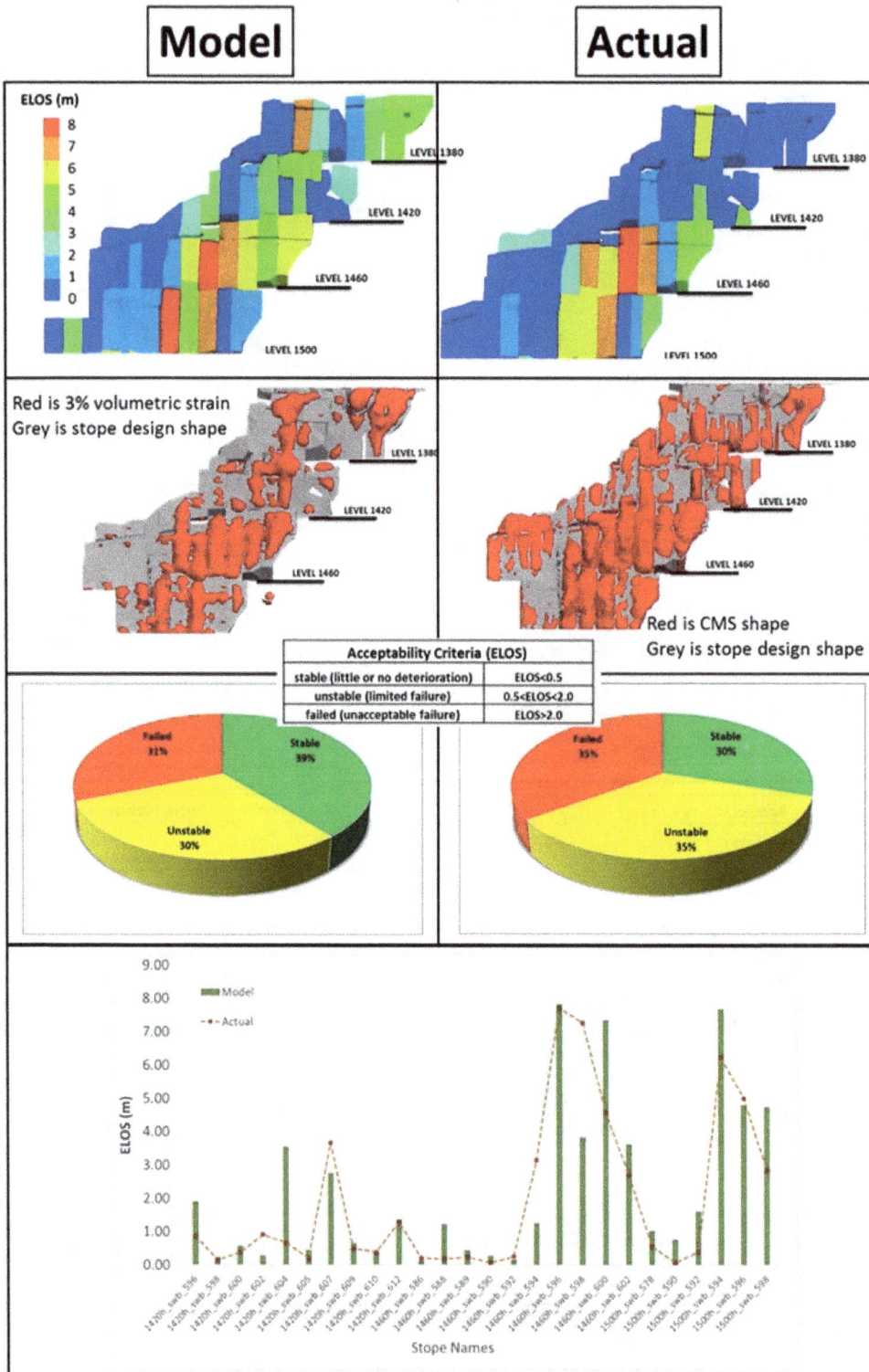

FIG 5 – Validation of the predictive model through back analysis of historical stope performance.

The modelling method plays an important role when dealing with dilution assessment in high stress or more complex rock mass conditions. At the subject mine site, the predictive numerical models prepared previously were not able to capture the observed hanging wall failure mechanism accurately. This was found to be largely due to the limitations of the applied modelling method.

In this framework, Mining One uses the IUCM (Vakili, 2016) modelling method that can account for complex geomechanical failure mechanisms causing hanging wall dilution. This method was able to capture many of the observed mechanisms leading to increased dilution at the subject mine.

The model was validated against the historical dilution data but also several other geotechnical performance indicators such as drive damage mapping, hanging wall SMART – cable monitoring and the seismic data.

STOPE SIZE OPTIMISATION

Following the initial investigation and review of historical stope performance, the validated dilution model was used to find the optimised stope layouts to minimise and control dilution. The following questions were raised to be answered through further analysis:

- What is the optimised stope strike length to manage unplanned dilution without any compromise on production rates?

- Considering the observed stress shadowing effect, is it necessary to have reduced stope strike length for all the planned stopes? Alternatively, is it possible to reduce the strike length only for the first stope on each level?

- Is it possible to optimise the excavation geometry of the hanging walls to manage the expected failure mechanism better?

To answer these questions, Authors, together with the site personnel, constructed and analysed several models and scenarios. These different scenarios were compared quantitatively, as shown in Figure 6, and the following conclusions were obtained:

- Reducing the strike length of only the first stope showed a significant reduction in dilution without any significant compromise on production rates.

- It was shown in the models that reduction of the strike length for the first stope on each level had a minimum impact on stress-shadowing effect and its benefits. Subsequent stopes on the level still showed considerably smaller overbreak than the first stope.

- Model results suggested that a de-stressing slot in the hanging wall of the first stopes can be very useful in minimising dilution, as it can create the same stress shadowing effect.

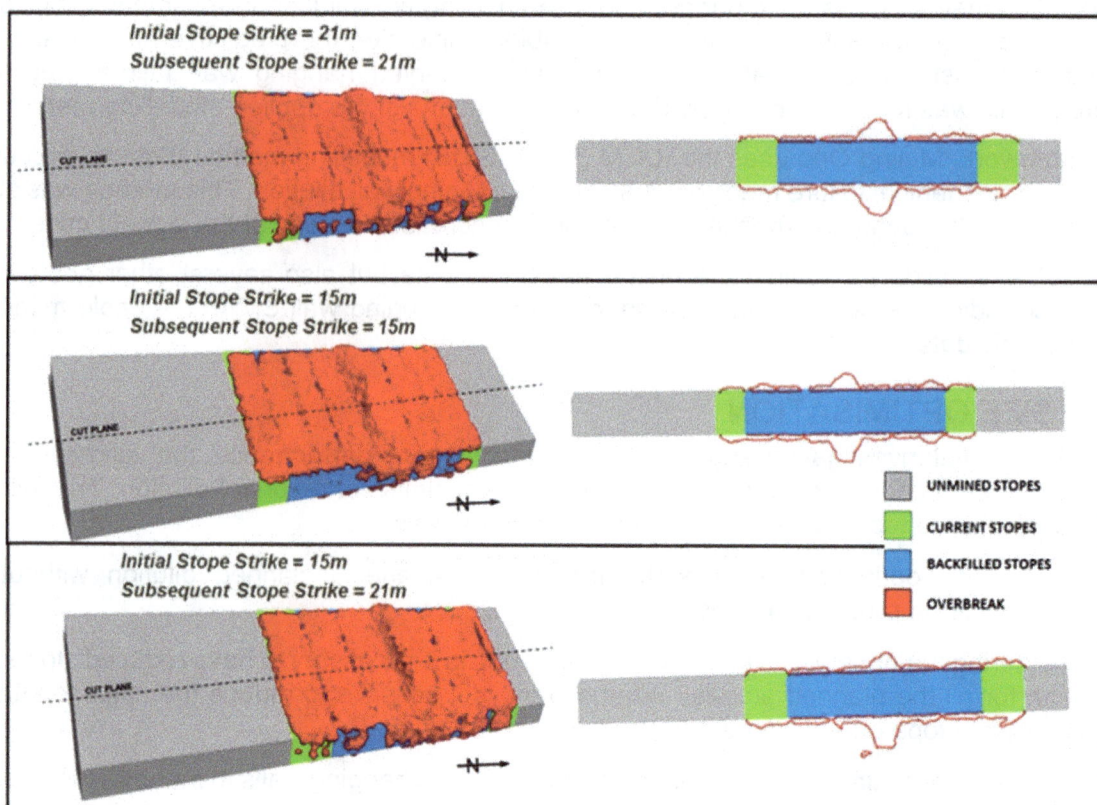

FIG 6 – Stope size optimisation analysis.

SEQUENCE OPTIMISATION

Historical mining identified two sequence-related factors that were considered to affect hanging wall dilution. First, south to north retreat, which is necessary when a centre-out chevron sequence is adopted. Secondly, due to the southerly plunge of the orebody and in order to maintain a centre-out down dip chevron sequence, the retreating front needed to be offset every two to three levels. The chevron offset was previously trialled which resulted in some adverse geotechnical conditions including hanging wall overbreak for stopes mined on the northern front.

From a purely geotechnical perspective, a north-to-south single end-retreat sequence was previously considered to be the most effective sequence to control hanging wall dilution. However, the single end-retreat has some shortcomings from a mine planning point of view, including less flexibility and fewer production rates.

As a result, it was decided to carry out a series of analysis to see if the chevron offset retreat sequence can be engineered or optimised. The aim was to consider a range of lead-lag options for the start of the offset, as well as different sequences regarding the formation of the closure pillar, extraction of the crossover stopes and subsequent mining of the remainder of the chevrons. Analysed options and comparisons with an end-retreat sequence are shown in Figure 7.

The following conclusions were made:

- The model with an offset distance of 42 m showed the lowest total dilution.

- Opening stopes on the offset level generated excessive dilution.

- An optimised chevron offset sequence can generate comparable results to a single end-retreat sequence.

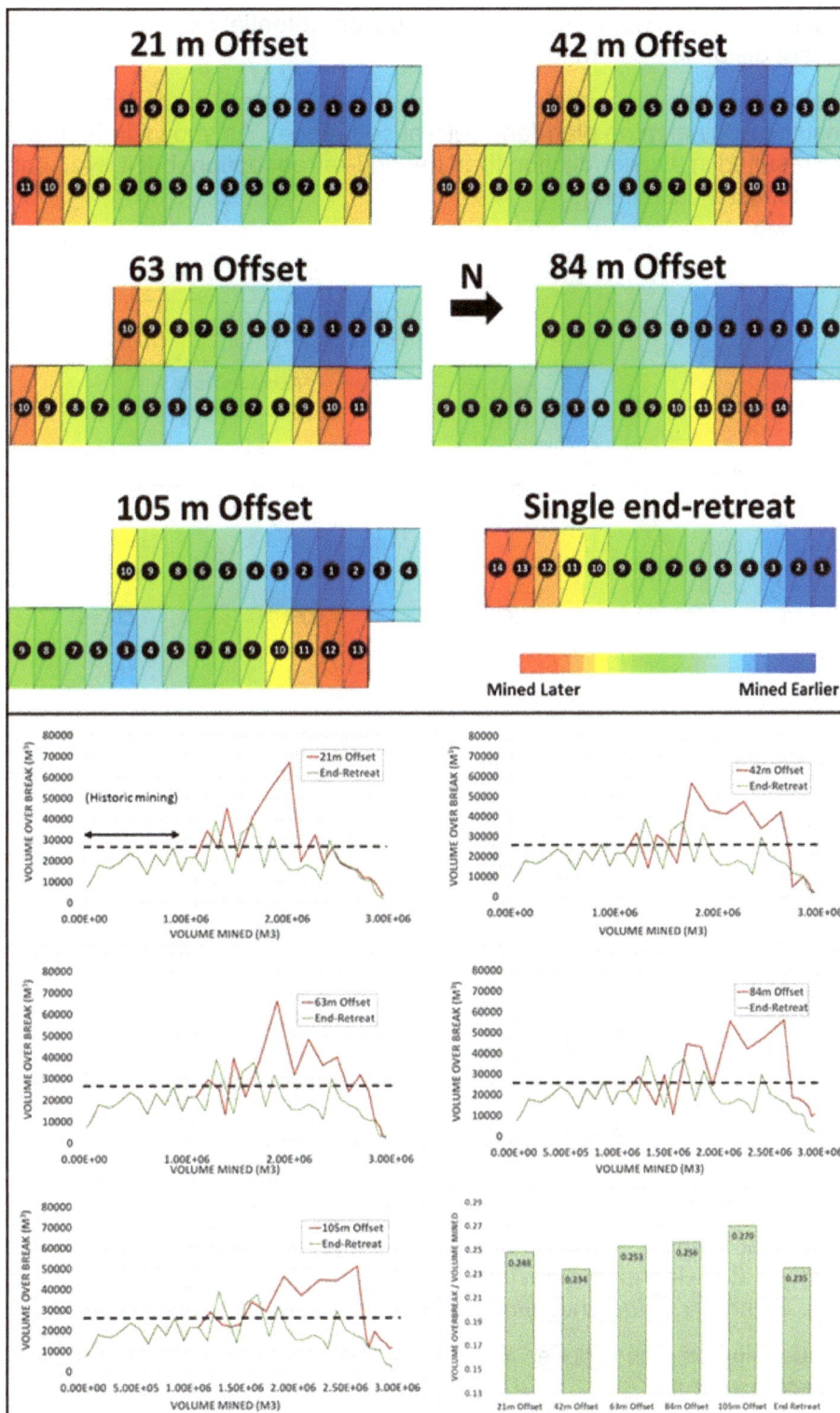

FIG 7 – Sequence optimisation analysis.

OPTIMISED MINING SCENARIO

As a result of sequence and stope size optimisation analyses, the mine site's planning team implemented some mine design measures to manage dilution better but also maintaining a sustainable production rate. These measures were included in an optimised mining scenario which included:

- a reduced strike length of the opening stope on each level from 21 m to 9 m
- a de-stressing slot in the hanging wall of the opening stopes
- a reduced strike length for the northern stopes (north of the opening stope) from 21 m to 15 m

- a chevron offset sequence, for maximum production potential from dual fronts on each level, using an offset distance of 33 m.

This optimised strategy was also analysed, using the calibrated numerical model, and it was compared against the original mine plan (end-retreat sequence). As shown in Figure 8, not only the optimised design resulted in a lower dilution than the end-retreat option, but it even showed reduced dilution compared to historical mining.

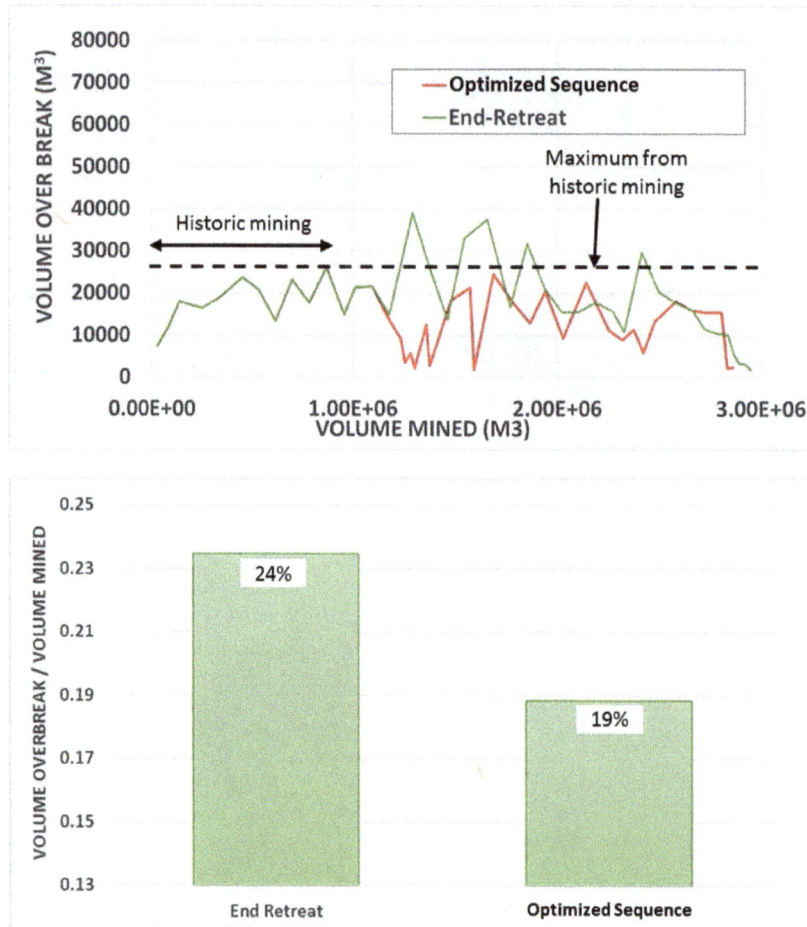

FIG 8 – Optimised mine design.

PROJECT OUTCOMES

The optimised mining scenario has been implemented at the mine site from year three on and results have been auspicious. Considerable reduction in dilution has been achieved, which was in line with model predictions. Summary of the critical improvements achieved are listed below:

- Overall annual dilution was reduced from 16 per cent in year two to 6 per cent in year three and 5 per cent in year four.

- Production was increased from 214 oz in year two to 267 oz in year three and 265 oz in year five.

- Cash operating costs were reduced from 690 (A$/oz) in year two to 609 (A$/oz) in year three and 592 (A$/oz) in year four.

Figure 9 shows the stope hanging wall conditions at the end of year four when the first offset level was successfully formed without any notable geotechnical problems.

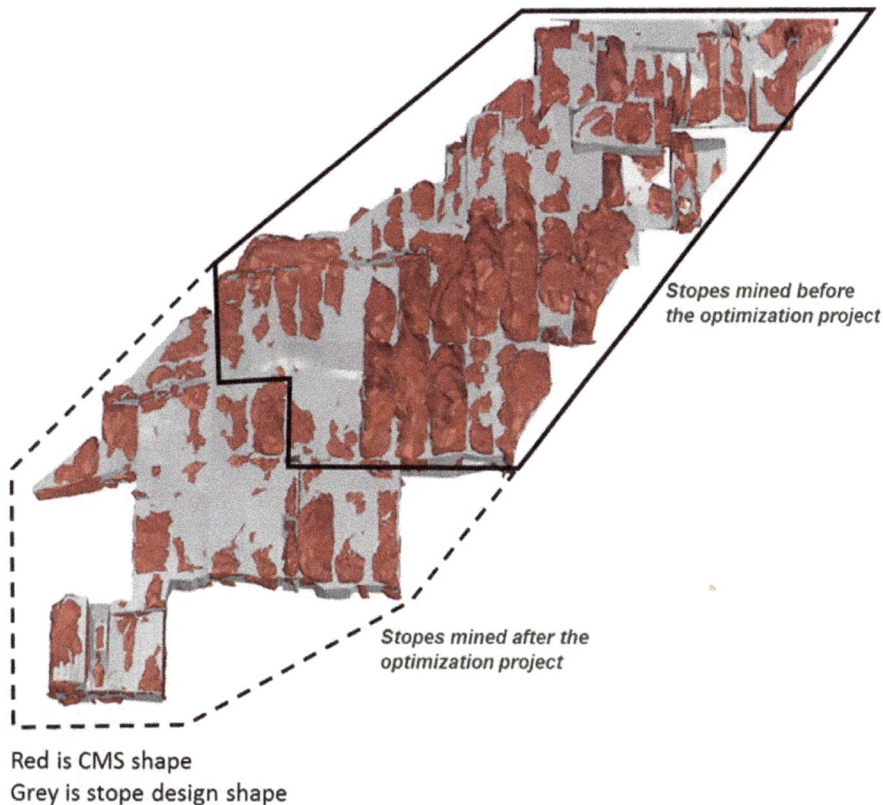

Red is CMS shape
Grey is stope design shape

FIG 9 – Stope hanging wall performance at the end of year four.

CONCLUSIONS

Dilution is a significant source of direct and indirect costs in open stoping operations. The ability to predict dilution quantitatively enables the economic risks associated with unplanned dilution to be managed much more effectively. Depending on the width of the orebody and the depth of mining, the economic performance of operations can become even more sensitive to dilution.

The empirical, statistical and benchmarking methods or the outdated analytical tools are way too simplistic to enable effective prediction of future stope performance. These conventional methods, at best, are suitable for understanding the factors controlling the historical dilution and have several shortcomings when used for forecasting future dilution.

This paper presented a proven method that makes use of advances in computational geomechanics in a more practical way. Mining One believes that application of this approach in other open stoping mines can bring about significant economic and safety benefits. However, this requires a fundamental shift in management culture and avoiding over-reliance on outdated design methods.

Mines are developing into deeper and more challenging geotechnical environments. It is impossible to expect that benchmarking and empirical methods, which were originally derived from case histories in considerably different geotechnical conditions, to be able to guide us for future mine design.

REFERENCES

Cavroc Pty Ltd, 2017. StopeExamine software plug-in, version 1.0. www.cavroc.com/stopeX.html

Clark, L and Pakalnis, R, 1997. An Empirical Design Approach for Estimating Unplanned Dilution from Open Stope Hangingwalls and Footwalls, 99th annual AGM-CIM conference, Vancouver, BC.

Itasca Consulting Group Inc., 2012. FLAC3D — Fast Lagrangian Analysis of Continua, software, version 5.0, Itasca Consulting Group Inc, Minnesota.

Stewart, P C and Trueman, R C, 2008. Strategies for Minimizing and Predicting Dilution in Narrow Vein Mines – The Narrow Vein Dilution Method, in *Proceedings of the Narrow Vein Mining Conference 2008*. (The Australasian Institute of Mining and Metallurgy: Melbourne).

Vakili, A, 2016. An improved unified constitutive model for rock material and guidelines for its application in numerical modelling, *Computers and Geotechnics*, 80:261–282.

Challenges and opportunities in accelerating electrification of underground heavy haulage equipment

J Webb[1], S Cribb[2], R Derries[3], M Keegan[4] and M Ratcliffe[5]

1. Practice Lead Mining Technology, South32, Perth WA 6000. Email: jayde.webb@south32.net
2. Program Director Net Zero Emissions (Acting), Newcrest, Melbourne Vic 3000. Email: siobhan.cribb@newcrest.com.au
3. Unit Manager: Innovation and Technology, Gold Fields, Perth WA 6000. Email: rob.derries@goldfields.com
4. Project Manager, Slate Advisory, Perth WA 6000. Email: michelle@stateofplay.org
5. Program Manager, Slate Advisory, Perth WA 6000. Email: madi@slateadvisory.com

ABSTRACT

The mining industry is relatively niche and complex compared to other global industries in its safety and operating environment. Whilst some of the technology required to decarbonise the mining industry will emerge from the mass market (eg batteries, energy storage solutions and renewable power generation) where exogeneous external investment and innovation will drive supply, bespoke applications such as material movement will be constrained by industry limitations.

Underground load and haul equipment burns the majority portion of underground mine diesel, and typically 25 per cent of scope one and two emissions. Transitioning away from incumbent diesel engines offers a considerable health benefit to all mine workers, as well as decarbonisation achievements, and through trials there is emerging evidence of more productive equipment and realistic expectations of lower operating costs of fleet. However, new equipment and material movement systems that support these industry goals are currently approached with caution due to the pivotal role they play in delivering productivity goals, and the lack of experiential data and certainty to mitigate the supplier development risk during this technology transition period, where all resources are required to meet a booming industry.

A group of mining companies and suppliers have the aligned aim to accelerate the understanding and availability of heavy haulage mobile equipment through collaboratively piloting equipment and working with suppliers and each other to aggregate market demand, encourage investment and resolve the mine design and technology choices through their involvement in the Electric Mine Consortium (EMC). This paper will outline the major challenges the industry must overcome in sourcing and adopting a zero-diesel particulate and zero-emission mine, drawn from cases and insights across ten EMC mining company members.

INTRODUCTION

Formation of the Electric Mine Consortium (EMC)

The Electric Mine Consortium was created in 2020, driven by a group of like-minded mining companies, with a goal to achieve a zero diesel particulate and zero carbon emission mine site. All at the start of their journeys, these companies recognised the opportunity to achieve their goals faster and as one large team via collaboration.

At that time, the EMC was comprised of five mid-tier miners and contractors, including South32, Gold Fields, OZ Minerals, IGO and Barminco. Since that time, it has grown to ten mining companies and 12 technology companies as outlined in Figure 1.

FIG 1 – Electric Mine Consortium (EMC) members, 2022 (Stanway, 2022).

EMC goals and working groups

The overall goal of the EMC members is to accelerate understanding and progress towards a fully electrified zero diesel particulate and zero carbon emission mine. As a group they defined the methods to achieve this, by:

- resolving technology choices
- shaping the supplier ecosystem
- influence policy makers
- communicating the business case to industry.

The members are assembled around six core and six enabling workstreams, as listed in Figure 2. These have been defined and are led by the mining companies in the consortium and are based on the big challenges they see needing to be addressed to achieve a zero diesel particulate and zero carbon emission mine. Work is completed by all members to achieve the workstream objectives and goals.

Challenge 1
Energy storage

Mine scale energy storage technologies are not yet operationally or economically proven in mining

Challenge 2
Electrical Infrastructure

Lack of understanding on the supporting infrastructure requirements for all electric equipment and vehicles

Challenge 3
Mine design

Traditional asset design does not enable the realisation of the full benefits of mine electrification

Challenge 4
Light BEVs & ancillary equipment

Economic and operating assumptions for light BEVs & ancillary equipment on site are unclear

Challenge 5
Underground haulage

Zero carbon load and haul equipment is not yet commercially available or technically viable underground at Australian scale

Challenge 6
Surface & long-road haulage

Large-scale, zero carbon surface and offsite haulage vehicles are not yet commercially available or technically understood

| Data | Valuation | Carbon | Skills | Policy | Communication |

FIG 2 – Electric Mine Consortium challenge areas (Stanway, 2022).

Electrification benefits

The benefits of electrification are multi-faceted, delivered through operational cost savings, emissions reductions and importantly a reduction in diesel particulate matter (DPM) exposure.

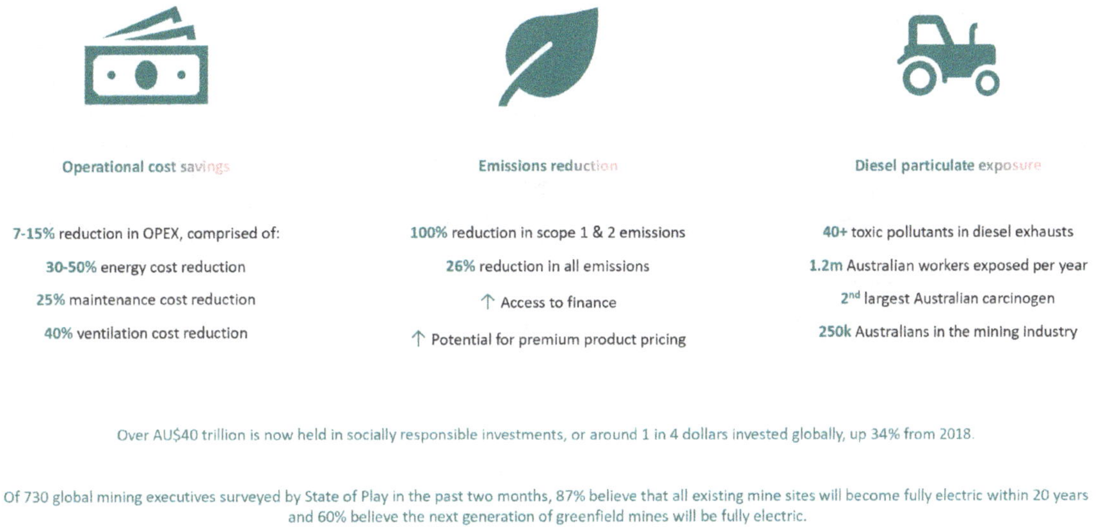

Operational cost savings
7-15% reduction in OPEX, comprised of:
30-50% energy cost reduction
25% maintenance cost reduction
40% ventilation cost reduction
Emissions reduction
100% reduction in scope 1 & 2 emissions
26% reduction in all emissions
↑ Access to finance
↑ Potential for premium product pricing
Diesel particulate exposure
40+ toxic pollutants in diesel exhausts
1.2m Australian workers exposed per year
2nd largest Australian carcinogen
250k Australians in the mining industry
Over AU$40 trillion is now held in socially responsible investments, or around 1 in 4 dollars invested globally, up 34% from 2018.
Of 730 global mining executives surveyed by State of Play in the past two months, 87% believe that all existing mine sites will become fully electric within 20 years and 60% believe the next generation of greenfield mines will be fully electric.

FIG 3 – Size of the prize from mine electrification (Stanway, 2022).

Typical underground mine emissions across the value chain

In line with the rest of the industry (and in some companies more aggressive than industry), the EMC members have an aligned range on emission reduction targets of 30–50 per cent by 2030 or 2035 and are progressing to carbon neutrality by 2050.

Within a typical underground gold mine, approximately 35 per cent of emissions are generated in mining, through the combination of drill and blast, load and haul and supporting services. While in a typical open pit gold mine, the mining emissions contribute only 20 per cent of total emissions.

Majority of emissions in gold mining come from crushing/grinding
Scope 1, 2 and 3 CO_2 emissions example: Gold; Australia
Other
Electricity
Diesel
Surface
Underground
Drilling
Blasting
Rock reinforcement
Loading
Hauling
Other Equip
Conveying
Crushing / grinding
Benficiation
Total
Conveying/ Hoisting
Mining
Processing
McKinsey & Company

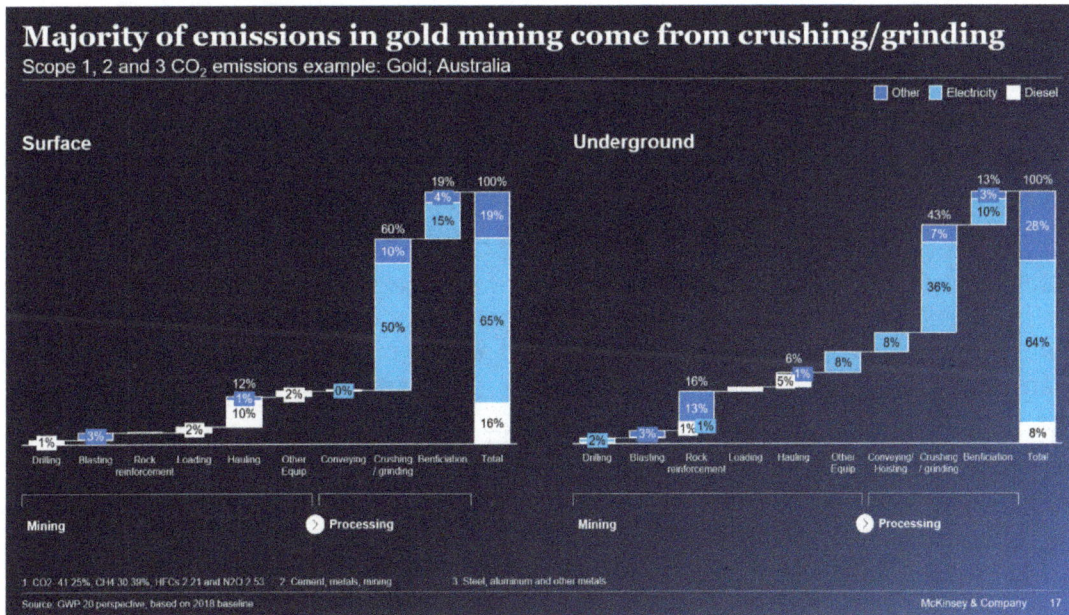

FIG 4 – Carbon emissions in a typical underground and surface gold mine (McKinsey and Company, 2021).

MAJOR CHALLENGES IDENTIFIED IN ACHIEVING ADOPTION IN THE UNDERGROUND ENVIRONMENT

Development timelines and impact on supply

In the underground environment, as previously identified, material movement (and associated services) presents one of the largest opportunities for DPM and carbon emissions abatement. However, within the Australian industry and driven by common equipment sizes being larger than other regions, battery electric loaders are not readily available to match the productivity at the mines ready to embrace the transition. The largest available 21 t loaders, commonly used in decline accessed mines in Australia, are a great example whereby currently the largest capacity battery electric loader available is the Sandvik LH518B (18 t). Similarly, the haulage needed at many Australian sites is a 63 t truck, which is only matched with the supply of the largest available battery electric truck being the Sandvik TH550B (50 t) and at time of publishing the TH665B (65 t) prototype is travelling the globe. Whilst original equipment manufacturers (OEMs) are working to develop larger loaders and trucks, the development timelines mean these may not be available to all mines until the late 2020s. In the interim, mining companies will continue to trial available electric vehicles with a higher technology readiness level (TRL) in order to begin learning and to transfer learnings into mine design and operating practices. This is, until the adoption is available of battery electric vehicles and electrical support structures across all operational applications.

Transition technologies

At the larger end of the scale there is a potential transition technology, Caterpillar are currently trialling the R2900XE diesel-electric hybrid loader through their site field follow trial process. Their claimed benefit is ~40 per cent less diesel fuel, while also delivering faster acceleration and deceleration. The lower diesel use is expected to deliver a proportional diesel particulate matter and carbon emissions reduction (to be measured during field trials), which puts this technology positively on the pathway to net zero emissions. In addition, Komatsu produce hybrid diesel-electric underground loaders, and field trials with the 22 t model have been completed in Western Australia by an EMC member and a further trial of an 18 t model is planned.

Charge system interoperability

As trials progress, charging infrastructure interoperability (or systems that offer a common interface) is emerging as another challenge to solve. Whereby the goal is to achieve charging from a single charger or charging station for battery electric vehicles (BEVs). The standardisation of OEM charging systems is not yet available, but is seen as critical to achieve infrastructure that is OEM agnostic and scalable to reduce the complexity of charging across any mine site. Together this would allow the ideal mine owner scenario of standardisation and optimisation of charging infrastructure that can both replace diesel and charge a mixed BEV fleet.

With the growth of vehicle providers underground, and a desire to adopt the ideal vehicle for the operation regardless of provider, it is seen that charging system interoperability will be necessary to facilitate the integration of a mixed OEM fleet desired across underground mines globally.

Future proofing of standards including larger current capacities and upstream communication protocols, is crucial to ensure readiness when new and greater capacity battery technology and BEV designs mature over time. Whereas in parallel the transmission cables are not expected to change in the short-term.

One of the outcomes from the Charge-On Innovation Challenge is the emerging mining taskforce within the well-established CharIN automotive group to develop the mining requirements ready for OEM integration. However, at this early stage it is important to recognise that the underground market is a small financial sector of some OEMs and in terms of standardisation could be skipped over if it does not elicit adequate participation from the underground mining subject matter experts.

Skills and capability

Electrification will bring about a number of implications for the mining workforce and its capabilities. Many of these will reinforce those resulting from the broad digital transformation trend, with some specific additions. Both require a new mindset, new leadership approach and new skills at multiple levels within businesses.

A major challenge lies in transitioning to a new operational and maintenance skillset required to run and sustain an electrified mine site. On face value this is the largest organisational change, demanding a sharp increase in electrical and electronics technicians and data-based diagnostics capability. Resolution will require retraining and close attention to apprenticeship programs, on which the industry is currently very focused.

TABLE 1

Mindset shifts associated with electrification (Stanway, 2020).

Diesel mine →	Electrified mine
Siloed →	Systemic
Point-to-point →	Networked
Centralised →	Distributed
Intuitive →	Analytical
Equipment focussed →	Service oriented

Commercial hurdles and total cost of ownership (TCO) comparison

Capital investment is traditionally assessed by the consideration of hurdle rate, whereby value-accretive projects not meeting the required rate are not progressed. In the future, decarbonisation projects will need to include penalties for carbon emissions, increasing fossil fuel price forecasts, green premiums, and a lower cost of capital for sustainability linked finance. Even when these are considered, the net zero investment may not meet the current investment hurdle rate. Decision-makers will then need to maintain their net zero commitment and internal/external stakeholder expectations through direct action projects which do not meet the investment hurdle rate or by carbon offsetting.

UNDERGROUND TRIALS UNDERWAY AND PLANNED

Across the EMC, over 40 electric vehicle trials are planned or underway, across 15 different sites globally.

Underground heavy haul trials in Australia and Canada

With limited vehicle supply across the ideal fleet conversions, the consortium has adopted a philosophy of trialling to learn across both hybrid and electric fleets, across vehicle providers. This includes hybrid loaders, battery electric loaders and battery electric trucks. With a typical fleet unit size in Canada much smaller than in Australia, the supply has been met earlier by suppliers who have begun their electrification journey in smaller vehicles. Given the global nature of the consortium, this has still meant access to information on fleet albeit smaller.

Light vehicle and auxiliary equipment trials

Despite having limited carbon emissions impact, the light and auxiliary vehicle category was seen as important to the EMC for electric vehicle adoption to enable a shift in mindset, easy learning at low cost and with reduced impact on production.

However, with very limited OEM supply across this category, the EMC members sought the option to trial OEM machines where available (eg Normet) and begun working with battery retrofit providers to achieve the same outcome, where not available.

Two significant areas of battery retrofit options were for services vehicles (Integrated Tool Carriers, ITs) and light vehicles. The EMC begun working with 3ME Technology in 2020 to deliver a retrofit of a Volvo IT L120, and with Zero Automotive in the retrofit of Toyotas.

Additionally, during the last two years, the EMC members have also ordered or commenced trials of alternate purpose built vehicle types to move people around a mine. In this process, they have uncovered and now working with Safescape in Australia to deliver the Bortana EV and Rokion in Canada.

Total trial status

With over 40 vehicle trials and conversions either in planning or in place across the globe, this represents a significant database of information to learn from and to begin to extract new operating assumptions for new mine builds or brownfields expansion. Current trials underway or planned are as outlined in Table 2.

TABLE 2

Electric Mine Consortium trials underway globally underground.

Vehicle category	In operation	Trial complete	Active trial	Awaiting delivery
Trucks	Newcrest (5) Sandvik Z50 Evolution (2) Epiroc MT42		Gold Fields Sandvik Z50	*Barminco* Sandvik TH665B Newcrest (3) Sandvik Z50
Loaders (LHD)	Evolution Epiroc (2) ST1030	*Gold Fields* Komatsu HD22 hybrid	South32, Newcrest, Barminco and MMG Caterpillar R2900 XE (Field Follow) Gold Fields Sandvik LH518B	Gold Fields Komatsu HD18 hybrid Evolution Epiroc ST14
Auxiliary/Support Equipment		*Barminco* Normet Charmec	*Gold Fields* BIT Volvo L120 Retrofit *Barminco* Sandvik longhole drill	South32 and Barminco BIT Volvo L120 Retrofit *Barminco* MacLean SS5 Sprayer
Light Vehicles	*Evolution* Rokion (2) OZ Minerals Zero Automotive ZED70Ti	*Evolution* Rokion Note: Several mines tested the first alpha prototype Bortana EV prior to the EMC	*Gold Fields* Murray Engineering eLV *Barminco* Zero Automotive ZED70Ti Newcrest Tembo eLV	South32 (3) and Oz Minerals (1) Bortana EV Barminco (additional) & Newcrest Zero Automotive ZED70Ti

INITIAL RESULTS AND DISCUSSION

Skills required and supporting frameworks

A zero diesel and zero emission mine will operate differently to traditional mines. The widespread technology and operating model changes that come from electrification will cause significant shifts in people, capability and skills requirements. Modelling conducted by the Victorian Automotive Chamber of Commerce estimates there is only currently 500 qualified electric vehicle technicians in the whole of Australia (Ratcliffe, 2021). The EMC recognises the need to collectively develop these capability needs and attempt to overcome the challenge of skills shortages and gaps for the future. The EMC is in the planning stages of establishing an industry-led, not-for-profit electric mine skills foundation. This will start as an initial three-year pilot, open globally to source providers and focused on rapid, micro-accreditation to accelerate the rate of technology change and skills pipeline. The focus will cover everything from trades through to university undergraduate and post graduate qualifications, including collaboration with OEMs.

OEM business models

The majority of the economic and environmental benefits of electrification lie with the operator, rather than the OEMs, who need to invest heavily in R&D and production. There is currently little incentive for OEMs to make the transition. However, shareholder and stakeholder pressures are beginning to change this – particularly for European manufacturers.

Incumbent OEM business models rely heavily on spare parts and maintenance. An electric vehicle engine has 20 moving parts, whereas a diesel vehicle has around 20 000. With 90 per cent fewer parts to service, the transition away from diesel equipment to electric equipment will drastically reduce parts and servicing requirements, a significant proportion of revenue for OEMs.

The nature of battery technology is driving a shift towards new ownership models wherein the supplier maintains ownership to manage the risk and maintenance and replacement over the life of the agreement (eg leasing models such as battery as a service, BaaS).

Mining company commercial or business models

Firstly, it is the mine owners responsibility to set the specifications upon which they wish to operate their own asset from a safety, schedule, technology maturity and training philosophy, including the decarbonisation strategy or increasing weighting on many Environmental, Social and Governance (ESG) criteria and/or commitments. Then a fair distribution of risk and reward should be considered as within the control of the parties involved. Where experience and benchmarking do not exist, or technology is not mature, this is where the commercial models, partnerships and innovation programs come into effect.

CONCLUSION

Collaborative design process within the EMC to virtually and physically trial

At such an early stage in the electrification transition within the underground mining environment, the collaborative design approach in the EMC has enabled the founding members to accelerate their progress, while attracting new mining companies to join to both accelerate their learning by accessing the information from existing trials and progress faster.

By establishing an operating methodology and associated trusted framework, the opportunity to accelerate electrification has been realised, by making it possible to virtually learn where physical trials could not be established and to trial the identified challenges in parallel at different mine sites, thereby also sharing the risk.

Clear demand picture to OEM suppliers

With a significant reduction in carbon emissions planned and with a safer work environment targeted across both EMC members and the wider mining industry, this has exacerbated the gap between OEM supply and industry demand, and so challenged OEMs to rapidly accelerate their delivery of vehicles to meet these needs.

Through the EMC, and while significant vehicle trials and adoption are in progress, the desired vehicle trial and adoptions plans have not been fulfilled. This is driving a requirement to work more closely with OEMs to articulate our requirements and risk appetite. It is clear that increasing transparency of demand is necessary by the mining industry in order to deliver on the required vehicle transition plan to achieve safety and carbon emission targets.

Key success criteria in achieving electrification

The opportunity in industry currently is to collaborate both with other mining companies, with OEMs and with non-traditional providers.

The options are to either wait for traditional OEMs to deliver a roadmap or commercially viable electric vehicles at the appropriate size, or we can define the areas of interest and match those technical equipment requirements with local retrofit companies, to meet the desired timelines.

The electrification journey is new to all parties, and demonstrations in the mining environment require patience and a risk tolerance to support new companies in all aspects of operation, service, support and training.

FUTURE WORK

Accelerating industry step change in the electrification journey for underground mining will require a collective of mining companies supported at the CEO level. In order to achieve this, the EMC will work across CEOs in three important areas:

1. Credible commitment:
   o To work together with suppliers on the design, trial and supply of vehicles.

- To work together to build case studies with simulation capability, to more rapidly define the clean sheet electric, zero carbon mine.

2. Unlocking skilled resources:
   - To develop frameworks with funding to develop the additional skills and capability needed in the mining industry to safely deliver on electrification.

3. Influencing policy:
   - To provide a clear and industry aligned voice to support low emissions mining.

REFERENCES

McKinsey and Company, 2021. Perspectives on mining – The haul road to net zero, McKinsey and Company.

Ratcliffe, M, 2021. The Electric Mine Consortium: A Case Study on Transformative Collaboration [online], Slate Advisory Pty Ltd. Available from: https://uploads-ssl.webflow.com/60529923ea318257ccfcadee/61fb7e4932a1517cb555 1dd0_EMC%20Report%2022%20WEB.pdf [Accessed: 1 June 2022].

Stanway, G, 2020. State of Play – Electrification Report 2021 [online], Slate Advisory Pty Ltd. Available from: https://uploads-ssl.webflow.com/60529923ea318257ccfcadee/60e16c5371e8111a0bada33d_1%202021%20SOP %20-%20Electrification.pdf [Accessed: 1 June 2022].

Stanway, G, 2022. Electric Mine Consortium Presentations. Slate Advisory Pty Ltd.

AUTHOR INDEX